PUBLIC

WATER

SUPPLY:

Data, models and operational management

To Pauna and Ann
for their patience, love and support

PUBLIC
WATER
SUPPLY:

Data, models and operational management

Dušan Obradović and Peter Lonsdale

Taylor & Francis

Taylor & Francis Group

LONDON AND NEW YORK

First published 1998 by Taylor & Francis
2 Park Square, Milton Park, Abingdon, Oxon OX14 4RN
52 Vanderbilt Avenue, New York, NY 10017

First issued in paperback 2020

Taylor & Francis is an imprint of the Taylor & Francis Group, an informa business

Copyright © 1998 Dušan Obradović and Peter Lonsdale

Typeset in 10/12pt Sabon by Scientific Publishing Services (P) Limited, Chennai

British Library Cataloguing in Publication Data
A catalogue record for this book is available from the British Library

ISBN 13: 978-0-367-65950-9 (pbk)
ISBN 13: 978-0-419-23220-9 (hbk)

CONTENTS

Introduction

The late twentieth century has seen a change possibly more significant than the Industrial Revolution. The 'information revolution' has touched millions of lives and generated an urgent need for new skills in understanding and using the new technologies. This book is intended to provide practical guidance in developing such skills for those providing safe, wholesome and reliable public water supplies. It shows how the traditional skills of the water industry can be the foundation for new skills to benefit from the new technologies.

The book is intended as a text for all sections of the industry, directed not only to managers, engineers, planners and consultants, but to the technicians, maintenance staff and operatives involved in the day-to-day effort to provide an efficient service. As well as being of interest to a broad range of personnel involved in the complex supply systems of the developed world, the book should also provide insights into improving relatively unsophisticated supply systems. It should also interest students considering a career in one of the most important and satisfying of industries – the water services.

The book has four principal themes and shows how they might be woven together to improve the effectiveness of the service.

- The first theme is that data in massive quantity can now be available to water companies almost in real time and at low cost. In theory, the more data that managers have available, the more efficiently they can operate, but the sheer volume of data means that properly trained personnel are needed to analyse data and elevate them to the status of useful information.
- The second thread is that the data streams delivering their cargoes to the water companies have, historically, been developed for a variety of purposes, but the information that can now be easily distilled from the data can be integrated to broaden the range of benefits for the utility.
- The third is that system models can be a central tool in integrating the databases of the utility and have the potential to act as sentinels of the systems, linked to telemetry, geographical information systems and other data systems, to monitor minute-by-minute activities of the real systems.
- The fourth and perhaps most important theme is that there are two groups of 'experts' involved in the information

systems of a modern water company, and the two need to develop areas of common ground to make the best use of the new technologies. The groups might be loosely labelled 'information technology experts' and 'water supply experts'. Although there is a mutual benefit in developing 'common ground', user-friendly software and the water system knowledge of the water specialists mean that it is better to train them to use the modelling techniques than the converse. The models themselves are ideal tools to train water staff quickly in skills that previously may have taken years to develop.

The book discusses telemetry, data collection and processing, modelling and potential future developments in those fields, but does not delve into the minutiae of their technicalities. Its primary purpose is to explain to those within and supporting the industry the potential benefits and shortcomings of the technologies and, conversely, to explain the needs and aspirations of the water industry to the information technology experts. To this end, most of the content is devoted to describing and explaining practical problems. The book includes some 445 illustrations, the majority based on real systems and illustrating real problems. The chapters are organised as follows.

Chapter 1 examines the data requirements to understand the supply network and the systems used to collect them. It examines how the data might be processed, stored and accessed, and looks at how the database thus created might be usefully integrated with other information held by the utility.

Chapter 2 looks at the quality of data harvested by telemetry. It describes the limitations of the data collection system, examining a number of common faults and ways of detecting and coping with them. The chapter looks behind the raw data to detect the important messages and ends by suggesting a key tool for using on-line data systems to benefit the organisation – mathematical modelling of the supply system.

Chapter 3 briefly describes the principles of mathematical models, describing how the physical elements of a system are modelled. It discusses the pitfalls awaiting newcomers to model making. In particular, the chapter suggests what should or should not be included in models. The longest section relates to pumps and pumping stations and, in particular, how they are controlled, in the real world and in the model. Much of the chapter will be of interest to anyone – modeller or not – involved in the operation of supply systems.

Chapter 4 examines a difficult aspect of system models, demand. It discusses factors affecting demand and the creation of demand patterns to be applied in the model. As well as daily, weekly and seasonal patterns generated by a range of user types, long-term trends are examined. The chapter shows how losses might be treated in the model and the relationship between leakage and pressure. Although written in relation to system modelling, the chapter will interest those concerned with demand management and forecasting.

Chapter 5 reviews the mathematical basis of modelling. Even the mathematically squeamish should follow the arguments quite easily but can bypass them if they wish, without losing the thread of the rest of the book. The chapter reviews the development of network modelling (or network analysis) from the early manual methods of the 1930s via the steady-state models developed in the 1960s and 1970s to the quasi-dynamic models that are the basis of most modelling software currently available. The chapter concludes by suggesting points to be taken into account in selecting software and ways in which the package could be sensibly linked to other databases.

Chapter 6 gets down to the actual business of creating models, offering practical advice on the steps to be taken by the novice, to the point at which a model is ready to run. The chapter lists data needed to construct the model, with more hints on what should or should not be included and descriptions of common configuration faults to be avoided. It includes examples of common solutions to hydraulic problems and suggests how they should be modelled. Having constructed a model and persuaded it to 'run', the chapter takes the new modeller through the proving steps necessary to be satisfied that the model is a reasonable representation of reality.

Chapter 7 looks at the three primary ways in which models can be used to address real-life problems of water supply: long-term planning, planning operational changes and operational management. It goes on to suggest ways in which modelling can be introduced to a water company with minimum pain. A section is devoted to the use of models to optimise pumping costs through control policy adjustment and tariff changes, and the chapter ends with a review of model usage in Europe and the USA.

The book includes a comprehensive list of references and further reading for information on particular aspects of the subject. Appendices include information on interfacing of databases, the pipe flow formulae and detailed data on friction and minor loss coefficients, pump data, valve characteristics and a selection of demand patterns covering a wide range of user types.

Acknowledgements

This book is the result of a quarter of a century of work in water supply and computer science, more precisely, in those areas where the two subjects overlap. The intention was to apply the potentially powerful technique of mathematical modelling to complex water supply systems, initially in the planning and design phase, and later for the benefit of operational management. Similar work has been undertaken by many other researchers and engineers throughout the world, and I was always eager to learn from my colleagues and to exchange information freely with them.

My special thanks go first to my older colleague and teacher, Milos Kordic, a brilliant engineer and designer of many water supply systems in Yugoslavia and elsewhere. He has showed me that analysis of water supply systems can be a fascinating subject.

The other important person is Rajko Cavor, my younger colleague and associate, who joined forces with me in this task some 10 years ago and provided much needed energy, imagination and drive for implementation of our ideas in practice, using his outstanding skills in and knowledge of computer science.

I owe special thanks to Ron Huntington, Ken Manley, John Snoxell and many other members of the staff of Wessex Water plc, UK, for their friendly support and cooperation, and the opportunity to work in the high-tech environment of that company. I am also grateful to my colleagues from the Istrian Waterworks (Croatia), Belgrade Waterworks (Yugoslavia), Energoprojekt (Belgrade, Yugoslavia) and other water companies, where I have learned so much. I regret that space will not permit me to name every individual, but a few, Darko Kranjcic, Predrag Uskokovic, Branislav Kujundzic and the late Vlatko Sulentic, are the names that come first to me.

Last but not least my thanks are due to the late Professor L. Huisman, The Netherlands, and the late Professor McPherson, USA, both well-known experts who have given me their support and encouragement at the times when I needed it most.

Dušan Obradović

In the late 1950s the late Professor R.J. Cornish arranged for his students to attend lectures by Professor Hardy-Cross, during a visit by the latter to Manchester. Professor Cornish imposed many laborious manual calculations on his students to teach the principles of network analysis, but did so with the amiable demeanour that caused him to be held in such great respect and affection. I thank him for his persistence.

In the early 1970s Norwich Water Department created an R&D section. I would like to thank the Engineer and Manager at that time, the late R.J. Bell, and his Deputy, Ken Rowe, for having that foresight and giving me and my colleagues the opportunity to explore the potential of network analysis.

From 1977 I worked for Wessex Water during an exciting period of development for the industry. Wessex played a leading role in many of the technical developments. I would like to thank Wessex and in particular Ron Huntington for that experience and the foundation it provided for my contribution to this book.

P.B. Lonsdale

As well as providing the platform on which many of the ideas in this book were developed, the Wessex Water plc information systems were also the primary sources of data on which the vast majority of examples and diagrams in the book are based. We would like to thank the company and in particular Colin Skellett for their generous permission to use that information in this book.

We would also like to thank Rajko Cavor and Aquaware Systems Ltd for their permission to include 'Standard interface formats' as Appendix A in the book

Dušan Obradović and P.B. Lonsdale
Belgrade and Wimborne,
August 1997

Notation

Symbol	Units	Variable/parameter
A	m^2	Flow area
C	–	Flow constant
c	$mg\ l^{-1}$	Concentration
D, d	mm	Diameter
F, f	–	A function
g	$m\ s^{-2}$	Gravitational acceleration
H	m	Hydraulic grade line
H_p	m	Pump head
K	–	Unit resistance of a pipe
k_h	–	Demand (load) factor
k	day^{-1}	Decay rate
L	m	Length
n	rpm	Rotational speed of pump impeller
n	–	Mannings number
n_q	–	Pump specific speed
P	kW	Power
p	Pa	Pressure
Re	–	Reynolds number $(= vd/v)$
S	%	Valve opening
T	s	Period of time
t	s	Time
v	$m\ s^{-1}$	Flow velocity
x, y, z	m	Cartesian coordinates
Y	$J\ kg^{-1}$	Pump head in SI units $(= gH_p)$
Z	m	Elevation
Δ	–	Increment
δ	mm	Internal roughness height
ε	–	A small quantity
ζ	–	Local (minor) loss coefficient
η	%	Efficiency
λ	–	Friction coefficient
v	$m^2\ s^{-1}$	Kinematic viscosity
ω	s^{-1}	Angular speed
ρ	$kg\ m^{-3}$	Density

Important SI units

The 'Système International d'Unités' (International Unit System) (SI) was agreed by the '10e Conférence Générale des Poids et Mesures' (The Tenth General Conference on Weights and Measures) in 1954. Since then it has been legally adopted by many countries throughout the world.

Physical dimension	Formula symbol	SI unit	Acceptable alternatives	Recommended units	Remarks
Length	l	m (metre)	km, dm, cm, mm	m	Basic unit
Volume	V	m^3	dm^3, litre, cm^3	m^3	–
Flow	Q	$m^3 \ s^{-1}$	$m^3 \ h^{-1}$, $l \ s^{-1}$, $Ml \ d^{-1}$	$m^3 \ s^{-1}$ and $l \ s^{-1}$	–
Time	t	s (second)	min, h, d, year	s	Basic unit
Rotational speed	n	$r \ s^{-1}$	$r \ min^{-1}$	$r \ s^{-1}$ and $r \ min^{-1}$	–
Mass	m	kg (kilogram)	–	kg	Basic unit
Density	ρ	$kg \ m^{-3}$	$kg \ dm^{-3}$	$kg \ m^{-3}$	–
Force	F	N (newton)	kN, mN	N	$= kg \ m \ s^{-2}$
Pressure	p	Pa (pascal)	bar ($=10^5$ Pa)	Pa and bar	–
Torque	M, T	N m	–	N m	–
Energy, work	E, A	J (joule)	kWh, W s, kJ	J	$= N \ m$
Head	H	m (metre)	–	m	–
Power	P	W (watt)	kW, MW	kW	$= J \ s^{-1}$
Temperature	T	K (kelvin)	°C (degrees Celsius)	K	–
Kinematic viscosity	v	$m^2 \ s^{-1}$	–	$m^2 \ s^{-1}$	–

Basic conversion factors

	English units to SI	SI to English units
Length	1 inch = 2.54 cm	1 cm = 0.393 701 inch
	1 ft = 0.3048 m	1 m = 3.280 84 ft
	1 yd = 0.9144 m	1 m = 1.093 613 yd
	1 mile = 1.6093 km	1 km = 0.621 39 miles
Mass	1 lb = 0.4536 kg	1 kg = 2.204 586 lb
	1 ton = 1016.047 kg	1000 kg = 0.984 21 ton (USA: long ton)
Pressure	1 psi = 6894.24 Pa	1 Pa = 0.000 145 psi
	1 psi = 0.068 9424 bar	1 bar = 14.505 psi
	1 inch H_2O = 2.4909 mbar	1 mbar = 0.401 46 inch H_2O
	1 ft H_2O = 29.8907 mbar	1 mbar = 0.033 455 ft H_2O
Energy, work	1 ft lb = 1.3558 J	1 J = 0.737 57 ft lb
	1 BTU = 1.0558 kJ	1 kJ = 0.947 15 BTU
Power	1 ft lb/s = 1.3558 W	1 W = 0.737 57 ft lb/s
	1 Hp = 0.7457 kW	1 kW = 1.341 Hp
Area	1 $inch^2$ = 6.4516 cm^2	1 cm^2 = 0.155 $inch^2$
	1 ft^2 = 0.092 903 m^2	1 m^2 = 10.7639 ft^2
	1 yd^2 = 0.8361 m^2	1 m^2 = 1.196 yd^2
	1 acre = 4046.86 m^2	1 hm^2 = 2.471 acre
Volume	1 $inch^3$ = 16.387 cm^3	1 cm^3 = 0.061 024 in^3
	1 ft^3 = 0.028 317 m^3	1 m^3 = 35.314 67 ft^3
	1 US gal = 3.7854 litre	1 litre = 0.264 173 US gal
	1 UK gal = 4.546 litre	1 litre = 0.219 97 UK gal
Velocity	1 ft/s = 0.3048 m s^{-1}	1 m s^{-1} = 3.280 84 ft/s
	1 ft/min = 0.005 08 m s^{-1}	1 m s^{-1} = 196.85 ft/min

	English units to SI	**SI to English units**
Flow	$1\ ft^3/s = 28.3268\ l\ s^{-1}$	$1\ l\ s^{-1} = 0.0353\ ft^3/s$
	$1\ US\ gal/s = 3.7854\ l\ s^{-1}$	$1\ l\ s^{-1} = 0.264\ 173\ US\ gal/s$
	$1\ UK\ gal/s = 4.546\ l\ s^{-1}$	$1\ l\ s^{-1} = 0.219\ 974\ UK\ gal/s$
	$1\ US\ gal/min = 0.063\ 09\ l\ s^{-1}$	$1\ l\ s^{-1} = 15.8504\ US\ gal/min$
	$1\ UK\ gal/min = 0.075\ 77\ l\ s^{-1}$	$1\ l\ s^{-1} = 13.1978\ UK\ gal/min$
Density	$1\ slug/ft^3 = 515.363\ kg\ m^{-3}$	$1\ kg\ m^{-3} = 0.001\ 9404\ slug/ft^3$
	$1\ lb/ft^3 = 0.0160\ kg\ dm^{-3}$	$1\ kg\ dm^{-3} = 62.50\ lb/ft^3$
	$1\ lb/inch^3 = 27.6799\ kg\ dm^{-3}$	$1\ kg\ dm^{-3} = 0.036\ 127\ lb/inch^3$
Kinematic viscosity	$1\ ft^2/s = 0.0929\ m^2\ s^{-1}$	$1\ m^2\ s^{-1} = 10.764\ ft^2/s$
Temperature Conversion	$t_C = (5/9)(t_F - 32)$ t_C is in degrees Celsius	$t_F = 32 + (9/5)t_C$ t_F is in degrees Fahrenheit

1 Data management systems in water supply

1.1 Water supply at the millennium

Public water supply, until very recently, attracted little or no public attention. The service was maintained with few difficulties for the customers, at relatively low cost.

That happy state is changing rapidly. The industry's management, its relationship with its customers and the general public, its relationship with the environment, its costs and so on are coming under increasing scrutiny. Some of the factors that are fuelling the dramatic changes in the industry are:

- a steady and universal rise in demand for water;
- a strong demand for operators to conform to and be judged against clearly defined standards in water quality and quantity, and quality of service;
- transfer of services from the public to the private sector – privatisation, in its various forms, has created pressure to increase financial efficiency by cutting costs and prices;
- increasing scarcity of resources, particularly in areas containing large urban populations;
- increasing sensitivity to the impact of public supply services on the environment – in particular, the effects of abstraction of water and the discharge of potential pollutants.

To meet the wide range of conflicting demands on an increasingly high-profile industry, there is a clear need to take advantage of the benefits of another high-profile area of modern life, information technology. The water operator needs to develop and implement a modern information technology system to achieve:

- close control of the water supply system;
- maximum use of resources and technology;
- speedy response to the changes demanded by society of an important public service.

New technology has allowed great advances in the collection and processing of data essential for efficient control of water supply systems. The reliability and accuracy of measuring devices, remote sensing systems, communications and processing facilities have improved significantly, contributing to the improved efficiency of water supply systems. In the late 1970s and early 1980s, new methods and procedures were developed around mainframe computers (Fallside, 1977, 1981; Fontaine, 1983; Kanbayashi *et al.*, 1977; Kashiwagi and Matsumoto, 1980; McPherson, 1975; Takagi *et al.*, 1983). Progress was not without practical problems, some of which will be discussed later, nor were the results always satisfactory (Cullen, 1987). Results did not always meet expectations, largely because complex supply systems are subject to influences that are difficult to quantify and control.

The twin benefits of further technological improvements and falling costs during the late 1980s and early 1990s have presented opportunities for improved control. In particular, the development of micro-technology, powerful personal computers and improving sensor and communication reliability are changing the information technology environment in the water industry yet again (Bost, 1986; Gotoh *et al.*, 1993; Halpern and Pascal, 1987; Huntington, 1990, 1993; Mitchell, 1993; Nguyen, 1994; Scherer and Phebey, 1995).

Operational data are now regularly captured, processed, displayed and archived, giving dramatic insight into the historic and current state of the system and permitting various analyses of particular events. The emphasis is on supervision and control in real time. There have also been proposals for expert systems by Al Hames (1987), Alla (1985), Chase and Ormsbee (1993), Cosgriff *et al.* (1985), Huntington (1984, 1993), Morita and Arakawa (1989), Nguyen (1994), Rance *et al.* (1993a) and Schulte and Malm (1993). The first expert systems under development were announced by Alla (1988), Kado and Itoh (1987), Santoni *et al.* (1987) and Walker (1988).

This book will not attempt to describe and review all these achievements in detail. The book does, however, examine in some detail key elements of comprehensive information systems within the context of water supply, in particular mathematical (network) modelling, and shows how it can be integrated into the information system to overcome some of the problems that plague staff in every water company.

1.2 Managing a water supply system

Modern water supply systems are large and complex, with hundreds of reservoirs, pumps and valves, thousands of kilometres of pipes and tens or hundreds of thousands of customers, while the process of capturing, purifying, transporting and distributing water involves many specialist trades and professions. Figure 1.1 shows a simplified management cycle.

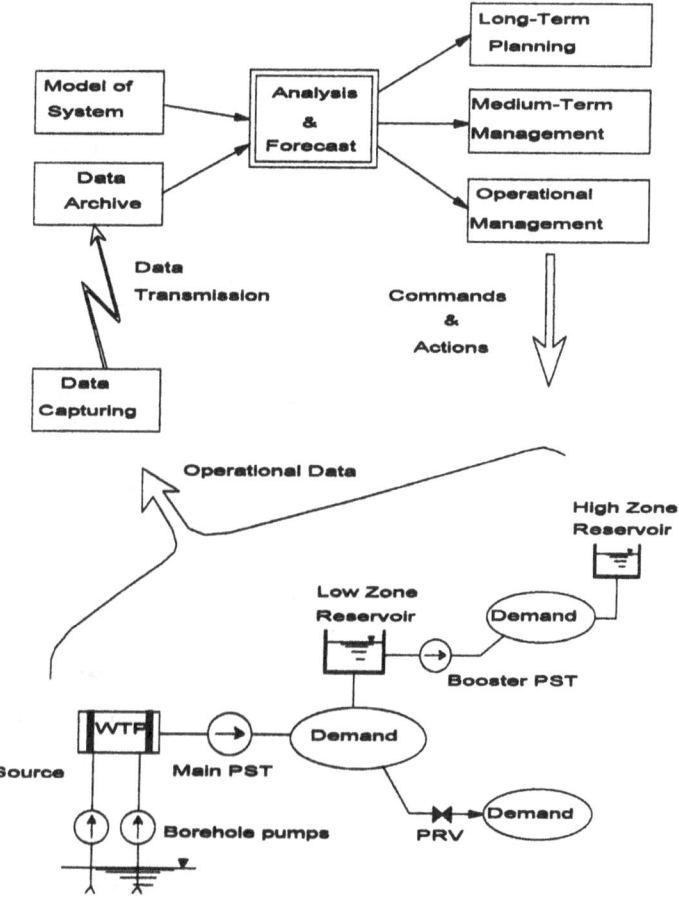

Figure 1.1 Overall control and management.

Data describing the current state of the system have to be captured, processed and transmitted to the water company. This information is analysed – preferably by using a model of the system – and evaluated. A forecast of future developments is then made and appropriate action taken by managers at various levels. Examples are:

- long-term management – planning of new resources to meet future potential shortfalls;
- medium-term management – changing operating regimes to take advantage of cheaper power tariffs;
- short-term management – redeployment of men and equipment to reduce leakage.

Some typical jobs are shown in Fig. 1.2. Note that all jobs are interrelated. Successful execution of the work relies on experience and knowledge acquired over a lifetime in the industry.

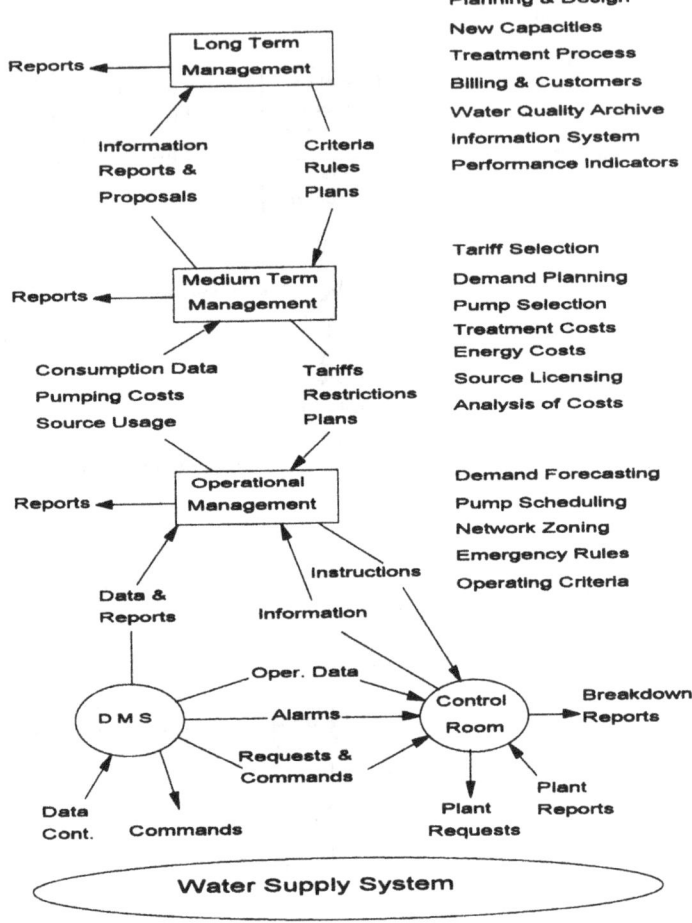

Figure 1.2 Jobs in a water company.

Management retains overall control but leaves details and technicalities to local staff, and the system operates largely under a form of self-momentum.

The reliance on experience and the apparent inherent momentum has changed with the introduction of new technology, which provides information to everyone, all the time, almost in real time and at an acceptable cost. Apart from human problems (people have a sense of 'possession' over data they collect), the benefits to the company are obvious.

The ever-present problem for the management is to keep a firm link between the people on the ground and the staff in the office; two-way information flow needs to be established and maintained, to ensure the safe, smooth and efficient running of the system.

Water balance

One of the most important tasks in every water company is to keep track of the water as it flows through the system: a form of 'water book-keeping'. Figure 1.3 shows the main items in the balance sheet. The balance shown in the schematic is on a company level, but similar balances should be made for lower levels. The items that need to be accounted for in the balance are:

- raw water abstracted from sources with exports (RWE) and imports (RWI);
- treated water, water used within treatment (TOU) and source works losses (SL);
- distribution input – net treated water plus treated water imports (WI), minus exports (WE);
- water delivered to local distribution networks – distribution input minus trunk main (TL) and service reservoirs losses (RL);
- water delivered to customers, plus distribution operation use (DOU), distribution main leakage (DML) and communication pipes leakage (CPL);
- water billed – measured (by meters at individual properties) and unmeasured plus supply pipes losses (SPL) minus various public uses (PU) and water taken unbilled, both legally (fire fighting, for instance) and illegally (theft and unauthorised use) (WTU).

The data necessary for such a balance sheet may be collected by telemetry, data-loggers, or manually – but usually a combination of all three methods is involved.

Figure 1.3 Water balance.

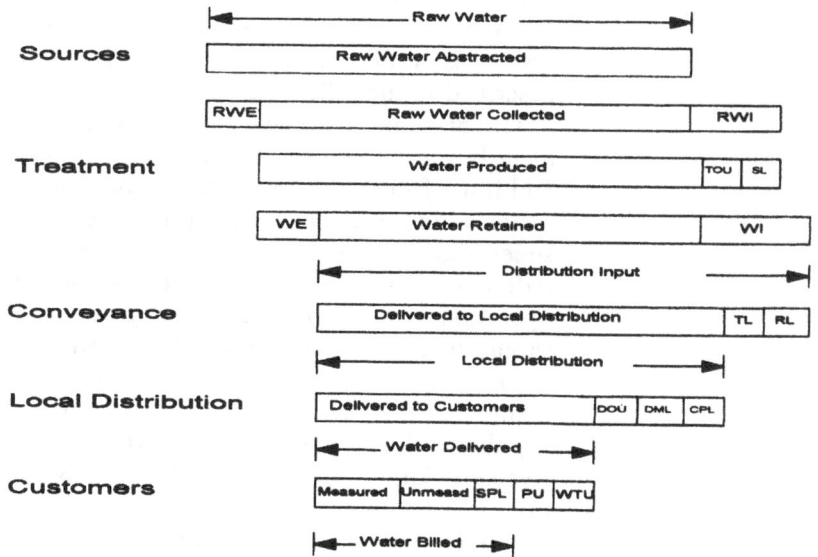

1.3 Data required for operational management

The very size of areas served by individual companies means that, even though the unit cost of the elements of a telemetry system (field instruments, outstations, power supply, accessories) is relatively low and continues to fall as the technology develops, the total initial costs of a system can be quite significant. To the initial, capital, cost should be added operating and maintenance costs and the costs of data transmission and processing. This means that the number of sensing devices and their location should be carefully selected to achieve the best cost/benefit ratio. How this should be done in a given case is a matter for separate study and cannot be prescribed generally; but some pointers might help.

It is obvious that a large supply system cannot operate without the constant influx of accurate data from all important points in the system: sources, reservoirs, pumping stations, main trunks and some of the major users (e.g. large factories). The first task is to *capture* the data at source, as accurately as possible, then to *transmit* the data to the water company quickly and reliably, *receive* the information and *interpret* it correctly. The most efficient way to do all this is through a telemetry system; other options are data-loggers and manual readings, which are cheaper in capital investment, but more labour-intensive. A control system is rarely fully telemetered – the price would be excessive. Generally, just the most important measurements (output of main water sources, for instance) are telemetered. The rest of the data are captured by data-loggers (which may be connected to the telephone network or isolated) or by instruments with local readings. A well designed control system is a mixture of all three. These options are briefly described in the following text.

The first task is to study carefully the water supply system, identify the key points and decide what should be monitored and how. Figure 1.4 shows the major elements of water supply systems and the most important measurements needed for supervision and control. Note that none of the instruments are redundant – they are all needed if a good insight into system operation is required. Figure 1.5 gives an idea how the data should be captured and with what frequency.

The individual instruments could be of very different reliability, quality and accuracy, and purchased from various suppliers. Some could be a part of the telemetry system (typically larger flowmeters and water-level and pressure gauges), while others are purely local (like water meters at customer's premises). The most important ones are those

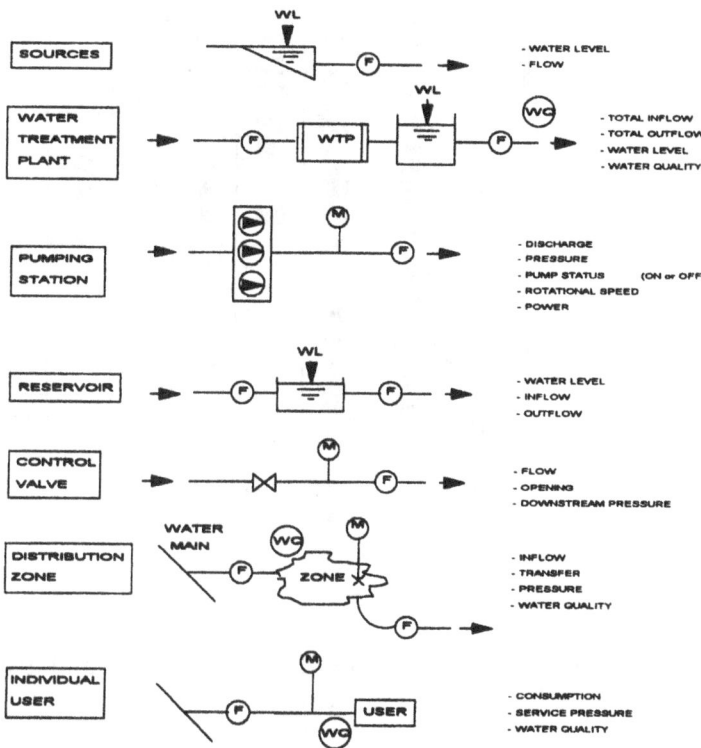

Figure 1.4 What data to collect?

monitoring flows, pressures and water levels at key points of the system. Further information on the types of instruments and some pointers as to the choice of particular instruments are included in Chapter 2.

Apart from flows, pressures and water levels – the 'primary' parameters of a supply system – other values might be usefully monitored, such as pump speed (for variable-speed units), power use, electric voltage and current, water quality parameters (e.g. turbidity, residual chlorine), etc.

Data captured by an instrument can be displayed locally, for the benefit of field staff, but in most cases the data are converted from an analogue signal (typically 4–20 mA) to a digital value for further processing. Digitised data are easier for handling, archiving and transferring.

Note that there are variations in accuracy, frequency of measurement, importance and cost of individual instruments installed for various purposes, which are:

- alarms (low storage, low pressure, pump failure);
- warnings (dangerous trends, e.g. depletion of storage or increasing level of leakage);

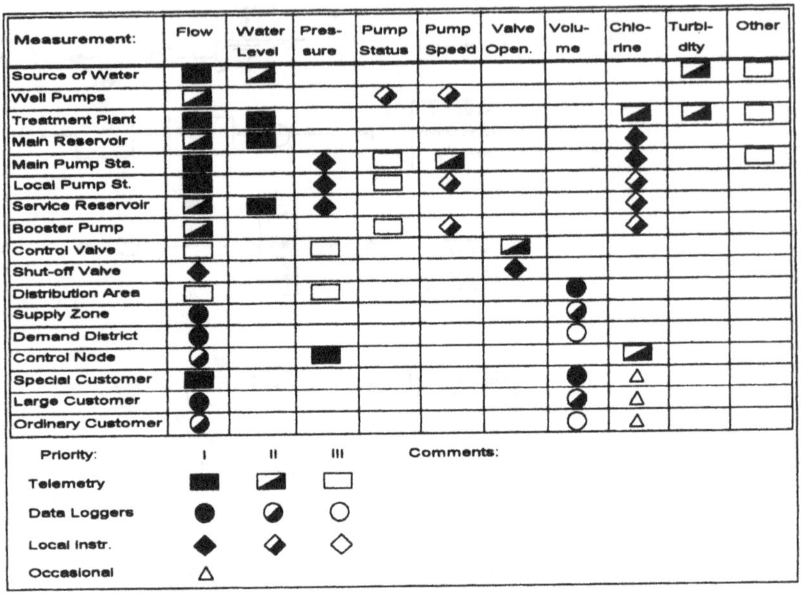

Measurement:	Flow	Water Level	Pres-sure	Pump Status	Pump Speed	Valve Open.	Volu-me	Chlo-rine	Turbi-dity	Other
Source of Water	■	◪							◪	☐
Well Pumps	◪			◈	◈					
Treatment Plant	■	■						◪	◪	☐
Main Reservoir	◪	■						◆		
Main Pump Sta.	■		◆	☐	◪			◆		☐
Local Pump St.	■		◆	☐	◈			◈		
Service Reservoir	◪	■	◆					◈		
Booster Pump	◪			☐	◈			◈		
Control Valve	☐		☐			◪				
Shut-off Valve	◆					◆				
Distribution Area	☐		☐					●		
Supply Zone	●							◑		
Demand District	●							○		
Control Node	◑		■						◪	
Special Customer	■							●	△	
Large Customer	●							◑	△	
Ordinary Customer	◑							○	△	

Priority:	I	II	III	Comments:
Telemetry	■	◪	☐	
Data Loggers	●	◑	○	
Local Instr.	◆	◈	◇	
Occasional	△			

Figure 1.5 Typical measurements in a water supply system.

- data capturing for analysis and planning at all levels of management;
- billing the customers.

Raw water system

A simplified raw water collection system is shown in Fig. 1.6. It consists of a surface source (impoundment) and two sources with three and two wells, respectively. Water flows to a treatment plant by gravity from the impounding reservoir and by pumping from the wells.

It is obvious that the flow of raw water needs to be monitored if any supervision and control is to be done; this is achieved by meters RW-1 and RW-2. The contribution of individual wells does not have to be monitored with the same quality instruments; in many cases it would be quite enough to monitor flow from groups of wells – here GM-A and GM-B. Note that a simple balance could be established, i.e.

$$Q_{RW-2} = Q_{GM-A} + Q_{GM-B}$$

giving a check on the accuracy of all three instruments; if any one fails, this will be quickly discovered and rectified.

Water conveyance system

In Fig. 1.7, water has to be transported from the treatment plant to areas of demand. Water is pumped from the treatment plant to reservoir A and then flows by gravity to reservoir B, then to reservoir C, close to the town. Note that certain quantities are diverted to satisfy local demands A, B and C.

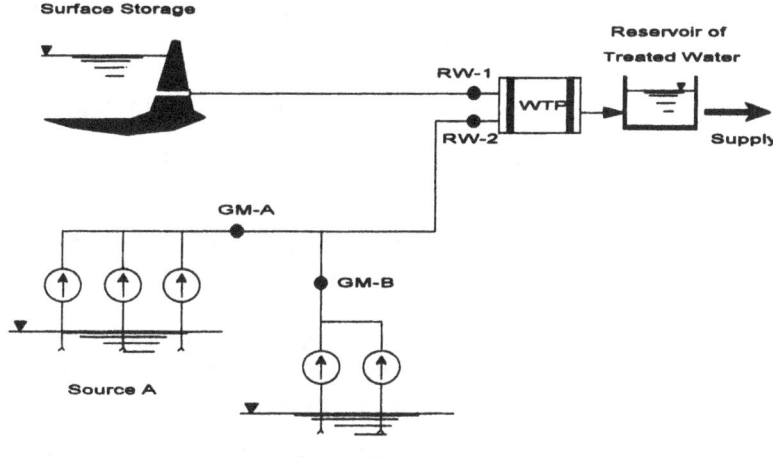

Figure 1.6 Raw water system coverage.

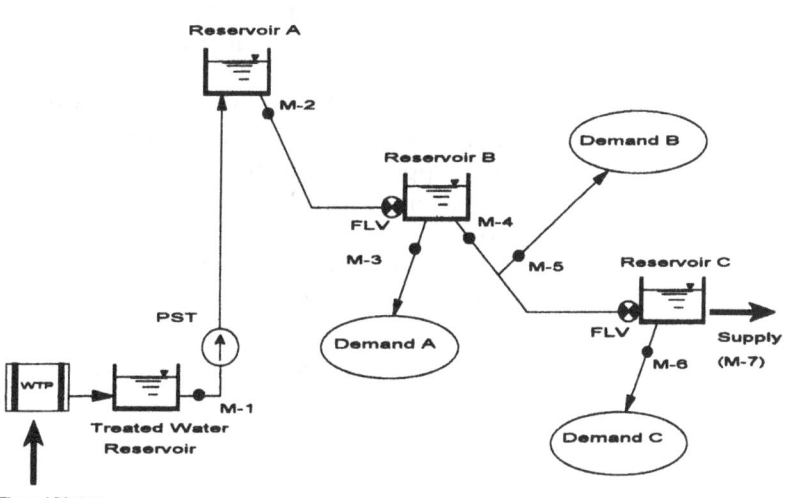

Figure 1.7 A conveyance system coverage.

Ideally, all the flows should be monitored, as well as levels in all the reservoirs, so the state of such an important system will be constantly known.

Note the possibilities for cross-checking; the total volume delivered by the pumping station has to be approximately equal to the volume registered by flowmeter M-2:

$$V_{M-1} = V_{M-2}$$

and also further downstream

$$V_{M-2} = V_{M-3} + V_{M-4}$$

$$V_{M-4} = V_{M-5} + V_{M-6} + V_{M-7}$$

If a section of pipe is very long (for instance between reservoirs A and B) then flowmeters might be installed at each end of the pipe, to show immediately any significant leakage (such duplication is not shown in Fig. 1.7).

Distribution system   The control of even a relatively small network may be difficult to achieve due to conflicting requirements:

- A system with many loops will provide better *security of supply* because customers can be supplied from two or more directions.
- A branched system will be easier to *supervise and control* because flows cannot change direction and the number of important pipes is reduced.

Moreover, many distribution systems have two basic functions: to supply local customers *and* to transport water to those further off; this is plainly visible in the next example. Figure 1.8 shows part of an urban water distribution system. Water flows from the main reservoir A to the lower one, reservoir B, by gravity. A PRV (pressure-reducing valve) controls the inflow in reservoir B. Several districts of the lower-pressure zone (A to D) are being supplied from this trunk. A booster station pumps water through the higher-pressure zone, district E, towards reservoir C.

All district inflows and outflows are monitored. As there is no storage, direct comparison between flows is possible in several places:

$$Q_{DistA} = Q_{M-2} + Q_{M-4} - Q_{M-7} - Q_{M-8}$$

$$Q_{DistB} = Q_{M-3}$$

$$Q_{DistC} = Q_{M-7}$$

$$Q_{DistD} = Q_{M-8}$$

$$Q_{DistE} = Q_{M-5} - \Delta V_{ResC}/\Delta t$$

Note that in the last case (district E) changes of storage in reservoir C must be taken into account because there is no flowmeter between the reservoir and the district.

Note also that the balance for the main trunk is:

$$Q_{M-1} = Q_{M-2} + Q_{M-3} + Q_{M-4} + Q_{M-5}$$

permitting checking of the meters and – even more important – an early warning if a serious burst occurs anywhere between the meters. Possibilities to set one meter against others are very useful for everyday supervision and control, and in many cases justify the cost of additional equipment.

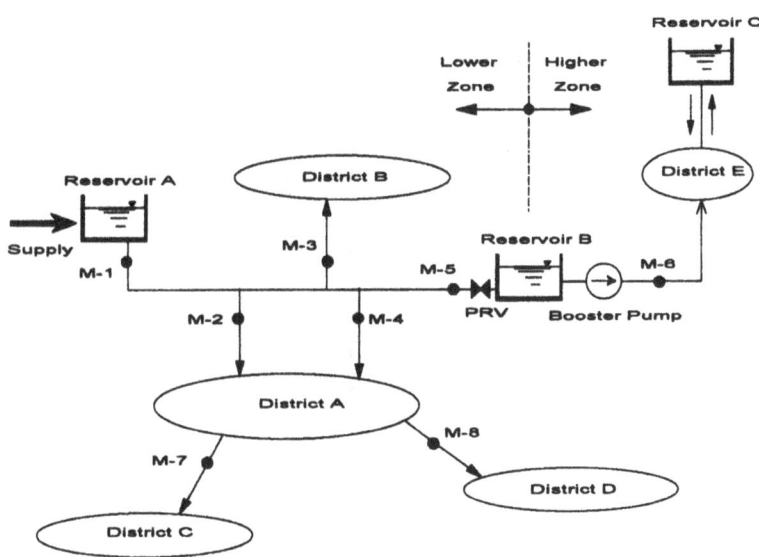

Figure 1.8 Distribution system coverage.

Parts of the distribution network can be organised into demand *districts*, as shown in Figs 1.9 and 1.10. The task might be difficult in urban systems, where the network is complex, with many loops and interconnections. In the simplified case shown in Fig. 1.9, the utility might have to close valves Val-2, Val-3 and Val-5, install a PRV in place of Val-1, and add two meters in pipes where Val-1 and Val-4 control the flow. This

Figure 1.9 Urban distribution system.

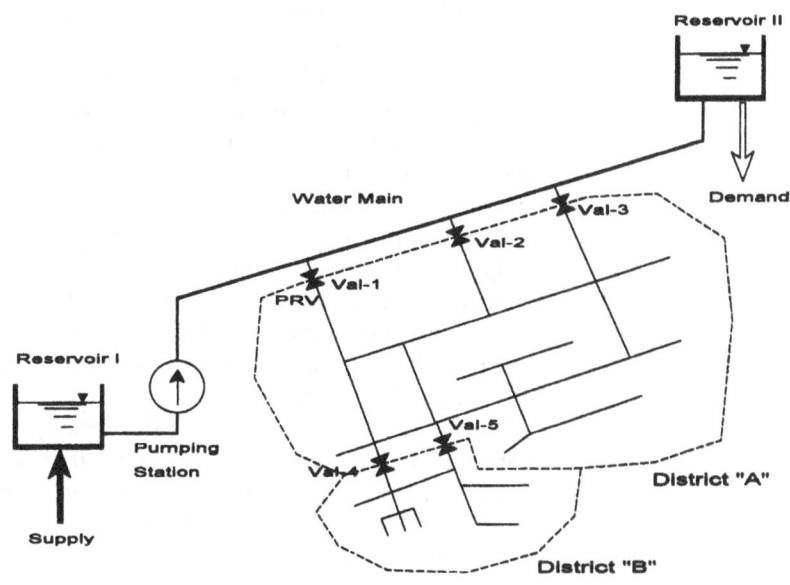

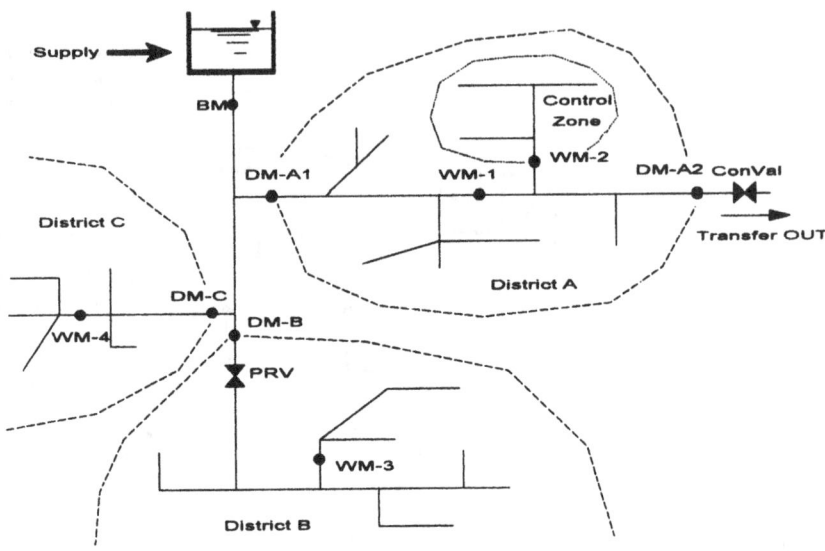

Figure 1.10 Rural distribution system.

will clearly delineate districts 'A' and 'B', and make overall control simpler. The closed valves could be reopened in emergencies or if the districts need to be changed. Note, however, a conflict in good control and good hydraulics – several loops have to be cut. Note also that demand in district 'A' is the *difference* between flow through Val-1 and Val-4, which means that both instruments must be operational and well calibrated, not just one as in the case of district 'B'.

The conflict between good demand data (control) and the hydraulics of the system are usually easy to resolve. The original system is still intact in that no pipes have been physically cut and operational staff will always be able to restore useful loops in an emergency. The loss of hydraulic capacity is normally insignificant in reality, as most pipes laid in streets, certainly in developed countries, are over-capacity in hydraulic terms. The major cost in laying the small-diameter pipes is in the excavation and reinstatement costs, and the pipe cost is usually an insignificant part of the total. This being the case, most distribution engineers would provide slightly oversize pipes, to avoid relatively early replacement and to meet what has certainly been seen over the past three or four decades as an inevitable rise in demand.

Rural distribution systems are usually of branching type, without loops, because the customers are few and dispersed over larger areas. Overall control is therefore simpler, as illustrated by Fig. 1.10. Water flows directly from the reservoir to the distribution system, with no other reservoirs or pumping stations. The network is subdivided into three districts, covering (presumably) separate villages or hamlets. The demand in each district is monitored directly:

$$Q_{DisA} = Q_{DM-A1} - Q_{DM-A2}$$

$$Q_{DisB} = Q_{DM-B}$$

$$Q_{DisC} = Q_{DM-C}$$

In many cases, even the districts are too large for efficient pressure control and leakage monitoring, so smaller territorial units, called here *control zones*, may need to be introduced – see Fig. 1.10. Note that the demand in this small area is monitored by flowmeter WM-2; note also that there are other flowmeters within the area monitoring local flow without a specifically assigned area (e.g. WM-1). It is desirable to establish a clear balance at district level, but this is not always possible or practicable (i.e. district demand equals the sum of demands in its control zones).

A few customers in district B in Fig. 1.10 are shown schematically in Fig. 1.11. Their consumption is measured by individual water meters installed either in the customer's premises (meter CM-A) or outside it (meter CM-B). This is an important detail because the total losses in supply pipes – due to corrosion and traffic damage – could be significant. The supply pipe, between the distribution main and the customer's meter, is, of course, the responsibility of the water company.

Note that service pressure in this district is controlled by a single PRV; note also flowmeter WM-3 monitoring demand in the corresponding control zone.

Figure 1.11 Metering of individual customers.

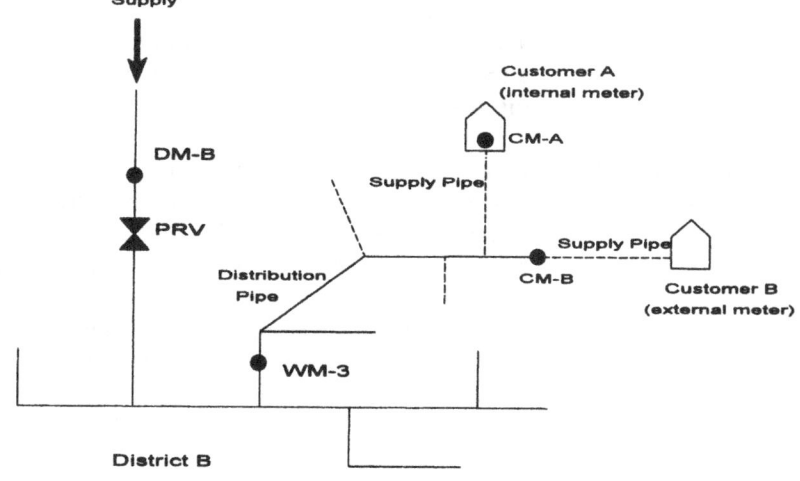

Supply

Customer A
(internal meter)

DM-B

CM-A

Supply Pipe

PRV

Supply Pipe

Distribution
Pipe

CM-B

Customer B
(external meter)

WM-3

District B

The consumption of metered users is regularly measured either by a special task force ('readers') or by remote sensing devices. The time interval is typically a month, quarter or half year for manual readings. With remote devices, reading frequency could be changed as needed. The price of such systems might appear to be prohibitively high at present, but a number of studies have considered the possibility of joint systems with other utilities, such as electricity and telecommunications, which have the facilities to transmit data through their systems (see next paragraph). These data are used primarily for *billing* purposes, but would also provide valuable information for various planning, design and modelling tasks.

The task of reading customer's meters is tedious and labour-consuming due to sheer numbers. The meters must be simple and relatively cheap, and are therefore not as accurate as the larger meters used for monitoring bulk flows. A major problem is how to measure low flows (usually during night hours), which might pass unrecorded. There have been attempts to quantify unrecorded flow (Bessey *et al.*, 1994). In recent years better instruments have been introduced, notably combination meters: two meters of different size with a controlling valve. However, the most important change may well be remote reading of meters via either public telephone lines or short-wave radio links (Gotoh *et al.*, 1993; Haddon, 1994, 1995; Pocock, 1992). Automatic meter reading (AMR) will open new possibilities for detailed analysis of water consumption at individual premises, and provide valuable data for modelling, design, planning and other jobs.

The billing data have to be allocated to respective districts in a simple and straightforward manner. This can be a difficult task if tens or hundreds of thousands of customers are involved. As the coordinates of individual houses may not be readily available, a public information system should be used for allocation of users to the districts. This important question is discussed below.

Difficulties in real-life systems

In practice, the main difficulty is to separate the conveyance system from 'pure' distribution, and to define a sensible set of districts acceptable to all branches of the water company – robust enough to be easily updated through the changes that are inevitable in real systems.

However, this task must be done before a data capture system can be designed. The whole territory of a water company should be divided into relatively small districts. The guiding principles are:

- A district should not contain reservoirs or pumping stations (otherwise the management of pressure will be difficult or even impossible).
- All flows to and from a district must be monitored by adequate meters.
- The boundaries of a district should not be frequently changed by operating its boundary valves – if this needs to be done regularly, then the district should be redesignated.
- A district should be small and homogeneous in terms of population and territory.
- Full use should be made of any existing system for territorial subdivision and addressing, such as the Postcode system in the UK or the ZIP Code in the USA.

These principles should be amended as required by local circumstances in each case. Note that this difficult task can be and must be carried out by local staff who have specific, detailed knowledge of the system.

The next two figures show real-life examples of demand management districts in urban (Fig. 1.12) and rural (Fig. 1.13) areas. Note boundary valves in Fig. 1.12, and also the lack of one in the place indicated by a circle – this pipe intersects the boundary, yet one is not sure whether it is closed (by using sector valves) or not. Such difficulties can only be resolved by involving local staff and their local knowledge. Fortunately, such problems do not usually arise in rural areas (Fig. 1.13).

The best solution when drawing the dividing line between demand districts is to avoid cutting across distribution pipes, as in the example in Fig. 1.14, but in many cases this may be

Figure 1.12 Demand management district in an urban area.

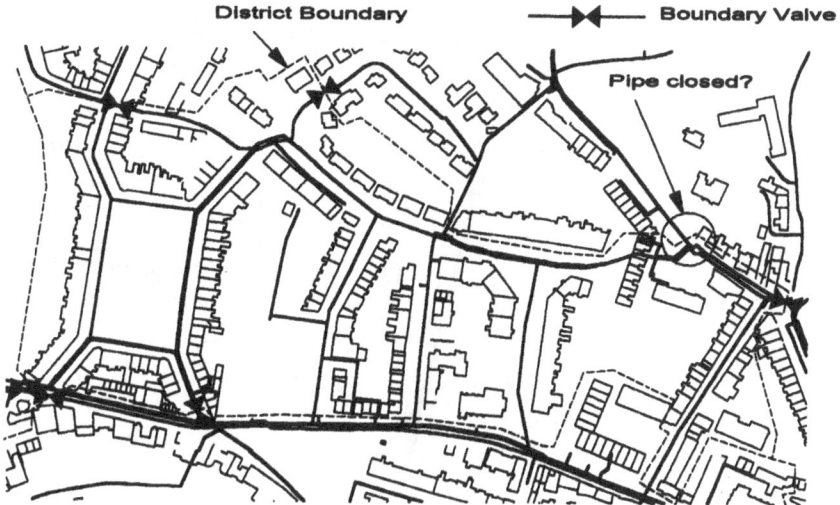

District Boundary   ▶◀ Boundary Valve

Pipe closed?

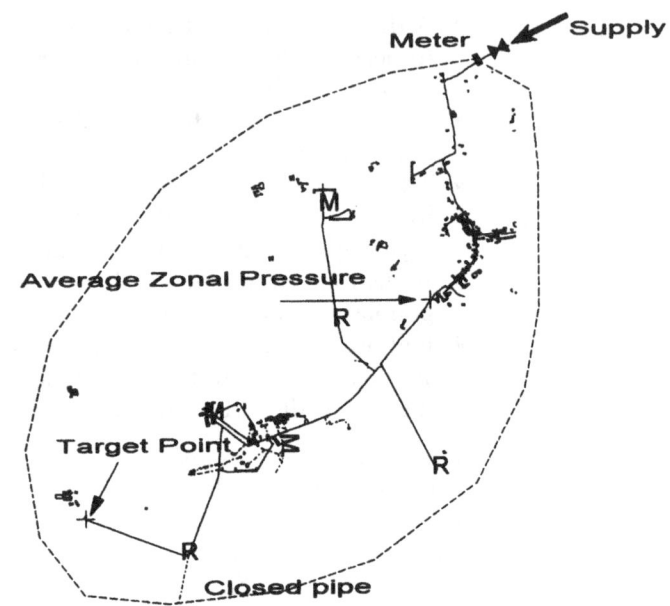

Figure 1.13 Demand management district in a rural area.

unavoidable, as in Fig. 1.15. Then meters should be installed at boundaries, in this case at points 'A' and 'B' – note that pipe 'I' just carries water to a distant area and supplies no customers in the two districts on the map. Meters should be installed in pipe 'II' on the district boundary at point 'B' and on the branch from pipe 'II' at point 'A'. No meters are needed on pipe 'I'.

Figure 1.14 Two districts are separated.

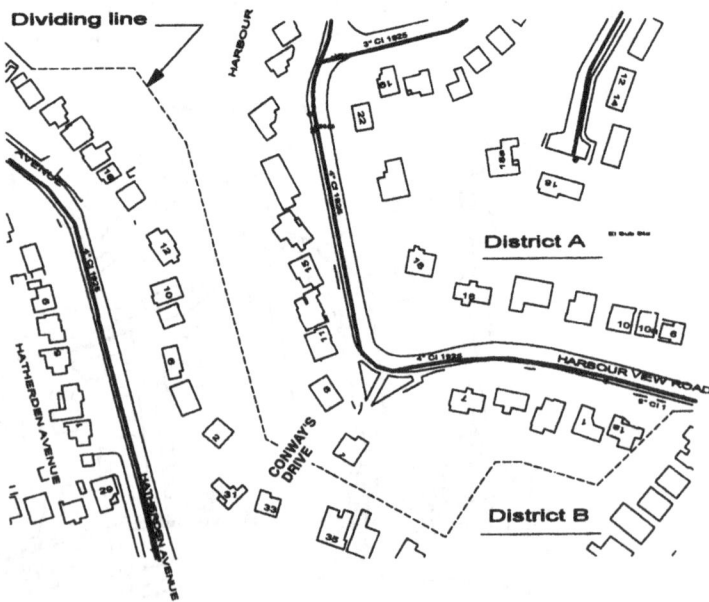

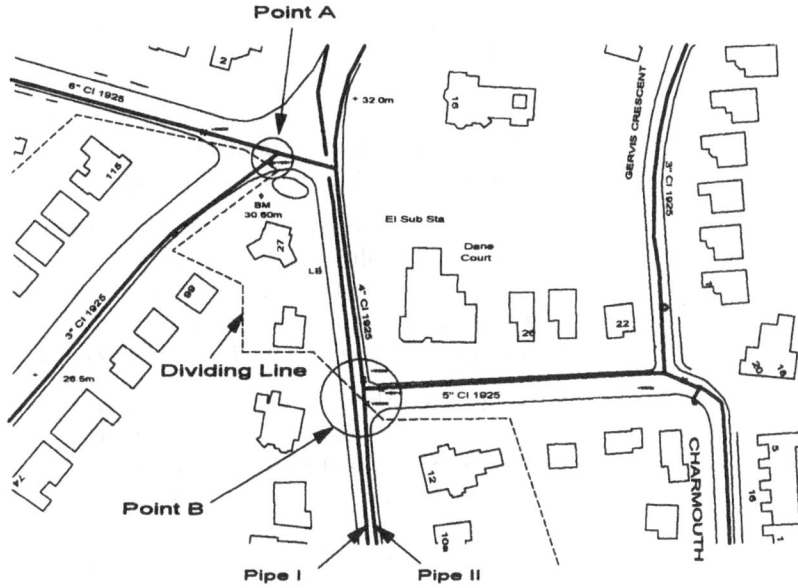

Figure 1.15 Meters installed at boundary.

This small example illustrates the importance of the careful delineation of demand management districts in practice – and how expensive it might be.

1.4 Data capturing system

Telemetry system

This is the best – but also the most expensive – system for monitoring a water supply system. It was developed in other industries much earlier, coming to water supply relatively late, in the 1970s. Since then it has become a standard feature of advanced water companies. The elements required in a complete telemetry system (Bland and Townend, 1987; Grombach, 1986; Huntington, 1990; Mitchell, 1993; Olner, 1985; Rance *et al.*, 1993a, b; Snoxell, 1993a, b) are the following.

- Instruments (sensors) that measure flow, level, pressure, pump status, valve opening and quality of water
- Analogue/digital converters, which change an analogue signal (voltage or current) into a digital signal – more suitable for transfer at a distance
- Outstations, which collect data from several sensors and control the local process
- Transmission – modems, communications links (radio, telephone lines, signal cables, etc.) and accessories
- Process (telemetry) computers
- Data processing computers with peripherals

A simplified scheme of a telemetered control system is shown in Fig. 1.16. Note the hierarchical structure from the host computers downwards. Such a system, if properly designed and used, is a priceless asset for efficient operational management and control of large and complex water supply systems. The benefits of such a modern and sophisticated system are many:

- excellent control over the water supply system;
- easier handling of emergencies;
- significant reduction of operating and maintenance costs;
- significant reduction of losses of water and energy;
- improved level of service to the general public.

There are other, intangible, benefits, like the increased competence and confidence of staff, easier introduction of new technology, and better information at all levels, from the top management down, which all pave the way for further progress.

However, there are some drawbacks, too. The price of such a system can be high. Moreover, it needs specially trained staff for operation and maintenance, expensive spare parts, constant surveillance, etc. It is also quite inflexible and all changes – inevitable in real-life systems – can be made only by specialised companies, incurring further costs and delays.

The development of a full telemetry system – desirable as it might be – cannot be completed overnight. The process is one of steady development and improvement, typically over several months or even years, and cannot be accelerated.

New technology, servers, powerful PCs and Windows-based software have all opened new possibilities. The host computer

Figure 1.16 A telemetry system.

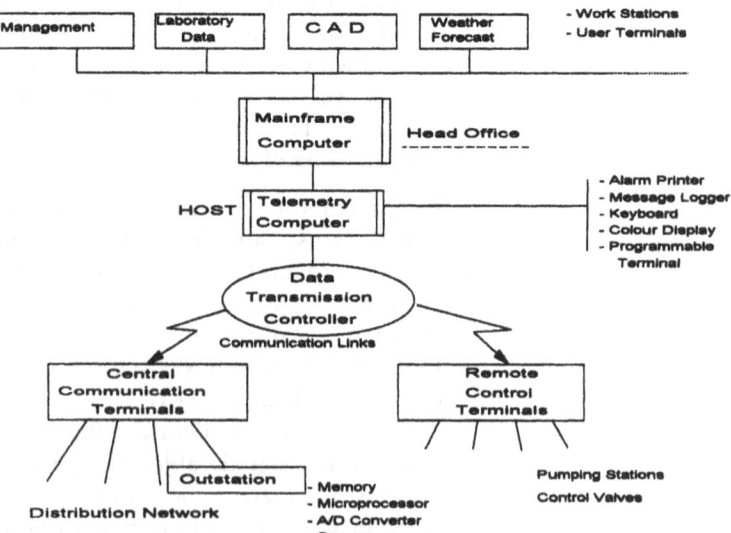

can now be replaced by computer networks such as WAN (wide-area network) and LAN (local-area network) – Fig. 1.17. The computing power of the latest PCs is such that many tasks previously done by a mainframe can now be entrusted to them, so the whole system is radically decentralised. Staff can have an active role in development of control software, so all subsequent changes could be dealt with in-house. The tendency is to use 'off-the-shelf' equipment and software, which is available at ever lower prices. The monolithic control rooms are being replaced with networked systems of PCs, used by people familiar with Windows-based applications (Lawson, 1996).

The new technology was initially used in the industry to improve local control of facilities such as pumping stations (Fig. 1.18) and pressure-reducing valves (Fig. 1.19). A modern PLC (programmable-logic controller) can collect local data and control the functioning of pumps and valves.

Selected operational data are regularly collected by telemetry that monitors the facilities, but telemetry will not be employed to interfere so long as the situation is normal (i.e. all important parameters are within the prescribed range). In emergencies the telemetry system will permit remote control over facilities. In this way the staff can be relieved of constant surveillance over facilities in the field and other repetitive (and boring) tasks. This is particularly important in systems with several PRVs – compare centralised control (Fig. 1.19a) with local control (Fig. 1.19b).

Figure 1.17 Decentralised telemetry system.

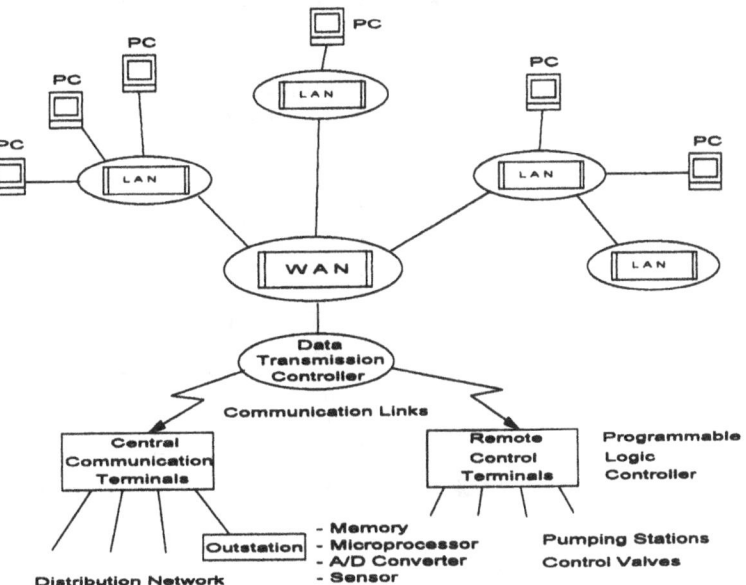

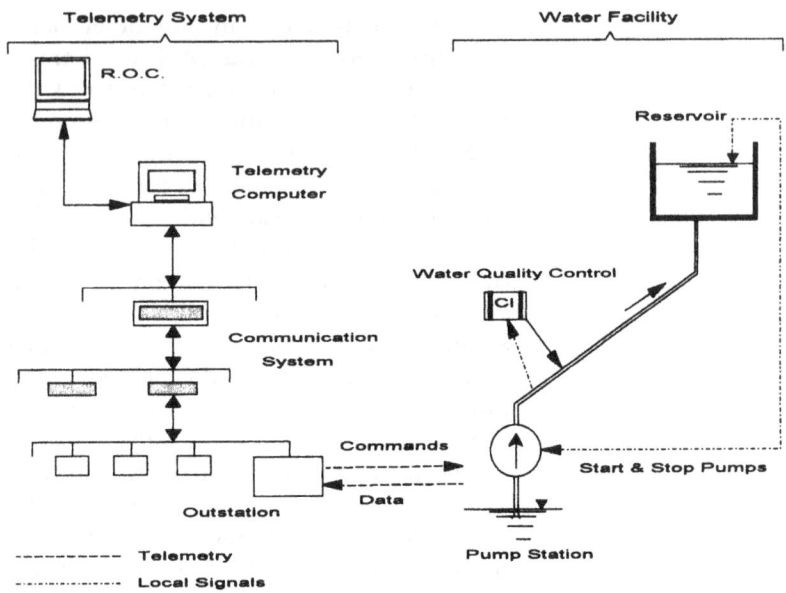

Figure 1.18 Control
over a pumping station.

The first task of a telemetry system is to monitor the supply
system and permit a measure of control (especially in emergencies), but there is a second one: capturing of operational data.
This information is valuable to many in the water company, and
should be made available as easily and seamlessly as possible.

Figure 1.19 Control
over a PRV: (a) centrali-
sed control; (b) local
control.

The transfer of data captured by the telemetry system to other
off-line applications, such as computation of water balance,
leakage analysis, modelling, etc., is rather difficult because of the

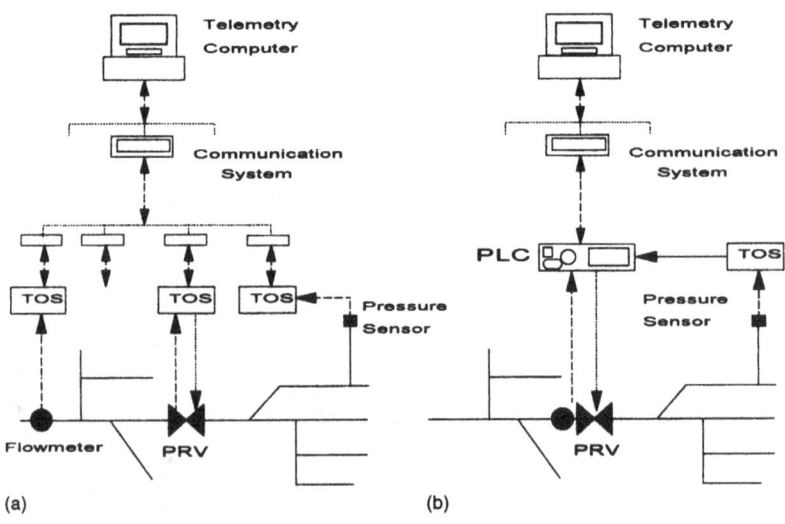

sheer volume of raw data and incompatibilities in hardware and software. The problems can be overcome by developing a software interface, and allocating sufficient resources to this task.

Data-loggers

After a telemetry system, the next best data capturing systems are based on data-loggers. These are instruments that collect and store data from primary instruments *in situ*. Every logger must first be prepared for active duty: loaded with data related to its next site, time and date, etc., prior to installation. This can be done on a personal computer using simple software.

Data collected during the logger's activity have to be downloaded to a host computer, either via a telephone link or by direct link with the office computer. More expensive loggers have their own peripheral devices: printers, displays, even portable microcomputers. The instruments used in conjunction with data-loggers are shown in Fig. 1.20, showing the great potential of this device.

Compared to a telemetry system, loggers have the advantage of being portable from site to site, while retaining almost the same level of precision and reliability. On the negative side, the data record they provide is not as complete or continuous as one from fixed-site telemetry. Therefore, the best policy seems to be to use loggers to complement the telemetry system. This means that the loggers should be used throughout the distribution system to cover:

- demand management districts,
- control zones within a district,
- important and large users.

Moreover, loggers are used very often to check the accuracy of telemetry instruments, or to examine more closely a part of the system after, say, a rezoning or a change of control policy, etc.

A data capture system could be based solely on data-loggers, as shown in Fig. 1.21, but more often it just complements the telemetry system. A rule of thumb is as follows.

- The telemetry system covers the sources and conveyance system (main reservoirs, water trunks, pumping stations), the so-called *production system*.
- Loggers cover the *distribution system* – from service reservoirs to customers.

Data from a logger can be downloaded to a lap-top computer and analysed on the spot, and/or transferred to the office computer to be analysed later (the option in Fig. 1.21a). Note the time lag (trip to the site and back cannot be made every day).

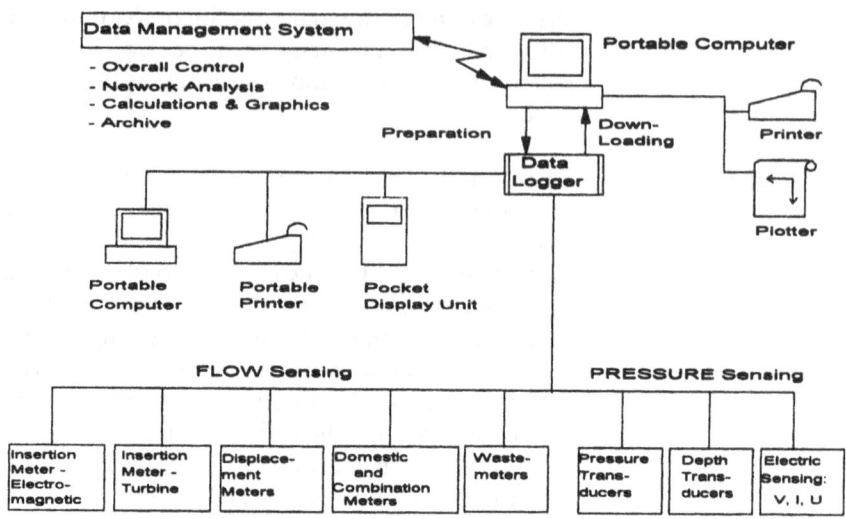

Figure 1.20 What a logger can do.

A more efficient solution is to send the data via radio or telephone line to either the district office or the central control room (the option in Fig. 1.21b); the data can be downloaded at any moment – usually once per day. Note that the data are not available before the downloading procedure has been completed – as opposed to the telemetry system, where current data are

Figure 1.21 Logger-based data collection system: (a) local interrogation; (b) district telemetry scheme; (c) central telemetry scheme.

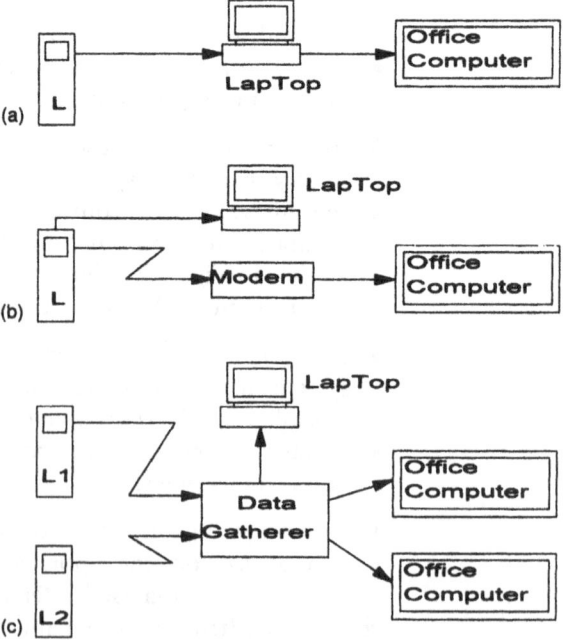

always available. On the other hand, the price of a logger-based system is relatively low.

The tendency is to move closer to a telemetry system – many data-loggers are being converted to telemetry units simply by connecting them to the public telephone system via a dedicated line – these are *dial-up loggers* (Fig. 1.22). This is, effectively, a telemetry system (the option in Fig. 1.21c). The user can call the logger at any time and download data, collected since the previous download, up to the current time. The procedure can be fully automated and controlled by a computer (Gotoh *et al.*, 1993).

Modern, well equipped water companies make good use both of telemetry and of loggers to great advantage.

Instruments with local readings are the cheapest solution in countries where labour is cheap and plentiful (and modern equipment too expensive). The water company should organise regular readings on-site; the data could be logged in a portable device or in a logbook and later sent to the control room. Such a control system cannot be as efficient as those just described, but nevertheless may be a necessary stage in development, giving time and opportunity to introduce gradually more sophisticated devices like data-loggers, etc. The historic data are very useful to compute trends and long-term effects (for an example see Obradović and Filip, 1986). Moreover, even in the most advanced water companies there are some locations that continue to be supervised in this way, for one reason or another.

The choice of an appropriate instrumentation and control system is a never-ending task. The criteria for selection are:

- ergonomics – size, weight, portability, visibility, ease of use, complexity, connections, etc.;

Figure 1.22 Dial-up meter.

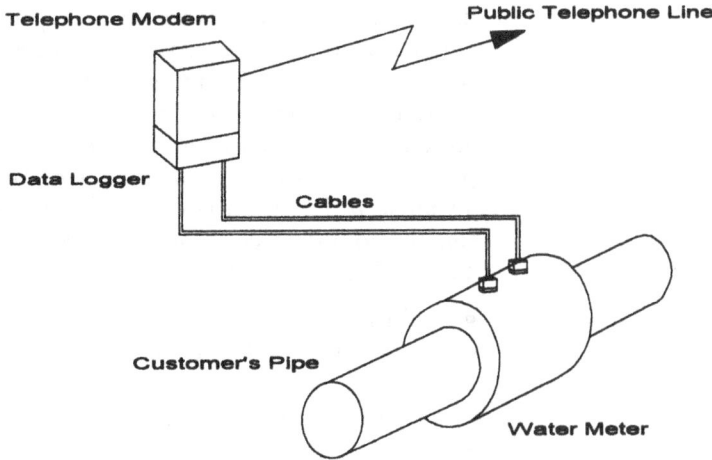

Telephone Modem

Public Telephone Line

Data Logger

Cables

Customer's Pipe

Water Meter

- operation – data handling characteristics, adaptability, memory capacity, ease of data retrieval, programming facilities, etc.;
- accuracy – linearity, hysteresis, repeatability, stability of instruments and their components;
- reliability – ruggedness, waterproofness, no interference with radio frequency, thermal stability;
- hardware and software requirements for host computer;
- price and terms of delivery;
- support – production range, quality control, training facilities, technical support, marketing, plans for improvement of this product.

All these questions are important when making decisions, which will have far-reaching effects, on how to create a cost-effective control system.

Billing data

Billing data are generally collected, manually, from water meters installed at customers' premises. The main purpose is to obtain data on which to base a bill and generate revenue, but the data are also irreplaceable information for leakage control, design and planning, and many other jobs. The problem is that data are slow in coming, often half-a-year old, inaccurate, prone to error, difficult to check and adjust, etc. Attempts have been made to employ hand-held data entry computers (HDET), but these are not a complete solution as they still depend on a relatively high use of manpower.

A new technology – automatic meter reading – is a potential answer to the problems (Gotoh et al., 1993). The meter – of a much improved design – will be controlled from a remote site, either through public telephone lines (Fig. 1.22) or radio links (Fig. 1.23). Data from individual meters will be transmitted to a data concentrator linked to a repeater. The repeater will communicate with a process computer in the company. The routine can be programmed and needs no human intervention apart from supervision. The data are immediately available to all concerned: customer service, meter shop, control room, planning and design department, etc. Another solution is to have a vehicle with the short-wave radio station that regularly collects signals from data concentrators (Haddon, 1994, 1995; Pocock, 1992). The benefits for the company are:

- real-time pricing,
- load control and analysis,
- alarm reporting,
- automated on/off service,
- planning of maintenance and replacement of water meters,

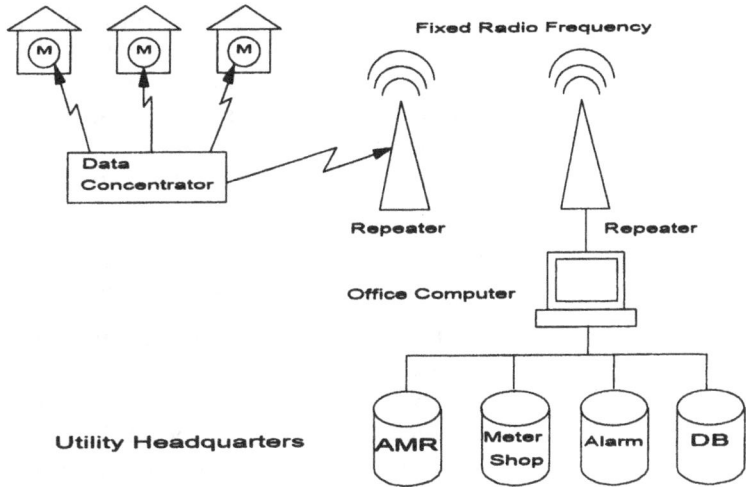

Figure 1.23 Radio links – another solution.

- detection of tampering with a meter, etc.

Further potential benefits that may well develop in the near future are: electronic funds transfer, customer service enhancement, electronic billing and many others.

The characteristics of the different systems outlined in the above three subsections are summarised in Table 1.1.

1.5 Data management system

Territorial sub-division

The steady inflow of operational data has to be well received and used in the water company. It involves data processing and archiving, as well as dissemination of information to all concerned. All this is done by a data management system (DMS). The simplified structure of one such system is shown in Fig. 1.24.

Table 1.1 The characteristics of the three different systems

	Telemetry	Data-loggers	Billing system
Numbers	Small	Average	Huge
Frequency	Minutes	Minutes	Months
Location	Known and fixed	Known and changeable	Difficult to pinpoint
Accuracy and reliability	Excellent	Good	Average to poor
Flexibility	None	Good	None
Gaps in records	Few	Many	Average
Installation costs	Very high	Low	Average
Operation and maintenance costs	High	Low	Not applicable
Required knowledge	High	Average	None

The *region* covers the whole territory supplied by the water utility. It could be divided into administrative *areas*, covering large tracts of land. One area could be subdivided into several *supply zones*, each having its own sources and distribution network. The transfers to or from individual supply zones must be identified.

Each supply zone then has three branches:

- sources,
- transfers (both to and from the zone),
- demand management districts – smaller territorial units.

Each *district* can have one or several *control zones* – a block of houses as an example – where demand and leakage losses are supervised constantly. The boundary of a control zone could be changed frequently, by opening/closing of control valves, while other territorial units remain unchanged for long periods, if not permanently.

The data describing all the items – their location, boundaries, area, population, important customers, etc. – are kept in a GIS (geographical information system) (Glasbrook, 1993; Scherer and Phebey, 1995; Snoxell, 1993a, b; and others). This is a computer system designed to capture, manage and process all information linked to a specific location (hence the term 'geographical'). The data are either graphical (such as maps and schematics) or textual (pipe diameter, reservoir volume, population numbers, etc.). The equipment is highly specialised and quite expensive compared to PC technology. It can only be operated by trained people. A further difficulty is the relative

Figure 1.24 DMS configuration.

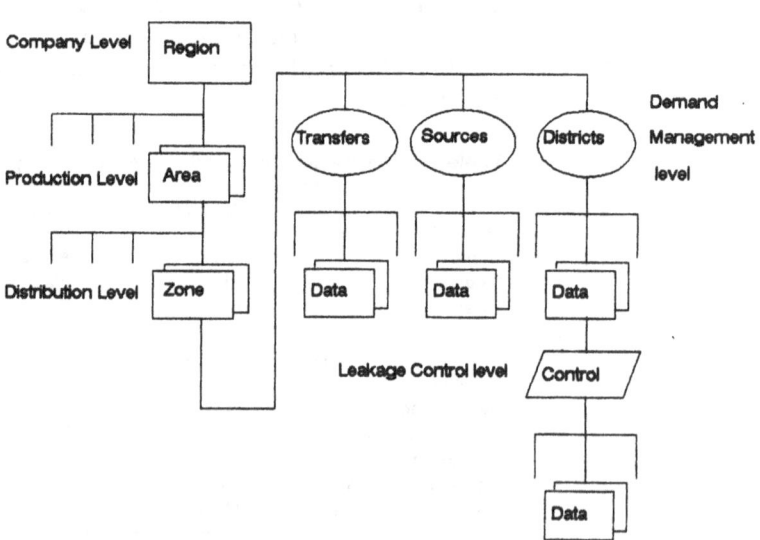

inaccessibility of captured data, except to those in the GIS department.

Despite these shortcomings, GIS is rapidly gaining ground in water utilities, because the potential advantages outweigh the problems. The technology is rapidly acquiring new tools and systems such as GPS (Global Positioning System), which permits instantaneous reading of global coordinates everywhere in the world, using a network of satellites.

A real, operational system is outlined below. In this system, information about the physical assets of the company (pipes, reservoirs, pumps, valves and so on) in a variety of formats (graphical and narrative) are held in the GIS system, and operational data (flows, pressures, pump states, reservoir levels, etc.) are held in a separate database. However, users of all this information, from either source, are able to access, manipulate and use it through their own network, stand-alone or portable PC. What the 'end-user' cannot do, without appropriate authorisation, is to alter the two 'information banks'.

The GIS department keeps vital data about all the physical assets of the company. Physical changes in the water supply system (e.g. a new main or a pump replacement) are regularly reported to the GIS operators and processed. The updated information is kept in the GIS database, and accessed through GIS hardware and software. To improve the accessibility of these vital data, various solutions are possible; one is shown in Fig. 1.25. It is based on the fact that GIS can export data in a standard format, like CGM (Computer Graphics Metafile) for instance. At regular intervals (say, once per week) the updated files are converted into CGM format and transferred to the central file-server, from where they are disseminated to several local servers (Fig. 1.25).

Figure 1.25 Regular update procedure of GIS data.

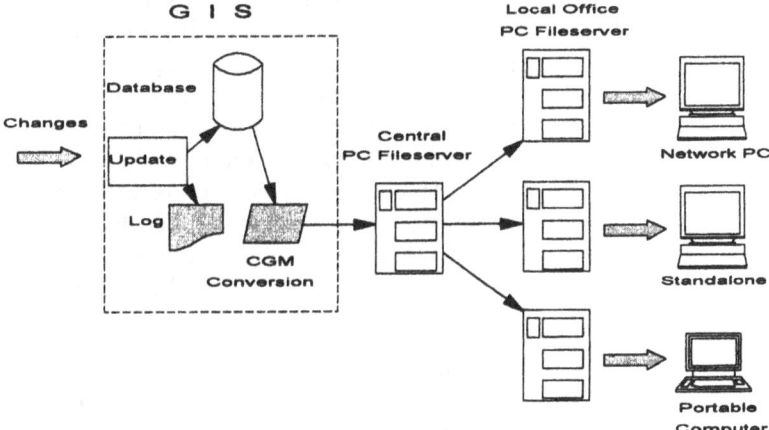

Operational data from the system are kept at the appropriate place in the operational database, with the configuration files (explaining the structure) at the root. Note from Fig. 1.24 that all lower-level items are completely contained in the higher one, with the exception of control zones – the sum of districts makes a supply zone, etc.

The files contain operational data as captured, without any change – interpretation, correction and filling of gaps is left for later phases of analysis. The users always see data 'as captured' and can judge for themselves how reliable they are, rather than wonder what the effects of any intermediate data processing have been.

Another feature of this system is that all data are kept in a standard format, and can be used easily by outside users for importing to spreadsheets, word-processing or utility programs, apart from having quite well developed analytical tools built into the basic information system.

Fresh data come from four main sources:

- telemetry system,
- data-loggers installed at selected places (portable equipment),
- billing system,
- manual readings of instruments.

Data come in continuously, through various channels and at various frequencies. Some data are taken primarily for tasks other than operational management, e.g. billing data for charging the customers. These primary functions should not be interfered with in developing links and data transfers to other data systems. Figure 1.26 shows how the telemetry data are processed.

The telemetry system could be very large, collecting thousands of signals. All operating data are collected in central telemetry computers (duty and standby mainframes) and refreshed continuously. Operators in the regional control centre (ROC) keep close control over the system. Other users can have access through dial-up computers, as in Fig. 1.26. Every few minutes (typically 15 min) the required data (mostly flows, pressures and water levels) are transferred to a PC file-server, where they are available for all analysis and archiving purposes.

The telemetry covers practically all sources, transfers between supply zones and main distribution trunks, so these data permit computation of:

- source outputs,
- total demand (consumption and losses) in the region, areas and supply zones.

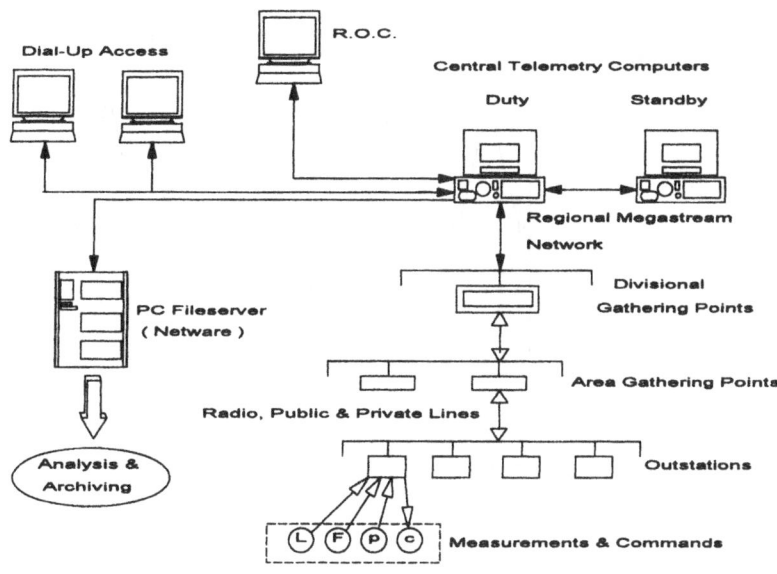

Figure 1.26 Telemetry configuration.

Fresh data are cross-checked whenever possible by setting one reading against others. Note that telemetry is a large, independent and notoriously self-sufficient system, designed primarily to supervise and control the water supply system – data capturing and archiving is usually not the first concern. The example shown in Fig. 1.27 illustrates this point.

A service reservoir is supplied by a pumping station. The pumps are level-controlled units, and therefore the current value of water level in the reservoir is known in the station (signal 'B'), as well as in the control centre itself (signal 'A'). If either of the two signals is lost, the control centre will know immediately and

Figure 1.27 Telemetry coverage of a reservoir.

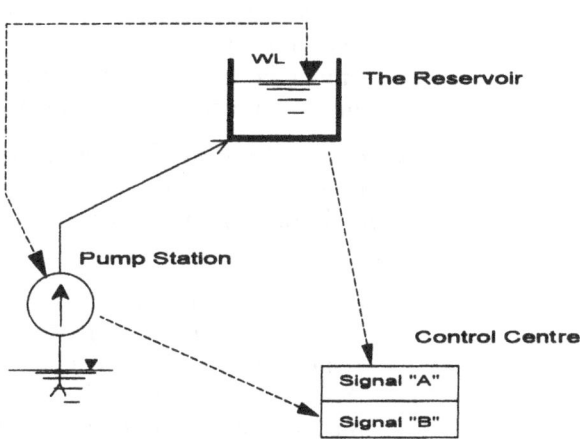

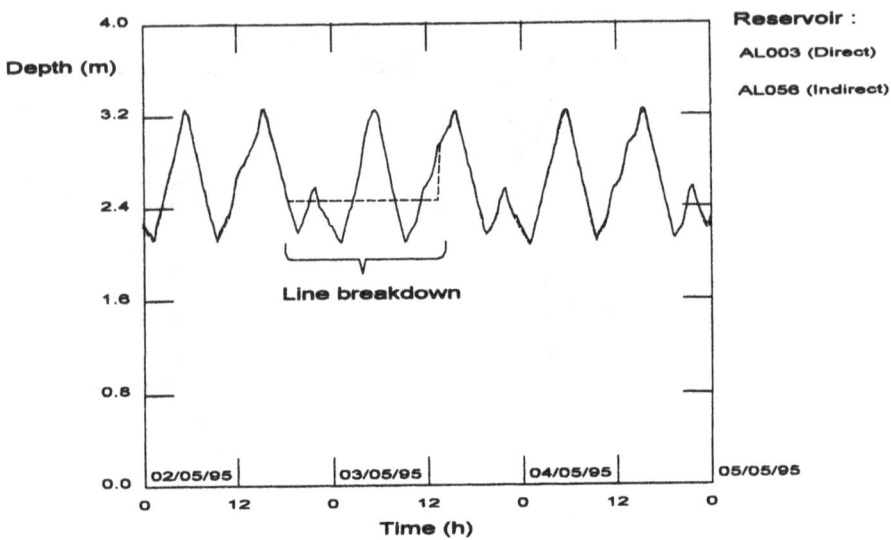

Figure 1.28 Telemetry
data system – water level
data: two signals that are
not identical.

react accordingly; both signals are treated as independent
variables, not necessarily identical. Sometimes, indeed, they
are not equal, as shown in Fig. 1.28.

**How should data
be merged?**

The overall picture is obtained by merging the data from all
these sources. This task is rather difficult – given the size of files,
different format and frequency of data capturing, and hardware
limitations – but has to be achieved. At present, there are no
standard or 'typical' solutions to the problem; however, one case
from real life is shown in Fig. 1.29.

Billing data are regularly transferred to the same PC file-server
receiving telemetry data. There are three types of files:

- monthly-read customers, the largest and the most important
  ones whose water meters are controlled (read) every month or
  even more frequently;
- commercial users, controlled typically twice per year;
- domestic users, also controlled twice every year.

One difficulty is in identifying the precise location of individual
customers and how they are connected to the network. There are
several solutions to this problem, none completely satisfying,
and all rather expensive in time and effort (Shore, 1986):

1 Link individual customers to the network records. This
   sounds very promising, but it takes enormous effort to
   achieve because even medium-sized utilities supply 50 to 100
   thousand customers. Moreover, the network itself is changing, with new customers appearing, etc.

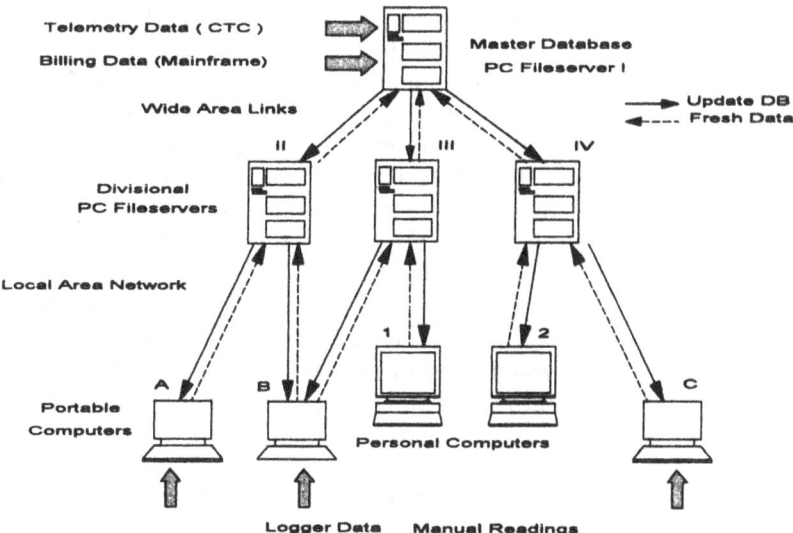

Figure 1.29 Merging operational data.

2 Link individual customers to streets. The public records contain data on streets and house numbers (in most cases), and the address of every customer is already part of the billing system. However, this solution has other drawbacks, such as not being unique (duplicate names of streets), being subject to changes from outside factors, and having many streets that are too long. In rural areas there might be no streets at all. Worst of all, the problem of linking the customer to the network remains open – the customer may not be supplied from a pipe lying in the same street.

3 Link each customer to a node of a model of the network or system. Again, the initial input and ongoing maintenance of such a system would be enormous.

4 Use a public system such as the ZIP Code in the USA or the Postcode in the UK. Thus the coordinates of individual users are supplied and updated by the state and local authorities. The customers must only be linked to that system, which initially takes a lot of time and effort, and has to be maintained later on.

In the case of the real-life information system described in section 1.3, option 4 was adopted, and the system is based on Postcodes.

The end-result is the coordinates of every customer. These coordinates are linked to the demand management districts through a GIS, as described below, so the total metered consumption can be computed for each district and hence for supply zones, areas and ultimately the region by simple summation (note, however, that this is *not* the total consumption of water because the system is not fully metered). Note that it is not enough to know the exact coordinates of a customer – who might not be connected to the nearest pipe to the property but a pipe outside the Postcode or ZIP Code area.

The data from telemetry computers and the billing system (another mainframe computer) are easily processed and stored in the master database, on the PC file-server I (Fig. 1.29), because they come from a single source. The other operational data come from other monitoring devices (hours run meters, the power company meters, and water meters located all over the region which are not incorporated into the telemetry system). Some of these instruments are equipped with portable data-loggers and others have to be read on the spot (manual readings). Attempts to organise centralised collection and processing of this valuable information have run into difficulties and high costs (the region in question is very large), so it was decided to organise distributed data processing instead. The staff were given portable computers (lap-tops) with adequate software: maps provided by GIS, as well as operational data for the last 100 days, for their respective area only, but in great detail. The user can then enter manual readings and/or download data from the data-logger, for all places he/she can visit during the day.

The members of staff report regularly (perhaps once per week) to the divisional headquarters, where a file-server keeps the database updated every night by the master database (Fig. 1.29). Once the user connects his/her own notebook computer to the divisional PC file-server, two tasks will be carried out automatically:

- fresh data from the notebook computer will be taken by the divisional PC file-server;
- the database on the notebook computer will be updated with new telemetry and billing data, and also data from other sites.

The process only takes a few minutes. The field data are kept in the divisional PC file-server and transferred each night to the master database for final processing.

Divisional PC file-servers provide additional capacity and flexibility for the system; even in the worst case – when all communications with the centre are broken – local staff can continue to work unhindered, and nothing will be lost.

The software on notebooks is very user-friendly and specifically tailored for field use. It is very robust and foolproof. No special training is needed. Note also that all the hardware is commercial and could be easily replaced.

1.6 The dissemination of operational data

All water companies always have a data management system (DMS) of some sort, to keep track of demand, water production and operation costs. This system may be quite rudimentary, based exclusively on manual data input and processing, or a highly sophisticated, computer-based system. A DMS reflects the general level of technology, local conditions, skills of management and of staff, tradition, etc. Given enough time, attention and resources, every DMS will progress and eventually reach a high level – but it cannot be rushed. Many attempts to leap over several stages of development have ended as costly failures. Therefore, it is important to review the experience of other companies and get a deeper understanding of the logic of development before starting with ambitious plans for modernisation.

Figure 1.30 shows schematically a modern computer-based data management system. The operational data flow through various channels, the most important ones being:

- The telemetry system provides information about source output, storage in large reservoirs, flow through main trunks, etc. (database 1 in Fig. 1.30).
- Data-loggers monitor local demand and operation of reservoirs and pumping stations not covered by the telemetry (DB 2).

Figure 1.30 Data management system.

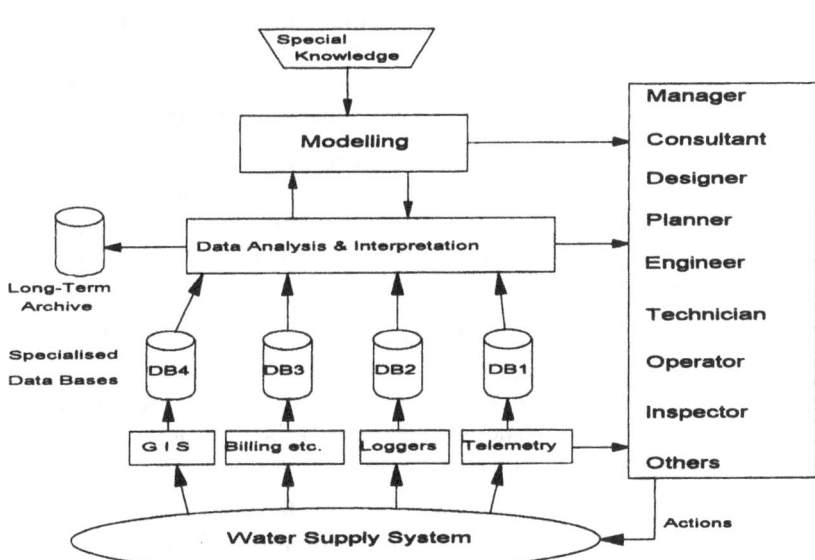

- The billing system contains the information about customers: who they are, where they are located, how much water they use, etc. (DB 3);
- The GIS (geographical information system) has digitised data about the capital assets: pipelines, reservoirs, valve chambers, etc. (DB 4).

The local database is usually of a 'rolling archive' type, which means that data are saved only for a limited period of time (say a month), to be overwritten afterwards by fresh data. Hence there is a need for a long-term archive, where the most important data, 'distilled' from the short-term archive, are stored indefinitely for any later use.

Software interface   Important considerations regarding the software interface are:

- Large organisations have a number of large, well established data systems, which are traditionally independent and self-sufficient, and are run by different sections within the company.
- Computerisation of the systems has happened at various times, and therefore they use different hardware and software solutions (e.g. the billing system is almost everywhere a mainframe application, while loggers are PC-oriented).
- Each piece of data has its own, unique identifiers (e.g. customer accounts or reference numbers in the billing system), which are unknown to outsiders, and subject to sudden changes without prior notice to other potential beneficiaries within the organisation.
- Additional attributes that may be of great benefit to other users are accepted and maintained reluctantly, and basic changes are obviously unacceptable.

The only answer is to establish an interface that will provide a link between these large systems by associating one set of unique identifiers with its counterpart. It has to be adopted by all concerned, and maintained by a special task force. There are two problems:

1 To make this interface in the first place, correct errors and establish confidence in the link.
2 To maintain the interface through all changes and reconstruction of one or other of the large systems to which it is linked.

It is very important to delineate responsibilities: Who allocates and can change identifiers and how will others be informed about the changes?

Software interfaces can be organised in plain text and graphic files accessible to all other systems – internationally accepted formats should be used to facilitate import and export of data files. A proposal for standard interface formats for telemetry, loggers and GIS, prepared by Rajko Cavor, is included as Appendix A.

Note that creating and maintaining a software interface, such as the one described above, is a separate job in its own right. An integrated information system is a tremendous advantage for the water company, enabling everyone to perform their tasks better and more efficiently than before (Beal, 1988; Bland and Townend, 1987; Bost, 1986; Cosgriff *et al.*, 1985; Fontaine, 1983; Huntington, 1990; Scherer and Phebey, 1995; Snoxell, 1994; Takagi *et al.*, 1983).

2 Analysis of operational data

2.1 The value of on-line data

The decade from the mid-1980s saw a shift from relatively primitive and labour-intensive data collection systems to quite sophisticated ones, collecting and transmitting the majority of data automatically to a central point, with little, if any, human intervention. Staff levels are reduced to a few highly specialised technicians, who supervise the process but do not intervene if operations are proceeding correctly. The state of the monitored system is always known by having frequently refreshed data available.

The efficient operation of a water company is impossible without a reliable database, accessible to all, at all times. An on-line database reduces the time lag between an event and its entry in the files; for example, typical scanning intervals of 5 min in a telemetry system mean that the database effectively represents the current state of the system. Such fresh information is extremely valuable to many in the company, from managers to operators.

Individual items of operational data (e.g. the current value of flow in a particular pipe) are of limited value – until put into the wider context of the whole system and compared with other data. Given the huge number of data, it is obvious that the flow of operational data must be carefully channelled through the company, as shown in Chapter 1. Data are gathered by telemetry, dial-up and portable loggers or manually processed by computer and stored in a database. The operators in a control centre use these data for evaluation of state, analysis, alarms and various reports, and use the same system to control pumps, valves and other devices.

It should be stressed at this point that the incoming data need to be interpreted correctly.

At first sight one may think that this is a nonsense: if all relevant data are always available, on well designed screens and/ or synoptic tables, it should be easy to make decisions if everything is shown in detail. This is not so. At least, it is not the case if chaos is to be avoided.

Flooded by data, a novice will soon feel lost and helpless. Numbers in themselves are meaningless, and only rise to the status of information if one understands how the system operates and uses the data to support that knowledge. Knowledge of the system, supported by reliable data, allows a soundly based judgement of the condition of the system and an understanding of the effect of intervention (or non-intervention).

It is not possible to learn how to control a huge water supply system by 'trial-and-error' methods – and no one would authorise this. Trained operators rely upon accumulated experience to control the system. They carry a mental picture of the real system, and new information just helps to confirm the picture.

The 'mental picture' is a kind of model, but is inaccessible to others. Individual operators may have excellent knowledge of the system, very good logic and other qualities – but these qualities cannot easily be transferred. Novices need months and months of training before the old-timers give them a chance to do anything on their own.

Conversely, reliance solely on experience has its dangers: if the data are not correctly or fully understood, one can make mistakes that pass unnoticed because no one outside the control room has access and/or enough knowledge to spot the error. But, in time, one may notice a steady deterioration of service, increasing losses of water and consumption of energy.

Ideally one needs to combine experience of the water system with an understanding of the opportunities and limitations of new technology. Achieving this blend of understanding and knowledge is difficult, and calls for a lot of patience and tolerance from all sides.

Everyone aspiring to monitor and control water supply systems, and keen to take full advantage of the information technologies available, should be familiar with the advantages and shortcomings of all aspects of the information technology – from sensors to the on-screen data.

The system operator, to fulfil his/her function, needs to understand that, however well the manufacturer constructs the sensing devices (a meter or a pressure sensor for example), they will be subject to limitations. An overlap of understanding – a common ground – is certainly needed, otherwise one may create a situation where the water company is 'data-rich, but information-poor' (Rees, 1996).

The data gathering system includes a number of key elements and is affected by a number of factors, many of which interrelate and affect the validity of data delivered into the database. The limitations of each element, and the potential cumulative effect of the various elements on the final data, need to be understood to allow sound judgement of the validity of information being monitored at the centre. This understanding is also essential if the most cost-effective data gathering system is to be developed. For example, very accurate instrumentation can be very expensive, but if the data processing elements of the system include 'rounding' or 'truncating' of data, it may be that less costly instruments are perfectly adequate for the circumstances.

The key elements and factors are:

- Primary instruments
- Transmission system, including signal converters and the transmission medium
- Reading, scanning and archiving frequencies

Instruments

The three principal measurements, essential in understanding the hydraulics of supply systems, are flow, pressure and level. A fourth, often used for determining flow, is velocity. Instruments have, historically, developed within a number of well defined groups or types, but the technologies developed over the last 20 years or so have resulted in a wide proliferation of primary instrumentation. A number of instrument types are listed below. It is not proposed to describe in detail the advantages and disadvantages of each type listed. There is adequate literature on that subject elsewhere. However, an understanding of the 'pros' and 'cons' of each of the instrument types is basic to an understanding of the validity of data being delivered to the information system. From an understanding of the strengths and weaknesses of the various instrument types should flow an understanding of the most appropriate type of instrument for particular situations.

For example, two common groups of flowmeters are differential pressure and positive displacement types. The former has no mechanical parts in the primary instrument, and tends to remain accurate over long periods of service. A disadvantage is that flow is proportional to the square root of the pressure difference across some form of impediment in the line (orifice plate or Venturi type of constriction), and is less accurate in the low range. In contrast, the positive displacement type of meter has moving parts, and tends to become less accurate as its mechanical elements wear with age. It follows that a differential pressure type meter will be more appropriate to measure major, steady flows from, say, a treatment plant or an impounding reservoir than a positive displacement meter. Conversely, a positive displacement meter might well be the appropriate choice to measure supplies to individual customers. Although they wear with age and tend to under-register in that condition, they are relatively cheap, easily refurbished and readily understood by customers.

Instrument types These may be considered under the four categories mentioned above.

1 Flow measurements
   - differential pressure (Venturi, Dall tube, orifice plate)
   - rotating mechanical (positive displacement)
   - electronic (magnetic, ultrasonic)
2 Point-velocity measurements
   - Pitot tube
   - electromagnetic velocity probe
   - insertion turbine
   - laser Doppler anemometer
   - ultrasonic Doppler velocity probe
   - insertion vortex
   - hot-wire anemometer
3 Pressure measurements
   - manometric techniques
   - flexing element (piezoelectric probe, etc.)
4 Level measurements
   - direct measurement (staff gauge, hook or point gauge, electrical depth, float-operated)
   - pressure-actuated (pressure transducers, air-filled and sealed system, bubbler systems)
   - electronic methods (capacitance probes, ultrasonic)
   - level detection (float switches, conducting or vibrating probes, radiofrequency, infrared)

See Fowles (1993) for more details, and also other sources like textbooks, manuals, handouts prepared by manufacturers, etc.

Instruments are man-made and not perfect. In selecting and using instruments, the user should pay attention to parameters like:

- discrimination
- accuracy
- range of measurement
- repeatability
- drift (stability)
- hysteresis
- linearity
- speed of response
- initial settings – zero offset and conversion factor (here 'human factor' comes into play)

Brief definitions of these parameters are given below, but, like the rather broad subject of instruments themselves, the reader should seek fuller explanations in other literature (e.g. Fowles, 1993), and further reading will be rewarding. For example 'accuracy' can be defined as the closeness of the instrument reading to the truth, but how is 'closeness' defined? Instrument manufacturers frequently define the accuracy of their products as either plus or minus a percentage of full scale or plus or minus a percentage of the reading. The difference may be significant or it may not, depending on the location of the instrument in the system, the relevance of the particular reading in respect of the whole system, and the cost of a more accurate instrument over a less accurate instrument.

Discrimination This parameter shows how finely an instrument can measure, i.e. the smallest changes of measured quantity that the instrument can recognise.

Accuracy This is a measure of the closeness of the reading to the true value. It can be expressed in relative terms, either as a percentage of the full-scale deflection, or as a percentage of the reading.

Range of measurement and turn-down ratio These are the range of the instrument over which the accuracy specification is met, and the ratio of the maximum and minimum readings that meet the accuracy specification. The range and turn-down ratios need to be understood in relation to claims of accuracy.

Repeatability and drift These are really two expressions of the same parameter. Repeatability is the ability of the instrument to measure the same quantity consistently, time after time. If it does not do this, then the instrument is displaying 'drift'. Good repeatability is not an indication of accuracy. It simply means that the same right or wrong answer is being presented each time the measurement is made!

Hysteresis Some instruments have a disturbing tendency to show different readings of the same physical value under rising and under falling trends. This is termed hysteresis and should be very small.

Linearity The output of most instruments – over the effective range – is normally linear. Good linearity is achieved when the calibration points are also linear. A brief definition is therefore that linearity is the departure of the calibration points from a straight line.

Speed of response This is the reaction time of an instrument, and might be important in transient conditions.

Zero offset This is the output of the transducer when the parameter being measured (e.g. pressure) is zero.

As indicated above, it is not our purpose to debate the problems inherent in the field of potential instrument error in detail. Instruments are subject to a whole raft of problems, which means that the truth of their readings is to some degree 'fuzzy' at the edges. We can and must live with those difficulties, but practitioners should be aware of them and their significance in particular circumstances.

Transmission

Once captured, data undergo several mutations before the information is presented to the end-user. First, sensor output is an *analogue* value, which must be converted into the *digital* value. The converted value is then sent through communication channels to the data gathering centre. The channels may include a number of stages. The initial stage is, of course, the instrument itself. The second stage is the creation of a digital signal from the analogue input. That digital signal may then be transmitted by private hard wire, radio or telephone lines to some local collection point, such as an outstation unit collecting a number of primary data streams, and so on until the data, eventually, arrive at the central control point. At each change, a device has to convert the input value to an output value, with a certain loss of accuracy, depending on the range and technical characteristics of the system involved. Errors are also introduced by *rounding* and *truncating* the data, resulting occasionally in serious deviations from the 'true' value.

Reading and scanning (or polling) frequencies

The analogue signal at the sensor is always present and constantly changing, because the sensing element, of necessity, responds directly to the parameter being measured (e.g. the diaphragm of a pressure cell) and is 'read' by the signal converter at intervals, typically of 30 s. All these readings are then communicated to either a data-logger or an outstation. A data-logger will not store all the data, only those current at predetermined intervals of time, up to say an hour; and an outstation will normally only retain the last incoming data that it received.

Neither outstations nor data-loggers that have been linked into a telemetry system communicate their data as they receive them, but wait to be interrogated from the central control. The centre can handle only a limited number of channels (communication lines); therefore the central telemetry process computer *scans* (or *polls*) each outstation in turn, usually every few minutes (1 to 6 min is a common range). Data are there, being replaced typically every 30 s or so, all the time, but the user will get only those which happen to be present when that particular line is 'open', i.e. when the centre is calling and directly connected to the outstation. Anything between scans will be lost. This is a minor drawback in water supply because, after all, the changes are slow. However, some quick events might be overlooked – for example, the precise time a pump actually started. Data-loggers are handled in a significantly different way. They are effectively autonomous and record data with a user-prescribed *scanning* (or *polling*) *frequency*; the only limitation is their storage space. Data-loggers that are connected to telemetry can deliver all the data they have been pre-set to acquire, in one interrogation from the central control. For example, a logger might be set to record pressure and flow at 15 min intervals and be scanned by the telemetry control at, say, 12 h or daily intervals. When scanned in this way, the logger will 'deliver' $12 \times 4 = 48$ items of data, or 96 items if scanned every 24 h.

In some cases the telemetry cannot reach all outstations due to practical difficulties (high cost, usually) and other local solutions are applied – like manual readings, short-wave radio link, public telephone network, etc. The reading frequency can drop to as low as one reading per day, or even worse, with a corresponding loss of reliability and accuracy.

Archiving frequency

Telemetry data are *collected* frequently, at intervals equal to the *scanning frequency* (1 to 6 min), and data retrieved during the scan replace data collected during the previous scan and are displayed as 'live' data on screens linked to the telemetry system – in the control centre, the manager's office or on any other screen temporarily or permanently accessing the system. However, the data are usually *saved* at a different rate, called the *archiving frequency* (Gwynne et al., 1996). The archiving frequency is typically between 15 and 30 min but can be a longer interval. Consequently, some data are being lost, which might be significant. The reason for the timing difference is that the collections are for two different purposes.

The primary function of telemetry is the monitoring and control of the minute-by-minute operation of the water supply system. To fulfil this function, it needs to have a relatively frequent update of its information. In a pressure control loop, for example, pressure changes may require a valve operation within a few minutes. Data being archived are being collected as a historic record for future analysis, used, in the short term, to investigate operational anomalies, carry out investigations of emergency situations and so on. In the longer term, the data may well be 'distilled' into a long-term archive in which, for example, daily demands rather than instantaneous demand will be of greater importance, and used by planners, designers and others for the longer-term development of the supply system.

The various reading, scanning and archiving frequencies involved in the collection and storage of data can lead to a slightly incorrect picture of the supply system, and the potential for distortion certainly should be understood by those needing to merge high (water) experience with high technology.

Other difficulties also exist. The communication lines might suffer from occasional breakdowns, or be influenced by other electrically operated devices (called 'noise'), or vandalism, etc. Examples are given below of the identification and interpretation of data losses from such causes.

The potential for error in the data collection and transmission system is considerable, and such errors have given rise to a certain scepticism about new technologies, especially in operators trained before the arrival of the high-tech systems. Rees (1996) lists the following reasons why the data are often distrusted by users:

- infrequent sensor calibration,
- poor location of sensors,
- lack of end-user involvement,
- inaccurate configuration.

The development of understanding of the potential for error and how to deal with it in the real world of data collection systems will be the key to gaining full benefit from the technology. Some suggestions on how to identify errors and overcome the problems they pose are given later in this chapter.

Interpretation of data

The mistrust of data described above is only too well known to managers. After great efforts and expenditure of time and money, staff are often sceptical about the benefits. Attitudes range from incredulity amongst the old hands to high hopes and blind trust amongst the new recruits, often leading to disappointment and deprecation of new technology. The staff must learn how to use it, and this is very difficult without a rational approach and proper training.

The correct interpretation of available data is not as easy as one may think, and the obvious conclusions are not necessarily the right ones; hence the need for training. Below are some guidelines on how to approach the problem of analysing data. Some of the guidelines assume that a dynamic, mathematical model of the network is available:

- Always make a plan of *how* to analyse this information. What should one investigate first?
- Always develop a *general idea* of how the system actually operates, and try to find the evidence that supports the general operational mode and that which contradicts it – find *'pro'* and *'contra'*. A briefing on the historical development of the system might shed some light on how the present system works.
- *Examine* a period of several time intervals (days, weeks or months) to spot irregularities or anomalies.
- *Zoom into* all singular events, by selecting a relevant, short time period.
- *Combine* the results in meaningful entities (like variations of WL in a reservoir and flow of its booster pump).
- *Compare* the results of simulation with observed data, spot differences and explain the causes.
- *Inform* the right person, efficiently and regularly, or ask for consultation and help if in doubt.

The main thing is to spot the problem as early as possible, to gain time for reaction.

The interpretation of data must be left to humans (at least for the time being) because of the complexity of the problem. For example, a sudden jump in the night line can be explained in four different ways:

1 failure of instrument and/or local calibration;
2 genuine increase in demand (fire fighting, holidays);
3 rezoning of the distribution network;
4 a pipe burst.

Obviously, urgent intervention is needed only in the last case – when the losses have to be stopped. Experienced users can deduce what has happened using small clues that are difficult to generalise; however, this might be overcome in future with the introduction of expert systems. For instance, an instrument failure (case 1 above) is indicated, usually, by a straight line, the disappearance of signal, or by strange shapes (see the examples below). If the instrument is in good order (and one of the other explanations is the reason for the increase), then:

- If the reason is 2, peaks will be higher, and the night line will remain more or less the same.
- If 3, the daily diagrams will be different, but still following the usual pattern.
- If 4, the whole diagram will be shifted upwards – this will be noticeable on the night line.

Operational data can be transferred to other departments via computer networks, but only after preliminary checks are made and obvious errors dealt with. The process needs to be controlled and supported by trained people, because it is far more difficult than it may seem at first glance. The mathematical model can provide substantial help in analysing and interpreting the operational data, as will be discussed later in this chapter.

Summary of the value of on-line data

A well designed and maintained data management system (DMS) is essential to any water company, not only for forecasting and planning capital investment but in everyday operations. This is true whether or not the company has highly sophisticated computer-based systems available to it. However, the effectiveness of all sections of the company is greatly enhanced if primary data are collected, analysed, archived and made accessible through a digital system. Everybody should have access to the same data, and access to a good database will raise general knowledge of the system throughout the organisation, to its general benefit. However, a note of caution is needed: every DMS left alone or infrequently used, without regular testing and critical evaluation, has a tendency to rot – with disastrous consequences.

A well constructed DMS can be used without modelling for many useful purposes like:

- regular balancing of the system at all levels;
- automatic leak detection;
- demand analysis and forecasting;
- computation of operational costs;
- basic information for planning and design.

The problems of data collection and processing

The influx of operational data is enormous even in medium-sized water supply systems; data come from different sources, with varying frequency, accuracy and reliability. Compare, for instance, billing data and telemetry data. The former are collected monthly or half-yearly, manually, while the latter come every few minutes, captured by the most modern equipment. However, this does not mean that either is completely reliable, and a measure of healthy scepticism is essential. All data have to be thoroughly checked before acceptance, to identify periods of missing data and reject (and flag) the faulty ones, before wrong conclusions are made.

The first problem is how to recognise and deal with wrong and/or incomplete data. The next few figures will illustrate a few typical cases. Two options are available for dealing with the problem:

1 Leave gaps as they are and fill them with the most likely values only when the data are used for analysis, leaving 'unadulterated' raw data.
2 Replace missing data with the most likely values, creating 'synthetic' data. Given the huge number of data, an automatic procedure will need to be devised and used, to reduce the manual editing of data to a minimum. This process needs to be considered very carefully.

The first approach is usually better because staff will have more confidence in data 'as they are' rather than 'as they should be' – but it must be said that this is arguable.

The key factor in dealing with the problem is that the period of missing data should be *flagged* in some way, and not processed through the DMS to perpetuate the impression that, during the periods of missing data, the parameter was actually zero, or incorporated into the database and then used to produce means or average quantities. The *data* were missing, *not* the flow or pressure or water in the reservoir or water being delivered to a customer!

Figure 2.1 shows an example of incomplete data: note gaps in the record during 26 and 27 January 1995, while the data for both the preceding and following days are complete.

Figure 2.1 Incomplete data.

Other data might be continuous, but obviously wrong. Examples are:

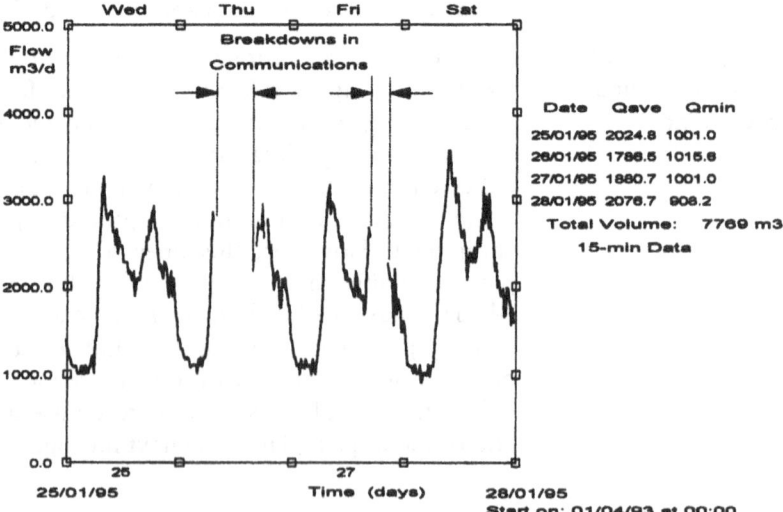

Data for District Meter: AF129 - Name: Water Tower Outflow

Date	Qave	Qmin
25/01/95	2024.8	1001.0
26/01/95	1786.5	1015.8
27/01/95	1880.7	1001.0
28/01/95	2076.7	908.2

Total Volume: 7769 m3
15-min Data

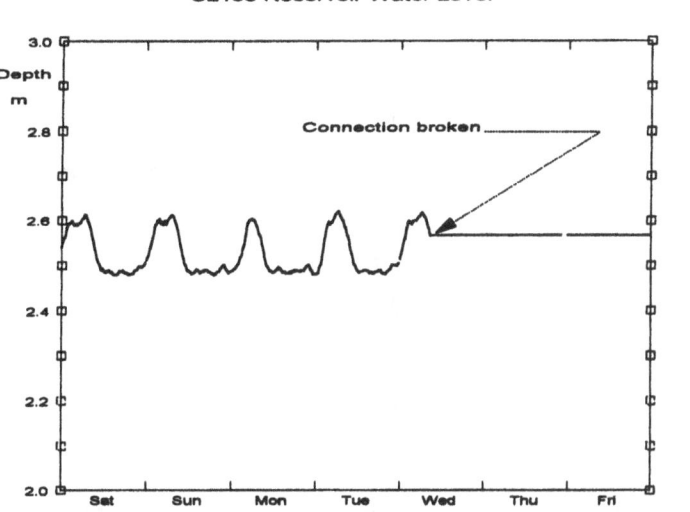

Figure 2.2 A constant value.

- a constant value – highly unlikely when measuring the flow in any pipe – probably a faulty primary instrument (Fig. 2.2);
- strange spikes – probably caused by malfunction of the transmission line (Fig 2.3).

In other cases, data might appear good but close examination reveals error. Figures 2.4 and 2.5 show flow in a pumping main and a distribution pipe, respectively.

Figure 2.3 Strange spikes – data error or unusual event?

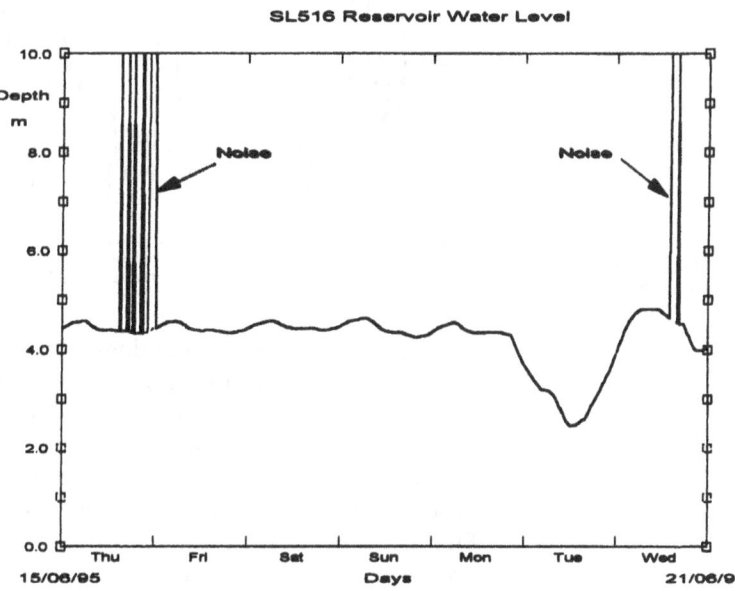

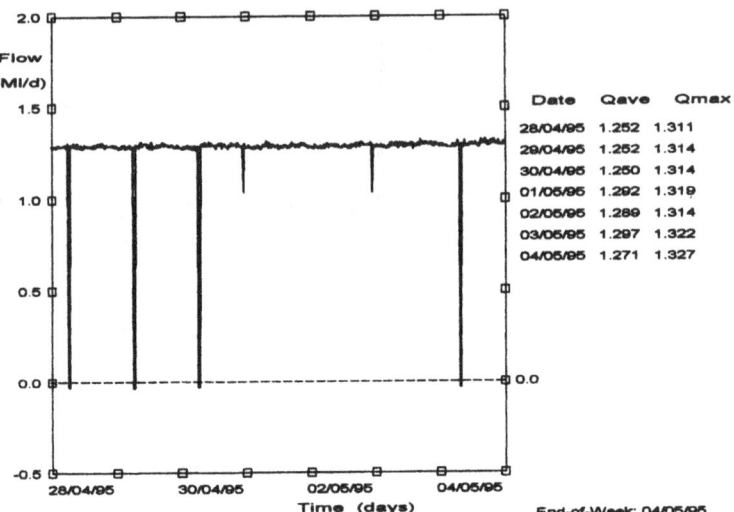

Figure 2.4 Acceptable data?

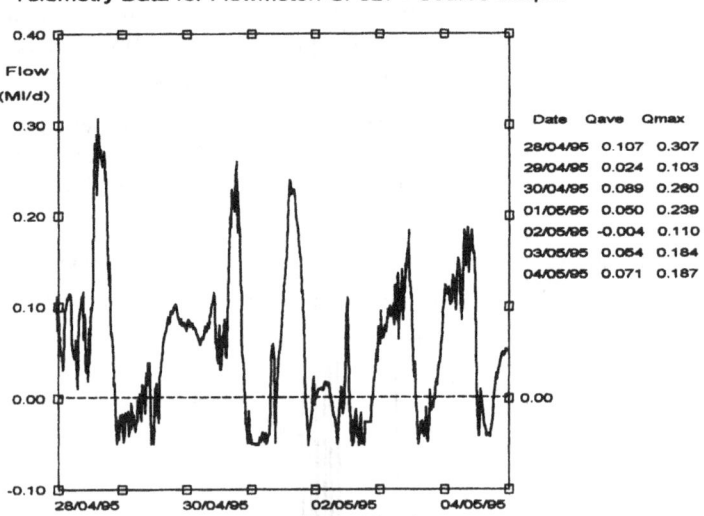

Figure 2.5 Graph just pulled down?

Although there are some dual-direction meters available, the overwhelming majority of sites only show unidirectional flow, so, with the caveat that a site *may* have negative flow, one can normally be confident that a negative flow is wrong. In these examples the conclusion is that the measuring range of the flowmeters probably needed adjustment. The whole graph in Fig. 2.5 is apparently 'pulled down', but a simple intervention on the site should put things right.

There is clearly a problem if the data are simply not feasible – as in the previous example. A greater problem arises if the flowmeter is out of adjustment but produces data that are still feasible. If the diagram in Fig. 2.5 is lifted by 0.05 Ml d^{-1}, it would show positive flow at all times but could still be significantly out of adjustment. The problem is overcome in the real world by regular recalibration of individual instruments and by linking data from the particular site with data from other related sites (see notes on Fig. 2.8 below).

Electronic and/or radio interference from outside sources can interfere with measurement transmissions and signal conversion equipment, producing strange peaks and breaks. Such interference is commonly described as 'noise' and illustrated in Fig. 2.6. If the effect is such that data are unduly distorted and the problem is persistent, some electronic screening might be considered to reduce or eliminate the problem.

Finally, the measuring range of the meter might be inadequate, as shown in Fig. 2.7. Note the sudden increase of flow on 23 June 1996, due to a pipe burst. The range of the meter was too small to measure such flows, and hence the peaks were clipped. The burst was spotted and repaired in a few days and afterwards flow was within the range. Circumstances and budgets will dictate any action to replace the meter with one capable of measuring extraordinary circumstances – in this case a loss of almost 0.5 Ml d^{-1} for 11 days!

Figure 2.6 Breakdowns in communications?

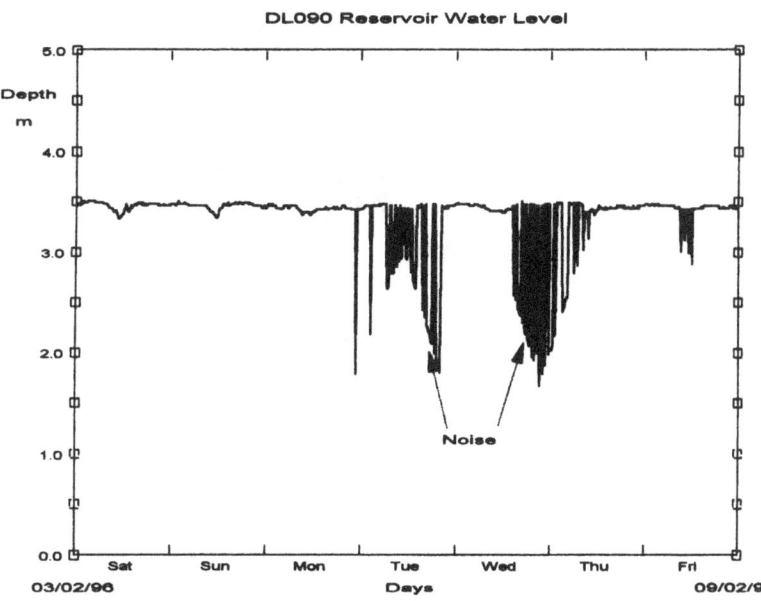

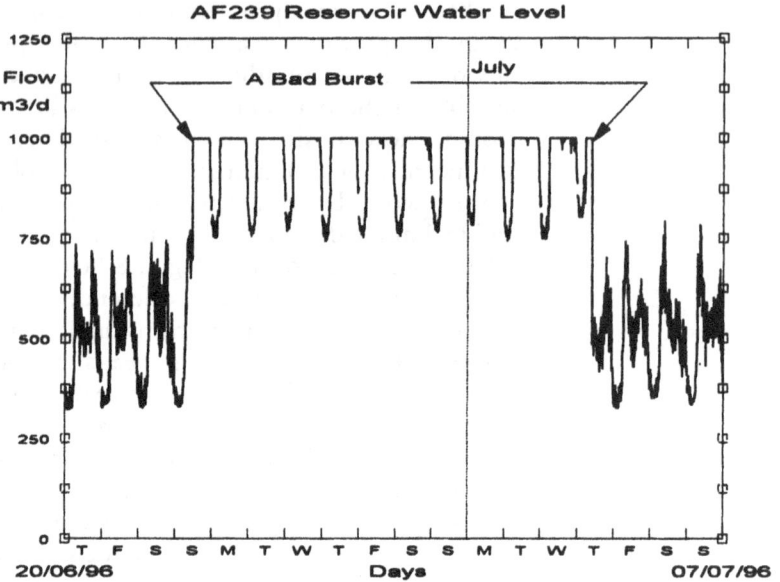

Figure 2.7 Measuring range was too small.

Figure 2.8 Range too small or reservoir overflowing?

In some cases it is not clear if the range of the instrument was exceeded or something else is wrong (Fig. 2.8). Water levels reach 4 m above the reservoir bottom but do not exceed this value. As it is, at the same time, the range of the sensor *and* the maximum depth, this could mean that water is overflowing – or that the sensor needs recalibration.

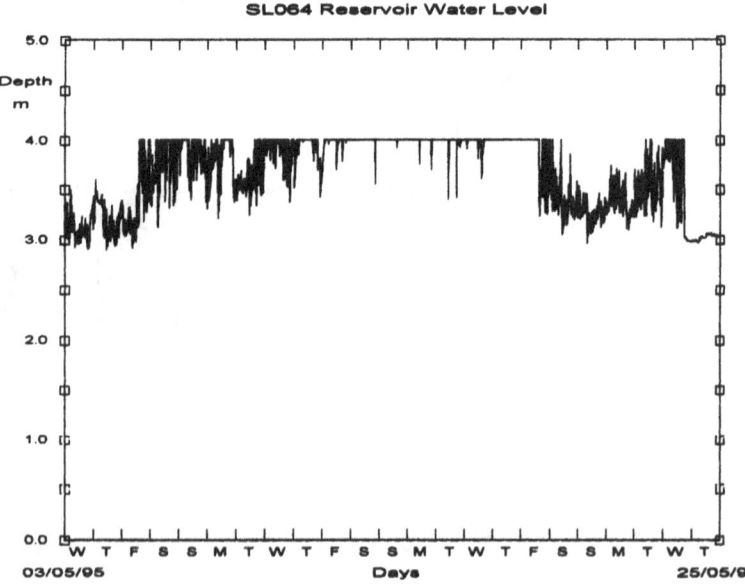

By careful analysis of data, one may notice step-like changes of a measured value, either flow or (more often) water level. As changes in reservoir level or flow are normally smooth and continuous, this pattern is obviously introduced by the data collection system (i.e. in the instrument, signal converter or transmission system). There are two common cases:

1 when the sensitivity level (threshold) of the sensor is too big – see the example in Fig. 2.9;
2 when the scanning frequency is too low, for instance once per day, as in Fig. 2.10.

Note that in the first case the 'height' of steps is constant, while in the other the 'width' does not change.

In all these cases the erroneous data could be easily spotted without further analysis. However, there are more difficult cases where the discrepancies can be found only by comparing data from one source with another. This means that there may be a surplus of data, collected exclusively for control – to be compared with other data. One example is described below.

A source has four borehole pumps, a contact tank (where water is chlorinated) and two relift pumps (Fig. 2.11). The output of each borehole pump is monitored, and also the inflow to the contact tank. The flow through AF222 should be equal to the sum of the flows through AF103, AF104, AF105 and AF106.

Figure 2.9 Effect of the threshold on data.

The difference between the combined output of the four borehole pumps (AF103 to AF106) and the inflow to the contact

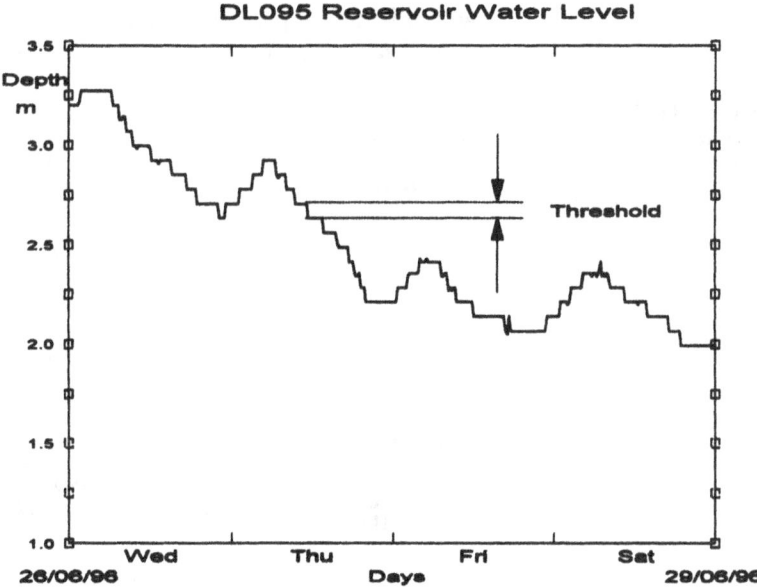

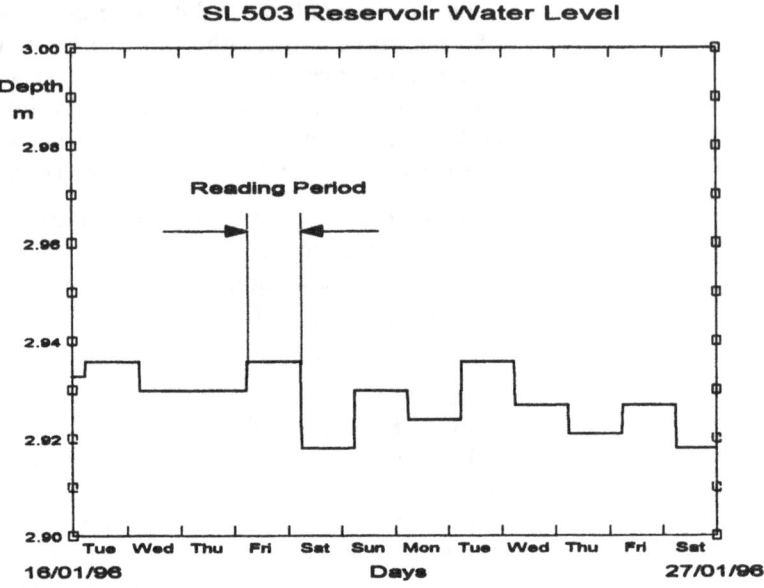

Figure 2.10 Effect of the reading period on data.

tank (AF222) for four days is shown in Fig. 2.12. Note that this difference is very close to zero most of the time. Occasional 'spikes' are due to the time lag between individual measurements (the telemetry system scans all instruments at the site in a cycle of 6 min).

Diagrams of the flows through the individual borehole pumps (not shown here) indicate several starts and stops, suggesting that the borehole pumps are not properly adjusted to the current flow conditions – probably they take more water out than the aquifer can replace, causing frequent trip-offs due to low suction

Figure 2.11 The schematic of the source.

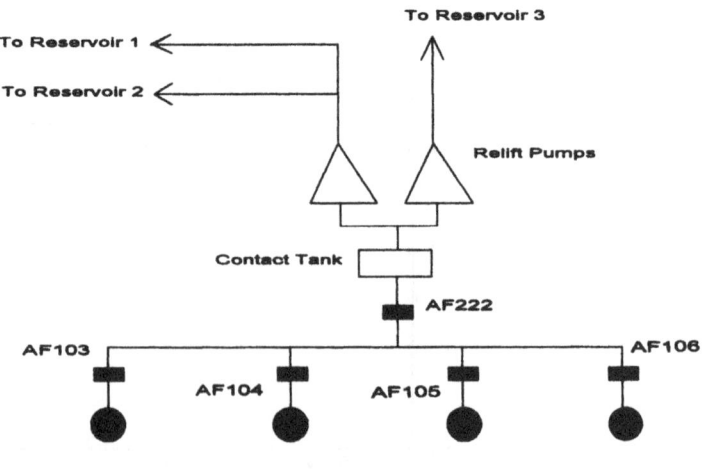

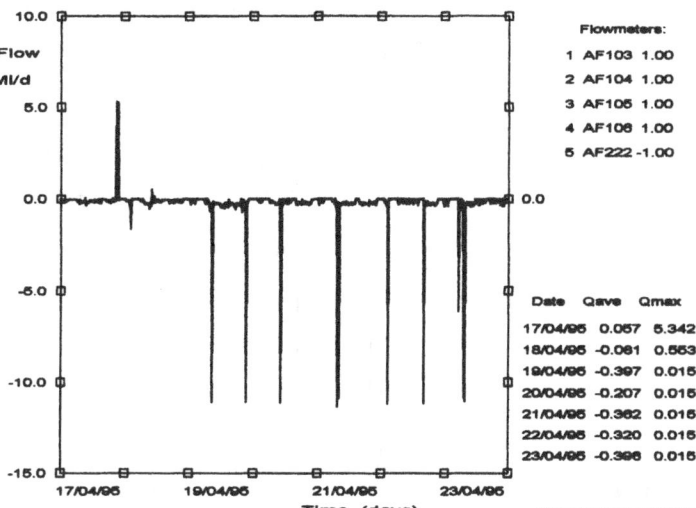

Figure 2.12 The difference in flows.

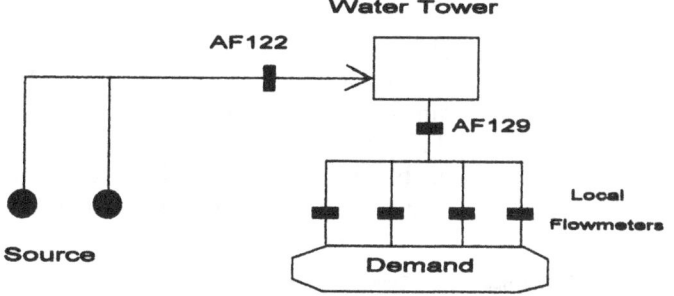

Figure 2.13 A simple, but complete, water supply system.

level. However, even in this unstable regime, the flowmeters operate quite reliably – the proof is the stable value of the difference (Fig. 2.12). The conclusion is that all five instruments are probably in order.

Another example is even simpler and permits quick computations and checks. The system is shown in Fig. 2.13. This is a small independent system consisting of two borehole pumps, one water tower and the distribution network (not shown in detail). Note that AF122 flowmeter monitors the output of the borehole pumps, i.e. the inflow to the water tower, while AF129 monitors demand.

The data for one day – 1 May 1995 – are shown in Fig. 2.14. Note the rectangular shape of flow through AF122, typical for pumping, and the two-peak daily diagram registered by AF129, typical for demand. The total volume of water in each case is similar: 2.51 Ml (AF122) versus 2.29 Ml. The difference is

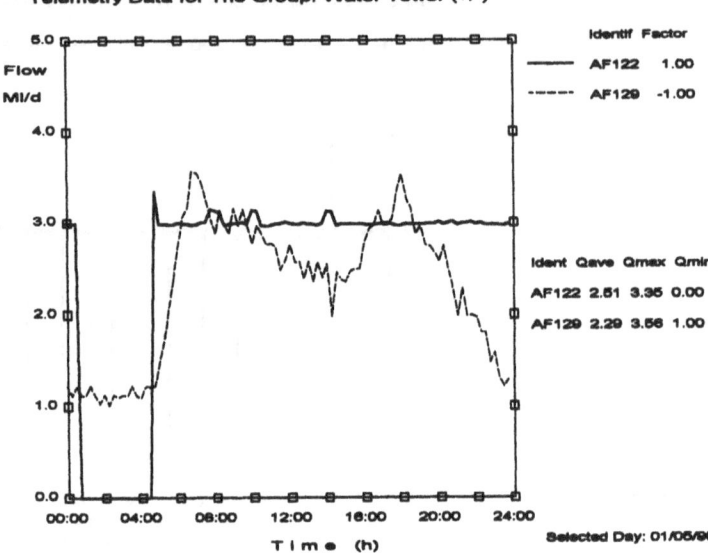

Figure 2.14 The balance on 1 May 1995 – good.

about 0.22 Ml or some 15% of the water tower's volume (1500 m³).

On another day – 29 April 1994 – the difference is greater, almost 50% of the water tower's volume (Fig. 2.15). The difference between inflows and outflows during several days is visible in Fig. 2.16 for the week starting on 28 April 1995, and in Fig. 2.17 for another week a year earlier, from 23 to 29 April

Figure 2.15 The balance on 29 April 1994 – bad.

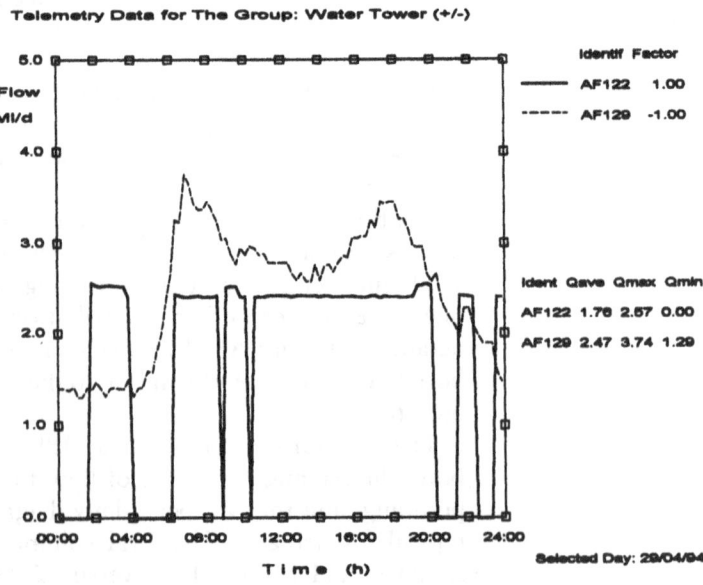

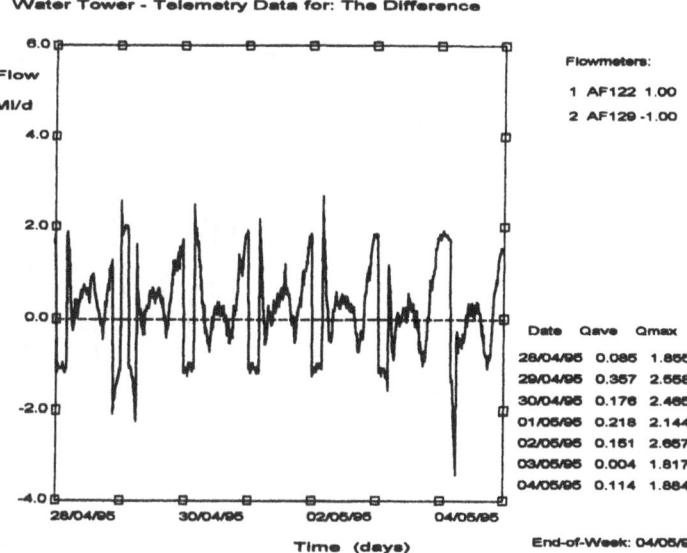

Figure 2.16 The difference, May 1995.

1994. Note the average values: in the first case they are smaller and positive, while in the second case they fall heavily on the negative side. Clearly one of these two instruments was in error; in this case AF122 was under-registering the inflow by 25% – the range was not properly set. Once discovered, the error was quickly fixed.

The problem is, clearly, to spot the error before the data are admitted further in the procedure. In some cases it is difficult to

Figure 2.17 The difference, April 1994.

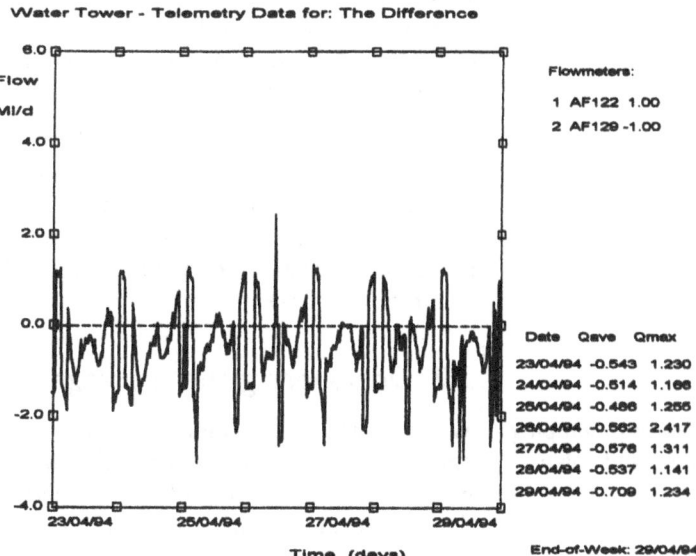

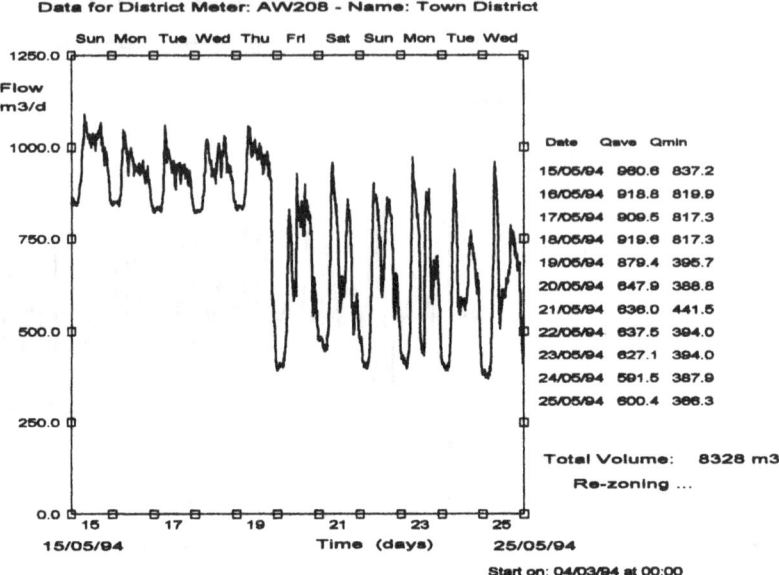

Data for District Meter: AW208 - Name: Town District

Date	Qave	Qmin
15/05/94	980.6	837.2
16/05/94	918.8	819.9
17/05/94	909.5	817.3
18/05/94	919.6	817.3
19/05/94	879.4	395.7
20/05/94	647.9	388.8
21/05/94	636.0	441.5
22/05/94	637.5	394.0
23/05/94	627.1	394.0
24/05/94	591.5	387.9
25/05/94	600.4	366.3

Total Volume: 8328 m3
Re-zoning ...

Start on: 04/03/94 at 00:00

Figure 2.18 Variations of demand – or something else?

decide what to do – see an example from real life in Fig. 2.18. This diagram shows variations of water demand in one district, with an abrupt change after 19 May. The diagrams before that date had exceptionally high night demand, while after 20 May they are back to normal shapes. Why? What was done on the night of 19/20 May 1994?

- Rezoning of the network?
- Repair of a bad leak?
- New PRV was installed?
- Or – the possibility can never be ruled out – was the flowmeter faulty?

As these values were captured by a data-logger, perhaps somebody mixed the files from two different instruments? This case illustrates the problems of real life. Whatever the difficulties might be, this task must be done as efficiently as possible, using the knowledge and expertise of the staff to create a reliable and accurate database.

2.2 Deciphering the message

Reservoirs, pumping stations and valves

Water level in a reservoir usually varies, over time, as a consequence of variation of both inflow and outflow. Normally, water level data need to be examined in conjunction with flow data to develop a meaningful understanding of the operational situation. However, if the level remains almost constant, or varies in a narrow band, it may be possible to glean some valuable information about the operation of the reservoir. Three cases are shown in Fig. 2.19, all based on real situations. These cases indicate (but do not prove!) some particular condition(s):

(a) Water level is almost constant (near 90% in this example) most of the day, but slightly lower during the peak demand time. Possible conclusions are:
   • A float valve is fitted with the upper limit of 90%.
   • Inflow exceeds outflow (except during peak hours), so the reservoir is lower than the hydraulic gradient.
   • There is spare capacity at this reservoir and it could be used to meet new demand.

Figure 2.19 Typical variations of WL in reservoirs. Possible conclusions are given in the text.

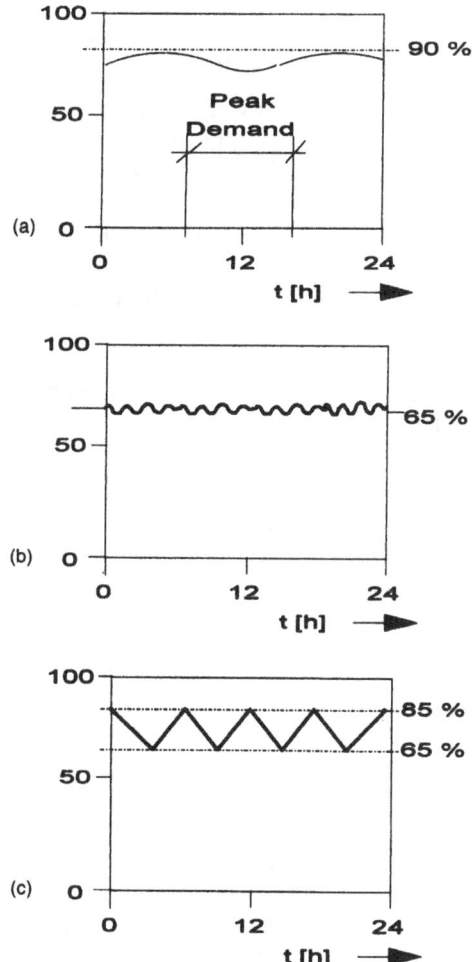

(b) Water level is fairly constant at 65%, with small oscillations. The conclusion is:
   • A variable-speed pump or a pressure-reducing valve keeps
     the level constant at 65%. This should be investigated
     further by the staff. Note that a constant line (without
     oscillations) will mean either that the instrument is
     inoperative or that the reservoir is out of use.
(c) Water level goes up and down in a sawtooth pattern, from
   65% to 85%. The conclusions are:
   • A pumping station with fixed-speed pumps either delivers
     water to the reservoir or takes water out of it (less likely).
   • Pumps are controlled by water level, apparently, with 'on'
     and 'off' levels set at 65% and 85%.
   • The balance between inflow and outflow may be com-
     puted easily, knowing the volume of the reservoir and
     reading time intervals from the diagram.
   • Frequent changes suggest that the reservoir may not have
     a sufficient capacity.

More examples from real life follow. Figure 2.20 shows changes
of water level in a reservoir with a float valve – note how the
FLV has prevented overspilling and waste of water.

Data for two reservoirs are shown in Fig. 2.21. Reservoir
DL109 operates with a standard pattern, reaching the maximum
early in the day and then gradually falling down until just before
midnight – nothing unusual, and all may be assumed to be well.

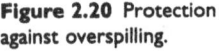

Figure 2.20 Protection
against overspilling.

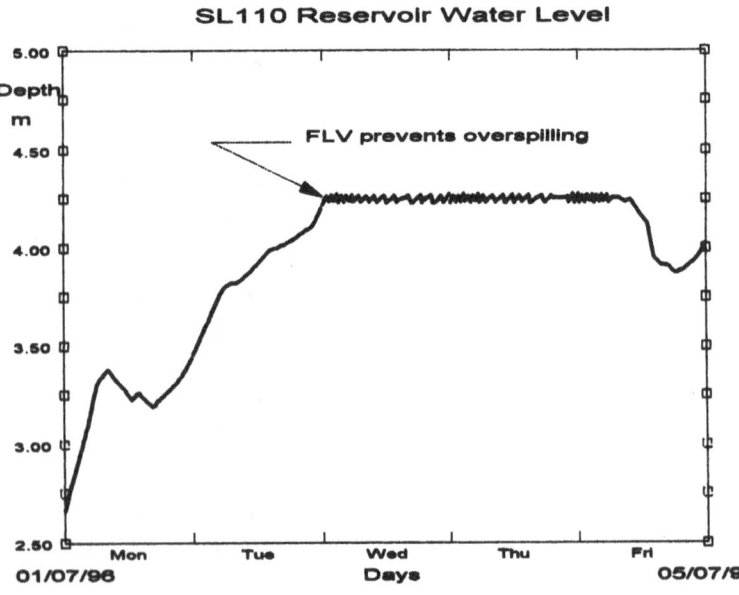

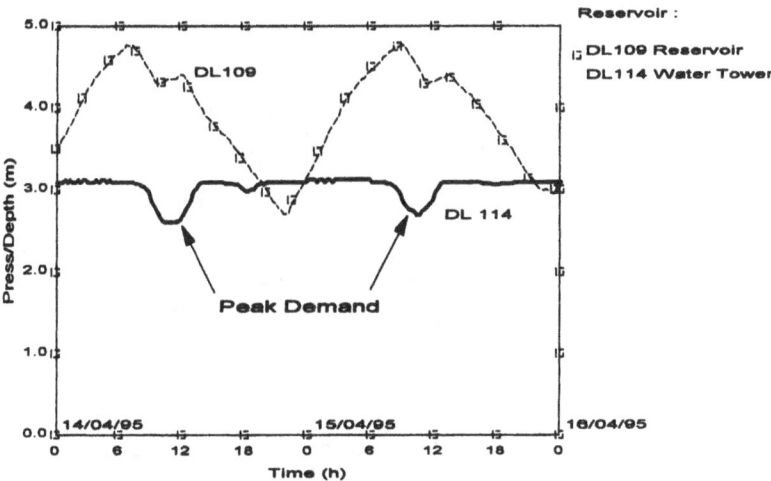

Figure 2.21 The standard pattern of DL109 and the excess capacity of DL114.

The example is included as something of a 'control' to compare with the rather unusual examples that follow. The other reservoir DL114 in Fig 2.21 is full most of the time, except during the morning peak demand periods. Why is it so? Further investigation might show that there is some potential for transferring load from other reservoirs and making better use of the capacity of DL114. It may even be possible to use the spare capacity to effect energy savings somewhere in the system by pumping only during off-peak electricity tariff periods or meet additional demand somewhere. It is clear that *spare* reservoir capacity exists and might be used to better advantage.

Two other cases are included here as Fig. 2.22. Water level in the small reservoir oscillates between set limits – probably fed by level-controlled pumps – while the large reservoir has erratic behaviour, as a part of a larger system. Both situations demand that operational staff investigate the causes of these features to further their understanding of the system and possibly improve the service.

In Fig. 2.23 the reservoir labelled as SL564 is displaying the classic sign of overflowing. During four periods from 23 March 1996 the reservoir was full, but the absolutely level line (clipped peaks) suggested that the reservoir was actually overflowing. (Was the decline in its level over the last three days of March due to the lost, overflowing water?) Towards the end of March a float valve was fitted (or refitted) to the reservoir, which managed to fill over the first two days of April. The float valve's presence is indicated from the night of 2/3 April. From that time when the reservoir fills, the float valve operates – indicated by an

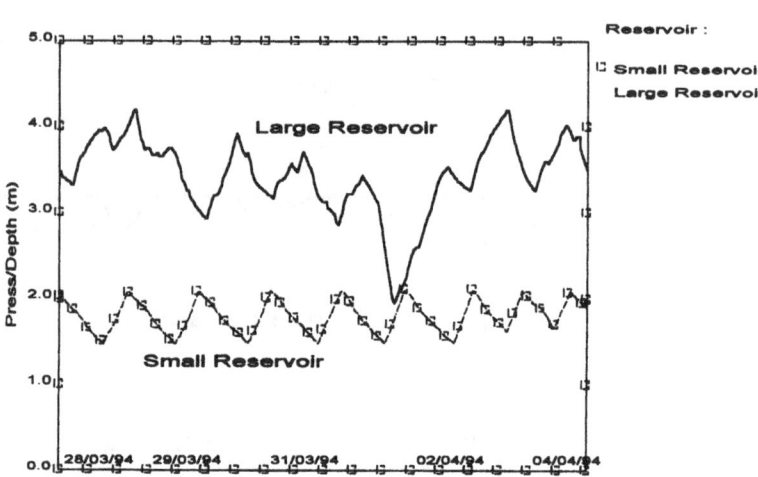

Figure 2.22 A reservoir that may be too small, and a large reservoir displaying erratic behaviour.

oscillation of water level. It will also be noticed that from the beginning of April, after the valve had been installed, 'demand' from the reservoir had reduced. Friday 29 March presents a much sadder picture to an operations manager than the picture the following week. This one diagram encapsulates a minor drama of leak detection and repair.

Figure 2.24 represents the flow through a pumping station. It can be concluded that this pumping station has at least three

Figure 2.23 Waste control in action – the fitting of a float valve.

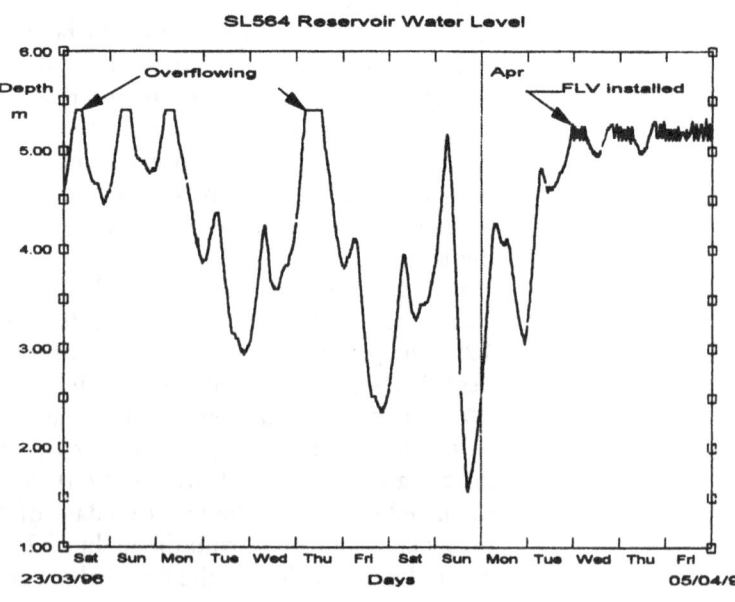

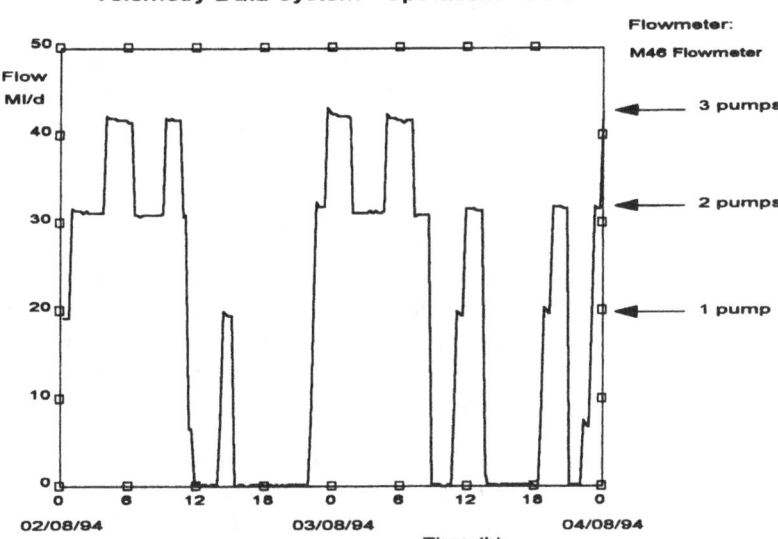

Figure 2.24 Pumping station with three units – typical diagram.

pumps. Note typical flow levels: about 20 Ml d^{-1} when a single pump is operating, slightly above 30 Ml d^{-1} for two units, and over 40 Ml d^{-1} for full capacity. Note the angular or square shape typical of pumping station flow charts.

The effects of a pressure-reducing valve can also be spotted on diagrams; see the example in Fig. 2.25. The oscillations of pressure in the network at point P01 suggest that pressure is not controlled. The patterns of pressure at points P20 and P23

Figure 2.25 The effects of a PRV.

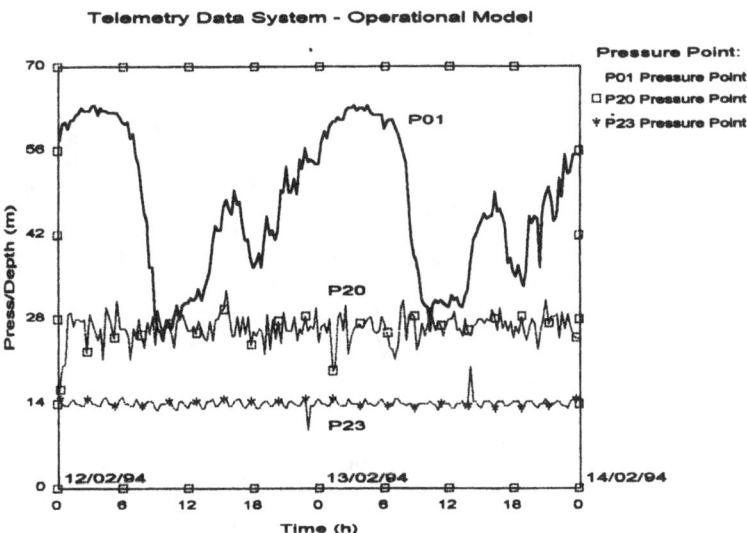

suggest a degree of pressure control, but the variations at P20 are greater than at point P23. If they are on the same system, this suggests that P23 is further away from the control point than P20 (the time lag is several minutes and the PRV servomotor continues to operate in the same mode – e.g. 'open' – until the new signal arrives).

2.3 The logic of data

Data gathered by a telemetry system may be wrong, and section 2.1 explored some reasons for this and means of identifying and coping with the problems. In other cases, data are valid but their true meaning is not understood. A simple case will illustrate this point.

Transfer or isolated demand?

Figure 2.26 shows variations of flow in a water main, and the pattern suggests a typical daily diagram of demand. Does this pipe supply an isolated zone only? This would mean that its downstream end was closed. It is possible that the flow includes a major transfer link, the demand of which could be constant up to 0.5 Ml d^{-1}, or alternatively high night-time demand could be the result of significant leakage. The question should be discussed with the staff – all sides could learn something.

Figure 2.26 Closed system? Major transfer at constant rate? Leakage?

After a careful examination of individual diagrams, one should try to discover the links between them. A few cases are discussed below, to illustrate the point.

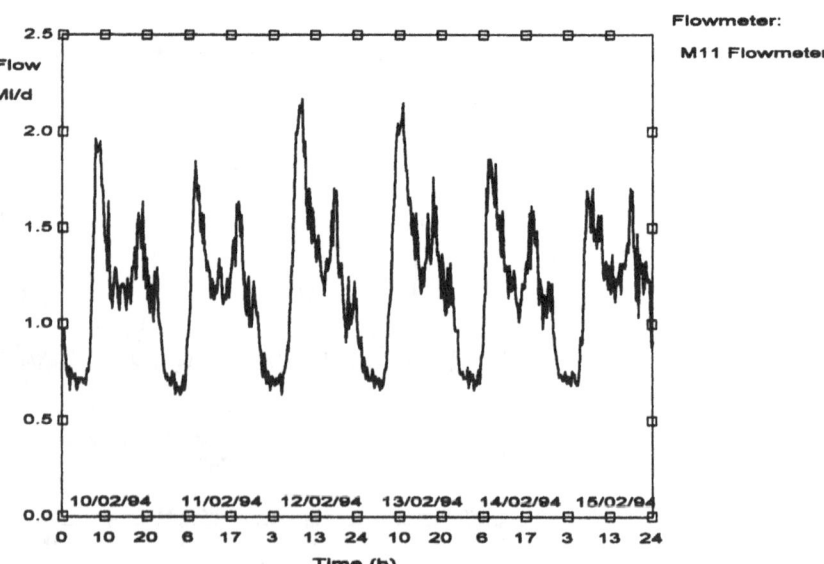

Telemetry Data System - Operational Model

Why does the
pressure change
abruptly?

Figure 2.27 (a) Sudden
drop of pressure
explained by (b) the start
of a downstream PST.

The next figure shows two diagrams from a regional water
supply system – Fig. 2.27a displays a sudden pressure drop
explained by examining Fig. 2.27b, showing the intermittent
operation of a downstream booster station. Note the perfect
timing of pump discharge and drop of pressure in the upper
diagram. The gauge P2 is nearer to the pumps, so pressure
changes are more apparent than in the case of P1.

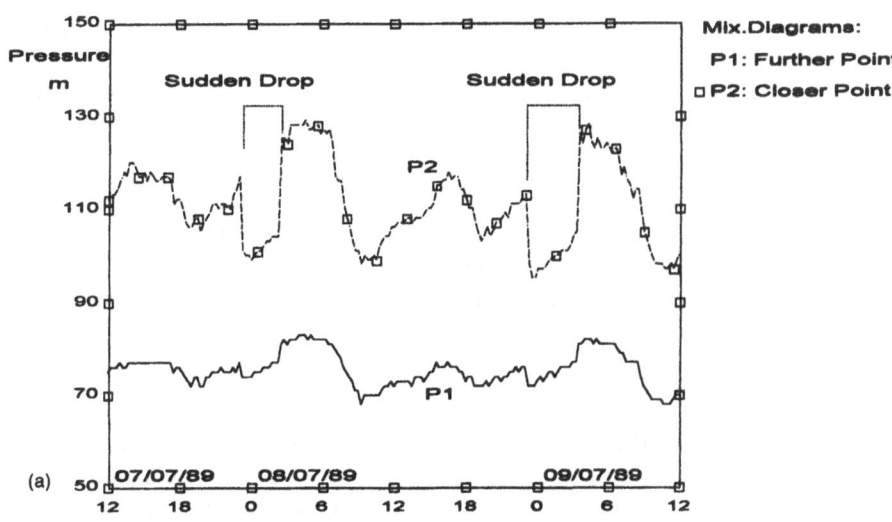

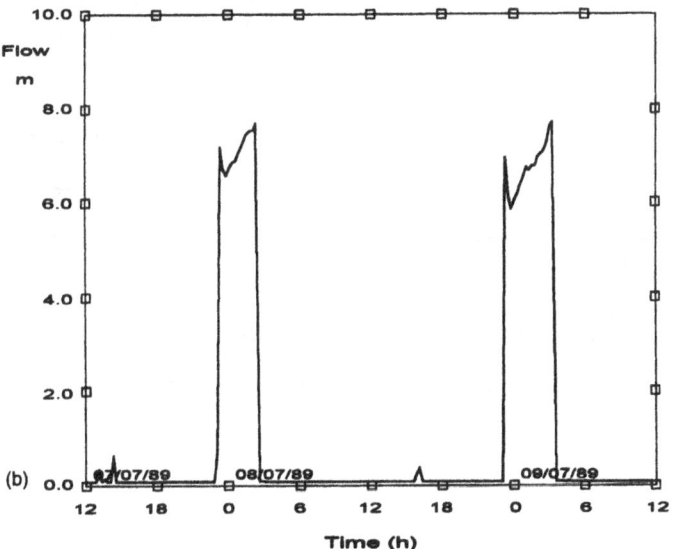

How is a pumping station controlled?

Another case where a booster pumping station is involved is seen in Fig. 2.28, which represents a small part of a more complicated distribution system. The puzzle is how the booster operates. Are the pumps level-controlled, pressure-controlled or do they follow some other regime? It is not an easy question because staff may change control policy seasonally or for some other reason (major works on roads, etc.) unknown to other persons. The answer can be found by studying telemetry data, using Figs 2.29 to 2.31.

The diagram of PST flow (Fig. 2.29a) shows that this station operates intermittently over the day. Variations of water level in both reservoirs are shown in Fig. 2.29b. One of them (the reservoir) is always full at 86–88% level, while the water level in the tower oscillates between roughly 70% and 90%. The conclusions are:

- Pumps are controlled by level in the tower (the higher of the two): 'on' at 70% and 'off' at 90%.
- The reservoir is placed too low (at least in normal conditions) and its float valve is closed to prevent overspilling.

This is valuable information for the final calibration of a model, which could not be obtained easily by other means. Note how the case is made more convincing by displaying relevant data on the same graph (Fig. 2.30): flows through the pumping station alongside water level in the affected reservoir.

Pressures are also affected by changes of flow in the network, as illustrated by the next example. Oscillations of pressure recorded by gauge P5 are shown in Fig. 2.31. Pressure *drops* when the pumps start, because gauge P5 is upstream of the pumping station (see Fig. 2.28), fitting nicely in the overall picture.

Figure 2.28 One booster PST, two reservoirs and three demand areas.

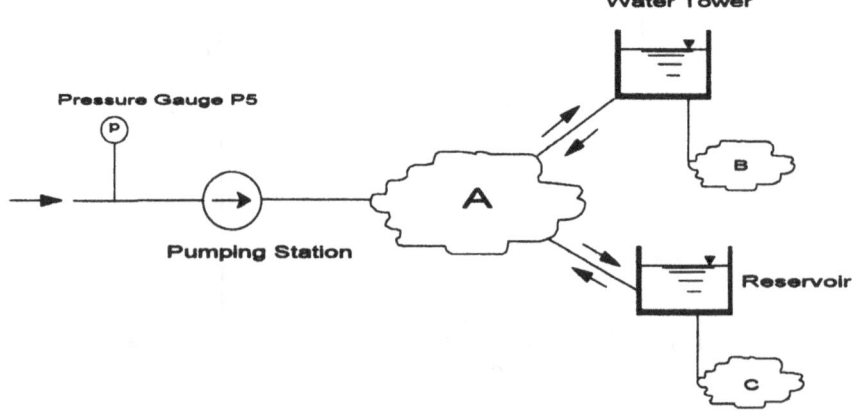

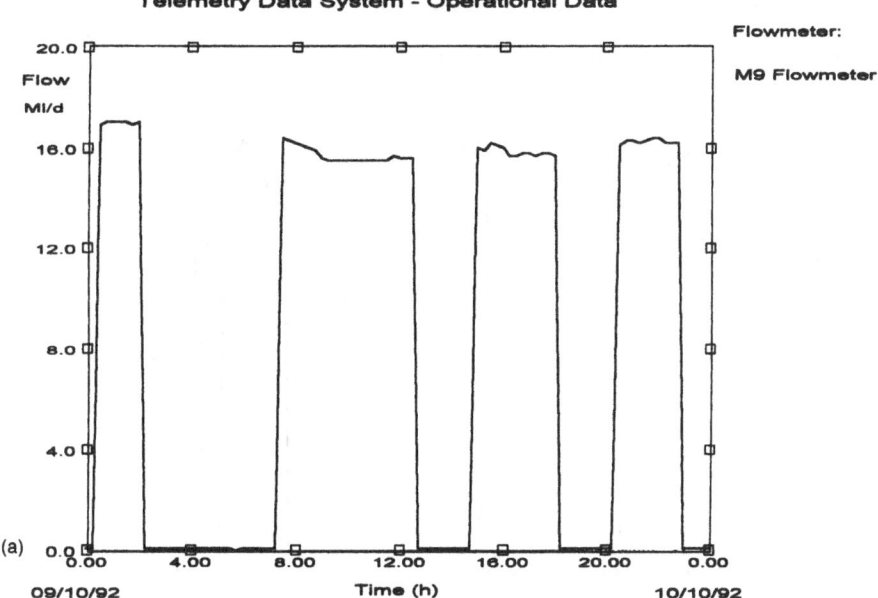

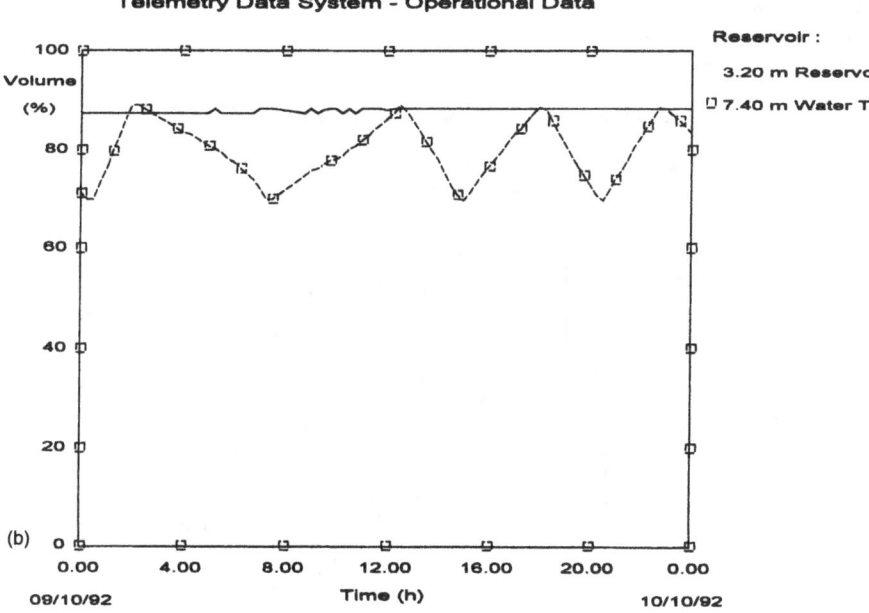

Figure 2.29 (a) Operation of booster PST and (b) variations of WL in reservoirs.

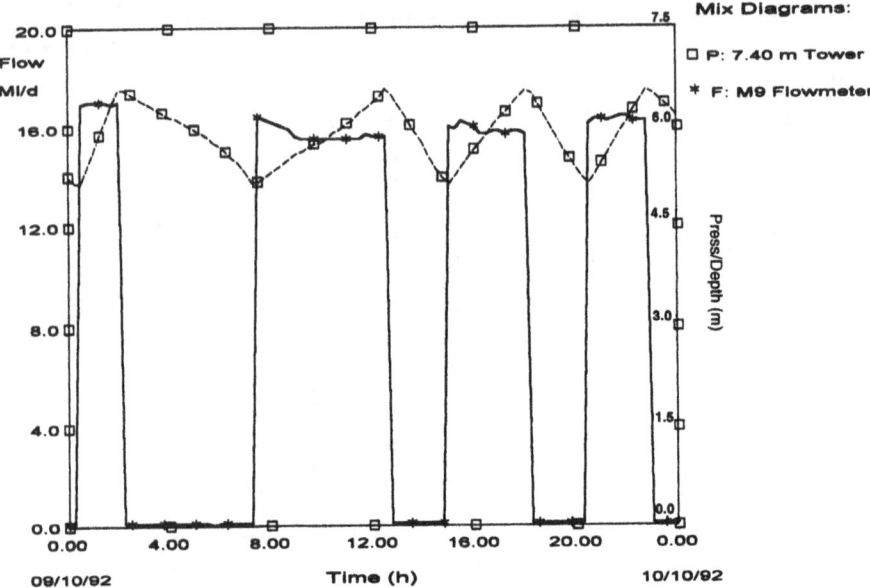

Figure 2.30 Pairing-up relevant data: pump flows and reservoir levels.

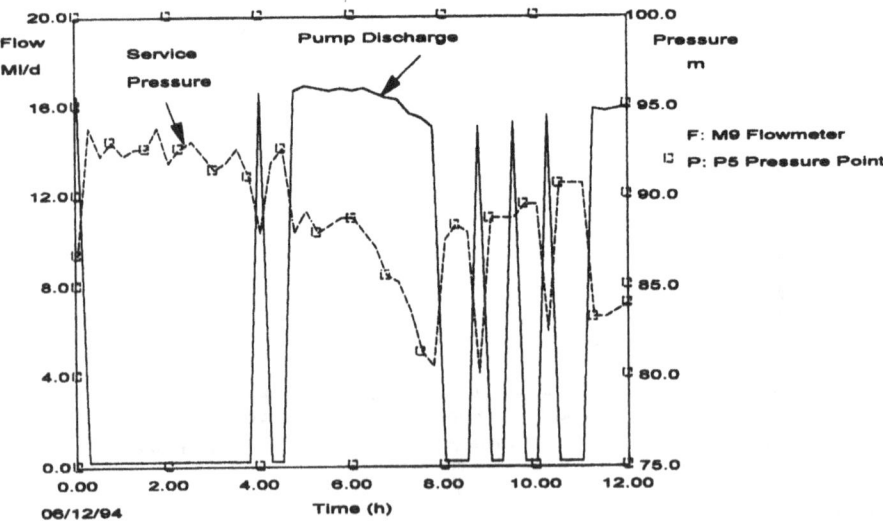

Figure 2.31 Oscillation of pressure due to pumping.

**Two reservoirs
operating in parallel**

Figure 2.32 shows variations of water level in a pair of reservoirs, operating in parallel, with a short connecting pipe between them. Note that the shapes are similar but not identical. The water levels above a common datum will, of course, be virtually the same if there is no significant flow between the two reservoirs.

**Gravity flow from
one reservoir to
another**

In the next case, mutual influence between two reservoirs is indicated. Both belong to a large regional water conveyance system and are connected through many links with sources, pumping stations and the distribution network of a nearby town. The scheme (very simplified) is shown in Fig. 2.33.

Water flows from the higher reservoir to the lower reservoir by gravity. The flow is measured in two places (M17 and M15), with a branch supplying some local demand in between. A float valve protects the lower reservoir from overflowing (it also has its own independent supply not shown in Fig. 2.33).

The next figure shows variations of flow (Fig. 2.34a) and water levels (Fig. 2.34b) during a typical day. One can notice a relatively sharp decline of flow after 2:00 a.m., undoubtedly due to closing of the FLV, because the water level in the lower reservoir is approaching the 90% mark. Later, when increased demand exceeds the inflow, even a small drop of water level in the lower reservoir is enough to open the float valve completely and let water flow freely from one reservoir to the other. The

Figure 2.32 Two reservoirs close together and effectively behaving as a single unit.

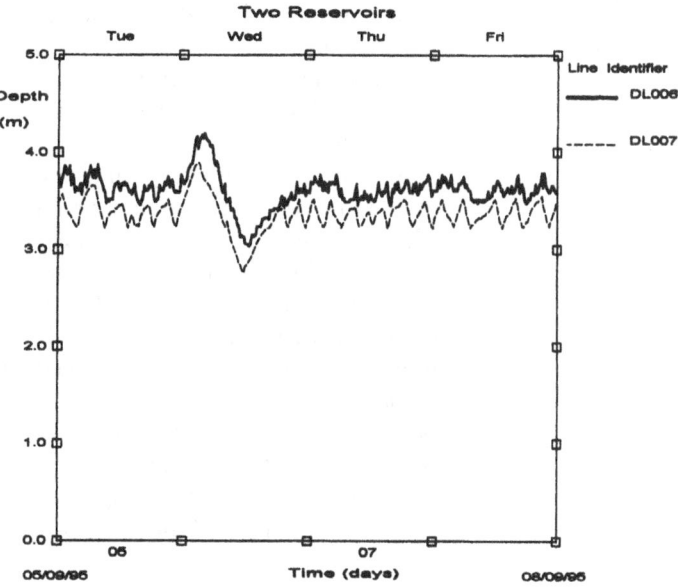

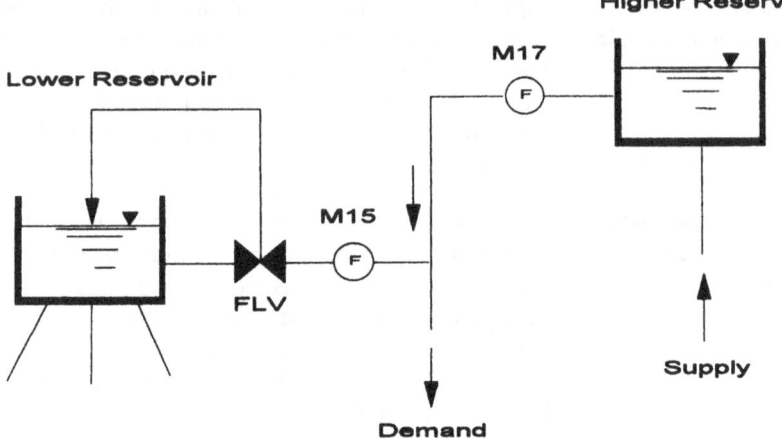

M17

Lower Reservoir

M15

FLV

Supply

Demand

Figure 2.33 Two
reservoirs in a system.

range in which the authority of this float valve is felt is quite
narrow, less than 0.5 m (it is probably a standard, lever-
operated, type of valve). All this is quite normal and should be
expected.

The point is that all pieces of information fall into place,
giving a proof of the reliability of the whole telemetry system.
One can continue to devise additional 'proofs', using other data,
to build confidence in the validity of the telemetry system and in
reaching operational decisions.

**Pumping station
between two
reservoirs**

Another case from a real system is shown in Fig. 2.35. This
represents only a small part of a much larger regional system. A
booster station transfers water from the lower to the higher
reservoir. Inflow, demand and flowmeters are all indicated in the
diagram. Water levels are displayed in Fig. 2.36a, and flows in
Fig. 2.36b.

At a first glance everything is quite normal – pumps start at
the 83% mark in the higher reservoir and stop at roughly 95%.
The increase of level in this reservoir is matched by a
corresponding decrease in the lower one. The picture is reversed
when the pumps are stopped: water level in the lower reservoir
rises and in the higher one declines. Why? Because the small
inflow (measured by M07) feeds the former and the small
demand (M05) empties the latter. The relative changes all have a
perfectly logical and legitimate explanation – but it would be
wise to check the non-return valve in the pumping station to see
if water escapes from the higher to the lower reservoir when the
pumps are stopped. One can also compute the change of storage

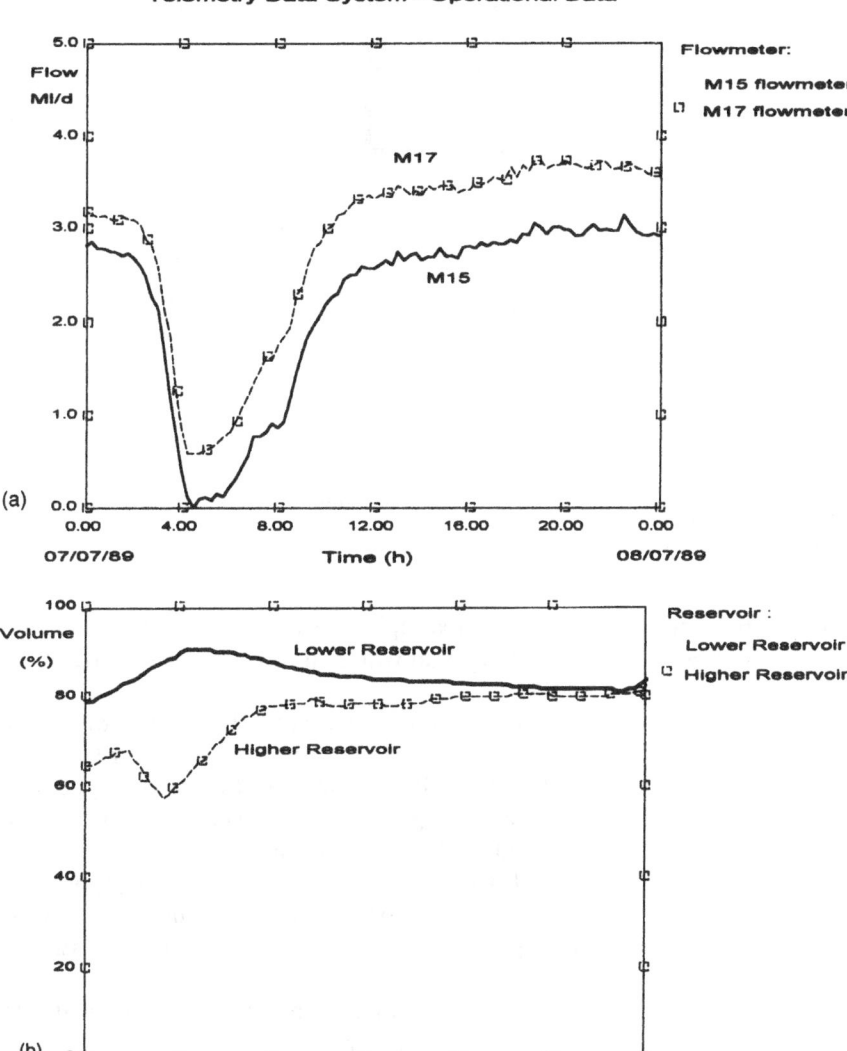

Figure 2.34 (a) The flow has declined because (b) WL is too high.

in both reservoirs to check the accuracy of the flowmeters – a relatively pain-free drop-test and a very good exercise indeed.

Careful analysis of accumulated information provides a splendid insight into the operation of the real water supply system, especially if one checks ideas on a model. The aim is to create a complete picture of the system: how it really operates and why. It is better to have a wrong picture than none at all – time and discussions with other members of staff will help in clearing up the errors.

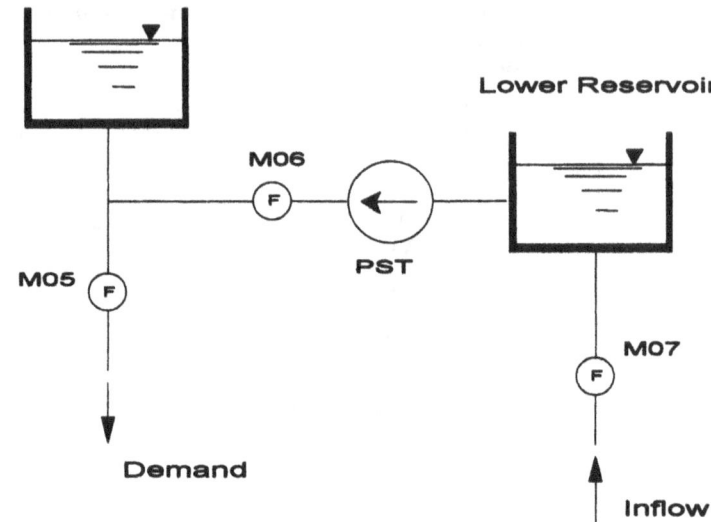

Higher Reservoir

Lower Reservoir

M06

PST

M05

M07

Demand

Inflow

Figure 2.35 A PST
between two reservoirs.

**Discovering the
control policy**

The next example shows two pumping stations belonging to a larger regional water supply system – a part of it is shown in Fig. 2.37.

Pumping station I supplies both reservoirs (I and II). Its pumps are controlled by water level in reservoir I, for which PST I is the only source. Reservoir II could also be supplied by PST II, and it controls the pumps in that station. Figure 2.38 displays flows through PST I and water levels in reservoir I. Note that pumps in PST I always start at the same level (approx. 1.8 m from the bottom) and stop at the same high level (approx. 3.3 m, till changed to 3.6 m on 6/12/90). This is a classic example of level control.

The pumps in PST II are controlled by water level in reservoir II (Fig. 2.39). However, this is not as simple a case as PST I: pumps start and stop at different levels and times (between 07:00 and 10:00, for instance). A Profiler device (see the subsection on tariffs in section 3.7) is installed with the task of keeping water level in the reservoir close to a prescribed daily profile. Points on the profile can be roughly identified by examining Fig. 2.39.

**How does a source
operate?**

A source of water is shown in Fig. 2.40. It has two boreholes (with one and two pumps, respectively), a small contact tank and two sets of relift pumps, each with three units, pumping to two different areas. The contact tank also receives a small gravity inflow from the regional system.

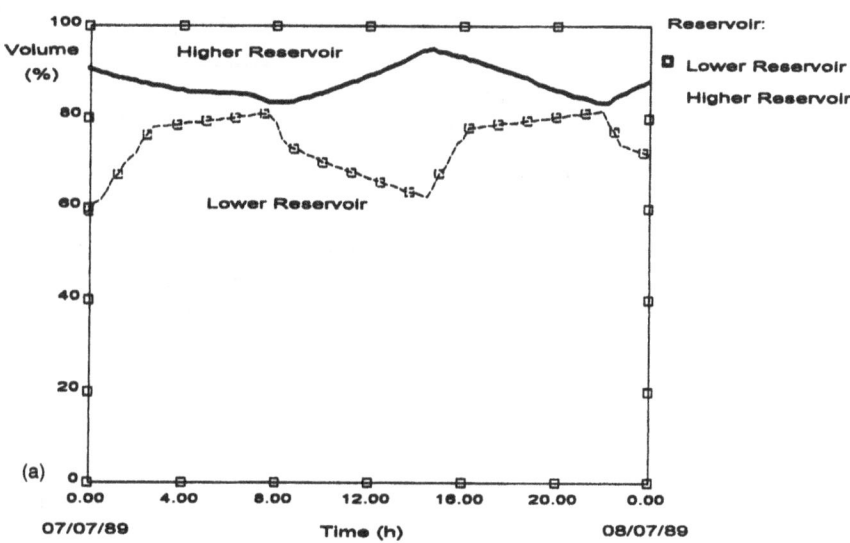

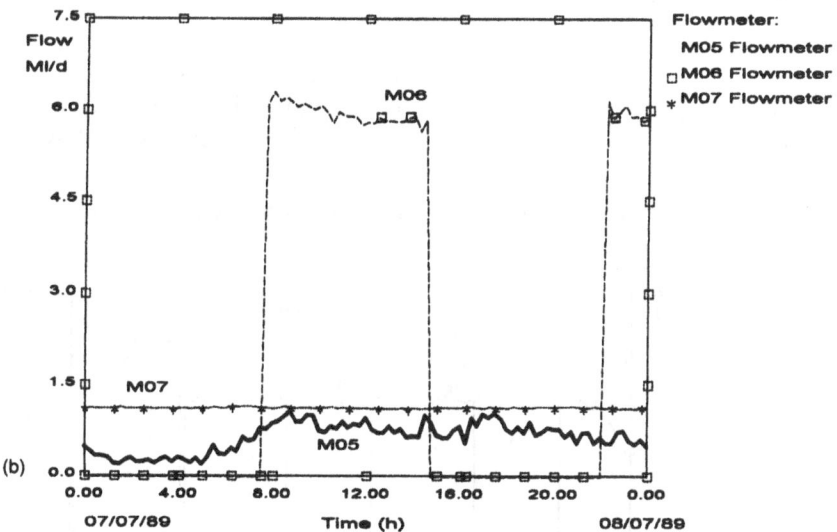

Figure 2.36 (a) Variations of WL was caused by (b) operation of a local PST.

The relift pumps supply distant places and the variations of flow are considerable – from zero to full capacity (Fig. 2.41). The regional system can provide a constant inflow, but this gravity flow is only a small proportion of the total demand on the site.

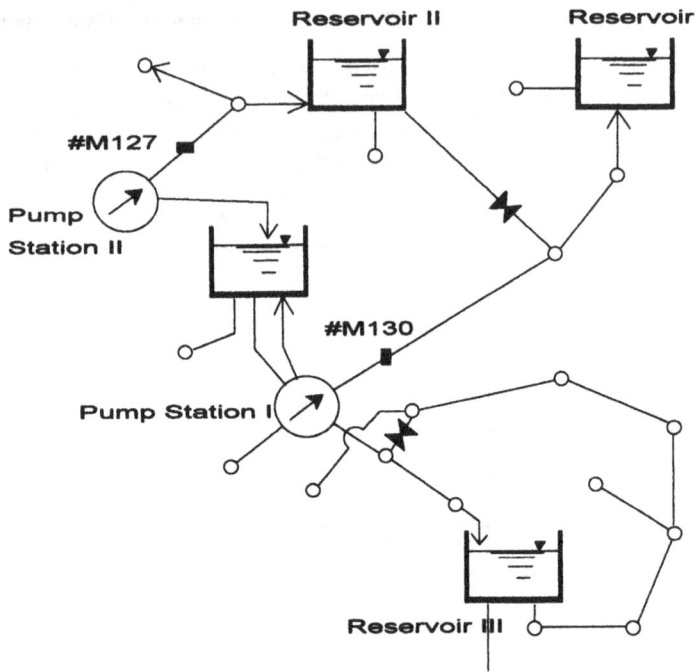

Figure 2.37 A PST in
the distribution system.

Figure 2.38 Level-
controlled pumps.

The contact tank is too small to moderate the oscillations of
flow, so the borehole pumps have to start and stop frequently.
To complicate the problem, there are a few customers close

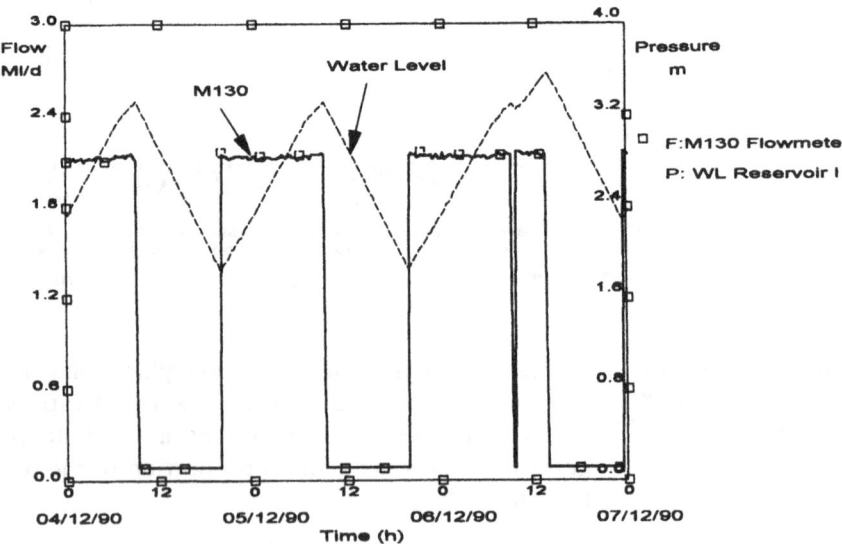

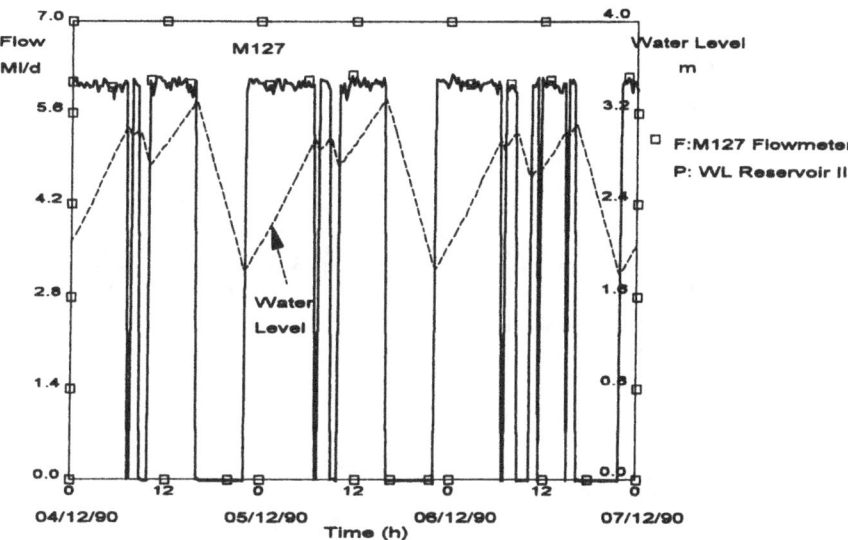

Figure 2.39 Profiler-controlled pumps.

to the site, so the tank must be kept as full as possible to maximise the chlorine contact time to avoid taste or odour complaints.

Meeting all these needs was achieved by using variable-speed pumps in both wells (one unit in each). These keep the tank about 90% full (Fig. 2.42). Note large changes of raw water

Figure 2.40 A source of raw water.

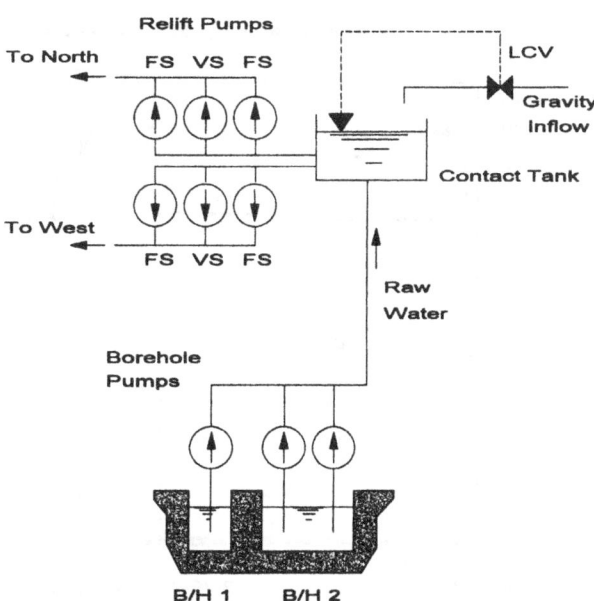

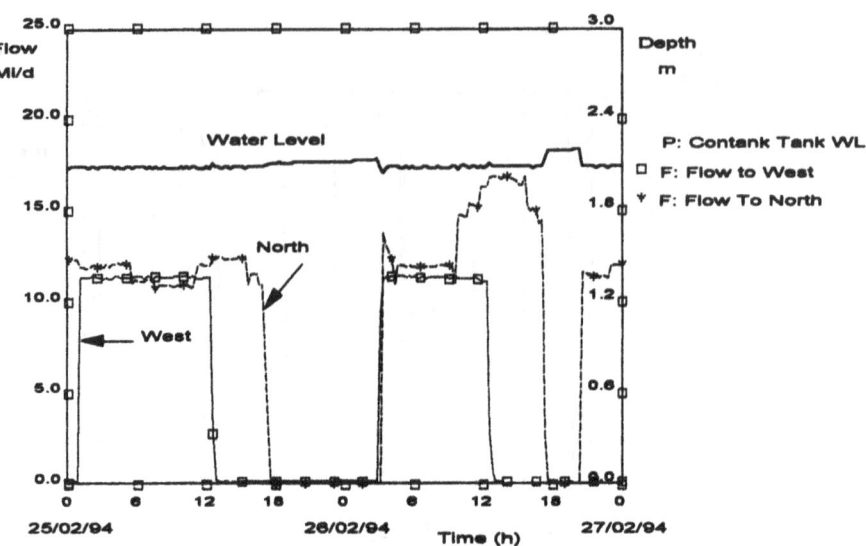

Figure 2.41 The output and WL in the contact tank.

Figure 2.42 Water balance at the contact tank.

inflow. A further, interesting feature of the diagrams is that when all output from the station stops at approximately 18:00 h on 26/02/94, and all pumping has stopped, a very short period of gravity inflow is sufficient to cause the contact tank float valve to close when the tank is 100% full.

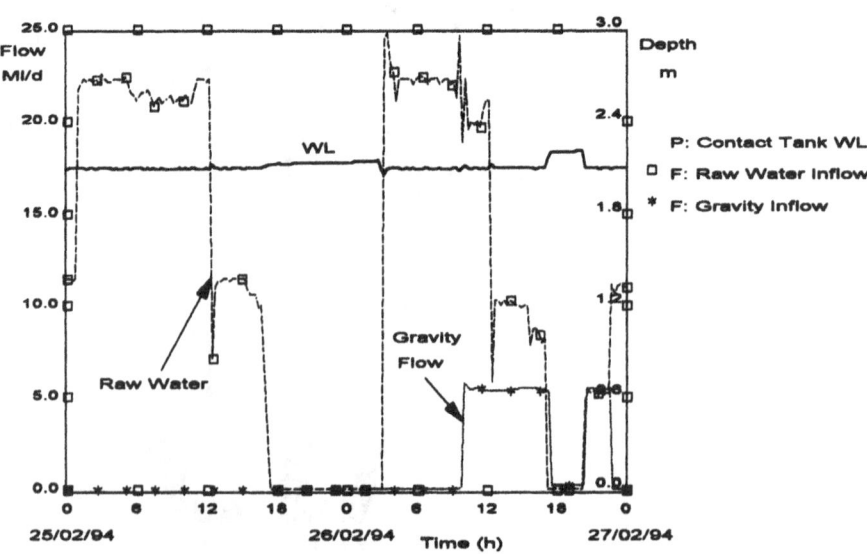

Closed or open valves?

The state of shut-off valves is not always known; even if staff claim that everything is under control, additional checks can be very revealing. In the next case the question was whether a valve between two pumps was open or not (Fig. 2.43).

The flow through each pump is monitored by flowmeters DF033 and DF034. A careful examination of recorded flows (Fig. 2.44) reveals that the valve in question is certainly open –

Figure 2.43 A pumping station.

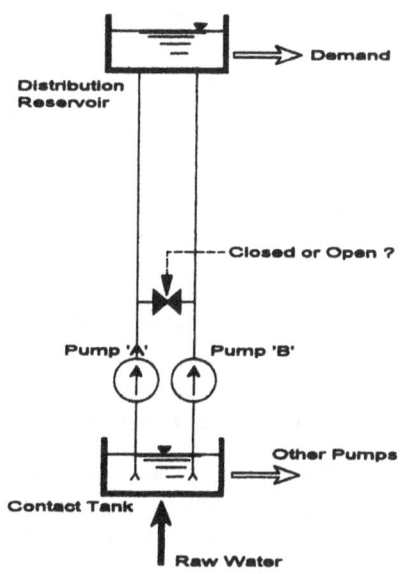

Figure 2.44 Proving that the valve is open.

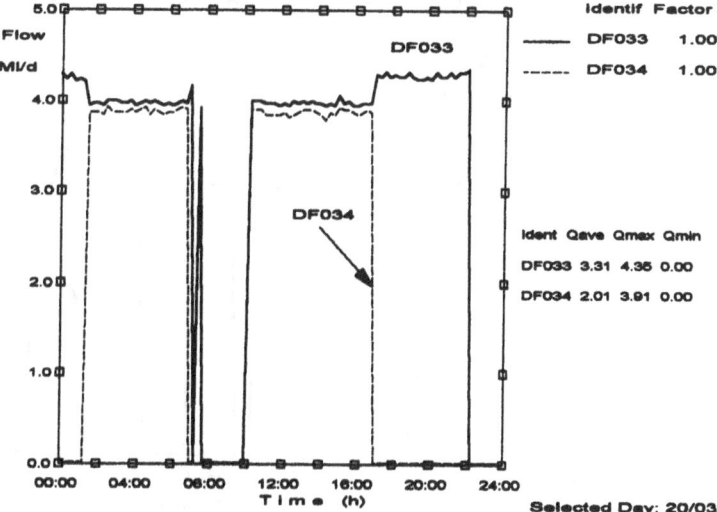

Telemetry Data for The Pumping Station Flowmeters

note the sudden increase of flow through pump A (DF033) when pump B was switched off. This can only be because the discharge pipework is open and the pumps are effectively operating in parallel – the valve must be open.

Where does the water come from?

Similar questions can be asked about the pumping station shown in Fig. 3.36 in Chapter 3. Do pumps take water from the contact tank (i.e. from the local source) or from the long suction pipe? The answer is visible in Fig. 2.45.

Flow through DF136 rises whenever the other pump is stopped, to fall significantly whenever the other pump starts. Pumps behave in this way when operated in parallel, and therefore it is clear that both take water from the common suction pipe. Any increase of flow through this long pipe causes the drop of pressure, i.e. the increase of pump head. If the pumps took water from the contact tank, very close to them, this effect would not take place.

Zone valve

The distribution network is usually split into several autonomous zones, separated by closing appropriate valves. In emergencies, these valves might be reopened, or valves may be temporarily closed to create temporary zones. The effect of such changes might be quite strong, as illustrated in Fig. 2.46.

Interference between pumping stations

Operation of one pumping station usually does not affect the regime in others, at least not considerably. The reason is obvious – staff wish to keep the system as simple as possible. In some cases this cannot be avoided – an example is given in Fig. 2.47.

Figure 2.45 Operation of two pumps.

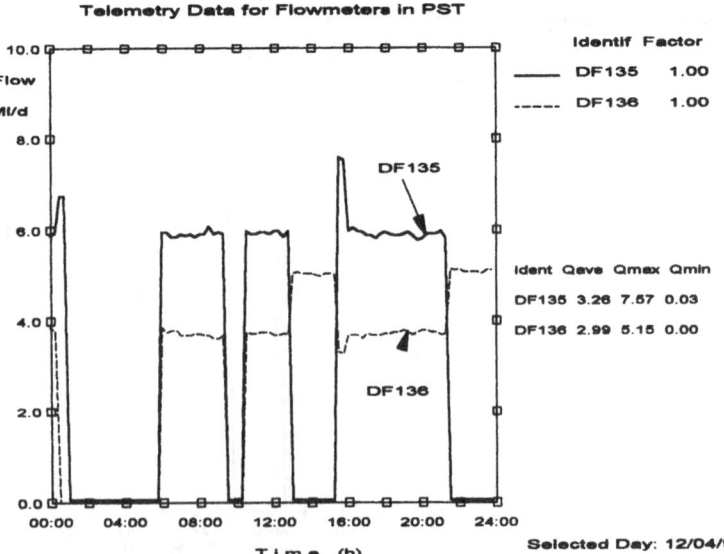

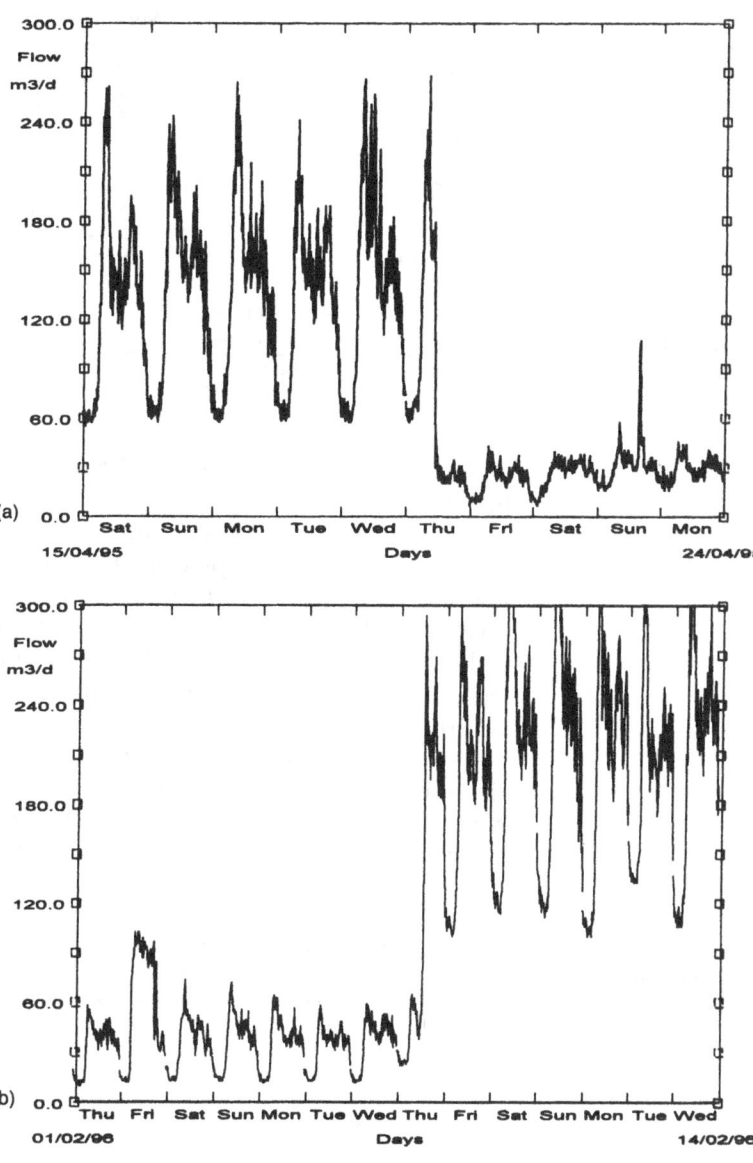

(a)

(b)

Figure 2.46 (a) Rezoning the beginning: a zone being reduced in size by valve closures. (b) Rezoning the end: the original zone being restored by valve operation.

The main pumping station pumps water from source 'A' to the main reservoir. The flow through its suction pipe – a rather long one – is monitored by flowmeter AF184. A smaller pump ('source pump') takes water from source 'B' and adds it to the main flow from time to time. When the pumps in the main PST start, flow through the source pump increases – because pressure in the common pipe will fall significantly. Note that the pressure side of the source pump is the suction side of the main PST. Conversely, the starting of the source pump will decrease the

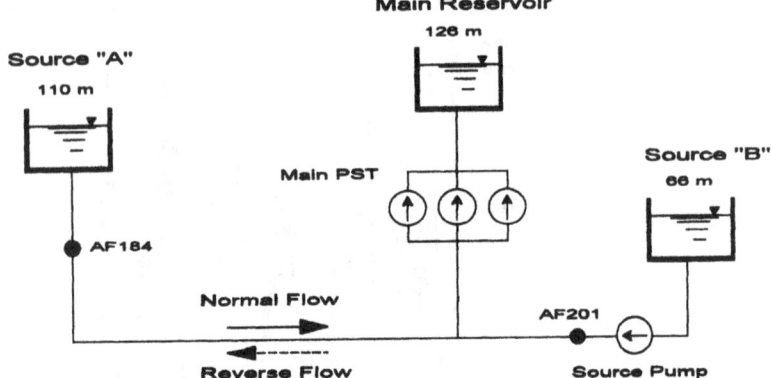

flow through the common pipe. These effects are clearly visible in Fig. 2.48.

The source pump operates whenever there is enough water in its suction tank, and stops when the water level there drops too low (Fig. 2.49). It is therefore a level-controlled pump, but in the reverse mode – it *empties* a reservoir (source 'B' in this case) rather than *fills* a reservoir. This is a fine example of so-called 'reverse control policy' (see the subsection on pumping station control in section 3.7).

Note also that, if all pumps in the main PST are stopped and the source pump still operates, its flow can only go *back to source 'A'*. Unfortunately, as AF184 is not a bidirectional flowmeter, the reverse flow is not recorded.

Figure 2.48 Flows through two pumping stations.

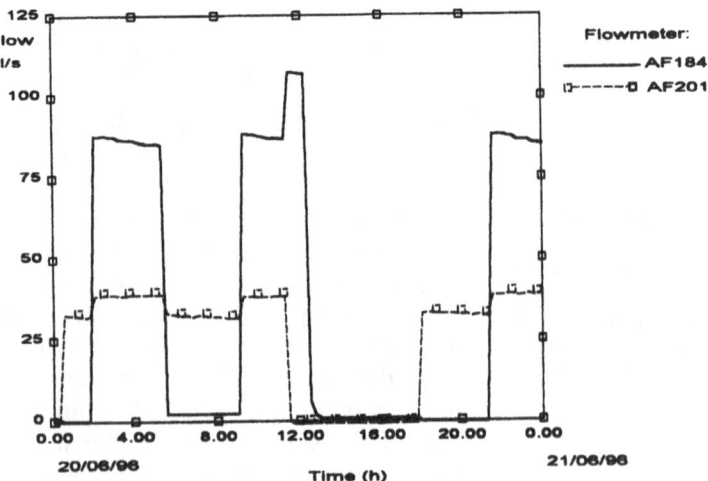

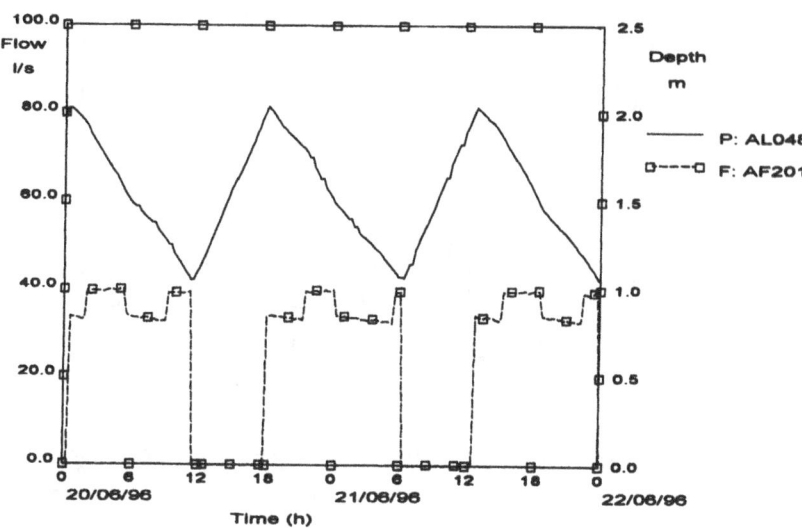

Analysing Control Policy ...

Figure 2.49 Operation
of the source pump.

Supplying an isolated
area

2.4 Analysing the balance

When all important values are monitored, it is possible to carry
out a variety of analyses, giving a better understanding of how
the system operates. In the next case both inflow to and outflow
from a water tower are monitored, as well as water level in the
tower (Fig. 2.50).

This favourable (and exceptional) situation permits analysis in
depth. Figure 2.51 shows the total balance for two consecutive
days, 20 and 21 April 1995. Note how the water level rises
whenever the inflow (AF147) exceeds the outflow (AF148), and
drops in the reverse case.

Figure 2.52 proves that the pumps are controlled by water
level. Note the 'off' level for the first (duty) pump and the second
(standby) pump; also the 'on' level for the second unit – the first
pump starts somewhere below that level.

Figure 2.50 Water
tower fully covered.

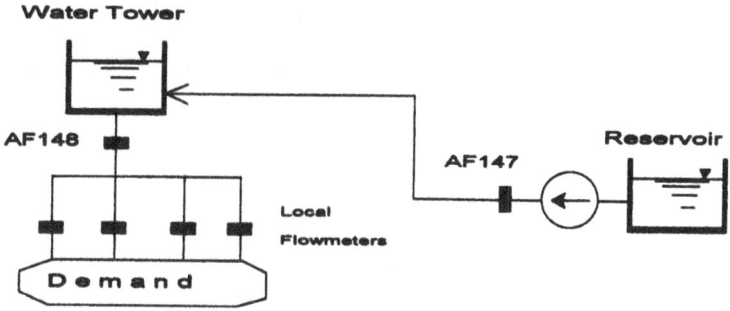

Analysis of operational data

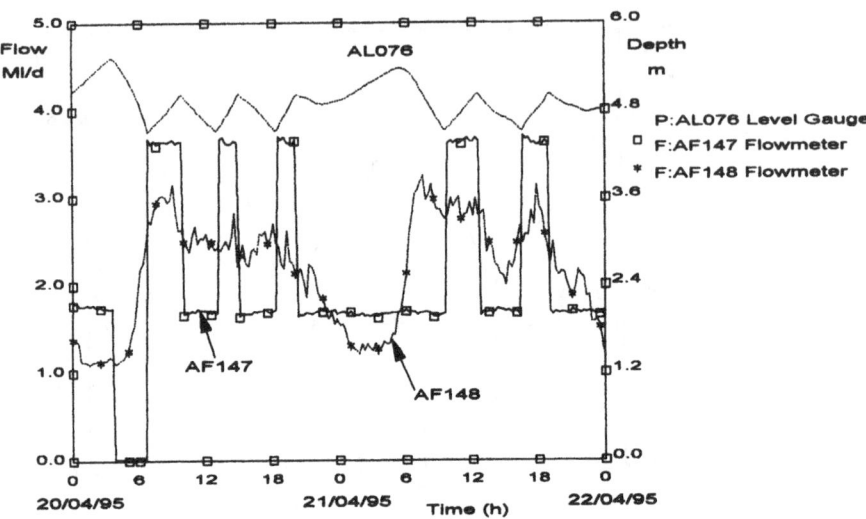

Figure 2.51 The balance
of inflow/outflow.

**Difference between
flowmeters**

Figure 2.52 Level-
controlled pumps.

The ideal situation is when the sum of all outflows (demands) is
close to the inflow in a given area, such as the one shown in Fig.
2.53. DW012 monitors flow into an apparently isolated
consumption area. Flow through branches further downstream
is recorded by four other meters, as shown in Fig. 2.53. The total
flow (monitored by DW012, see Fig. 2.54) considerably exceeds
the sum of the other four flows. Why?

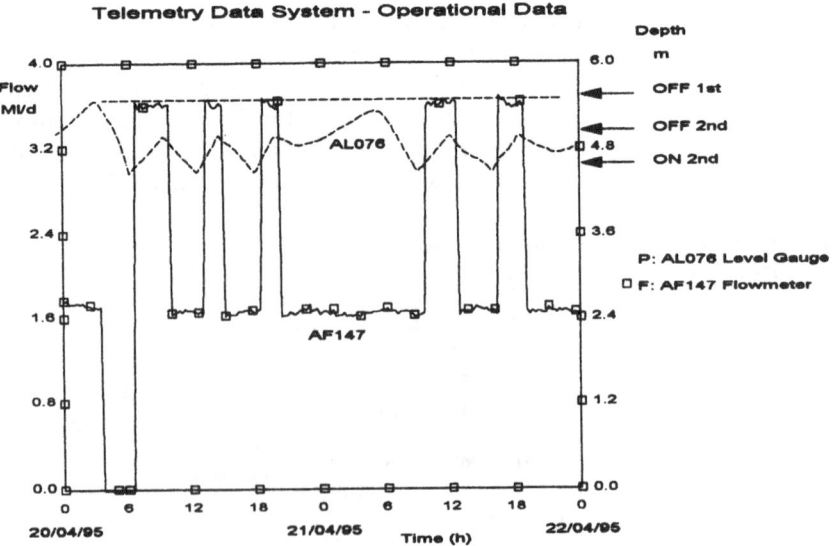

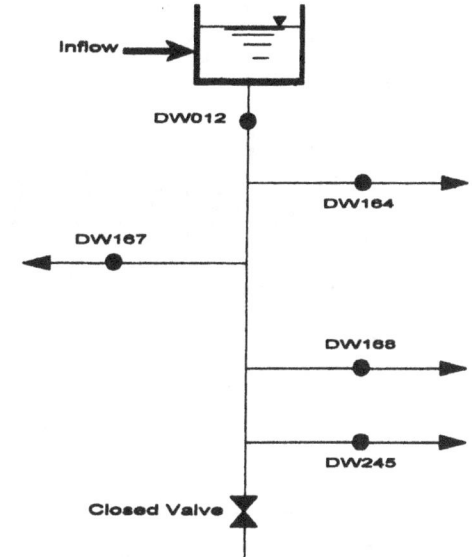

Figure 2.53 Analysing flows in the distribution network.

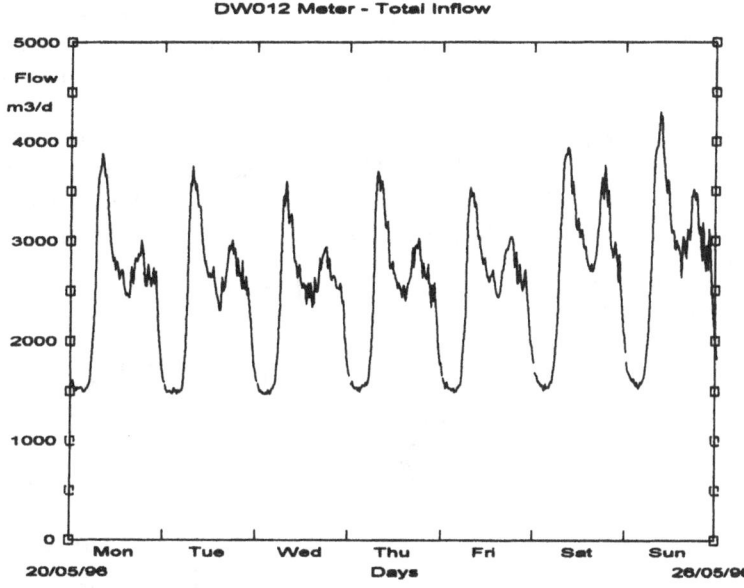

Figure 2.54 Total inflow in the area.

There are four possible explanations. Three of them are: faulty instruments; undetected leak and/or non-monitored consumption; and the 'closed' valve is not closed. However, by examining the pattern of the difference (Fig. 2.55), one can conclude that there are *certainly* one or more important consumers in the area who are not monitored. The evidence is a recognisable daily diagram – note lower night consumption

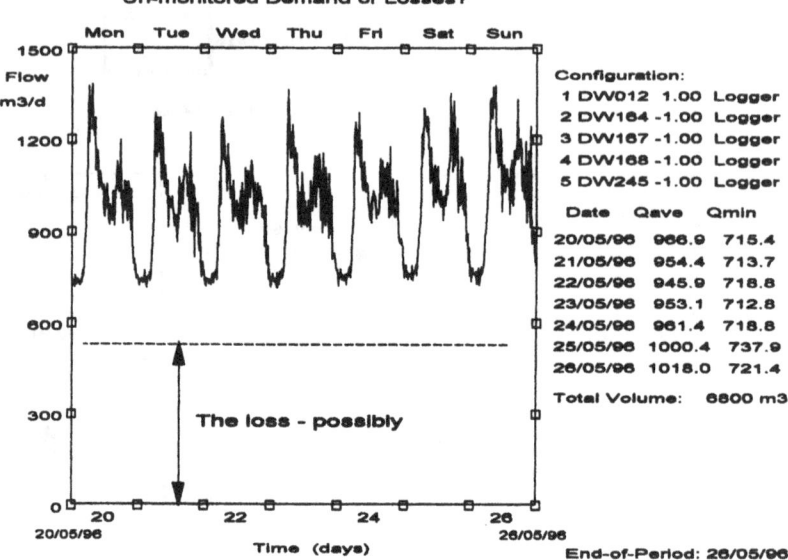

Figure 2.55 The difference between aggregate of outflows and total inflow.

and two peaks. As the night demand (in Fig. 2.55) is still rather high, the other three options cannot be ruled out, but cannot be totally responsible for such a large difference.

A change in control policy

In the next case the situation is not so favourable to simple analysis. There are several pumping stations and reservoirs, including a tower, in a medium-sized distribution network shown in Fig. 2.56 (part of a larger system). The pumps are

Figure 2.56 Part of a distribution system.

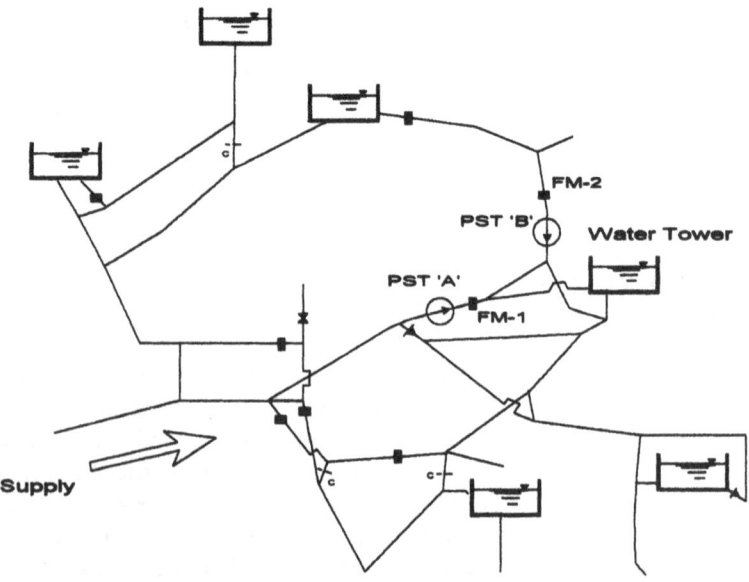

controlled by level in the water tower. There have been two control policies used for the system:

1 Under the original policy, the pumps started and stopped at pre-set levels regardless of the time (Fig. 2.57). Note that all 'ons' are at the same horizontal line – the same applies to all 'offs'.
2 Under the new policy, called Profiler, the pumps try to provide a predetermined water level in the tower throughout the day (Fig. 2.58). Note how 'ons' are *not* all at the same level as before – nor the 'offs'.

The profile developed to control pump sets needs to be designed with some care. The objectives of installing it should be clear. Is the purpose to minimise peak electricity tariff? Is it to contain demand on a limited source? Is the control range too narrow, with the potential to cause frequent (possibly damaging) switching of the pump? Figure 2.59 is an example of the latter problem. Here the need for re-examination of the adopted profile is obvious, because the pumps start and stop too often; the system is unstable.

Discovering a serious leak

Figure 2.57 Old control policy.

The last example shows how telemetry data were used to discover a significant leak just by analysing information already available. Figure 2.60 shows part of the distribution system of the city of Bath, UK. The inflow from an outside source is monitored as well as flow through each branch: one towards reservoir 'A' and the other supplying reservoir 'B'.

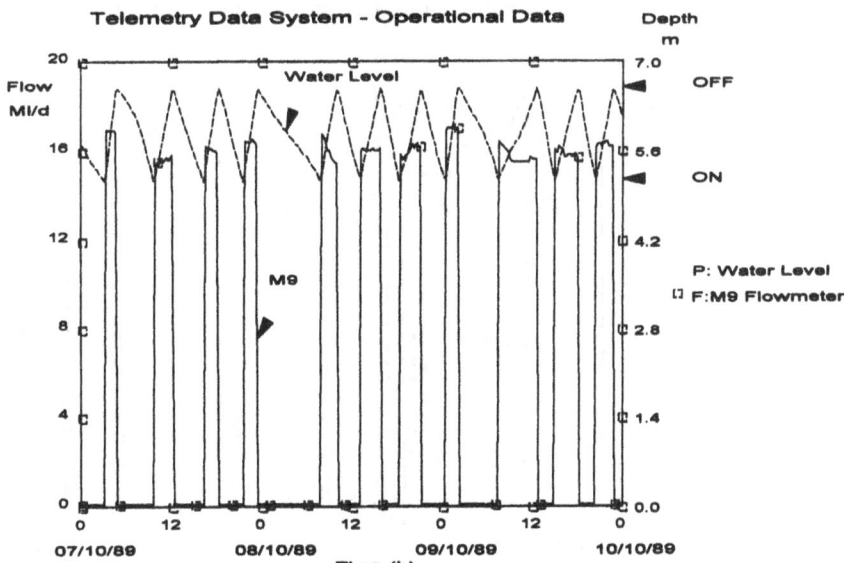

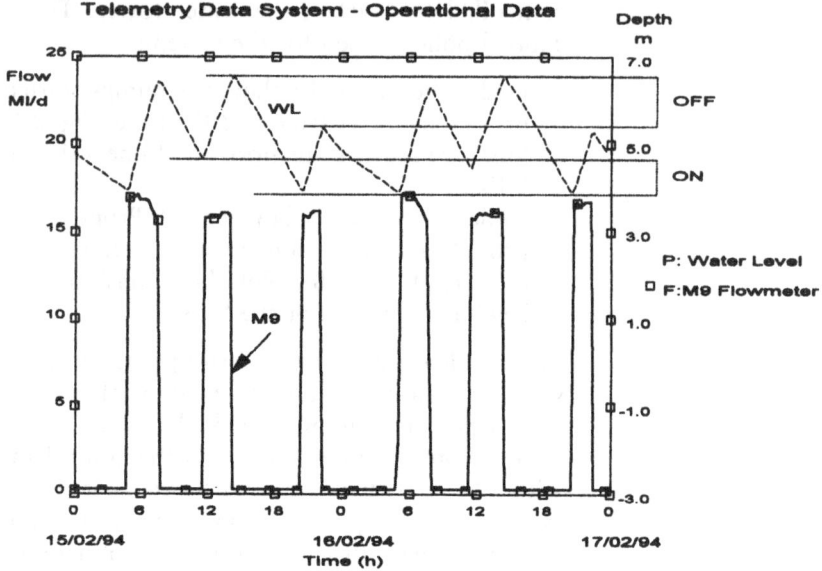

Figure 2.58 New control policy – Profiler.

Figure 2.59 Problems in real life.

All three flows are shown in Fig. 2.61. Note that the inflow is practically constant, while both outflows vary from time to time; when the flow in one branch increases, the other flow decreases – all perfectly logical at a first glance. But the *sum* of the two outflows is *smaller* than the inflow (Fig. 2.62). The difference is significant and could hardly be attributed to the inaccuracy of

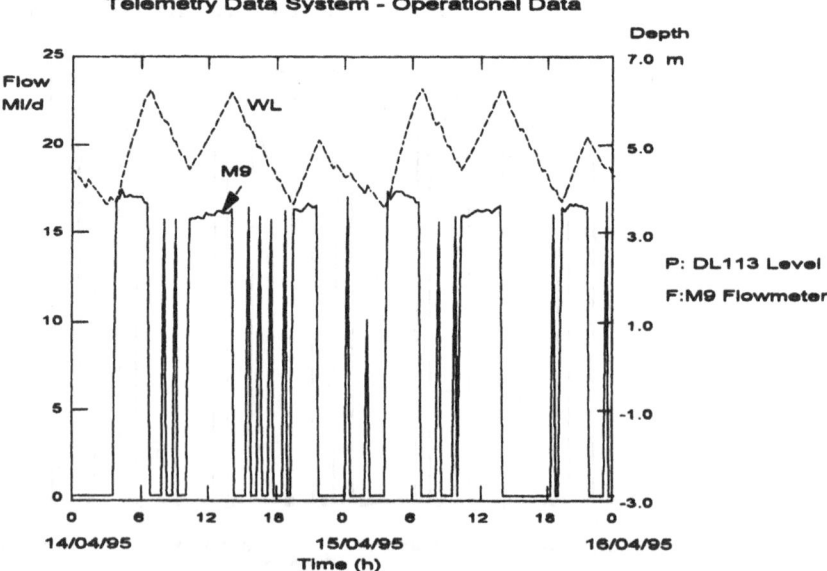

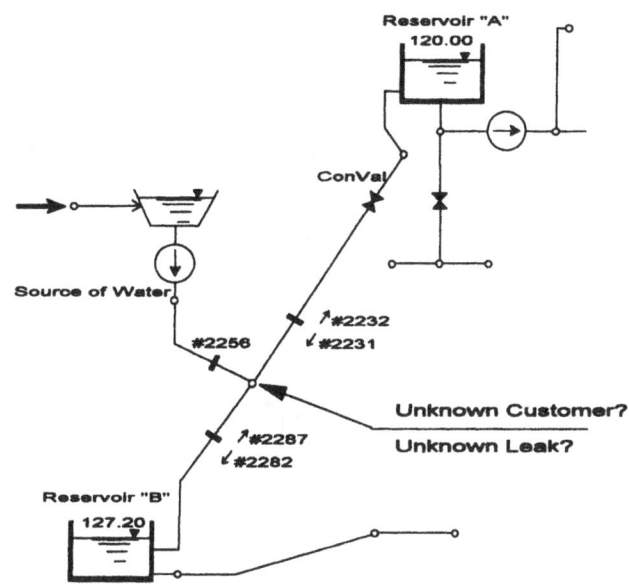

Figure 2.60 A part of a distribution system.

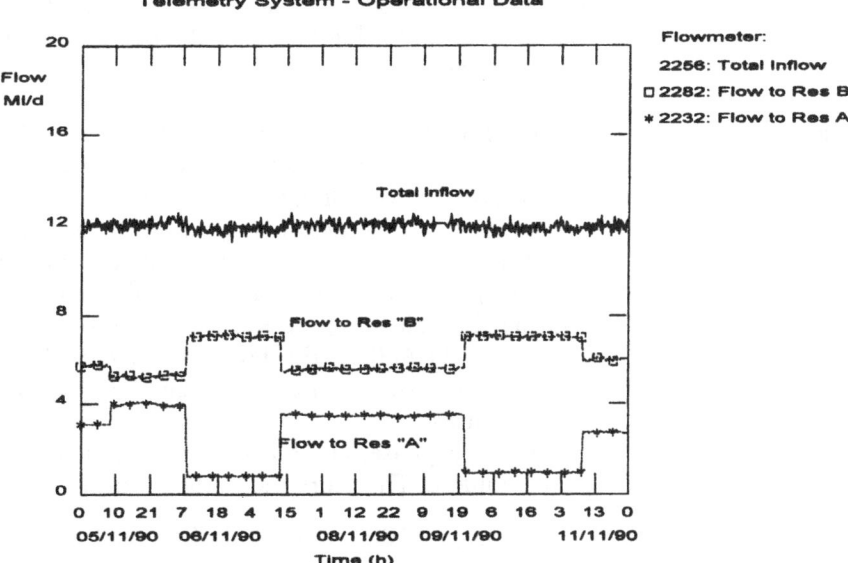

Figure 2.61 Total inflow and flow through branches.

three flowmeters. Moreover, this difference *increases* when the flow towards reservoir 'A' is throttled by closing a PRV in this branch (see Fig. 2.60), and reverts to the old level when this PRV is open again – as visible in Fig. 2.61. This is exactly how a leak flow would behave, if located somewhere upstream of the PRV (the pressure there will increase on closing the PRV). This

Analysis of operational data

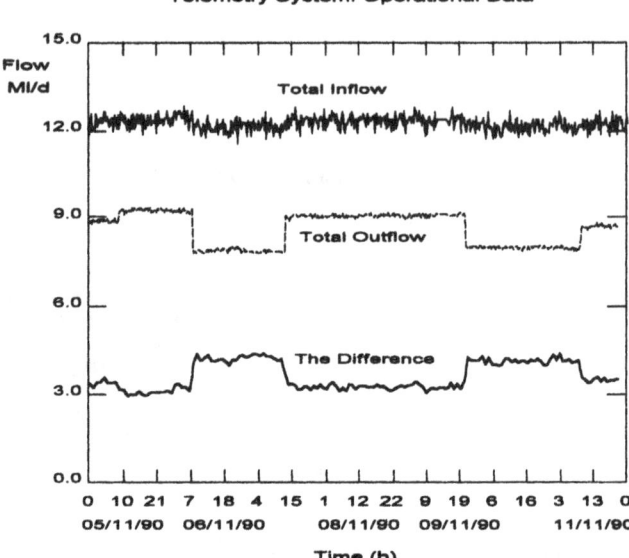

Figure 2.62 The difference between the inflow and total outflow.

evidence was sound enough to persuade the local staff to investigate, and a big leak was discovered near the junction of the three pipes.

Manual calculations

Figure 2.63 shows variations of level in a 150 m³ tank. Note that inflow occurs when the level falls below 1.25 m. The depth of this tank is, according to records, 3.0 m. A simple calculation shows that the real depth is significantly less than 3.0 m.

If the depth of the reservoir is 3.0 m and the volume is correct (150 m³), then the plan area of the tank must be 150/3 = 50 m². During the first filling period, when the pump is operating for 3.22 h (= 11 600 s), the level rises by 0.385 m, and the net increase of stored water must be 0.385 × 50 × 1000 litres = 19 250 litres.

The next phase, emptying, lasts 5.55 h (= 19 980 s) for the same volume. Therefore the outflow during this period must be

$$\Delta Q_{outflow} = 1000 \times 50 \times 0.385/19\,980 = 0.96\,\mathrm{l\,s^{-1}}$$

If we assume that the outflow rate from the reservoir during the pumping period was of a similar order, then the total volume pumped into the reservoir during the 'filling' period must have been 19 250 + 0.96 × 11 600 = 30 386 litres. But the pump was delivering water to the reservoir at a rate of 4 l s⁻¹ for 11 600 s, a total volume of 46 400 litres.

The actual depth of the reservoir is 2 m – not 3 m – and the plan area of the tank is 75 m³, not 50 m³. Note that 0.385 m

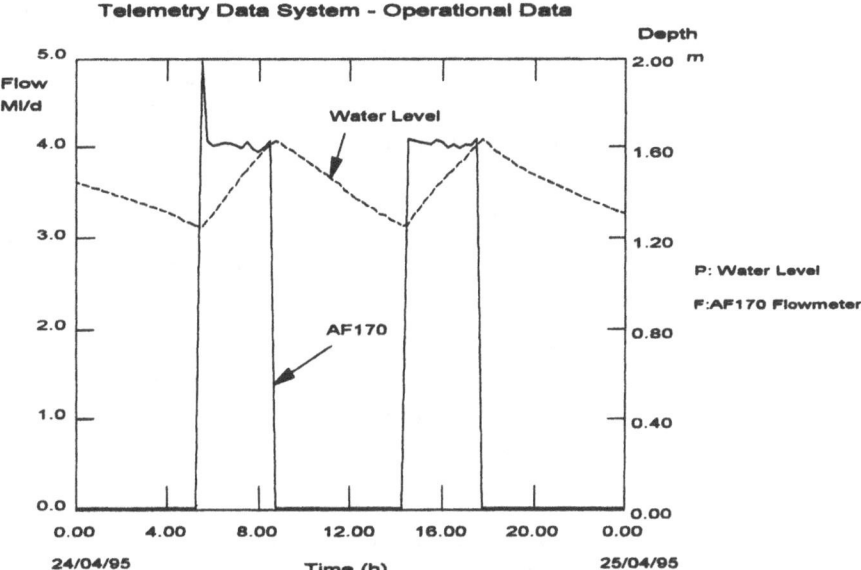

Figure 2.63 Manual calculation

represents $0.385 \times 75 \times 1000$ litres $= 28\ 875$ litres, not the 19 250 litres previously calculated.

The revised calculation of water delivered to the reservoir during pumping, based on change in water level, shows that $28\ 875 + 0.96 \times 11\ 600 = 40\ 011$ litres is a truer estimate. The calculation has a degree of uncertainty, of course, unless the outflow figure during the pumping period is available. However, the discrepancy between the calculation of 30 386 and 46 400 litres was too big to ignore, and further investigation confirmed that the recorded depth was incorrect.

2.5 The case for mathematical modelling

Operational data available on-line are a tremendous asset to every water company, and the cornerstone of any automated control system. Useful as they are, the data alone cannot *predict* (except in the crudest sense) and cannot *evaluate* options. The quantity and quality of operational data now available on-line are almost terrifying, presenting a real danger of staff being 'flooded by data, starved of information'. To place all these data in a proper context and draw the right conclusion, one must resort to tools like mathematical modelling (also known as network analysis).

Briefly, a mathematical model is an artificial entity that describes the behaviour of a real system in various situations. In the case of a water supply system, this means that the model can simulate variations of flow and pressure in a distribution network, operation of pumping stations, functioning of all devices like control, check and float valves, changes in water quality at service connections, compute costs, forecast demand patterns, optimise the use of pumps, etc. – and all this in any conceivable normal, emergency or catastrophic case.

Building and testing a model takes time, effort and significant cost. No one will do it without good reason, least of all a water engineer pressed with other urgent tasks. But good reasons do exist:

- The model will collate all known facts and data of the real system(s) with all known relationships between its elements. The very discipline needed to build a working model creates a comprehensive database about the components of the system being modelled.
- The model will permit the analysis of operational data in great detail and depth, helping to understand the complex process within the prototype system, identifying causes, effects and solutions.
- All interested parties – operators, designers, engineers and managers – can participate in the building of the model, give a contribution and later use it for various tasks.
- The current control policy can be tested on the model, verified or improved, without risky or expensive experiments with the real system.
- The response of the system in various normal, emergency and catastrophic situations may be analysed beforehand, and adequate strategies to deal with such situations developed and perfected.
- The model is a powerful tool for design and planning, greatly reducing the time and costs for such activities.
- The model is a powerful but safe and interesting device for the education and training of new hands, from operators to managers.

These are tangible benefits, but there are also intangible benefits in developing and using models in a water company: general knowledge about the system will be raised, myths dispelled, and basic data collected, validated and stored in a comprehensive and logical format.

These arguments, as well as the availability of computer hardware and software, have caused a proliferation of models and modelling techniques in the water industry over the last few decades. One can argue that modelling was always an integral part of the design process; the novelty was its application to whole water supply systems and not just to elements of the system. Nowadays the models have become a standard tool in water companies, a part of the broader information system, and used by staff at all levels. Modelling is no longer an exclusive academic exercise!

There are two main types of models:

- *simulation models*, which *describe* the behaviour of a real system as closely as possible;
- *optimisation models*, which *prescribe* what should be done in order to achieve a certain objective.

The first type is more popular in operational management, because the models are simpler, easier to understand, more flexible in use and there is no complicated mathematical apparatus or 'black box' necessary for optimisation. Moreover, staff can make simulation models alone, without special training, which is not the case with optimisation models. The problems of integrating the two types of model are discussed in Chapter 7. The models referred to in this and subsequent chapters, apart from Chapter 7, are generally simulation models.

The main applications of modelling are in:

- *operational management* of existing water supply systems;
- *design and analysis* of existing and future water supply systems;
- *education, study and research* in hydraulics and water supply.

Typical jobs for which the models are being used are:

- analysis of how the system really operates in normal and emergency conditions;
- assessing levels of service, e.g. locating pressure areas, how many people are affected, etc.;
- verification of current control policy;
- evaluation of storage reserves in emergencies;
- contingency planning;
- optimum pump scheduling;
- re-routing flows around pipe bursts;
- identifying 'bottlenecks' within the system;
- water quality studies through computation of flow paths and travel times;
- decisions on tariff selection;
- extension of telemetry system;
- design of district metering systems;
- design of pressure management zones;
- design of new facilities;
- design of enlargements and reconstruction of the system;

- design of a new control/telemetry system;
- 'war games' for the staff;
- training of novices on various levels, etc.

What is needed to make a start?

The principal ingredients of a model are a good program for hydraulic analysis, a complete database of operational and fixed data, and specific knowledge about the system in question. Note that only the first item can be bought, while the other two must emerge from the company. There are many up-to-date programs on the market, with more to appear soon, so a new customer will have a range of software to choose from. A few guidelines in making a choice are given here to help.

Modern software products are labelled 'user-friendly' because they do not ask for any special knowledge or training prior to use (usually only partly true). The characteristic features of a good water supply system modelling package are:

- It is extremely easy to learn and use, because it is tailored to suit user's demands.
- It is completely menu-driven, so the user does not have to look very often in manuals.
- It has a comprehensive system for capturing errors, with an explanation in plain language, so the user is protected against common blunders and mistakes.
- On-line help is available at any stage.
- It can model a wide range of possible control policies for pumping stations and control valves, but keeps all computing 'technicalities' invisible to the user.
- It has the ability to keep the user informed about the state of the system during a simulation run.
- It always provides an answer to the user regardless of the size or complexity of the case, and never loses control!
- It has good computer graphics available to the user for both input and output jobs.
- It can compare operational data with simulation results on the same graph.
- It permits users to create the output according to their particular wishes in a very versatile way.
- It is a very robust tool, which will not crash whatever the user might do through inexperience or negligence.

As for the user, it is assumed that:

- The user is well informed about the water supply system that he/she is modelling, but does not have to be an expert in mathematical modelling or computers.
- The user should describe the real supply system as it is, without the need to invent hypothetical elements, and the

program must be versatile enough to cope with the complexities of real supply systems.

In conclusion, it might be said that good mathematical models should have the following characteristics:

1 Desirable
- Embrace all known facts and relationships
- Put operational data in context
- Everybody can participate and contribute
- Control policy can be tested and improved
- Education and training of the staff
- Correct response in normal/emergency/catastrophic situations
- Invaluable for planning and design jobs

2 Feasible
- Basic relationships are (mostly) known
- Response of water supply systems is predictable
- Environmental impacts could be observed and monitored
- Basic data are already available in the company
- Hardware is not expensive
- Software is not expensive and does not require input from outside the company
- No special training is needed
- Could be introduced in stages

3 Benefits
- Better understanding of water supply system functioning
- Better overall control
- Better use of facilities
- Better use of resources and funds
- Substantial cut in unaccounted-for water levels and energy costs

4 Dangers
- A new link in the chain is introduced
- Dependence on technology is increased
- Human problems: old (experience) versus new (special knowledge)
- Overconfidence can lead to disasters.

3 Modelling water supply system elements

3.1 Structure of a hydraulic model

Any water supply system may be quite adequately represented by a set of nodes and links.

A *node* is a location within the system where either head or inflow/outflow is known. Nodes fall into one of three groups:

1 *Fixed-head nodes*, where water level remains unchanged during the simulation because the storage is infinitely larger than all withdrawals and/or inputs. Examples are: aquifers, lakes, large pools, springs, wells, etc.
2 *Variable-head nodes*, where water level rises if the inflow exceeds the outflow and vice versa. Examples are: reservoirs, water towers, contact and balancing tanks.
3 *Ordinary nodes*, where the total sum of all inflows and outflows is known and the head is to be computed. Examples are:
   - a transfer point, where a known quantity of water is either delivered to or taken from the system;
   - a junction (branching point) of two or more pipes, with or without some local consumption;
   - a service connection to an important user, where local consumption of water is known in advance;
   - any other point within the system that is important for any reason.

In both 1 and 2, water level is a known quantity, while outflows and inflows need to be determined through the computation. Examples of nodes used in modelling are shown in Fig. 3.1.

A *link* connects two nodes and water can be conveyed through it. There are three types of links:

1 *Pipes*, which simply transport water between two nodes.
2 *Pumping stations*, which add energy to water.
3 *Control valves* (any type), which regulate the flow of water by reducing its hydraulic energy 'throttling'.

These names are mnemonic: they describe the behaviour of one element of the model by using the name of the similar physical device. This does not mean that each real valve in a water supply system will be represented by a 'valve' in the model – this is

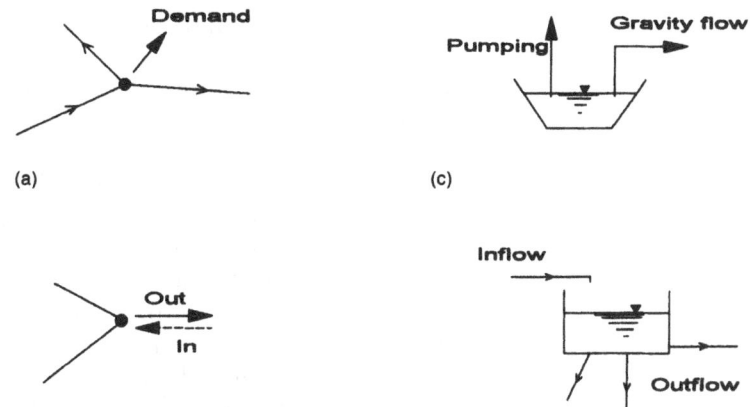

Figure 3.1 Nodes in modelling: (a) ordinary node; (b) transfer node; (c) fixed-head node; (d) variable-head node.

seldom practicable. Quite the opposite: only a few important physical valves (control valves, for instance) will be shown as such in the model – and the bulk (sector valves) will simply be included as minor losses in their pipes.

A Pipe in the model represents any conduit, tunnel, water main or other pipe – anything used to convey water from one place to another, under pressure but not through open channels. A single real pumping station that houses several pumps under the same roof may need to be represented in the model as several 'pumping stations' (in a modelling sense) if the suction or the delivery node is not common to all the pumps (see section 3.7).

Figure 3.2 Links of a model: (a) pipe; (b) valve; (c) pumping station.

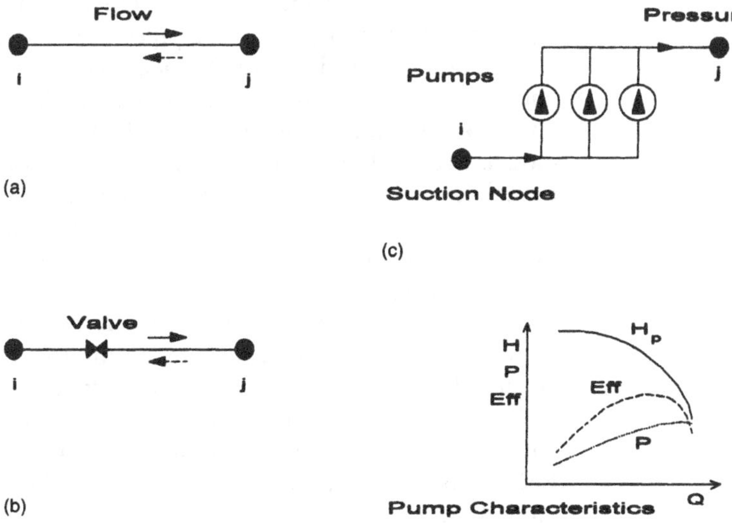

Examples of links are shown in Fig. 3.2. Note that hydraulic energy falls in the direction of the flow in the case of 'pipe' and 'valve', but increases in the direction of flow through a 'pumping station'.

It can be easily seen that all the elements of any water supply system can be described by these items – either directly or by suitable combinations.

Thus far it has not been defined whether the model will serve for analysis of steady or transient regimes. Further development will be limited to relatively slow changes – where inertial forces and compressibility of water can be ignored. This assumption is valid for all normal and most emergency situations in water supply systems, especially if longer time periods (up to several days) are being analysed. Of course, when analysing transient states – which last just a few minutes at most – inertial forces, elasticity of pipe walls and compressibility of water must be taken into account, using more sophisticated analytical tools.

A further simplification of the real system is about water withdrawals. It is clearly impossible to represent each tap in the model, so whole buildings and blocks of houses are represented as 'sinks' – nodes with aggregated demand (total consumption plus some losses). At the very best, this can be only a crude image of the real process, but it is still useful. Even with all these assumptions, the task cannot be solved without a computer.

3.2 Modelling real-life systems

Modern water supply systems have attained unprecedented complexity, size and importance. The distribution network of a small city is shown in Fig. 3.3. Only the pipes are shown. A small area of this city is shown in Fig. 3.4 – note road names, house numbers and other details. These details are omitted in Fig. 3.5, which shows only the pipes of that area.

Given the size and power of modern computers, all these pipes and house connections *could* be modelled, despite their numbers – tens or even hundreds of thousands. On the other hand, such models are too big and impracticable for operational use, so the model is usually limited to larger pipes, reservoirs, pumping stations and control valves. For instance, the macro model of the same city is rather small, as represented by the schematic diagram in Fig. 3.6.

The model consists of reservoirs, pumping stations and a few main trunks, but it can still represent the real system quite accurately, at least when the overall balance is sought. Note the difference between this schematic and the cartographically correct diagram of the distribution system shown in Fig. 3.3. Either type of representative diagram could be used as a basis for modelling.

Figure 3.3 Distribution network of a small city.

This case illustrates the first difficulty in modelling a supply system: what to retain and what to omit? There is no simple answer; much depends on the purpose of the model, the data, software and hardware available, and the preferences of the modeller. At this level of development, modelling is not solely an art and completely subjective, but neither is it a totally objective creation based entirely on mathematics and physics. However, if the model as a whole can be developed in one of several ways, none completely superior to the others, many of the parts or elements of the model can be created as objective entities by reasonably standardised procedures. The principal elements of any water supply system will be identified and briefly described

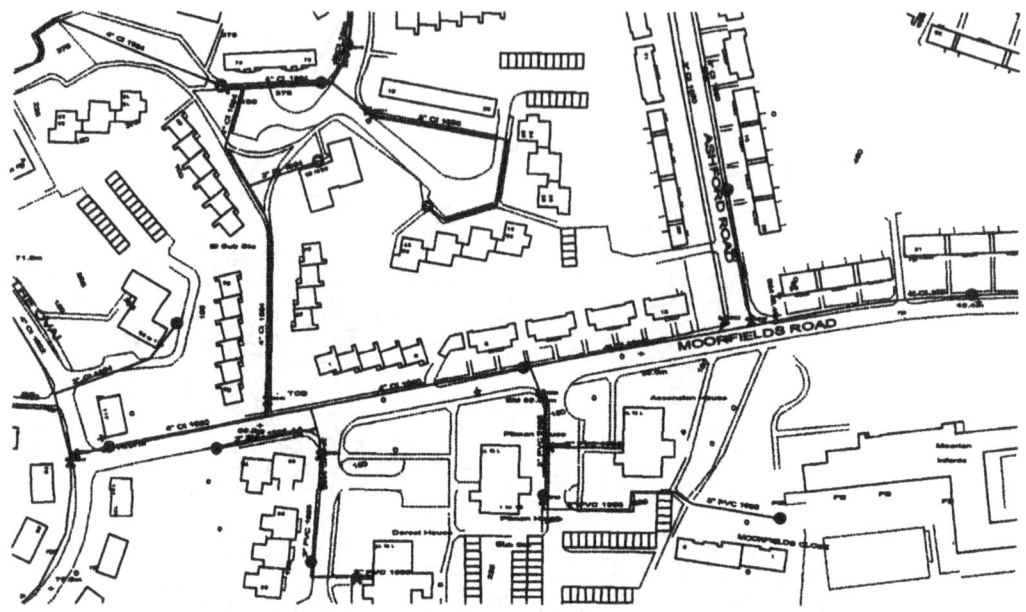

Figure 3.4 A detail of the distribution network.

Figure 3.5 Pipes of the network.

in this chapter. The text will concentrate only on aspects relevant to modelling, and hopefully avoid detail not relevant to the model.

Every real water supply system (WSS) has more features than it is feasible to include in a model of the system (just think how

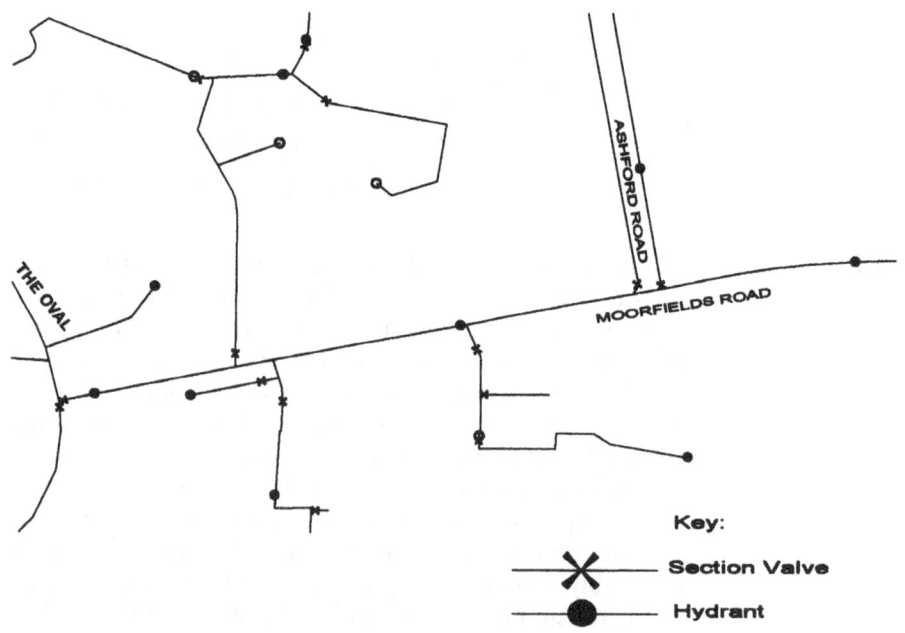

Key:

✕ ── Section Valve

● ── Hydrant

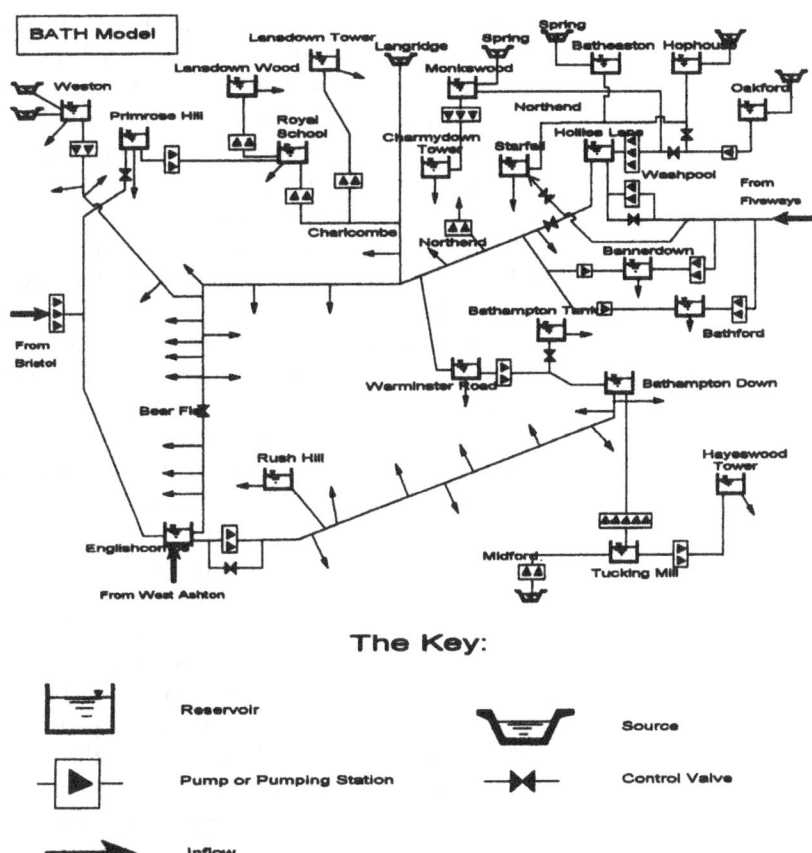

Figure 3.6 Macro model of the same city.

many taps there might be in even a modest sized public building). Any model must concentrate on the most important facets, or the task of modelling will be hopeless. This situation therefore poses the problem of identifying and describing *important* facts – and what to do with the rest? This whole chapter is devoted to this problem – as a prelude to mathematical modelling techniques.

The first task is to identify the basic elements of the system, those which will later be key 'building blocks' of the model. Those are:

- sources of water,
- pipes,
- points at which water is taken from the system (service connections, standpipes, etc.)
- water storage facilities (reservoirs and water towers),
- pumping stations,
- valves and control equipment.

All these elements may be present in an actual WSS, arranged in any order. All of them – and the system as a whole – may be represented by mathematical means to create a new entity: a mathematical model of the real water supply system.

It is interesting to note that a few items are omitted from the list above, for various reasons. First, all equipment for protection against pressure surges, such as air valves, air vessels, aeration pipes, etc., are ignored. As will be seen later, it will not be the purpose of this book to discuss the modelling of transient regimes. Secondly, water treatment plants cannot easily be represented as one element. The usual practice is to model these plants separately, not as part of the distribution system. Distribution system models usually start at the treated water reservoir, with a known inflow from a treatment plant. However, this is not an inflexible rule, and the final decision should be made independently in each case.

The first task is to collect data describing the physical characteristics of each of the items to be included in the model: reservoir, valve, or pump. In practice, the data might be incomplete, unreliable or even incorrect. It will be found to be relatively easy to identify such deficiencies and to insert sensible default values until better information is obtained. Some data on such default values are provided in Appendices B, D and E.

The next task is to understand how the system operates, and gather information about the control regime or policy for pumping stations and flow-regulating valves. Again, experience and engineering sense should be mobilised to check, augment and correct any suspect information.

The problem is to separate fact from fiction (even myths). It might be tempting to ignore facts that do not fit nicely into a preconceived idea about the system, but one must always be ready to revise the model in the light of the new, proven facts. One should not use 'tricks' in modelling – for instance, replacing a two-unit pumping station by nine pump sets – even if this might appear to be an elegant solution in meeting the proven data or a complex control policy.

3.3 Sources of water

The sources of water may be:

- a spring,
- a stream or river,
- a well,
- a lake,
- a storage reservoir, etc.

Some examples are shown in Fig. 3.7.

The quantity and quality of water that a source can yield usually varies with time, depending upon climate and local hydrologic conditions. These changes may also be analysed on mathematical models, but of a different kind, covering much larger areas and with a longer time step. For the purpose of developing our supply system model, one can introduce rather crude simplifications, assuming that either

- the water level at the source is constant, regardless of quantities taken, or
- a known inflow (which may or may not vary with time) is supplied from this source.

These simplifications do not make problems in a short timescale, up to a week or so. For longer periods of time, it seems that some interlinking between those two types of models will be needed – but this remains a task for the future.

Figure 3.8 shows the yield of a source during a period of four years. Note that the output is season-dependent; the minimum is always in September – before autumn rains.

Figure 3.7 Modelling the sources.

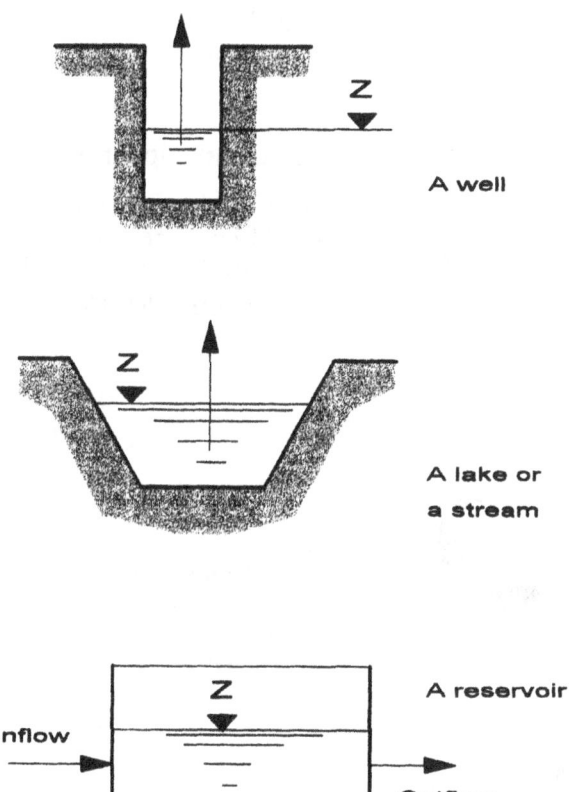

A well

A lake or a stream

A reservoir

Inflow

Outflow

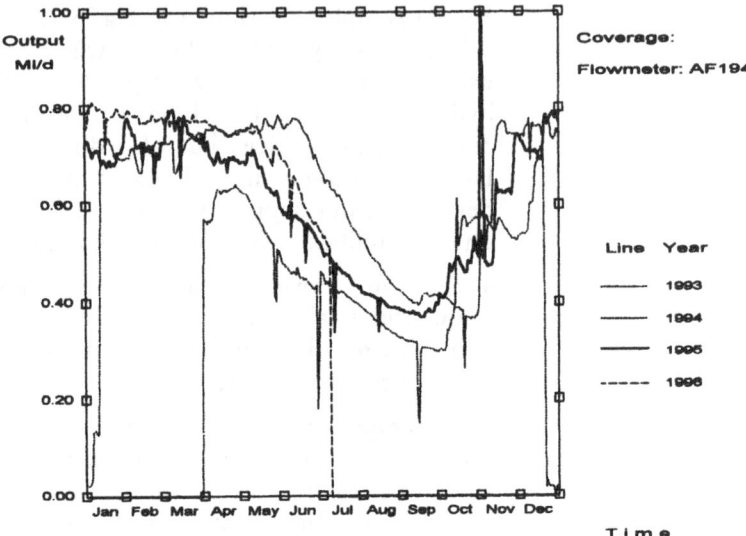

Figure 3.8 A source yield throughout four years.

Water from a source can flow either by gravity alone (Fig. 3.9a) or be pumped (Fig. 3.9b). All these elements should be modelled – this is better practice than to assume that the source will supply whatever quantity is required by the rest of the model.

3.4 Pipes and pipe networks

Figure 3.9 Modelling a source: (a) gravity flow; (b) pumping.

Urban distribution network records were, until very recently, usually held on large-scale hard-copy maps (typically 1:500 or 1:1000). Latterly, such records, including background detail, have increasingly been transferred to GIS (geographical infor-

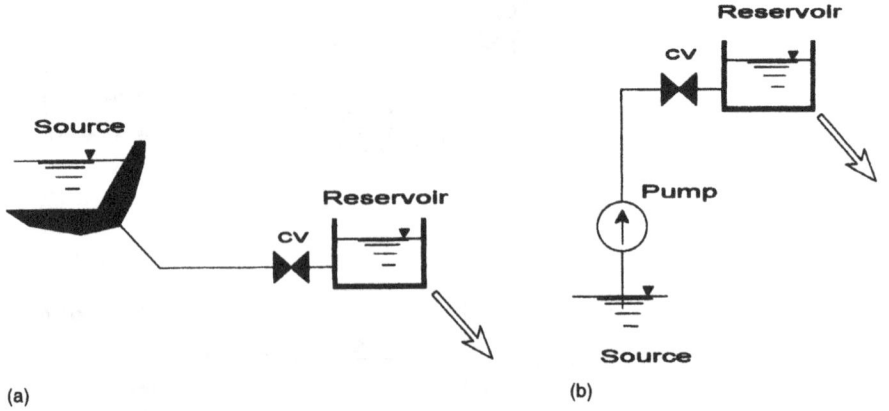

mation systems). These records contain a lot of details – an
example is shown in Fig. 3.10 (a very small part of a large
distribution network). The pipe between nodes 'i' and 'j' is a 4
inch cast iron pipe laid in 1905. Hydrants, section valves, streets
and individual public and private buildings are all indicated on
the map. Other details not shown in this particular example may
be held within a GIS, such as house numbers or names, other
public utility mains and services, and so on. All valuable data for
water company staff who have to operate and maintain the
water supply system may be included.

To collect all this information from all sources (paper maps,
sketches, investigations, interviews, etc.) and to keep it updated
is a formidable task, calling for powerful computers, disciplined
procedures and trained personnel. The investment in time,
money and effort is considerable, but the benefits to the water
company are also great and are enjoyed by all departments.

However, for hydraulic analysis of pipe flow, most of these
data are not needed. Relevant data are:

- pipe length, L (m),
- internal diameter, d (mm),
- internal roughness, k (mm),
- information about pipe fittings like section valves, bends,
  reductions/enlargements, etc.

Note that pipe length and diameter can be taken from Fig. 3.10,
but internal roughness has to be estimated (it could be

Figure 3.10 A pipe
within the network.

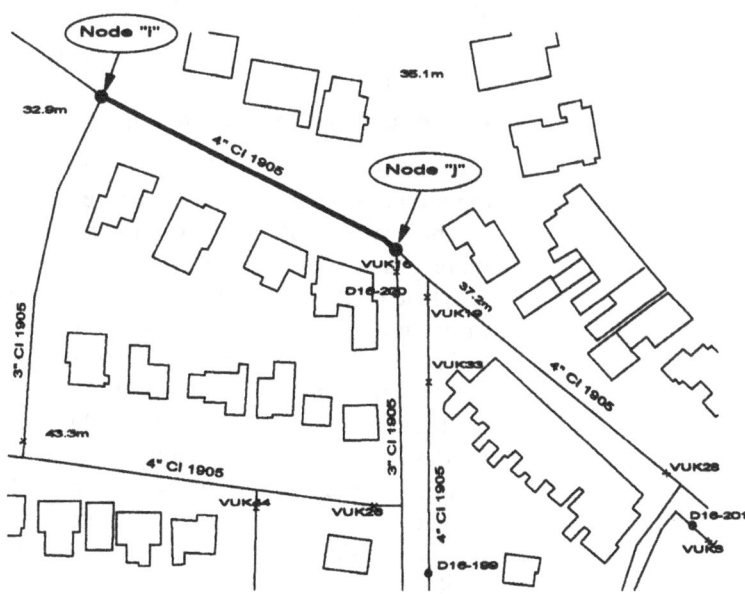

measured, but this is a very expensive exercise). The estimation of roughness may well be influenced by data held on maps or in the GIS. The material and age of the pipe from i to j for example, and included on the map, will allow an experienced water engineer to make an informed judgement on a suitable roughness value.

Computation of head loss

The four items of data listed above are needed to compute loss of energy in the pipe. It is equal to the sum of losses due to friction between water and pipe walls and local ('minor') losses caused by changes of direction and/or velocity of flow:

$$\Delta H = \Delta H_{fric} + \Delta H_{local} \qquad (3.1)$$

Several formulae for computation of *friction losses* can be found in the relevant hydraulics literature. However, the two formulae most frequently used are Darcy–Weissbach and Hazen–Williams – for others, the reader should consult hydraulics textbooks, such as Streeter and Wylie (1979), Miller (1978), or Idel'chik (1979).

The *Darcy–Weissbach formula* states that

$$\Delta H_{fric} = \lambda \frac{L v^2}{2gd} \qquad (3.2)$$

where λ (–) is the friction coefficient, L (m) and d (m) are the pipe's length and internal diameter, respectively, v (m s^{-1}) is the velocity of the water and g ($= 9.806\ 65$ m s^{-2}) is the acceleration due to gravity. Unfortunately, λ is not a constant but an implicit function of several factors, notably

- Reynolds number, $Re = vd/v$, where v is the kinematic viscosity of the fluid (m^2 s^{-1}),
- internal roughness of the pipe's wall, k.

The variation of λ in relation to Re and to the internal roughness is explained in many textbooks (see e.g. Streeter and Wylie, 1979).

The values of internal roughness, k, for different pipe materials (cast iron, ductile iron, steel, asbestos-cement, polyethylene or PVC) are given in pipe manufacturers' handouts and in textbooks, usually only for brand-new pipes.

These equations for λ are very accurate, being the result of many very precise experiments in several laboratories. The problem is that λ is a function of Re, which is itself a function of fluid velocity and is unknown at the beginning of the computation. Further unknowns are the actual roughness of the pipe, k, and the actual internal diameter of the pipe, d. Owing to deposits and/or encrustation after several years of service, d may well not be the original diameter. There is also the added complication that the actual internal diameter of a pipe may differ significantly from the nominal pipe diameter – pipe manufacturers' literature should provide the actual internal size of their products. Therefore, in many cases the use of implicit formulae is not justified and the old Hazen–Williams formula may give even better results!

The Hazen–Williams formula exists in many different forms. One is

$$\Delta H = \frac{10.675}{d^m} L \left(\frac{Q}{C}\right)^n \tag{3.3}$$

where Q (m^3 s^{-1}) is the flow, and the two empirical constants are

$$n = 1.852 \quad \text{and} \quad m = 4.8704$$

The coefficient C is, in a sense, a measure of pipe capacity. Most water engineers, particularly those in the English-speaking world, are familiar with the formula and can estimate the value of C quite accurately both for old and new pipes. More details may be found in Streeter and Wylie (1979). The relationship between λ and C is expressed by:

$$\lambda = \frac{1014.2}{C^{1.852} d^{0.0184} Re^{-0.148}} \tag{3.4}$$

Local (minor) losses are caused by valves, bends, reductions, sleeves and other fittings. Usually, they are small compared with the friction losses incurred by the distribution mains themselves. It is quite common in assembling data for distribution pipes to 'round' the scaling off from maps of pipe length and ignore minor losses due to fittings. There are, however, some circumstances in which the losses through fittings *are* relevant, and the possibility of minor losses having an effect within modelled systems should never be discounted. For example, minor losses within pumping stations in which there are a significant number of fittings and tight constraints on suction pressures might well be the key to understanding a particular problem. Similarly, an inappropriately sized differential pressure meter may be the key to a particular puzzle.

The expression for minor loss is

$$\Delta H = \zeta \frac{v^2}{2g} \tag{3.5}$$

where ζ (–) is a loss coefficient, whose value can be found from tables in appropriate manuals (e.g. Idel'chik (1979) or Miller (1978)).

Introducing equations (3.2) and (3.5) into equation (3.1), the following result will be obtained:

$$\Delta H = \left(\lambda \frac{L}{d} + \Sigma \zeta\right) \frac{v^2}{2g} = KQ^2 \tag{3.6}$$

where

$$K = \frac{\lambda \frac{L}{d} + \Sigma \zeta}{2gA^2} \tag{3.7}$$

Note that flow

$$Q = vA \tag{3.8}$$

where flow area is $A = \pi d^2/4$. Using the Hazen–Williams formula (instead of Darcy–Weissbach) one can derive similar relationships between pipe size, pipe length, flow and roughness.

The formulae, derived empirically, are the result of many years' work by some of the greatest scientists both in the laboratory and in the field. The formulae were hotly debated at the time of their publication and consensus agreement forged from the debates. The result is that contemporary practitioners only have to collect data and apply formulae to compute flow through any pipeline. However, there still remains an element of judgement in choosing appropriate figures from a range of acceptable values for the roughness coefficient and the size of older pipes, and the modeller may be at a loss as to what to select. A few researchers like Walski (1983, 1984), or Ormsbee and Wood (1986), have attacked this problem using mathematical tools. Yet in many cases the facts stubbornly refuse to fit the range; this will be illustrated in the following two examples from the authors' own practice. (See Appendix C for minor loss values of particular fittings.)

How to explain the difference

Figure 3.11 shows a pipeline between two reservoirs. The data were known and the maximum flow was measured by monitoring changes of storage in both reservoirs over a day. The flow was far below the expected value. Why? The reason was a heavy encrustation, which had reduced the pipe size quite significantly; there was also silt and other deposits in the pipe. This fact should be taken into account by reducing the pipe diameter and not by increasing the internal roughness to ridiculous values.

In the other case, the recorded flow exceeded the theoretical maximum by a 10–15% margin. The difference could have been attributed to inaccuracy of the meter and forgotten, but the overall balance suggested that the measured value was correct, so a closer examination took place. The pipeline in question was laid between two reservoirs (Fig. 3.12). The assumption was

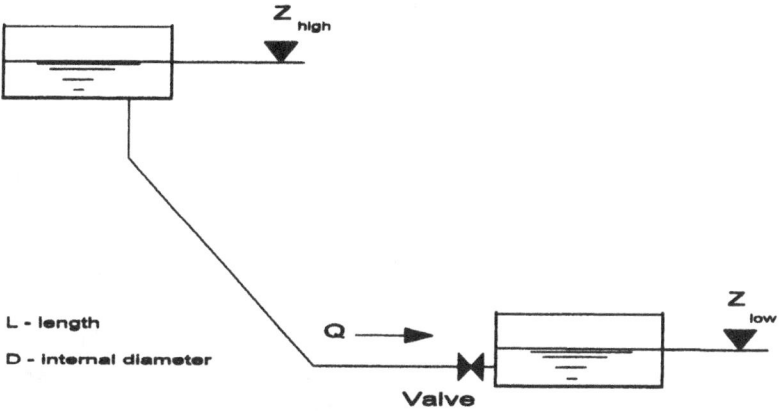

Figure 3.11 Maximum capacity of a pipe.

that both reservoirs were partly full and that the bypass valve around the lower one was always closed. In reality, the crew had to open this bypass to alleviate a shortage of water during hot days, letting the lower reservoir empty, being higher than the hydraulic gradient (a practice not known to the water company manager).

Closed pipes

These are very often important when modelling a real water distribution system. In design, they are often ignored, as the standard practice is to install valves at each end of a pipe, but they are assumed to be fully open in normal operation. During emergencies – repairs, for instance – these valves may be closed, and alternative routes used to maintain supply to as many customers as possible. Sometimes the temporarily closed valves become permanently closed due to a memory loss on the part of the repair crew! There may be a few pipes in the network that are closed intentionally; to separate pressure zones, for instance. Two examples in which closed pipes play an important role are listed below:

- A low-level reservoir is protected from excessive inflow by closing several pipes around it (Fig. 3.13).
- A small town is supplied by its own local sources most of the time, but in emergencies it can get additional water from a nearby regional system (Fig. 3.14).

Making a model of a distribution network

In developing a model, the first impulse is to include everything – each pipe and all the services. However, this is seldom (if ever) practicable because of the sheer size of such a model. Computers can handle very large models nowadays, but other problems inhibit the creation of such a comprehensive model. These include the difficulty of allocating demand to a great number of

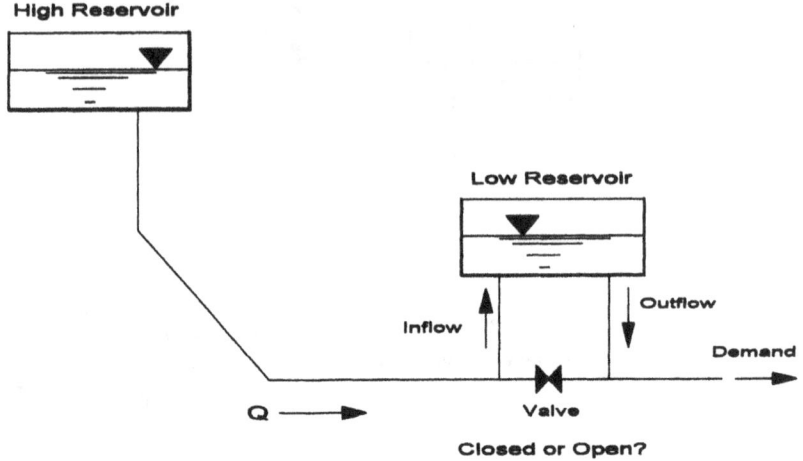

Figure 3.12 Why is the flow greater than the theoretical maximum?

nodes, trying to absorb the large volume of output data generated by the model, and the huge task of accurately entering fixed data such as pipe lengths, which may be considerable if services are at frequent intervals in a road or street. A degree of simplification is essential to bring the model to a manageable size.

Following the rejection of the idea of modelling everything, the next idea that usually occurs to modellers is to discard all pipes smaller than a certain limit (e.g. all pipes of less than, say, 100 mm (4 inches)) under the assumption that these small pipes are not significant to the overall hydraulic regime. This

Figure 3.13 A low-level reservoir.

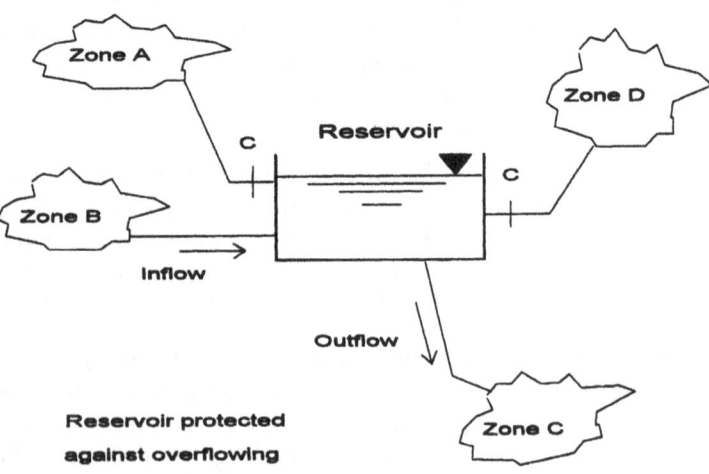

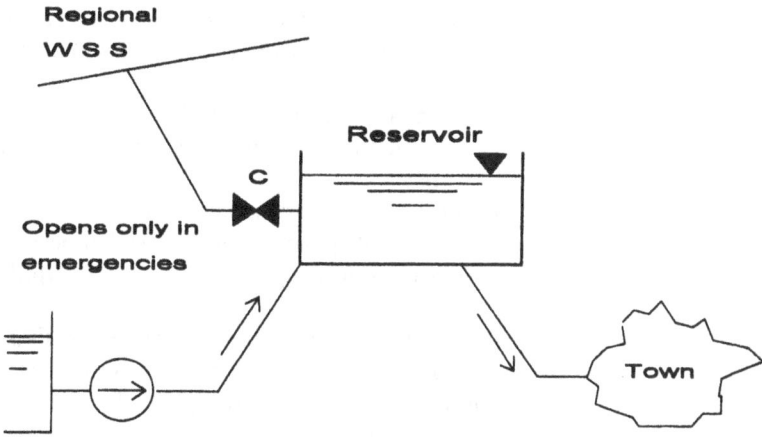

Figure 3.14 Emergency/seasonal supply.

approach can be dangerous because sometimes loops are broken, or the supply to a large consumer is lost – with significant consequences.

The best way is to approach the problem from a slightly different direction and adopt a number of criteria or ground rules for the modelling process, for example:

Figure 3.15 Part of a distribution network, showing loops 'A' and 'B'.

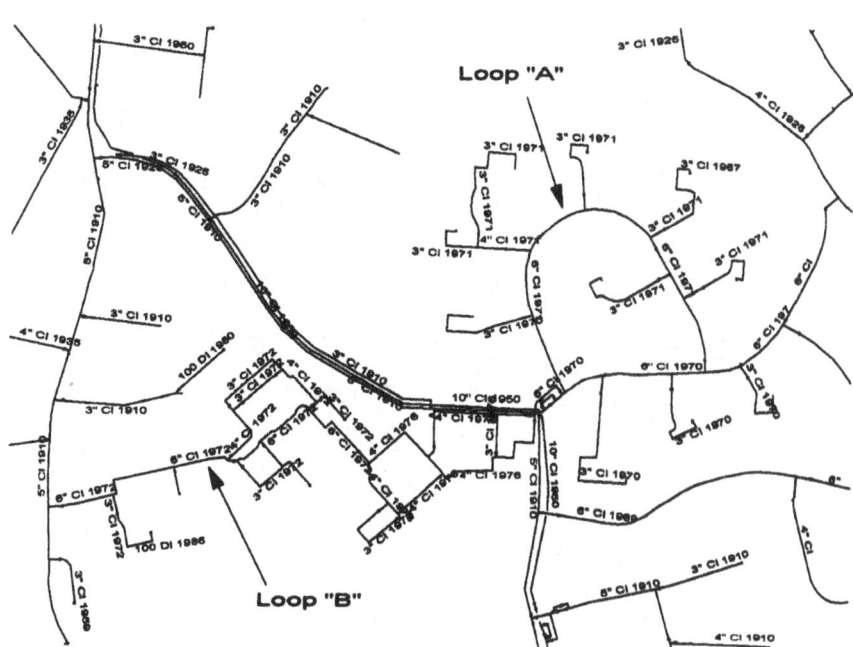

- identify important points – sources, pumping stations, reservoirs, large customers – and retain all pipes leading to or from them;
- delineate the area supplied by the pipe and allocate all the area demand to the pipe end nodes, in appropriate proportions;
- retain all pipes that are part of a loop;
- discard all permanently closed pipes;
- ignore smaller branches or those supplying no demand.

A case from real life is illustrated in Figs 3.15 to 3.21. This is part of an urban distribution network. All pipes belonging to the company are shown in Fig. 3.15, together with their data. The task is to make a model of the network, as simple as possible but omitting no important items.

Loop 'A' is examined first. Before any decision is taken, one should examine the demand area supplied through the pipes in question. Figure 3.16 shows dwelling houses in that area. It is possible to see how individual houses are supplied without elaborate investigation, and one of the demand sub-areas is delineated within the system. The full model of that part of the system, with all pipes and nodes, is shown in Fig. 3.17. It is

Figure 3.16 Defining a demand sub-area for loop 'A'.

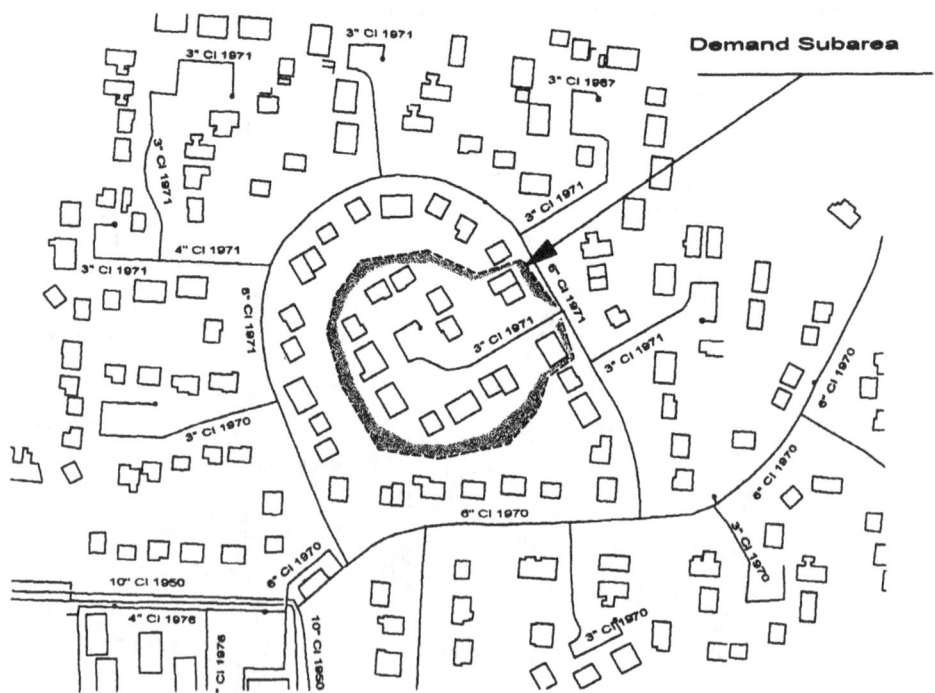

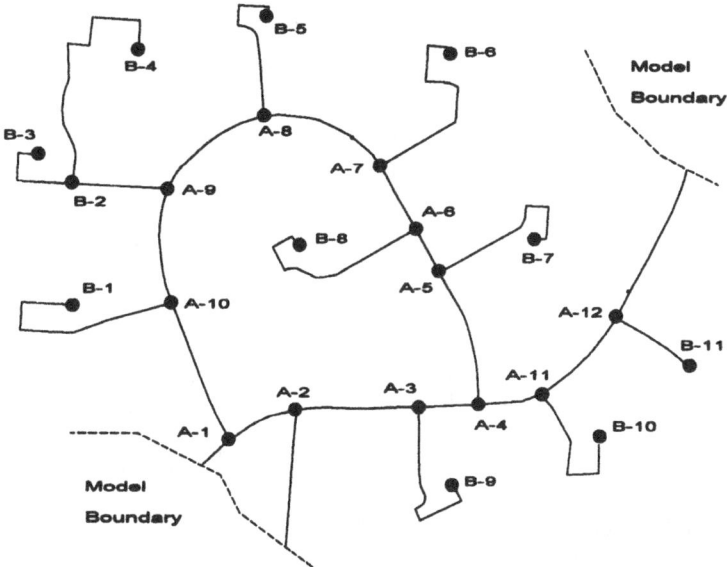

Figure 3.17 The full set of nodes for loop 'A'.

obvious that the hydraulic regime in the system will not be altered if nodes B-1 to B-11 with their connecting pipes are discarded from the model and replaced with respective demands at nodes around loop 'A'. This will produce a smaller model (Fig. 3.18).

Note that this simplification will not be appropriate if one needs to evaluate pressures at customers' premises – that is in the nodes B-1 to B-11 – or to study the changes of water quality in

Figure 3.18 Number of nodes reduced.

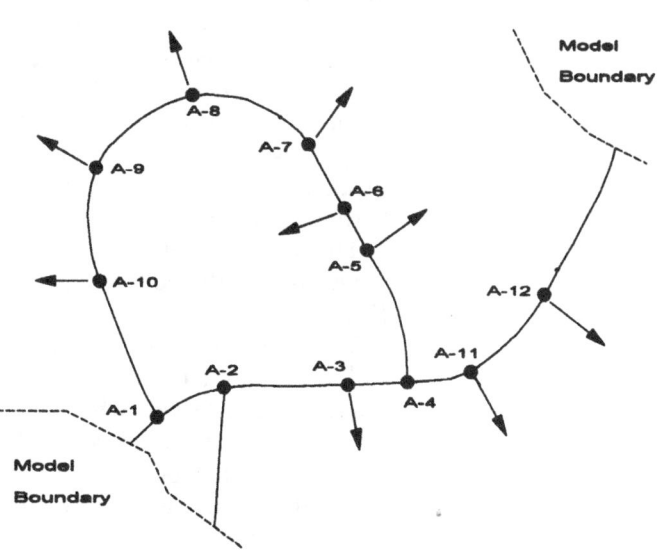

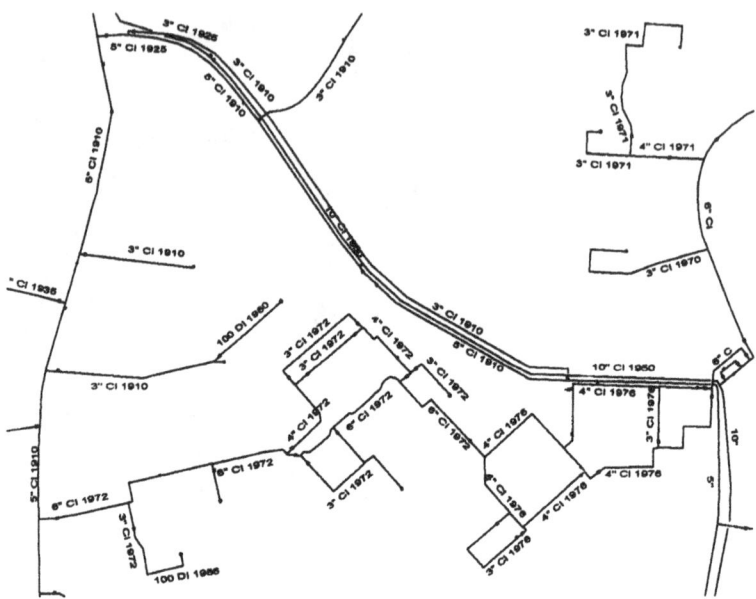

Figure 3.19 A more difficult task is to model loop 'B'.

the system (because the worst problems arise in the furthest points of the distribution system).

This case was relatively simple and straightforward, but there are more complicated ones, as modelling the other part of the same system, loop 'B', demonstrates (Fig. 3.19). These pipes supply groups of houses and serve as a part of a clearly visible large triangular loop. The smaller loops might be discarded – probably.

Again, the first task is to examine the demand areas, shown in Fig. 3.20. Three sub-areas could be established with their centres in nodes 'A', 'B' and 'C', respectively. The simplified model is shown in Fig. 3.21. This solution is certainly not the only possible one to simplify the model nor, necessarily, the ideal, but serves as a working hypothesis until proven – or a better one is proposed.

The final example is an 'island' – part of distribution network connected to the rest of the system through a single long pipe (Fig. 3.22).

The network data for this case are given in Fig. 3.23 and the full model in Fig. 3.24. This detailed model must be used – regardless of its size – whenever the analysis is centred on service pressure or water quality problems. In other cases, for example when the control policy over the macro-system is under scrutiny, this part of the network could be replaced by a single node without any loss of accuracy.

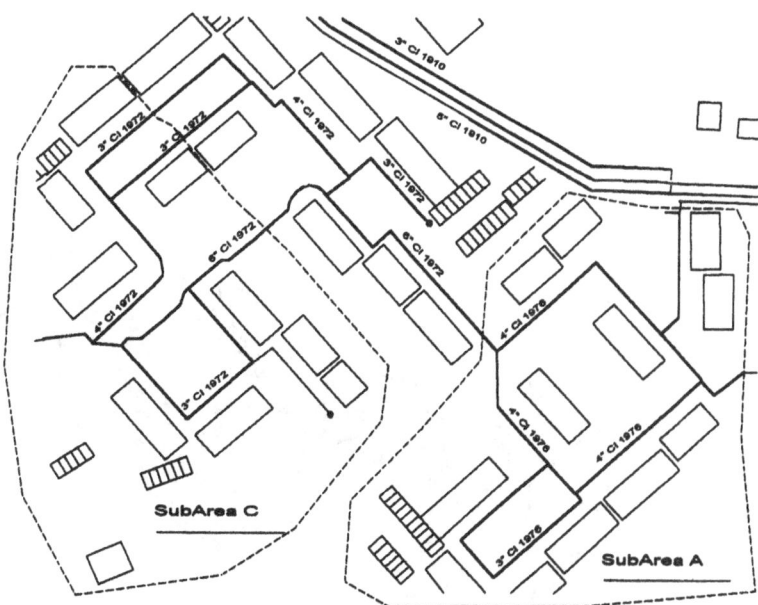

Figure 3.20 Demand sub-areas for loop 'B'.

These examples show how difficult it is to replace an actual distribution network with a simple and relatively crude 'spider's web' – the model. To keep the model within reasonable limits, one must sacrifice smaller pipes and introduce many simplifications, but experience has shown that the final results can still be good. Moreover, as larger and more powerful computers become available, the limits of modelling are rapidly expanding,

Figure 3.21 The result of modelling loop 'B'.

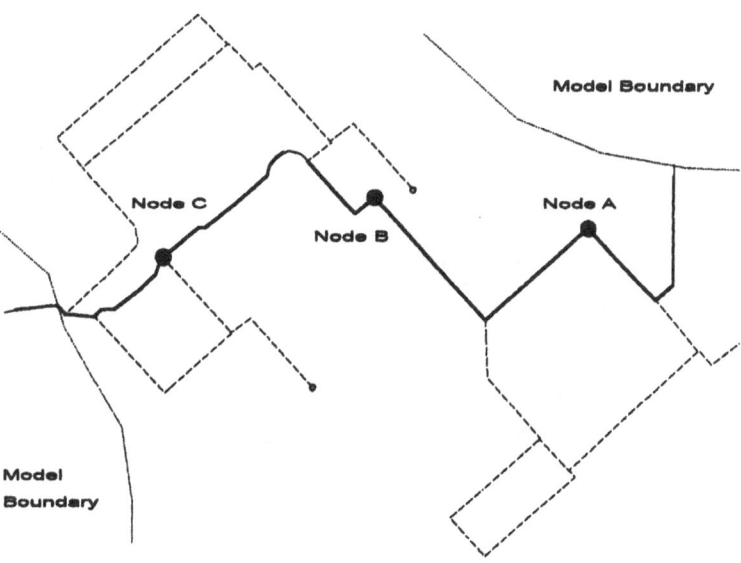

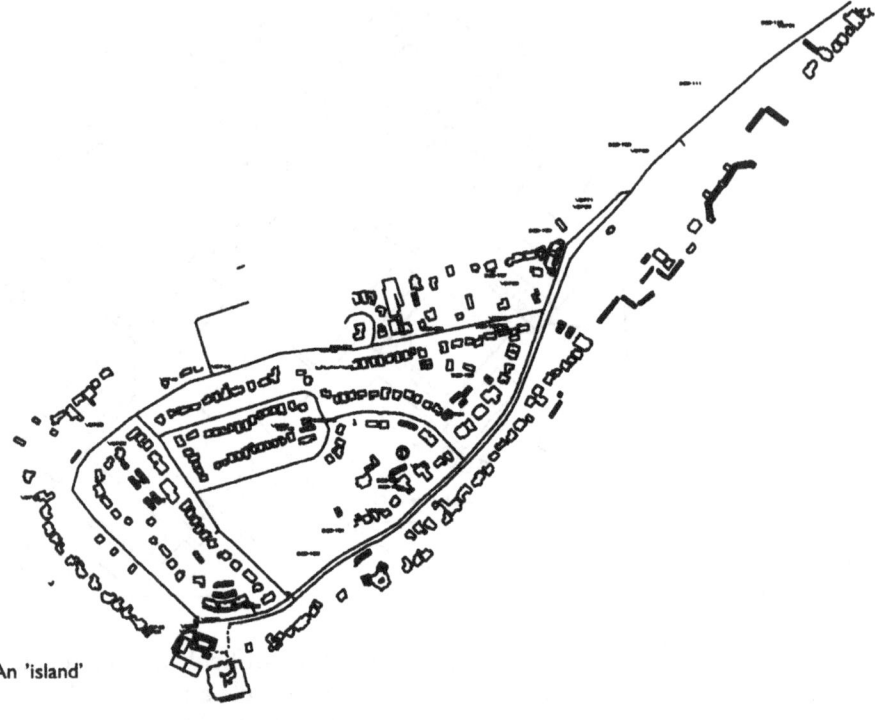

Figure 3.22 An 'island' in the system.

Figure 3.23 The network of the 'island'.

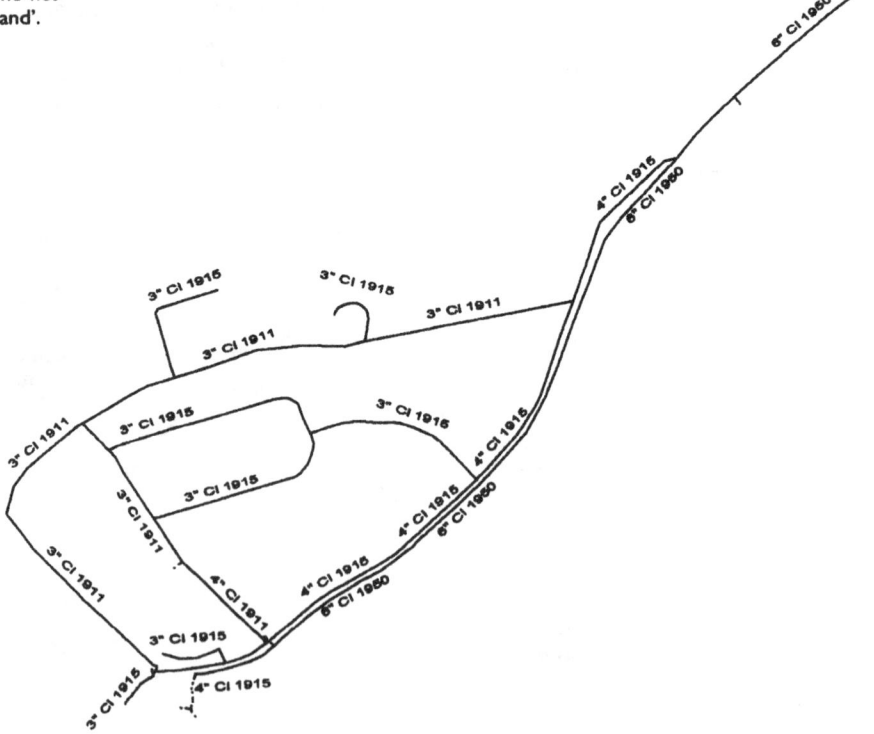

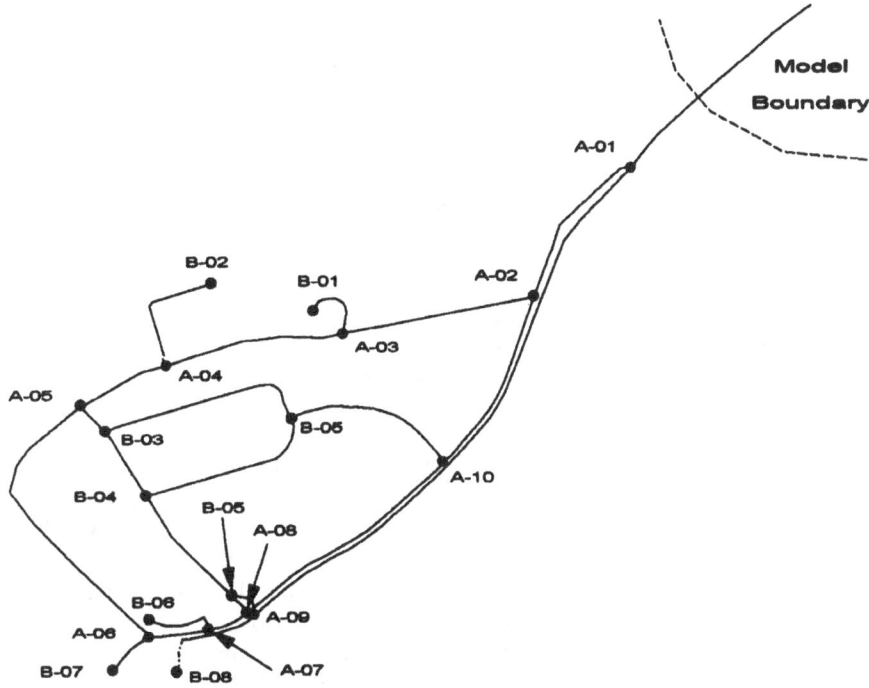

Figure 3.24 Finally the model of the 'island'.

permitting better modelling. The problem of 'loading' data may remain in some circumstances, but even that problem is diminishing as GIS develops and entering the physical details of pipes becomes a matter of loading a file into the modelling computer.

3.5 Service connections and public standpipes

Water companies supply individual customers through service connections. Customers include factories, hotels, offices, schools, public houses, individual houses, flats – down to taps in private yards. The consumption of water may be measured or not (Fig. 3.25). If it is measured, the 'sensor' will normally be a simple water meter.

The meter is read at regular intervals (monthly, quarterly, annually) and the data used for billing. This labour-intensive and expensive task is now being overtaken by technology – automatic meter reading (AMR) (see Gotoh et al. 1993).

The principal objective of the water company is to offer a supply to all its customers at all times, which meets the customers' needs in terms of quantity and complies with all the appropriate water quality criteria. Any shortage or breakdown in service must be mended in the shortest possible time. In some

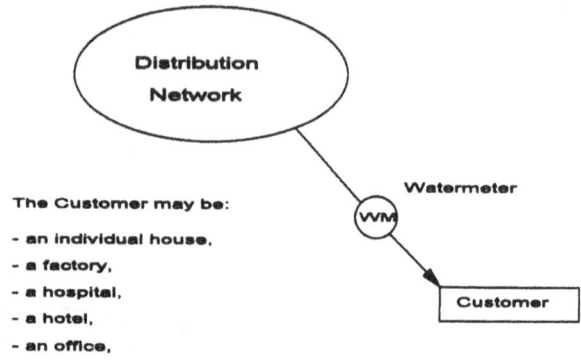

The Customer may be:

- an individual house,
- a factory,
- a hospital,
- a hotel,
- an office,
- a block of houses, etc.

Figure 3.25 The service connection.

countries, water companies are obliged by law to meet so-called 'standards of service', which can be very high.

The water company also has to provide water for public use like public taps, fountains, fire hydrants, street washing, watering of parks and lawns, flushing of water mains, etc. This use might be metered, but more often it is not.

The reader will find more on this subject in textbooks on water supply (e.g. Twort *et al.*, 1994). From the point of view of modelling, each connection (service or public) is a *sink* – a point where water is taken from the system. The total quantity leaving the system is the 'demand' and includes real, useful consumption and losses, either within the system (upstream of the water meter) or in customers' premises (downstream of the water meter).

Both components vary in time and are dependent upon local pressure in a rather complicated way, not yet fully understood, so several assumptions must be made, some of them very crude, as will be demonstrated in later chapters.

It is obvious that only the largest customers can be individually represented in a model. The bulk of the smaller ones have to be aggregated into demand sub-areas as shown earlier (Figs 3.15 to 3.21). Two examples are illustrated in Fig. 3.26:

- a large building (could be a factory, a hospital, a hotel or a large dwelling house),
- a block of houses.

The demand can be computed from water bill data, taken by one or several meters.

These meters could be very simple instruments (to keep the costs down) or relatively sophisticated, combining a small meter for low flows and a large one for peak demand. The data are either shown locally or transmitted to the company via

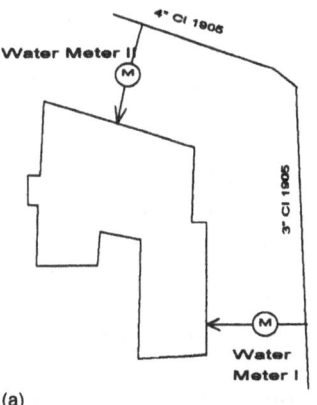

(a)

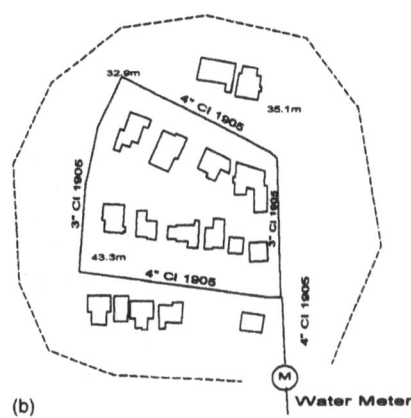

(b)

Figure 3.26 Examples of metering: (a) a large building; (b) a block of dwelling houses.

telephone lines or short radio waves. New technologies are emerging rapidly (see Gotoh *et al.*, 1993).

In determining the level of demand at nodes, it is clearly sensible to base it on the aggregation of individual metered demands if this is possible. It is also useful to meter zonal demands to check the extent of leakage in the company's network. In the absence of complete individual meter records, or if the metering and billing process is suspected to be less than accurate, it may be necessary to meter small groups of properties individually to generate some real data to be extrapolated to the whole system.

There are two possibilities to model the demand – one traditional, the other an emerging method. In the traditional method the demand is given as a function of time and does not depend on local conditions in the network (an oversimplification, perhaps, but widely used). In the new one the demand is a complex function of pressure and time, with an underlying diurnal demand pattern that is modulated by the effect of pressure variations during the 24 hour period. It should also produce a more accurate picture of the real situation if the leakage component of demand, both within individual properties and from the distribution network, is pressure-related too.

3.6 Reservoirs

What is a reservoir?

Reservoirs are facilities where a quantity of water is stored. There are two types of reservoir found in the water industry. Impounding reservoirs are designed to store raw water derived from surface waters in large quantities to meet seasonal variations in demand. Similar reservoirs are constructed to create storage for hydro-electric power generation and river flow modulation. In the context of this book, impounding reservoirs are normally treated as sources having a fixed level (see section 3.3), and although they may be rather dramatic physical entities they have little relevance in the context of most water supply system models.

The second type of reservoir in the industry are relatively small tanks (small compared with impounding reservoirs) used as balancing reservoirs in the distribution system, and contact tanks, balancing tanks and pump sumps in treatment plants and pumping stations. The principal function of most tanks or service reservoirs within a distribution network is to provide a buffer storage of treated water to meet diurnal fluctuations in demand. A secondary function is to provide a reserve of water to meet emergency situations – fire fighting demand or a burst main and so on. In the context of distribution system modelling, reference to 'reservoirs' normally implies service reservoirs or tanks used as pump sumps or treatment contact tanks. Service reservoirs used purely to provide a buffer against diurnal demand fluctuation are normally sited at the highest point within or close to their demand area. In situations where no suitable high ground is available, water towers may be constructed. The hydraulic function of water towers is just the same as normal service reservoirs.

A typical large service reservoir is shown in Fig. 3.27. This reservoir has two chambers (which may or may not be identical), to allow periodic cleaning, repairs, handling of emergencies and so on. Water enters from a main, via a valve chamber. The chamber contains three sets of pipes and valves: inflow, outflow and wash-out, the latter to allow emptying of

Figure 3.27 A large reservoir.

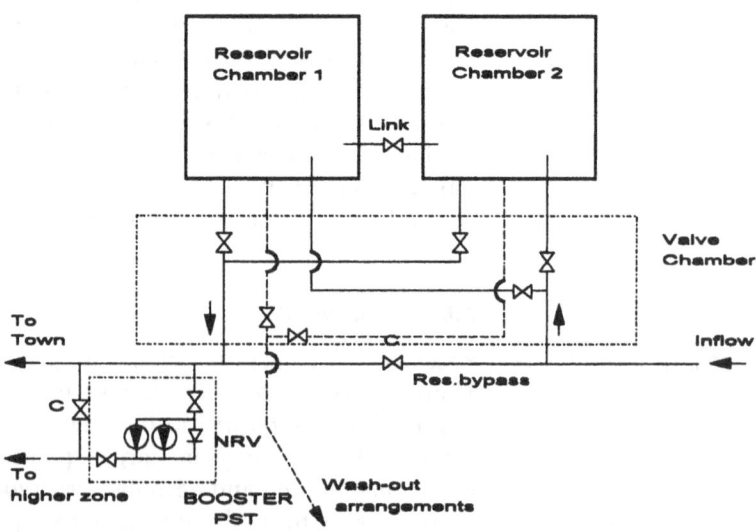

the reservoir for cleaning, etc. Only the important valves are shown in the schematic.

Water flows from the reservoir through another main pipe. It is always important to provide a bypass pipe around the reservoir – just for emergencies when it has to be emptied completely – for all sorts of reasons (reservoir leaks badly, or is seriously damaged, or its piping must be mended or replaced, etc.). The details important for modelling purposes are discussed below.

Shape and volume

Large reservoirs, constructed at or below ground level, usually have two or more separate chambers, normally prismatic or cylindrical in shape. Water towers, built above ground, can have more elaborate shapes. Some are plain spheres, others are like bowls or have toroidal shapes. A well designed and suitably placed water tower may be a recognisable feature of the urban landscape (as in Kuwait for instance). A few typical shapes are shown in Fig. 3.28.

For modelling purposes just a few data are important, namely:

- elevation, Z, of the current water level,
- possible range of Z: maximum (weir) and minimum (bottom),
- change of storage volume, V, with elevation, Z.

Figure 3.28 Different reservoir shapes: (a) a single-chamber reservoir; (b) a double-chamber reservoir of unequal depths; (c) a ball-shaped water tower.

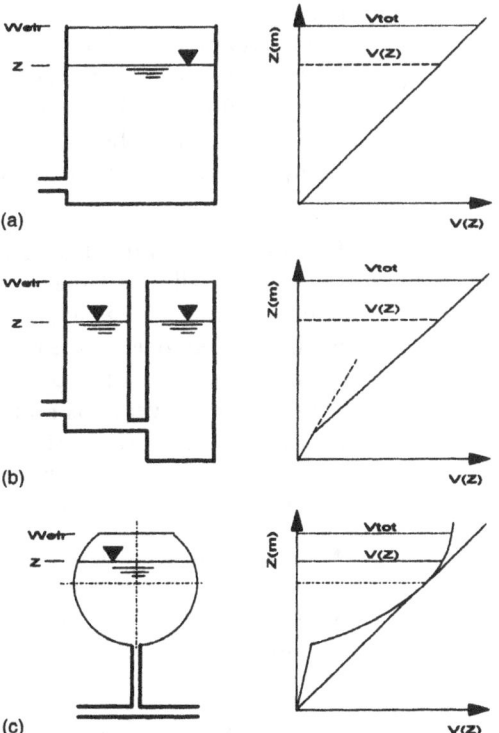

A multi-chambered reservoir (Fig. 3.28b), acting hydraulically as a single tank, could be represented either as one large reservoir (a better solution) or as several smaller ones, inter-connected by adequate pipes and fittings (a more complicated solution). The second solution is necessary whenever water levels (WL) are not the same in all chambers. This condition is unusual because the connecting pipes are normally left open, so the reasons for operating reservoirs of this type as independent units should be carefully investigated with the local staff.

It is always better to avoid, if possible, complicated solutions in modelling, e.g. several chambers with short large-capacity pipes between them. This can lead to problems in simulation, and the benefits are rather dubious! However, in some cases this cannot be avoided, for example when emergency regimes have to be analysed.

Inflow arrangements

Another important piece of information is how water enters the reservoir. Various arrangements are shown in Fig. 3.29. Note that in Fig. 3.29c no backflow from the reservoir is possible, while in Fig. 3.29d a non-return valve (NRV) permits backflow. Arrangements with devices that prevent overflowing are dealt with in the subsection on float valves in section 3.8.

The relative merits of these four systems depend on other features of the system, such as the following. Where are the sources? What are the length and size of the inflow pipe? Is there some demand to be met upstream of the reservoir? Is water pumped to the reservoir or flows by gravity?

Outflow arrangements

In most cases, water leaves the reservoir through a pipe that is submerged all the time – its mouth as low as possible, to maximise the storage available (Fig. 3.30a). In rare cases water is discharged over an elevated weir – see Fig. 3.30b. Water cannot enter the outlet pipe until its level reaches the top of the weir – then it overspills and leaves the reservoir. If the capacity of the weir is sufficiently large, the water level is effectively constant all the time. Such arrangements are applied only to satisfy special considerations:

- to provide a constant head in the system downstream of the reservoir (to operate filters, for instance),
- as a second outlet for a reservoir where both inflow and outflow vary very much, to save water that would otherwise be lost.

Peculiarities like this should not be neglected in modelling.

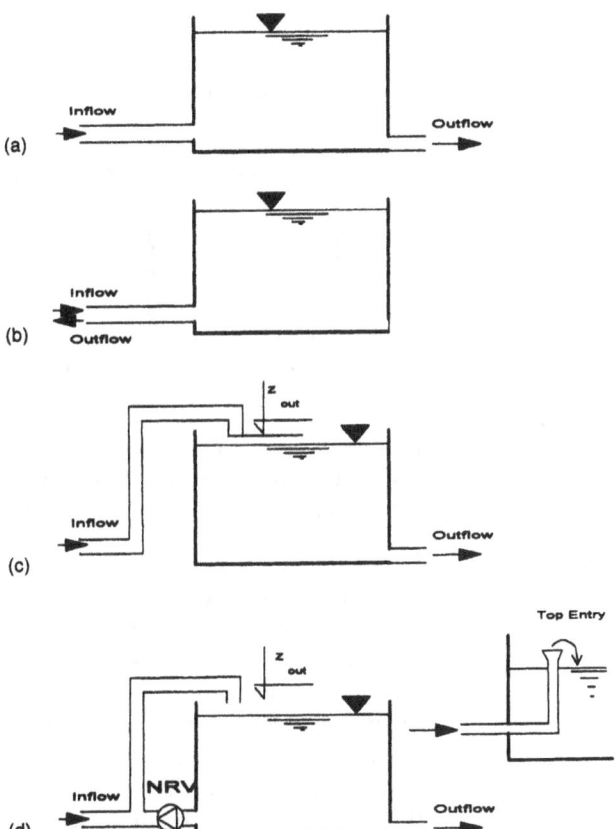

Figure 3.29 Different inflow arrangements: (a) inflow from below, with separate I/O pipes; (b) inflow from below, with one pipe for both inflow and outflow; (c) inflow from above; (d) inflow from above, with an NRV.

Internal walls

Figure 3.30 Outflow arrangements: (a) bottom outlet; (b) top outle.

Another feature that may cause problems in modelling – if unnoticed – is a partition wall in the reservoir (Fig. 3.31). Water enters the first chamber, fills it, then overflows to the second. In normal operation, the first chamber is always full, while the water level in the other one varies according to the balance between inflow and outflow. The effective volume of such a reservoir is reduced to that of the second chamber only. The rationale for such an arrangement may be to provide a longer retention time in contact tanks, or to maintain a positive head to

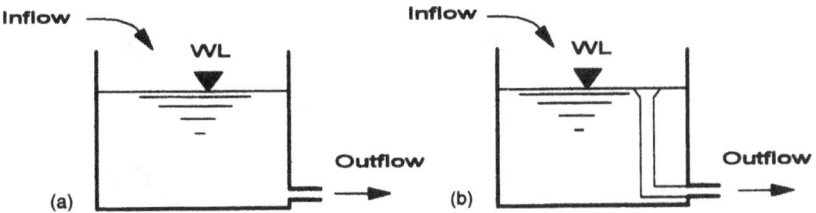

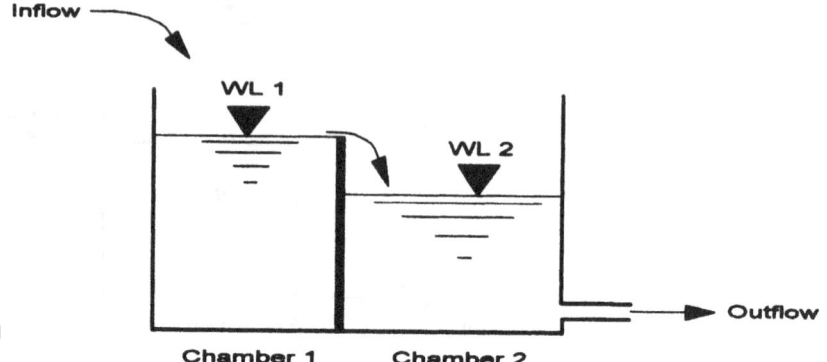

Figure 3.31 Internal
wall in reservoir.

pumps using the first chamber as a sump. In other places this is a
highly unusual feature (although the authors know of a few such
cases: remnants of incomplete reconstruction, etc.).

Bypass pipe

Every service reservoir is normally provided with a bypass pipe
to be available in emergencies. Typical arrangements are shown
in Fig. 3.32. The bypass pipe is normally closed by a valve
(Fig. 3.32a). All water must first enter the reservoir, before it
reaches the demand zone, a good policy to ensure that the water
in the reservoir is constantly replaced to avoid stagnation and
therefore water quality problems. However, if the inflow might
be interrupted and there are some users along the inflow pipe,
then the pipe could be provided with a non-return valve (NRV),
allowing backflow whenever the head in the reservoir inflow
pipe drops (Fig. 3.32b).

The bypass pipeline, normally closed (Fig. 3.32a), is usually
ignored in the modelling. However, the engineer should be

Figure 3.32 Bypass pipe
around reservoir: (a) with
a stop value; (b) with a
check valve.

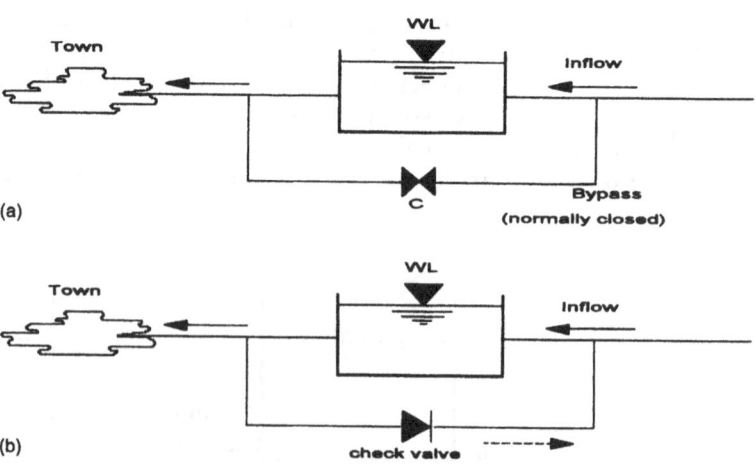

aware of its existence – just in case some unusual situations have to be analysed in the model. If the reservoir has to be emptied, then the bypass pipe must be reintroduced as a link between the two parts of the system.

How to model a
reservoir

Modelling a reservoir is not as easy as it may first appear. A reservoir similar to the one in Fig. 3.27, with several chambers and inlets, can be represented in many different ways, depending on the eventual use of the model. A few possibilities are shown in Fig. 3.33.

(a) Both chambers are represented as a single unit, inlet and outlet pipes are represented by single pipes, and bypass and other pipes are omitted.
(b) As in the first case, but the bypass pipe is added, complete with a control valve.
(c) The chambers are represented separately, each with its pipes and valves.

The degrees of complexity are different; model (a), the simplest, can be used for analysis of a normal operational regime, while the other two also permit the analysis of various emergency cases – involving the use of the bypass pipe and asymmetrical filling of reservoir chambers, etc. Note that all three solutions are feasible, correct and acceptable.

Figure 3.33 Modelling a reservoir: (a) the simplest model; (b) a model with a bypass pipe; (c) a model with two chambers and a bypass pipe.

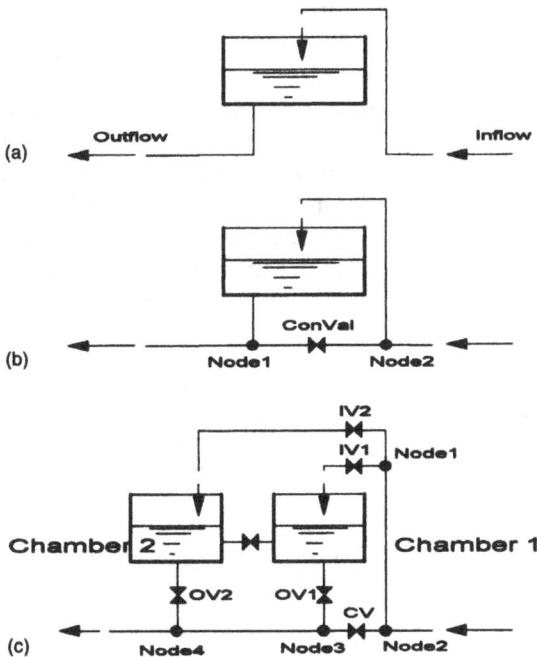

Control of water level

Water level in every reservoir can be closely supervised and monitored using modern technology. It has to hold appropriate reserves at all times and never overflow. Nor should the reservoir empty completely, because this almost inevitably causes problems: air will enter the system and may pollute the water and/or damage pipework (free air wandering throughout the network can cause severe surges). The importance of this point cannot be overstated. It may be said that the operating policy in every system should be directed at keeping all reservoirs reasonably full and under control all the time.

This end is achieved by both local and remote devices. The former are different kinds of valves, and the latter are pumping stations, control valves and the distribution system in general. Figure 3.34 shows the most common local devices:

- float valves, FLVs (Fig. 3.34a);
- pressure-reducing valves, PRVs (Fig. 3.34b).

A float valve only prevents loss of water through the reservoir overflowing. As water level rises, it starts to throttle the inflow, progressively, closing completely before the water reaches weir level. Its influence is only felt close to the reservoir top, in the last 0.5 m or so. Obviously, such a device cannot prevent emptying of the reservoir. More details can be found in the subsection on float valves in section 3.8.

A PRV can be used more effectively. It covers the whole height of the reservoir. If the water level exceeds the set level, a PRV throttles the inflow, and vice versa. The set level can be a constant or can vary with time, supervised by a small local processor. Note that such a device can be applied only when a sufficient head is available on the upstream side. There are more sophisticated devices (programmable controllers with microprocessors) on the market, which might be useful in many cases.

The devices described above control storage in a reservoir downstream of the device. In large conveyance systems, another

Figure 3.34 Control of water level: (a) just to prevent overspilling; (b) to keep WL constant.

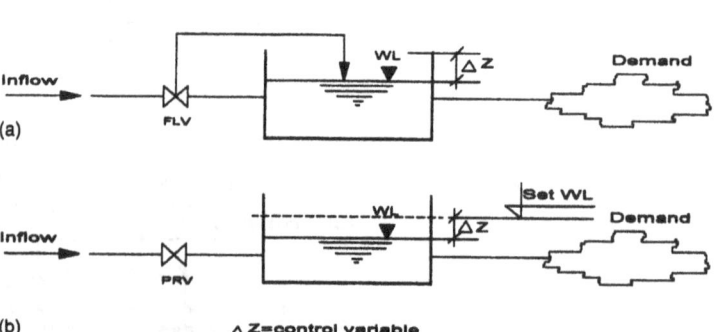

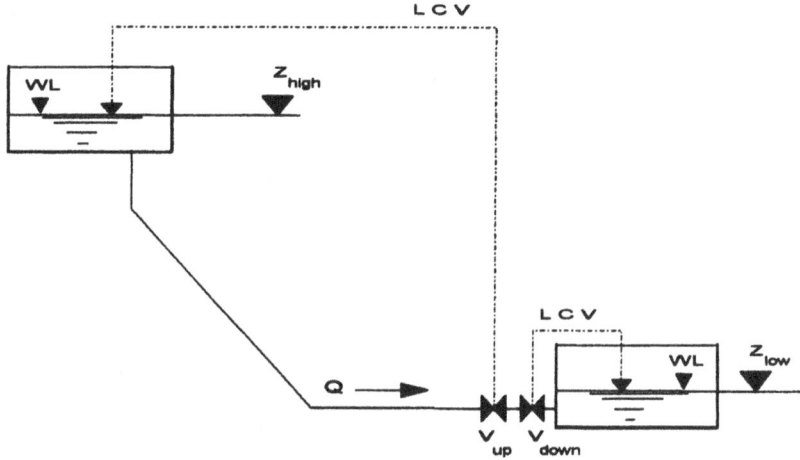

Figure 3.35 Control valves in transport.

valve may be placed downstream of the reservoir, with the task of preventing an upstream reservoir from emptying (Fig. 3.35). Here the valve 'up' closes down whenever water level in the upstream reservoir falls below a critical depth, thus keeping some water in the reservoir – and the pipe full. Note that this valve is not controlled by the downstream reservoir, which is very close to it, and is independent of the valve 'down', which prevents spilling of water from the downstream reservoir. This system would ensure that air is not drawn into the pipe or negative pressure suffered by it, thus avoiding the danger of pollution and damage to pipe walls.

3.7 Pumping stations

Description of a pumping station

Pumping stations in water supply are used for:

- raw water abstraction (e.g. well pumps);
- distribution of raw or treated water (e.g. relift pumps in treatment plants);
- control of service pressure within the distribution network (e.g. booster pumps).

Pumping stations are very important elements of the system, as they redirect and regulate the flow of water. Consequently, the proper modelling of pumping stations is very important. It is not necessary to be an expert in hydraulic machinery to be able to create properly modelled pumping stations, but a basic knowledge of pumps is needed.

A pumping station to a non-expert is a frightening maze of pipes, valves, pumps, surge vessels and lots of other peripheral equipment, which might – or might not – be important for the

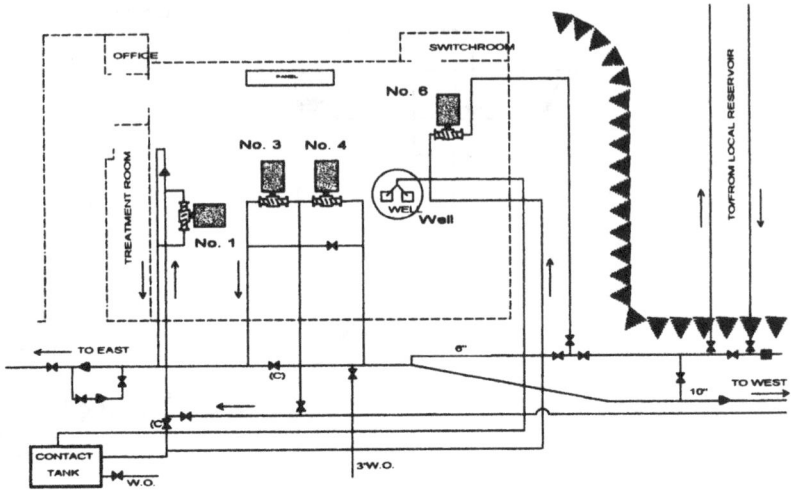

Figure 3.36 A pumping station.

modelling. Even a schematic diagram – if available at all – may not be very illuminating; see the example in Fig. 3.36.

Where to start? A good starting point is to identify first the *inlet* and the *outlet* points for the whole station and then of individual pumps. It may be that there are several 'pumping stations' under the same roof (as in the case above). In the modelling sense a 'pumping station' should have one inlet and one outlet only. Several pumps may be connected in parallel to common points on their suction and delivery sides (single units are relatively rare because a reserve pump is always needed), and such an arrangement should be regarded as a single 'pumping station'.

A crude sketch, such as the one in Fig. 3.37, is very helpful in understanding how a pumping station is actually arranged. The three pumps are arranged in parallel, each with a check valve (NRV) and a control valve on the pressure side, and also with several shut-off valves. Pumps are driven by electric motors. Two surge vessels are also visible, with a compressor to replenishment air in the surge vessels. A bypass pipe – with another check valve – permits a short link between inlet and outlet for emergencies.

Figure 3.38 represents a fairly general case. This hypothetical pumping station has three pumps, which are not necessarily identical. Pumps are configured to operate in parallel. Individual units may rotate with fixed speed or with variable speed. Note throttling valves, installed at the downstream side of each pump, and shut-off valves, which can isolate individual units. A bypass pipe, with a check valve (NRV), permits transfer of water from inlet to outlet side whenever the difference of pressure permits,

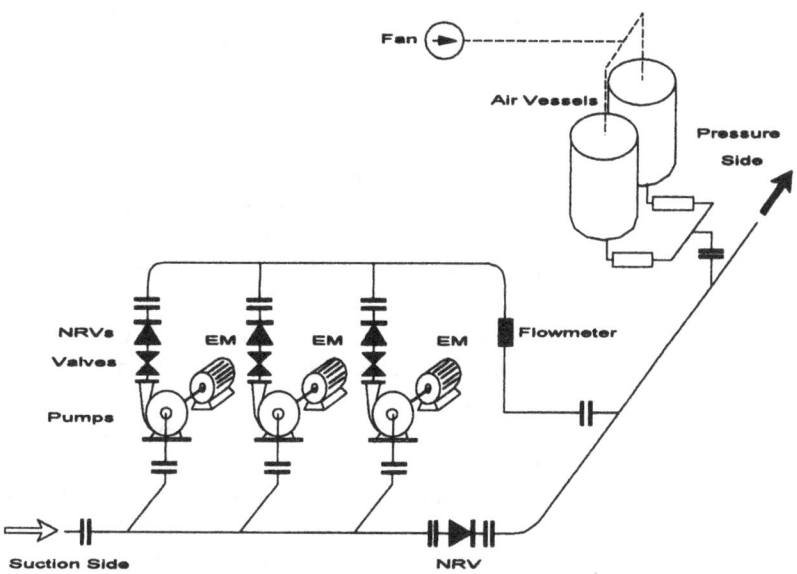

Figure 3.37 A view of a pumping station.

but prevents flow from the delivery to the suction side of the pumps.

Some of the pumps may be active (duty and assist units) while others are stand-by units. Note that some peripheral elements of the pumping station, shown in Fig. 3.37, are omitted – air vessels, for instance, because they are important only in transient regimes and are irrelevant to the modelling procedure.

Each pump has a valve in its pressure pipe (downstream side), which should be fully open in operation. However, in real life, staff very frequently use such valves to reduce the flow through individual pumps (causing considerable losses of energy!). This

Figure 3.38 The model of a typical PST.

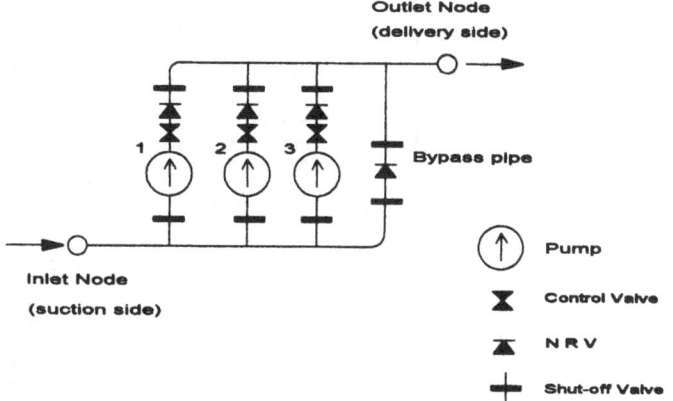

often occurs at stations with fixed-speed units, and therefore no alternative for controlling discharge. Minor losses in bends, pipe reductions, etc., may be either attributed to this valve or deducted from the head, H_p.

The PST represented in Fig. 3.38 could be modelled as a single link in the model, with obvious advantages in comparison with the more traditional procedure of modelling each pump separately. In the case represented in Fig. 3.36, the single, real pumping station needs to be modelled as four separate 'stations' (Fig. 3.39). This revised schematic diagram of the PST shown in Fig. 3.36 is almost self-explanatory. Water flows in from the trunk main, feeding pumps 1, 3 and 4 (unit 2 was discarded).

Units 1 and 3 are used to deliver water to the east – the two pumps are configured in parallel, but mechanical constraints mean that only one of the two can operate at one time. In the model, therefore, only one of these two units can be active, with the other on stand-by, but they can be modelled as a single pumping station, the first of the four.

The balance of water from the trunk main is directed towards the reservoir through unit 4. This single pump represents the second of the four pumping stations.

A well source, located within the real pumping station building, could be used in emergencies by operating either of the two pumps in the well to deliver raw water to the contact tank, where chlorine is added. The well pumps, effectively

Figure 3.39 Hydraulic scheme of the PST.

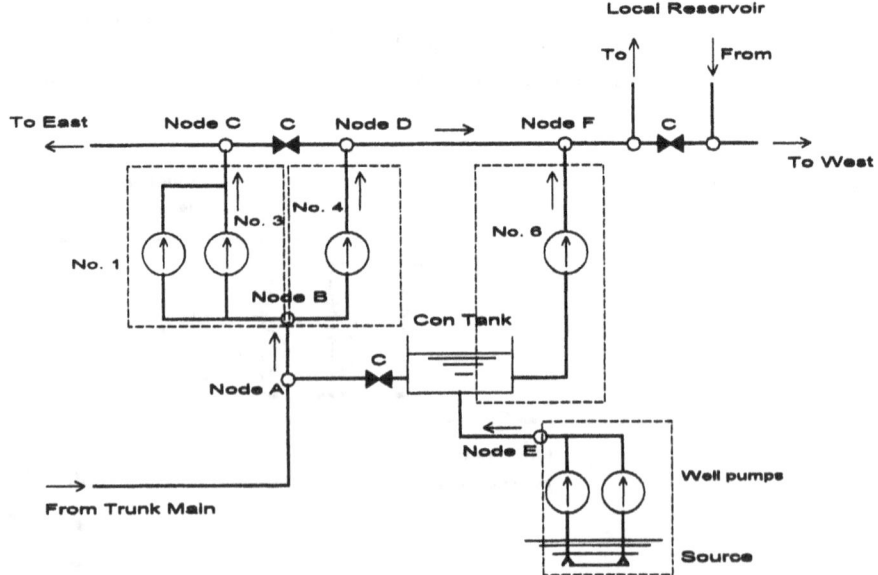

configured in parallel, constitute the third of our four pumping stations.

Water is taken from the contact tank, which doubles as the sump for unit 6 delivering water to the main taking water towards the reservoir. The single pump set, unit 6, constitutes our fourth pumping station.

Pumps 4 and 6 are not configured to operate in parallel; the reservoir is fed either by local water (from well pumps, via unit 6) or from the regional system (from the water main, via unit 4). A common delivery point therefore exists, but the suctions of units 4 and 6 are not common and are at different pressures. Note that a bypass pipe, normally closed, could let water from the main pipe *into* the contact tank – in emergencies.

In summary the four pumping stations, in a modelling sense, are:

- units 1 and 3,
- unit 4,
- two well pumps,
- unit 6.

These four groups should be modelled separately, and then interlocked as appropriate, by linking pipes and controls. The (normally closed) bypass pipes should also be included in the model for later analysis of various emergencies.

When a reasonably good schematic of the pumping station has been made, work can progress to details. The next difficulty is to find the data about individual pumps – the so-called Q–H–P characteristics. These data are normally supplied by the manufacturer, as a set of general data for the particular pump type or even as a set of test data for individual pumps. The latter data are the best for the purposes of modelling, but in many cases they are not available when needed for modelling purposes. However, the missing information can always be reconstructed with reasonable accuracy using data from other sources and a general knowledge of pumps.

The pumps – what data are required?

The main items in any pumping station are, naturally, pumps. This is in itself a huge topic – the reader may refer to several textbooks, like Dickenson (1988), or to catalogues provided by manufacturers – but only the characteristics important for modelling will be discussed here.

Pumps fall into one of two general groups. These are positive displacement (PD) pumps and rotodynamic (RD) pumps. Although both types are used to add pressure energy to a fluid, the PD type units do so directly in a type of batch process. They take a quantity of fluid, lift it, often into a pipe, and then some mechanism 'holds' the fluid in place whilst the process of injecting a further batch of fluid is repeated. PD pump sets are used in a wide range of applications in the water industry but not, usually, for the mass transport of water, and they are not of considerable interest in the context of this book. For further information about such units see, for instance, Dickenson (1988).

Rotodynamic pumps add pressure energy to water indirectly. High dynamic energy is created in the pump set by the inducement of high angular velocities, which are converted to pressure energy in accordance with the Euler theorem.

The principle of the rotodynamic pump is common to pumps having a wide range of duties. The physical configuration of pumps has evolved into a number of distinctive groups, which range from units having a relatively high capacity in terms of flow but only able to generate relatively low heads, to pumps delivering relatively small flows against quite high heads.

The relatively benign chemical nature of water and its specific gravity mean that pumps for water supply lack the special, distinctive features needed in other applications – for example, pumps for milk and juice, or for heating, or for boilers, or for liquid metals. Each kind of application has its own peculiarities. Consequently, it is easier to find a book on, say, pumps for dairies than one specialising in pumps for water supply, despite the large number of units employed in the industry – two exceptions are Monin (1984) and Frischherz and Sacher (1984). The text that follows will, hopefully, help to close this gap slightly.

Rotodynamic pumps used in water supply belong to three basic types (Fig. 3.40):

- *Axial-flow pumps* are relatively rare in water supply (usually only for raw water transport or within water treatment plants), but their high flow/low head character makes them ideal for land drainage applications.
- *Mixed-flow pumps* may be single- or multi-stage units.
- *Centrifugal pumps* also may be single- or multi-stage units, single- or double-suction types.

The three groups are listed in the order of their historical development, and all three types are used in water supply, and from time to time need to be modelled.

As a rule of thumb, centrifugal pumps provide higher pressures and lower flows than mixed-flow units of similar size. Mixed-flow pumps are similarly related to axial pumps, which generate high flows against low heads and were originally developed by encasing ship-type propellers in tubes.

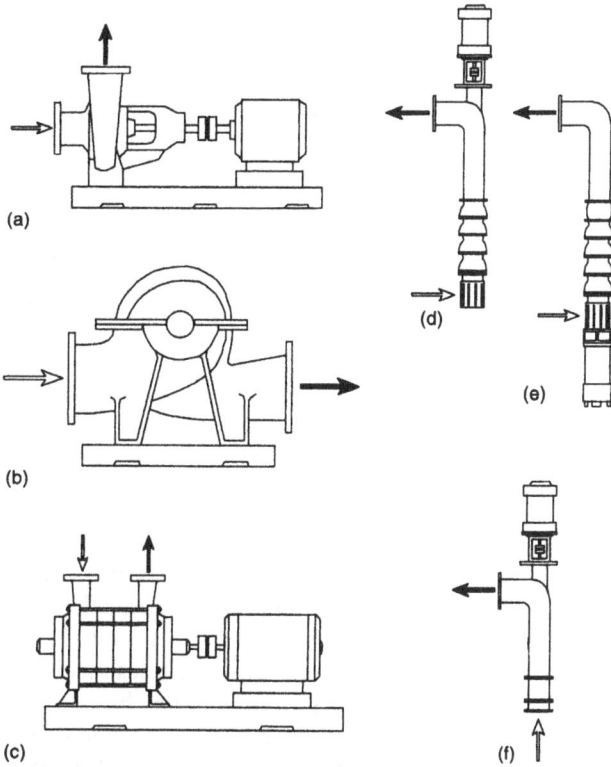

Figure 3.40 Pumps used in water supply: (a) single-stage centrifugal; (b) double suction centrifugal; (c) multi-stage centrifugal; (d) dry mixed flow; (e) wet mixed flow; (f) axial.

The information about a pump, needed for modelling, is:

- manufacturer's name and pump type ('trademark');
- rated flow, Q_0;
- duty head, H_p;
- rotational speed ω_0;
- optimum efficiency, η_0;
- number of pump stages (default $= 1$);
- pump characteristics, popularly known as the 'Q–H–P curve'.

The characteristics for each type of rotodynamic pump are shown in Figs 3.41 to 3.43. The lines in these figures represent:

- pump head, H_p (m),
- hydraulic power, P (kW), i.e. power delivered to the pump's shaft,
- efficiency, η (%).

Pump efficiency η is equal to

$$\eta = \frac{\rho Q g H_p}{1000 P} \tag{3.9}$$

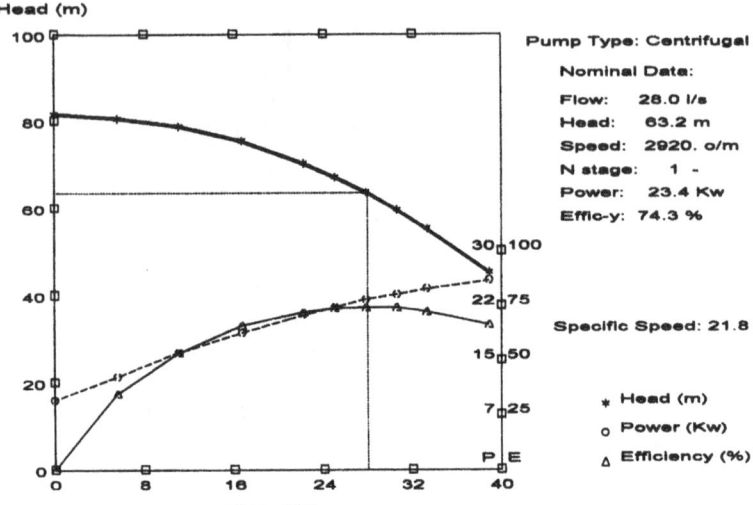

Figure 3.41 Standard centrifugal pump characteristics.

where, ρ (kg m^{-3}) is the density of water, Q (m^3 s^{-1}) is discharge, H_p (m) is pump head, g is acceleration due to gravity ($= 9.806\ 65$ m s^{-2}) and P (kW) is the power used by the electric motor (includes mechanical and electric losses within the electric motor). The characteristic corresponds to a single-stage pump operating at its rated speed.

The characteristics of a multi-stage pump could be easily computed by multiplying the pump head H_p and power P of a single-stage version by the number of stages.

Figure 3.42 Standard mixed-flow pump characteristics.

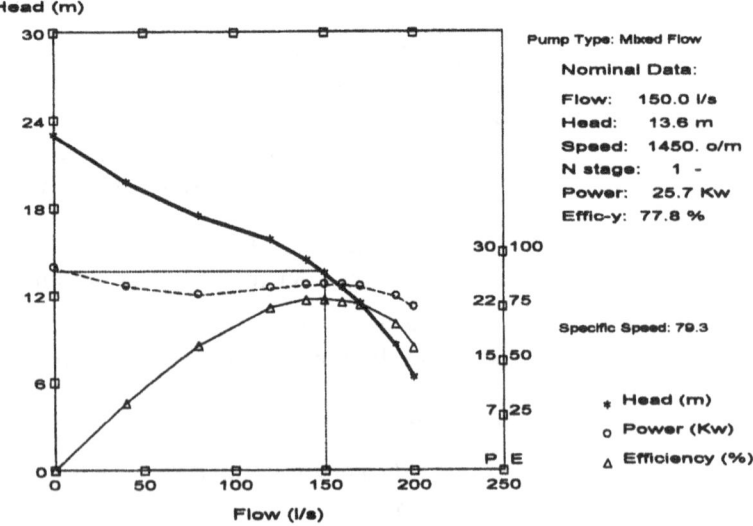

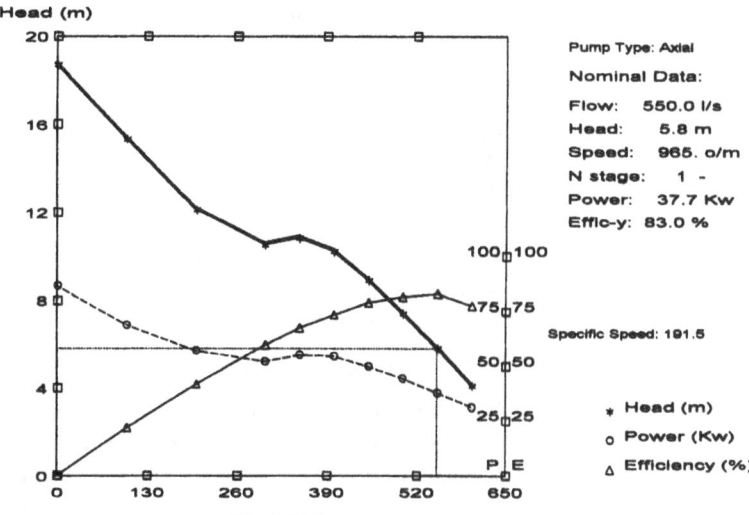

Figure 3.43 Standard axial pump characteristics.

Data from the literature and/or catalogues should be treated cautiously! The best information may be obtained by measurements *in situ*, using modern equipment like Yates meter – direct measurement of efficiency by a couple of thermodynamic probes. However, this is not always possible, so other sources must be used instead.

Individual pumps are classified by their *specific number*, which is equal to

$$n_q = \frac{n\sqrt{Q}}{H_p^{0.75}} \tag{3.10}$$

where n stands for the rotational speed (rpm). Note that the specific number is approximately 22 for the centrifugal pump shown in Fig. 3.41, around 80 for the mixed-flow pump (Fig. 3.42) and above 190 for the axial pump (Fig. 3.43). Note that for double-suction pumps the flow Q in this equation should be just half of the pump flow, and that for multi-stage pumps H_p should be the head provided by just one stage – not the total pump head (which is the sum of the heads provided by all stages).

The *specific speed* is a parameter still widely used, but it is more advisable to use the SI system, which is independent of units. By dropping all constants one can express specific speed as

$$\Omega_s = \frac{n_q}{52.9} \tag{3.11}$$

Table 3.1 Specific speed range for different types of pump

Pump type	n_q	Ω_x	N_{USA}
Flow/head units	$(l\ s^{-1})/(m)$	$(m^3\ s^{-1})/(J\ kg^{-1})$	(US gallon)/(ft)
Centrifugal	12–65	0.25–1.20	680–3300
Mixed-flow	65–175	1.20–3.30	3300–9000
Axial	185–400	3.50–7.50	9500–20000

In the USA the specific speed, N_s, is given in different units: flow is in US gallons and pump head is measured in feet. The relationship is

$$N_s = 51.6n_q = 2730\Omega_s \quad \text{(approximately)}$$

The typical ranges are given in Table 3.1.

Pump speed must also be known. The impeller inside the pump must rotate in order to transfer energy to the water. The rotational speed is determined by the engine, usually an asynchronous electric motor (exceptions are rare). The motor speed is a function of the frequency f (Hz) in the national power grid ($f = 50$ Hz in Europe, $f = 60$ Hz in USA) and the number of pairs of poles p (usually from 1 to 5 for water supply pumps). The speed is then equal to

$$n = 60f/p$$

Angular speed ω (rad s^{-1}) is equal to

$$\omega = 2\pi n/60$$

Typical values of pump speed are listed in Table 3.2. Centrifugal pumps usually have a speed of 1450 or 2930 rpm (smaller and/ or multi-stage units); if no other data are available, these values should be used as defaults.

The *inlet* and *outlet diameters* of a pump might be needed for the computation of net positive suction head (NPSH), for which inlet diameter may be important, and/or necessary throttling (outlet diameter). If this information is not available, one will not be very much mistaken if one assumes that water will flow into the pump with a speed of 3 to 4 m s^{-1}, and leave it with a

Table 3.2 Speed range for pumps

Pairs of poles	n_{Europe}[rpm]		n_{USA} [rpm]	
	Synchronous	Typical range	Synchronous	Typical range
1	3000	2800–2960	3600	3360–3550
2	1500	1400–1480	1800	1680–1780
3	1000	900–960	1200	1080–1150
4	750	670–735	900	800–880
5	600	540–570	720	650–680

speed of 4 to 6 m s^{-1}. This means that a pump delivering, say, 250 l s^{-1} will probably have an inlet diameter of 300 mm and an outlet diameter of 250 mm.

Maximum efficiency is another parameter that can be predicted with some accuracy. Dickenson (1988) relates it to the rated flow of the pump:

$$\eta_0 = 0.94 - \frac{1}{(13\,200Q_0)^{0.32}} \tag{3.12}$$

where Q_0 (m^3 s^{-1}) is the nominal flow. This formula was obtained by the least-squares method applied on a sample of 15 000 different pumps. The values derived from the formula are usually slightly higher than the real ones, at least in water supply. A method of using the formula to assess potential energy savings is described in Chapter 7.

Specific energy is that used by the pump to deliver 1 m^3 of water and is a parameter carefully monitored in every water company. It is equal to

$$e = \frac{P}{3600Q} = \frac{g\rho H_p}{3600\,000\eta}$$

and its value is in kWh m^{-3}. A simple case is included here as an illustration; the data are

- pump flow $Q = 150$ l s^{-1} = 0.150 m^3 s^{-1},
- pump head $H_p = 70$ m,
- hydraulic efficiency $\eta = 79\% = 0.79$.

The result is

$$e = 9.81 \times 1000 \times 70/(3600\,000 \times 0.79) = 0.24 \text{ kWh m}^{-3}$$

Electro/hydraulic ratio must also be considered. The parameters mentioned so far, power and efficiency, are related to the pump only; however, there are other losses of energy that must also be accounted for. These are mechanical losses (friction in bearings), recirculation losses (a small part of the flow never leaves the impeller and just circulates around) and electrical losses (heating of electric motor parts). All together, they are expressed as

$$R = P_{electro}/P_{hydro}$$

A typical value of this parameter is 1.20 to 1.40, getting worse (that is higher) as wear and tear on the pump increases with time.

The total energy used by a pump delivering V (m^3) of water is therefore equal to

$$E = RVe \quad \text{(kWh)}$$

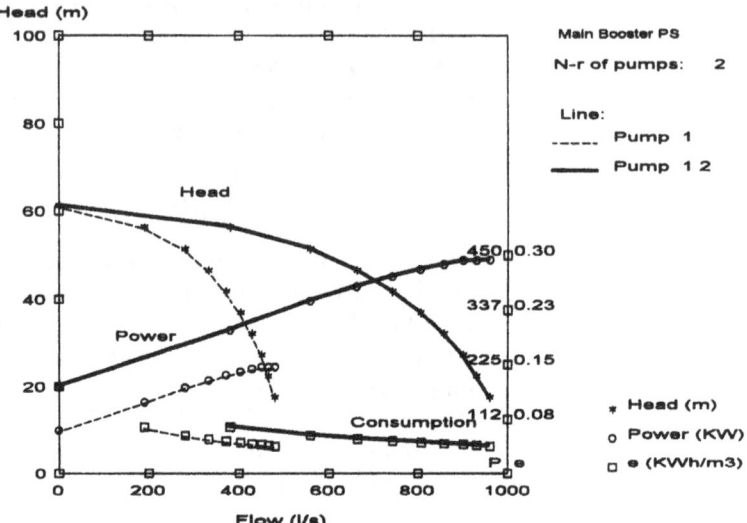

Figure 3.44 Parallel operation of pumps.

and that is the amount charged by the power distribution company.

Fixed-speed pumps

These pumps are by far the most frequently used units in any water supply system, alone or in parallel. Pumps in parallel need not be of the same type and size. The operating range of individual units depends both on the state within the network (inlet and outlet pressures) and whether other units are operating or not. This point is illustrated in Fig. 3.44 for two pumps operating in parallel. Note the operating range for a single pump and for two units together, and also slight overlapping in the middle range (flows from 400 to 500 l s^{-1}).

Things are more complicated when the station has several unit, operating in parallel. The complication arises because of the nature of the system characteristic. The head to be generated by the pumps has two components: the static head to be met plus the friction losses to be overcome for a particular flow. As the physical dimensions of the pipe system through which the water is pumped do not change, the friction losses increase as flow increases. This increase in friction head is not a linear relationship, and so the system characteristic is a rising curve in relation to increasing flows. A typical system characteristic is shown in Fig. 3.45 as the 'H' line.

In the case shown in Fig. 3.45, the PST has four identical units. When unit 1 operates alone, it can give the maximum flow of one unit. As other units step in, the total output of the

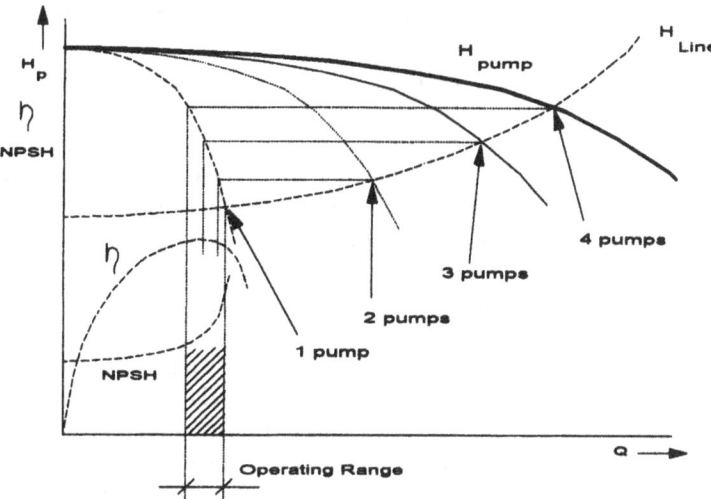

Figure 3.45 Four pumps in parallel.

pumping station does not increase in proportion to the number of pumps that are in operation. For example, if a single pump is operating, bringing in a second pump will not double the flow; it will be significantly less than twice the flow of a single unit. Similarly, a third pump being switched in will not triple the flow, and so on. The station capacity for one, two, three and four pumps, respectively, is the point on the system curve at which the $Q–H$ curve for the appropriate number of pumps crosses the system characteristic curve.

The shaded area represents the whole operating range for unit 1. Note that this pump, which operates alone, will use maximum power (this is true only for centrifugal pumps) – and may operate in the zone of dangerous cavitation (high value of NPSH), which roughly corresponds to flows above the BEP (best efficiency point).

This fact is usually known to the staff, so, when adjusting pumps for automatic operation in parallel, they throttle the first unit (the one which is likely to operate alone) 'just a little' in order to restrict its flow.

Note of caution The actual rotational speed of a pump may differ considerably from the rated speed, thus changing its characteristics. In a modern pumping station, the rotational speed of all units is monitored or logged. When preparing input for the model, the actual rotational speed should first be measured (once or twice) and the characteristics adjusted using the affinity laws given below. The corrections are not usually too significant – but there are some exceptions!

The characteristic of a pump is important for any analysis of energy consumption. As the price of energy rapidly rises everywhere, water companies regularly compute specific consumption of energy (expressed in kWh m^{-3}) and try to reduce it as to as low a level as possible.

Staff use such diagrams to determine the best operating range for different combinations of active pumps. This task is not simple for pumping stations with several pumps of different size and type, so a process computer is often used to perform this task continuously.

What can go wrong? The question 'Is this pump good enough?' is not the appropriate one. All pumps currently on the market are reasonably well designed and manufactured, with good operating characteristics – the competition is hard enough to ensure that. A question more to the point will be 'Is this pump good *for its intended use?*' Even the best pump could be misplaced and/or badly used, as will be illustrated in the next few cases (see also Walski, 1993; Tarquin and Dowdy, 1989). The important parameters are the following.

- The operating range of the pump (defined by minimum and maximum flow) should be fairly narrow and close to the best efficiency regime so the average efficiency will be relatively high (Fig. 3.46).

Figure 3.46 Operating range of pumps: good – operating around BEP.

- Low flows are not good because the hydraulic regime might be unstable (apart from low efficiency); also the efficiency of

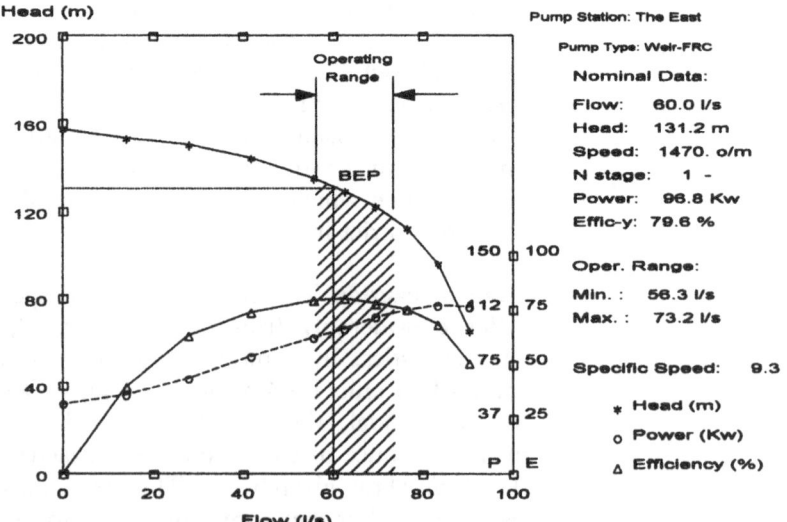

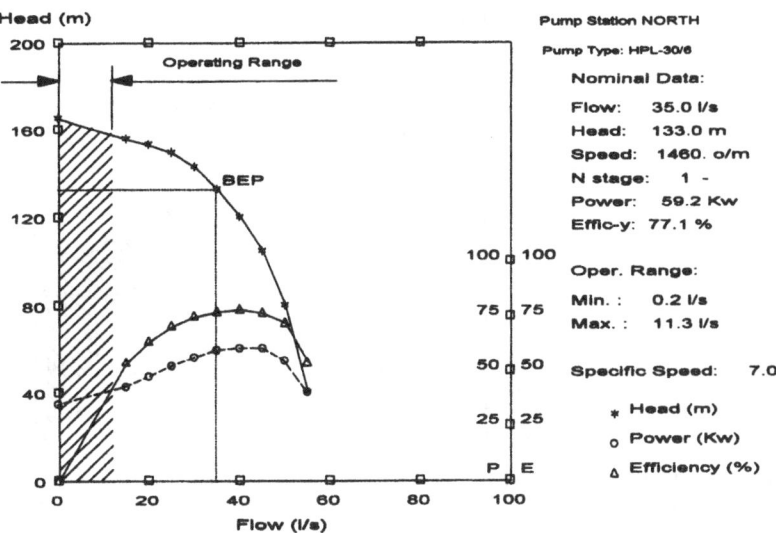

Figure 3.47 Operating
range of pumps: bad –
operating at low
efficiency.

the electric motor might be very low – if the power is less than
40% of rated value (Fig. 3.47).

- Too high flow may mean cavitation damage, especially if the
  pump operates for long periods (Fig. 3.48).
- The maximum power used by the electric motor should be
  checked, because it might exceed the capacity of the electrical
  network or transformers.

Figure 3.48 Operating
range of pumps: bad – the
flow is too high.

If the pump operates in the low-flow range, the remedy is to
reduce the resistance (the friction losses) in the system (perhaps

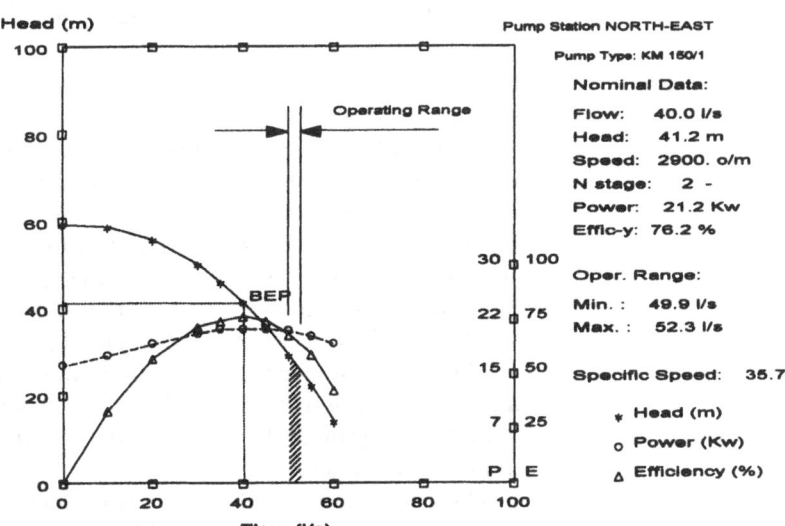

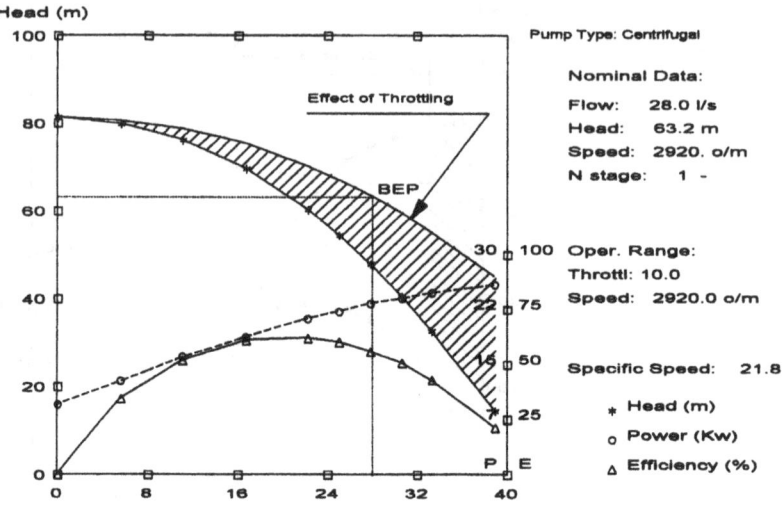

Figure 3.49 Pump characteristics: effects of throttling.

by opening some closed pipes and valves in its path), install a stronger electric motor, increase pump speed (if possible), install a larger impeller, or replace the pump.

If the pump operates in the high-flow range, the first remedy is to throttle it by (partially) closing its control valve – see the effects in Fig. 3.49. Do not forget – *throttling is always a loss of energy*. Other remedies are decreasing pump speed, installing a smaller impeller or replacing the unit with a smaller one. Generally speaking, in surprisingly many cases the pump is oversized because the designer was too cautious; in others, the system has been changed since pump installation.

Variable speed pumps

Fixed-speed pumps operate best in a relatively narrow range of flows, but demand within the system can vary widely. In some cases the difference is provided by water towers and reservoirs, but in many cases the pumps simply have to operate in unfavourable regimes. If another requirement is added – to keep service pressure more or less stable – fixed-speed pumps simply cannot meet the need. Variable-speed pumps provide the flexibility to reach the targets for both flow and pressure (see Hosho and Fukui, 1983; McNaught, 1993).

The effects of a speed reduction (from 2920 to 2700 rpm) on a pump are shown in Fig. 3.50. Note that pump head and power values are significantly lower, while the efficiency has remained almost unchanged.

The correct values of flow, head, efficiency and power for a different rotational speed could be determined by physical tests.

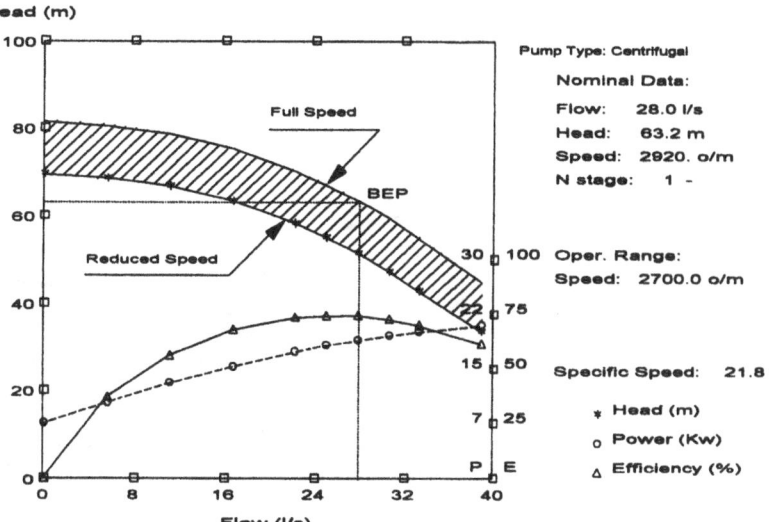

Figure 3.50 Pump characteristics: effects of speed decrease.

However, *affinity laws* give a precise answer to the problem. They state that flow, head and power depend on actual rotational speed as:

$$\frac{Q}{Q_0} = \frac{n}{n_0}$$

$$\frac{H_p}{H_{p0}} = \left(\frac{n}{n_0}\right)^2 \qquad (3.13)$$

$$\frac{P}{P_0} = \left(\frac{n}{n_0}\right)^3$$

while the efficiency remains much the same, regardless of the speed.

The best way to control discharge through a multi-unit PST is to combine one variable-speed pump with several fixed-speed ones, as in the case shown in Fig. 3.51. Note that the contribution of the second unit – the variable-speed pump – is likely to be small in comparison to the fixed-speed one.

The variable-speed unit is always running, alone or in combination with another, fixed-speed pump. The rotational speed of the variable-speed pump may go as low as 70% of the rated speed, and cannot be higher than the rated speed if using traditional devices (thyristors, etc.). New devices such as frequency converters can considerably enlarge this range, permitting speeds between zero and much greater than the rated speed.

Note of caution The decision whether an expensive variable-speed pump will be the best answer should be made only after

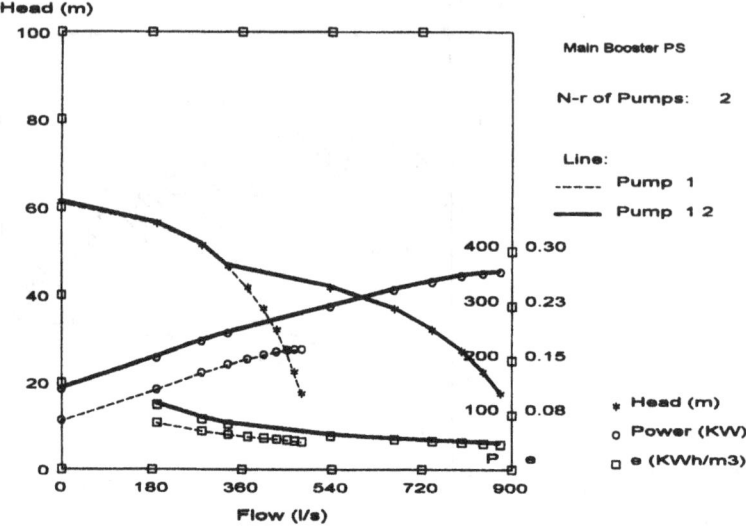

Figure 3.51 Combining variable-speed and fixed-speed pumps.

Figure 3.52 Problems with variable-speed pumps: operating range.

careful consideration. Pumps with a flat characteristic might be very inefficient if the speed is just slightly reduced, as illustrated in Fig. 3.52. If the pump head is fairly constant (other fixed-speed units will see to that), a relatively small decrease of speed forces the variable-speed unit to operate in a zone of low efficiency, thus increasing the total consumption of energy – as shown in Fig. 3.53.

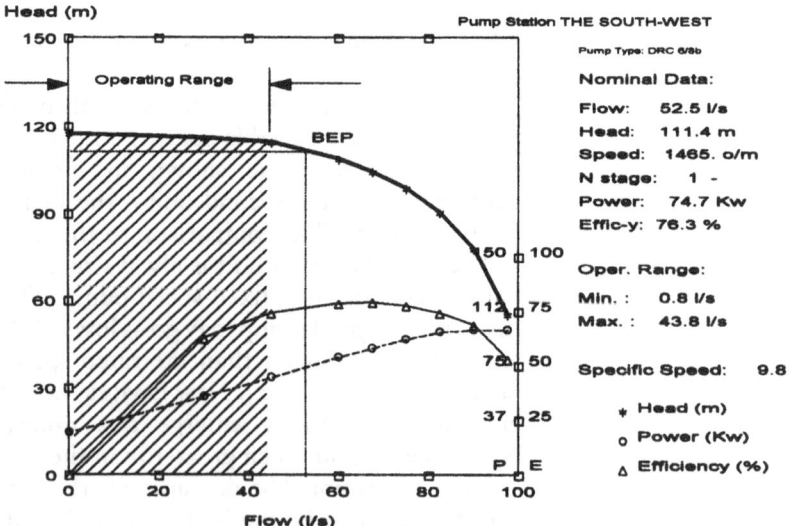

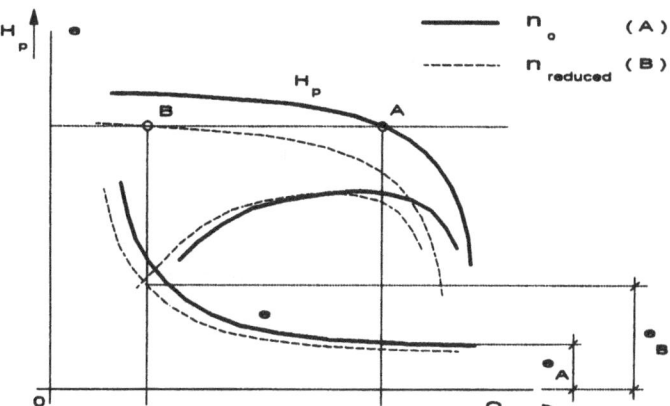

Figure 3.53 Inefficient operation of variable-speed pump.

Control of a pumping station

A PST operates either manually on local or remote control, or automatically. In the first two cases somebody has to switch 'on' or 'off' individual pumps according to an agreed schedule, or on their own initiative. In the last case, the PST is related to a particular location in the network. As the pressure goes up, pumps should switch off one by one, and vice versa. This 'control location' is usually a reservoir or a water tower, as shown in Fig. 3.54. For instance, pump 1 will be stopped whenever the level of water in this reservoir exceeds elevation 'IN', to be restarted again when the level falls below elevation 'OUT'. These control levels are measured from the reservoir bottom. Note the symmetry: pump 1 is the first to start, the last to stop; next comes unit 2 , etc. The last unit is in reserve – it operates only under emergencies. The idea is to keep water storage in the reservoir at sufficient levels all the time.

Figure 3.54 Keeping WL high enough.

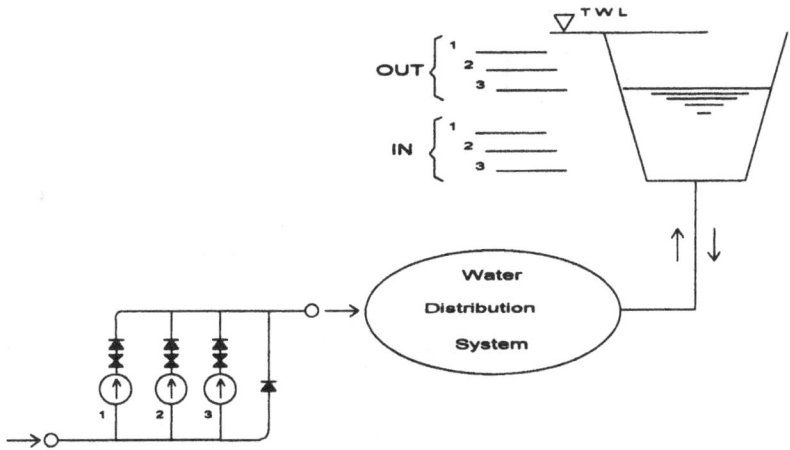

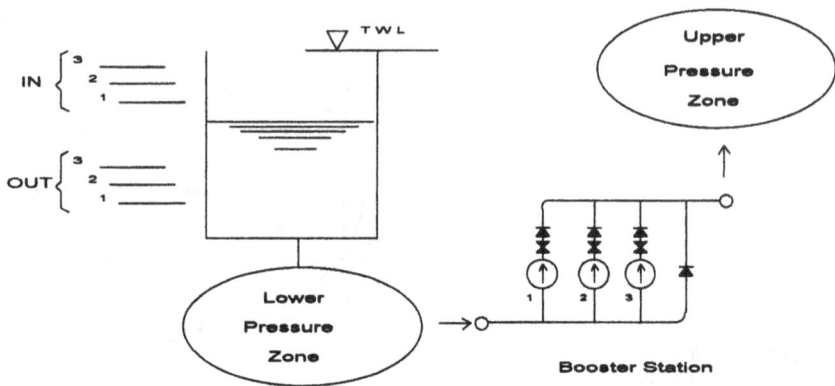

Figure 3.55 Keeping WL sufficiently low.

Similar logic is applied in special cases where a reservoir is placed too low and in constant danger of overflowing. Then the pumps are switched 'on' whenever the level is sufficiently high, and turned 'off' when it falls again (Fig. 3.55). In control jargon, this is 'reverse logic' because it is the opposite of the usual one (compare with Fig. 3.54).

An example from the real world, rather simplified, is included for illustration in Figs 3.56 to 3.58. High-head pumping station (AF172) keeps reservoir AL119 always full of water (Fig. 3.57), while the role of low-head pumping station (AF184) is to transfer water from reservoir AL126 to the main one (AL120) whenever the level there (AL126) is sufficiently high (Fig. 3.58).

A pumping station may be applied to control *pressure* in the network at a desired level (constant, or time-variable). This is

Figure 3.55 Keeping WL sufficiently low.

Figure 3.56 Different control policies.

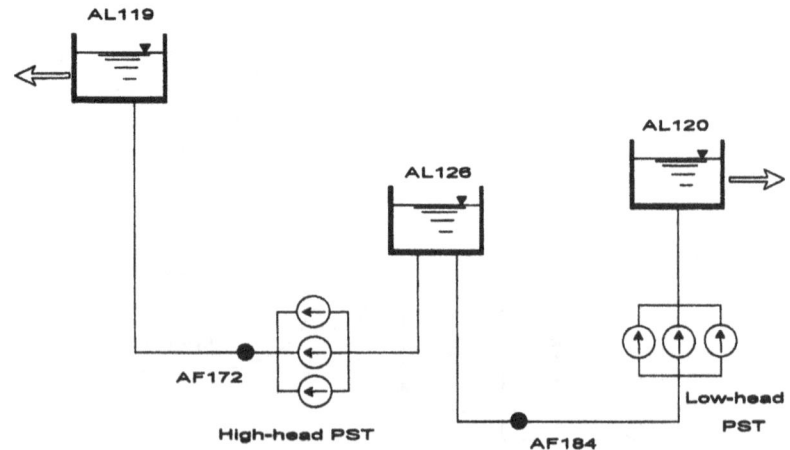

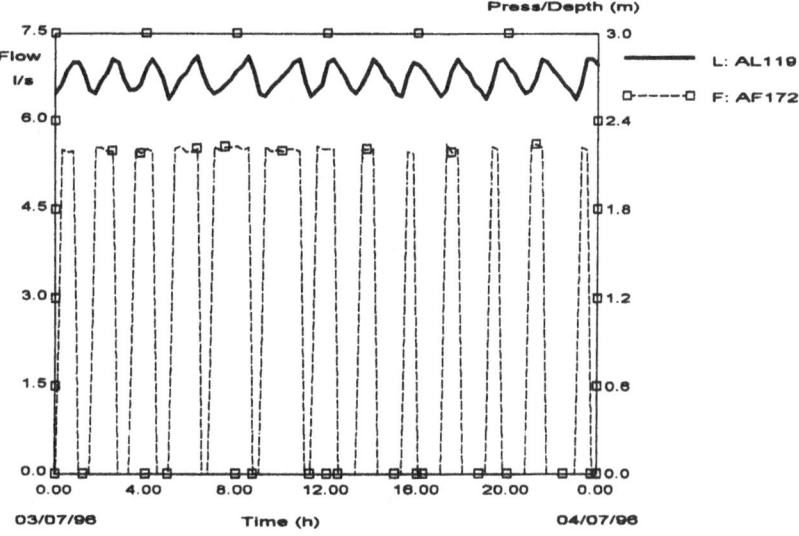

Figure 3.57 Level-
controlled pumps –
standard policy.

Figure 3.58 Level-
controlled pumps –
reverse policy (using all
available water).

usually achieved by variable-speed pumps, as shown in
Fig. 3.59. A sensor (4) monitors the pressure at a selected
point of the network and informs the controller (3) where this
value is compared with target pressure; if the pressure is lower
than required, the speed of electric motor (2) is increased, thus
boosting the pressure by speeding up the pump (1), etc. Target
pressure could be either constant or time-related.

This solution is frequently used for the supply to isolated
zones and smaller settlements, where no reservoir or water

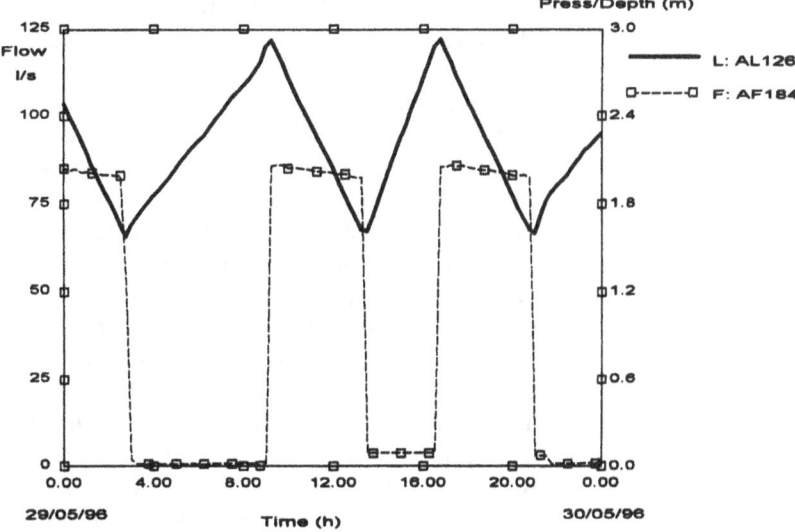

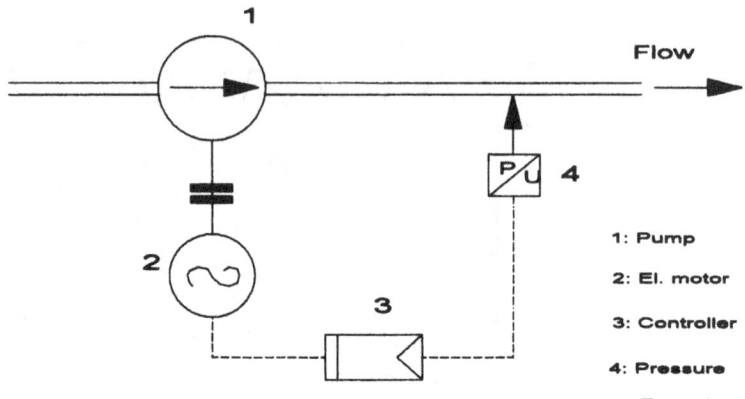

Flow

1: Pump

2: El. motor

3: Controller

4: Pressure
   Transducer

Figure 3.59 Pressure-controlled pump.

tower is available to maintain the pressure. The weakness of these systems is that they do not make allowances for tariff changes.

The electricity tariffs   The cost of energy is very high and, in many countries, electricity suppliers have elaborate tariff structures that take into account season, time of day, weekends and other factors. Two examples are given below.

The first tariff, 'A' (Fig. 3.60), charges less for energy used during the night (the power suppliers' 'night' is only seven hours) than for energy used during the rest of the 24 hours. Moreover, the user has to pay special, high charges on maximum power used for at least 15 minutes during the current month, during the winter.

The other tariff, 'B' (Fig. 3.61), differentiates between working and weekend days. In winter months, the energy used during afternoon peak hours is very costly – seven times the average cost! On the other hand, there are no charges on maximum power used.

It is obvious that any control policy which ignores these peculiarities of tariff will not be cost-effective. A new control device has been developed, called 'Profile Controller' (Profiler, in short). The basic idea is that the reservoir should be refilled during the period of cheap energy, and pumping reduced to a minimum during the rest of the day. Such a policy would result in a diagram for water level changes similar to the one shown in Fig. 3.62. The diagram could be planned in advance and realised through the Profiler installed in the pumping station. It operates as follows. If the actual water level deviates from the desired line, pumps will be started or stopped accordingly. It is the further development of the earlier 'constant-level' policy (compare with Fig. 3.54), but now control levels vary with time,

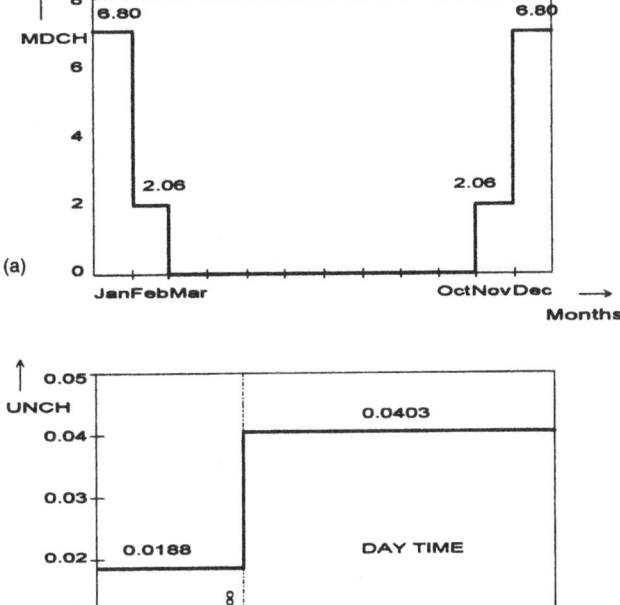

Figure 3.60 Tariff 'A' charges for power and energy: (a) maximum demand charges; (b) unit charges for energy.

following the desired water level changes. The savings in operating costs may be in excess of 10%.

Basically, the pumps will not be used during high-cost tariff, as shown in the example in Fig. 3.63a. Note that this is not an important consideration for a flat tariff (Fig. 3.63b) as the difference in costs is not significant.

A careless control policy can lead to very high costs (Fig. 3.64): in Fig. 3.64a the pump is operating round the clock and in Fig. 3.64b the pump was switched on just in time to incur high costs. To do justice to the staff, their first priority is to provide good, uninterrupted water supply and think about the costs later; maybe they did not have any other option.

In conclusion, it is obvious that control policies must take the tariff into account, as shown in the simple case of a PST with four pumps (Fig. 3.65). It is assumed that there is sufficient storage capacity in the network.

Note that if tariff 'A' was adopted, then all four units should not be operated together, to avoid incurring maximum power charge – and at least one pump should operate all the time. In the other case (tariff 'B'), pumps should be used to the full

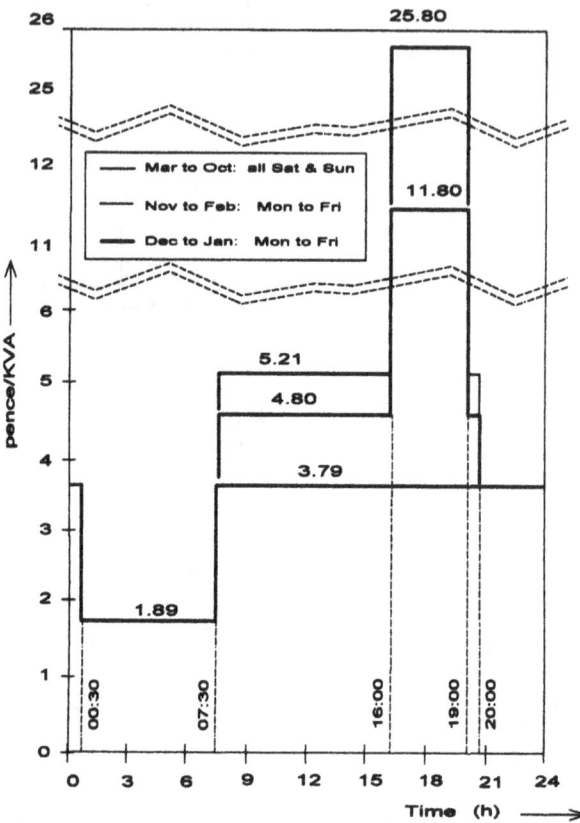

Figure 3.61 Tariff 'B' charges only for power (STOD).

Note: MDCH = 0, regardless of the season

capacity during night hours – to replenish reservoirs, storage permitting – and used very sparingly during the day; at peak tariff time (between 16:00 and 19:00) all units should be stopped and the customers supplied from storage with no pumped replenishment.

The final answer can be found only after a thorough analysis of the mathematical model. See Chapter 7 for further discussion on the problem of minimising power costs.

Protection of pumps   Pumps have to be protected against:

- entrapment of air (which happens when suction pressure drops too low);
- operating at zero flow in the unstable zone.

Both protections are shown schematically in Fig. 3.66. These protections always exist but they are seldom triggered.

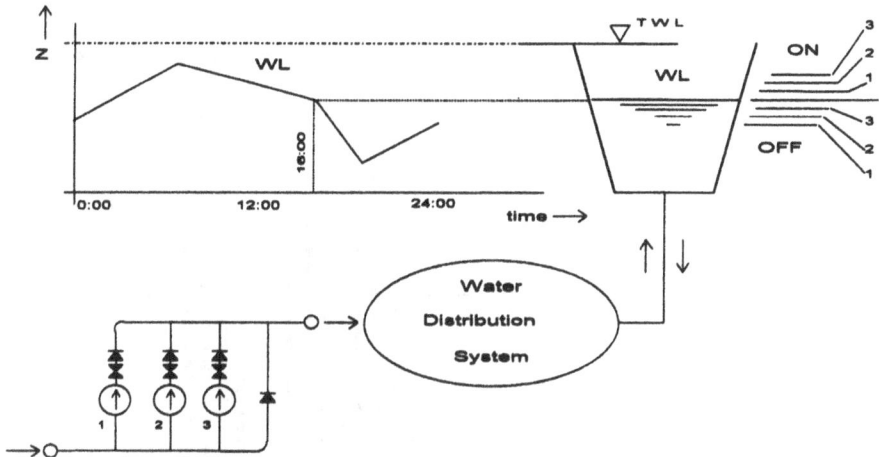

Figure 3.62 Profile Controller (or Profiler).

When (and if) water level in the suction tank drops too low, pumps are stopped – in this case, first unit 2 and then unit 1. This is an overriding command, of a higher priority than any request generated by the controlling system. Pumps cannot start again – whatever the need – before water level in the suction pool has recovered sufficiently (in some cases time relays postpone a pump restart). In this case the control signal is water level in the suction tank; in other ones, where pumps take water from a closed conduit, the signal is the suction pressure value, but the principle remains the same.

This system is also called 'low suction pressure protection'. The outlet pressure is also under control (Fig. 3.66). If this pressure exceeds a given limit, all active pumps will be

Figure 3.63 (a) STOD tariff can be 'dangerous' but (b) simple tariff is less so.

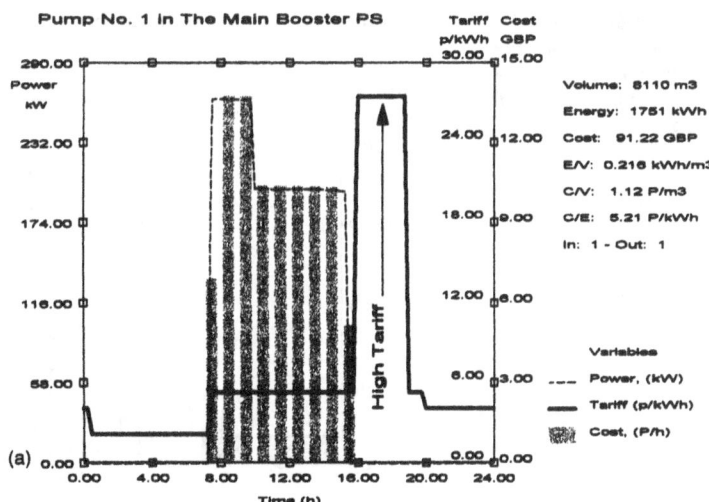

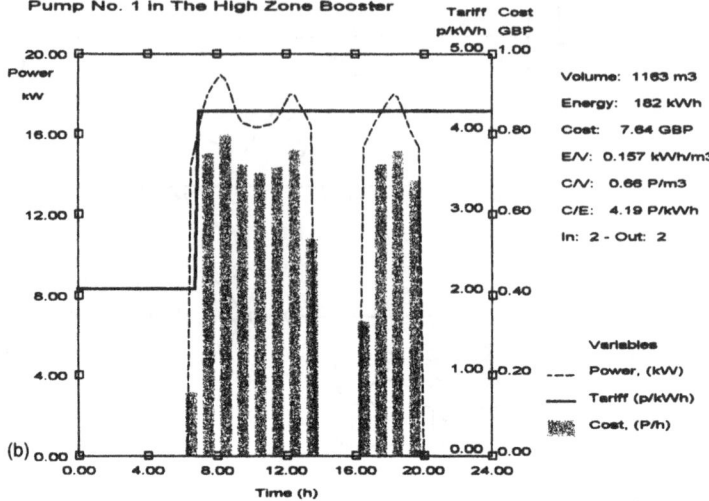

Figure 3.63 (b)

automatically tripped off (similar to the previous case). The units will be switched on later, when the pressure in the distribution network falls below the pre-set limit. This simple device prevents operation of pumps in the zone of low flow and low efficiency, and protects the network, too.

Bypass pipeline

Some PSTs have a bypass pipeline around the station, with a non-return valve (NRV) in it. The idea is to let water flow from the inlet side towards the pressure side at times when all pumps are 'off' but pressure on the inlet side is higher than on the pump delivery

Figure 3.64 (a) STOD and continuous operation and (b) errors in control policy can both incur high costs.

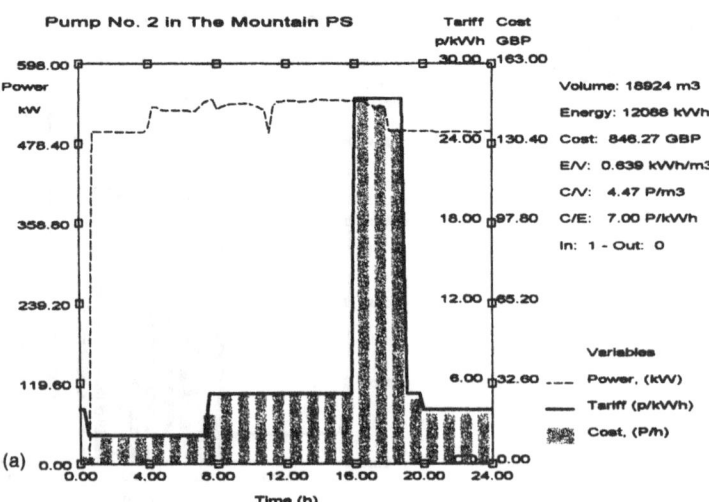

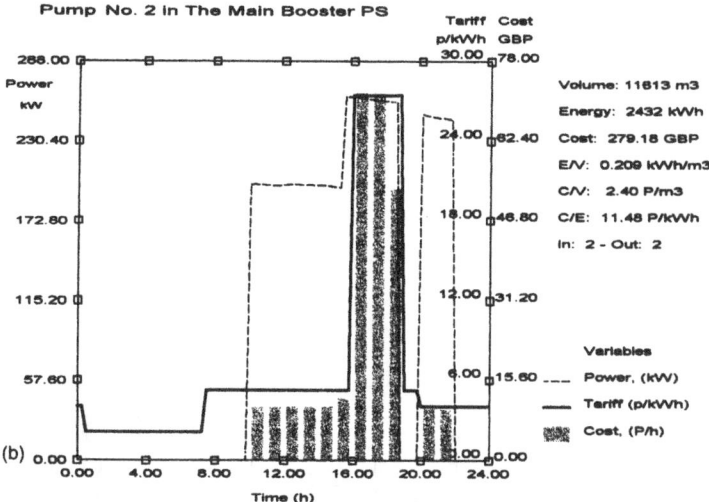

Figure 3.64 (b)

side. This is often the case with booster PSTs within a distribution network. Pumps are used only during the peak consumption time, but the flow through the bypass satisfies night demand.

The more important case is when the booster station (BST) has to increase the capacity of an existing pipe, usually part of a conveyance system (Fig. 3.67). This system can operate in two distinct regimes (Fig. 3.68): by gravity alone, or by boosting the flow. Even though incurring additional operating costs, this

Figure 3.65 Different control policies: (a) for tariff 'A'; (b) for tariff 'B'.

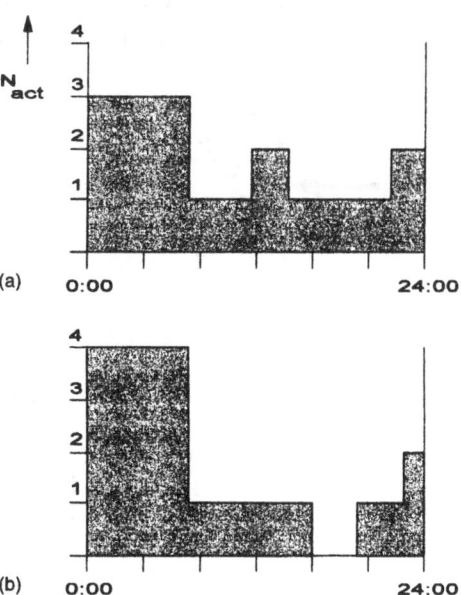

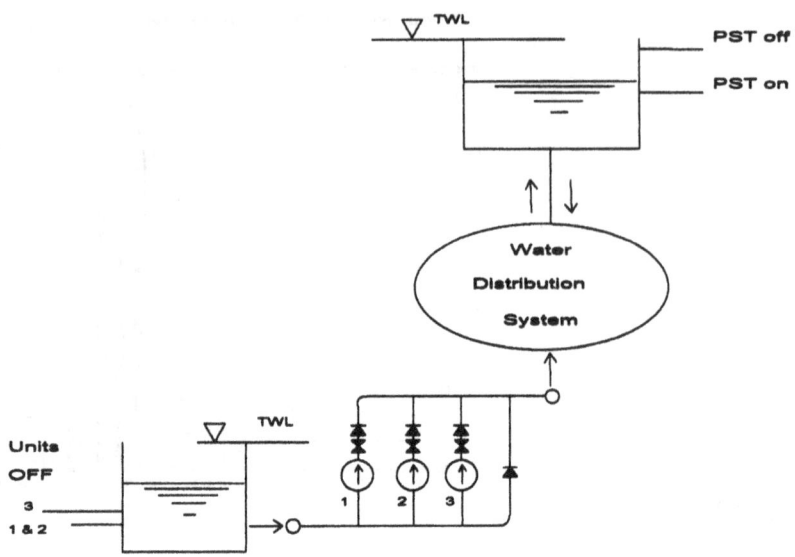

Figure 3.66 Protection of pumps.

option might be cheaper than laying another pipe alongside the existing main; it is also very safe in operation.

3.8 Valves

Valve types and their roles

Figure 3.67 Booster pump in the bypass.

In addition to the many, simple sluice valves found in water supply systems, there are normally many examples of 'special' valves, which have a wide range of forms and functions. Many different designs are used in practice, with new ones offered

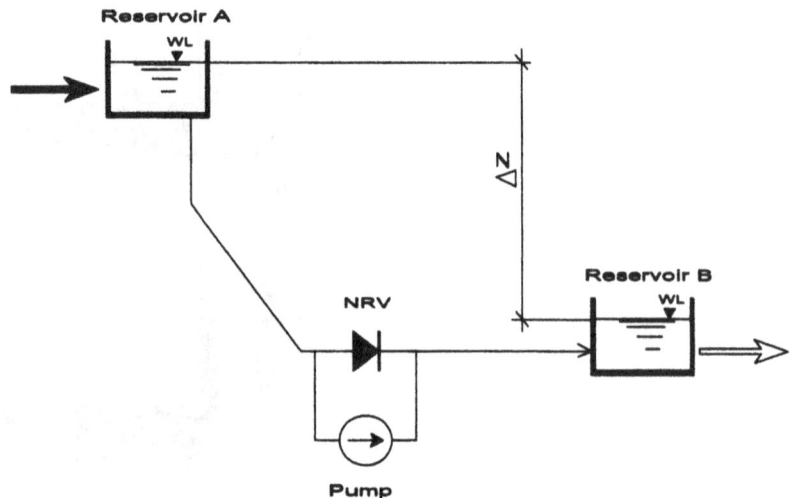

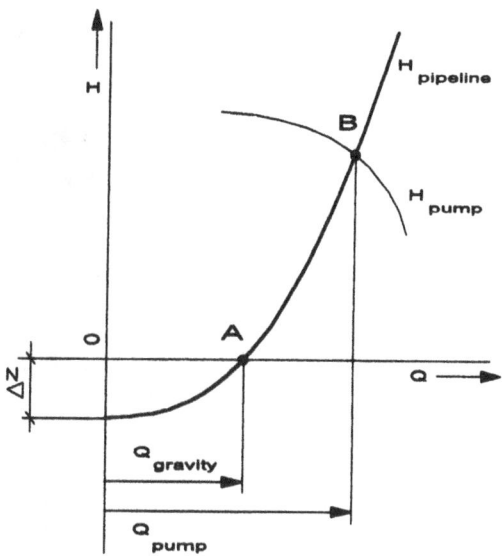

Figure 3.68 Operating range of a BST.

every year. All valves may be classified in three main groups (Fig. 3.69):

- *Linear motion* – gate (sluice), globe, needle valves
- *Rotary motion* – butterfly, spherical, plug valves
- *Elastic deformations* – valves with membranes

Each main type has its varieties; for instance, there are butterfly valves with symmetric vanes, eccentric discs or double ('sandwich') vanes, etc. Details can be found in *The Process Control*

Figure 3.69 Valve classification.

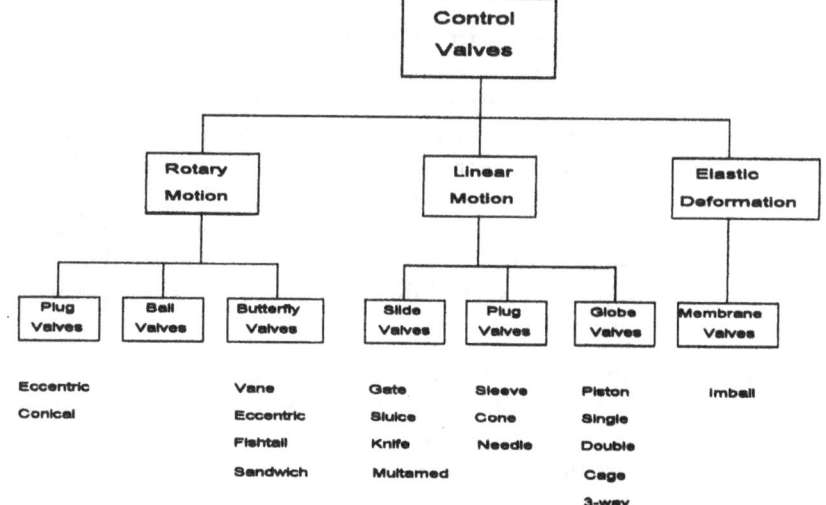

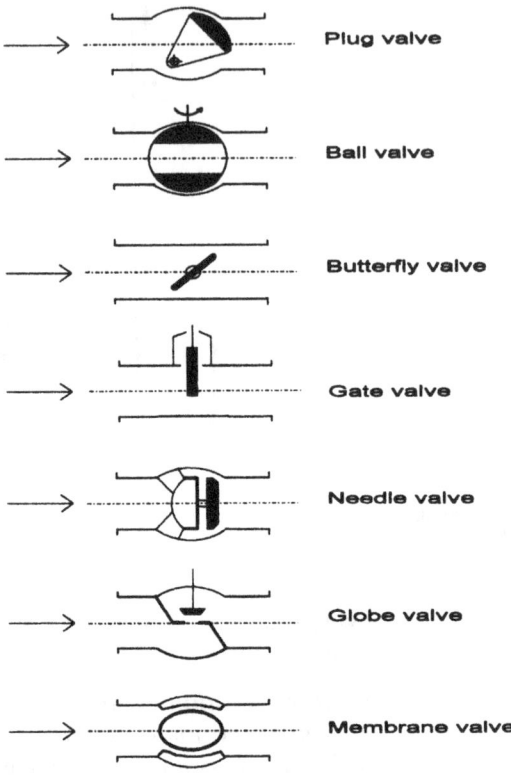

Figure 3.70 Different types of valves.

Handbook 1986–87 (CHW, 1986), and various textbooks. Sketches of these valve types are included here as Fig. 3.70. The valves may perform the following duties in water supply systems (as shown in Fig. 3.71):

- To isolate a part of the network or a pipe, temporarily or permanently, for various reasons, e.g. operation and maintenance jobs, forming zones, emergencies, etc. (Fig. 3.71a).
- To protect elements of the system, e.g. reservoirs, pumps, areas of the distribution network, from potentially damaging conditions or loss of water (Fig. 3.71b,c).
- To regulate flow and/or pressure within the distribution system, under either local or remote control (Fig. 3.71d).
- To perform special duties, e.g. make possible the washout of water mains.

Modelling of valves The main purpose of a valve is to throttle the flow – down to zero, if necessary. The relationship between its opening and degree of throttling cannot be computed but must be determined by measurements. The characteristic for a typical gate valve is shown in Fig. 3.72.

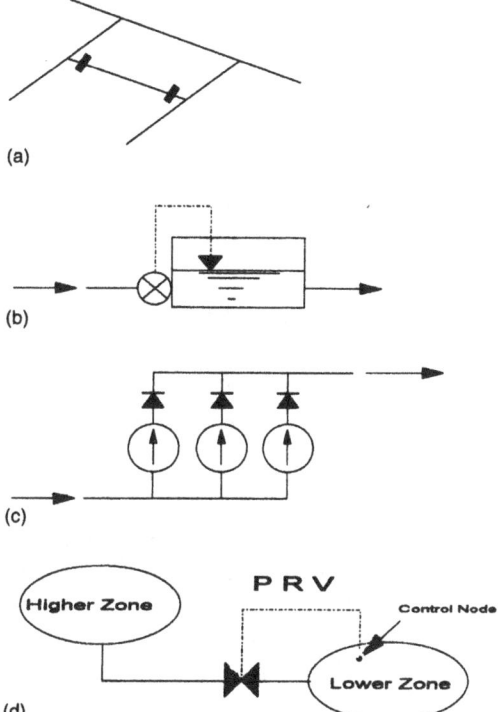

Figure 3.71 Valves in water supply: (a) section valves; (b) float valve; (c) check valves; (d) control valve.

The head loss in a valve is equal to:

$$\Delta H = \zeta \frac{v^2}{2g} \qquad (3.14)$$

where ζ is the coefficient of loss (–) and v (m s^{-1}) is the flow velocity in the valve. Sometimes another coefficient, k_D, is used instead of ζ:

$$k_D = \frac{1}{\sqrt{1 + \zeta}} \qquad (3.15)$$

The minimum information on each important valve (i.e. those which have some influence upon the flow regime) is:

- valve type and its nominal diameter;
- the coefficient of loss of energy in fully open position;
- its characteristics (like Fig. 3.72).

Other data for modelling depend upon the role of this individual valve in the supply system.

English units Manufacturers often use different parameters to show the operational characteristics of their products. For

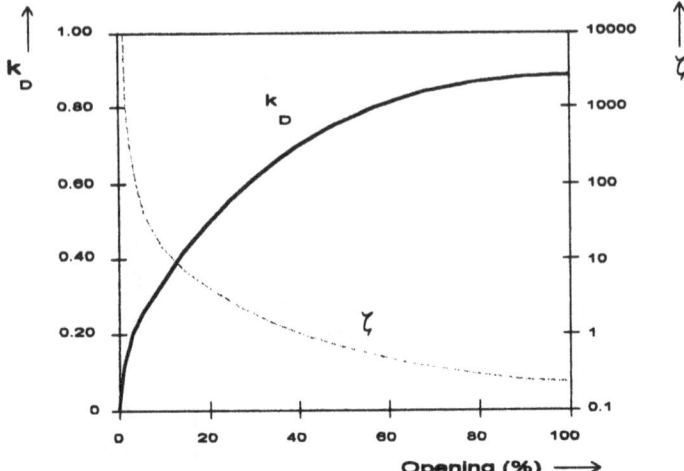

Figure 3.72 Valve characteristic for a typical gate valve.

instance, in the UK a popular parameter is *the flow coefficient* K_v defined as water flow through the valve in $m^3 s^{-1}$ causing a pressure drop of 1 bar. The relationship is

$$K_v = \frac{Q}{31.6}\sqrt{\frac{\rho}{\Delta p}} \qquad (3.16)$$

where Q is in $m^3 s^{-1}$, $\rho = 1000$ kg/m³, Δp is the pressure drop in bars. These numbers are relatively high. Table 3.3 gives data for a butterfly valve, size 300 mm, in all three systems.

Section valves

Normal practice provides each section of pipe in the distribution network with a valve at each end (Fig. 3.73). They serve to isolate the section when needed: repairs, new service connections, washouts, etc. Section valves are usually simple gate valves. In normal operation, they should be fully opened, and the loss of pressure is quite negligible. These valves should have just two states: fully open or closed (1/0). Most of these

Table 3.3 Energy loss coefficients for 300 mm butterfly valve

Angle (deg)	ζ	k_D	K_v	Comment
0	0.298	0.878	6600	fully open
10	0.363	0.856	5985	
20	1.069	0.695	3486	
30	3.833	0.455	1841	
40	9.283	0.312	1183	
50	29.73	0.180	661	
60	83.68	0.109	394	
70	334.7	0.055	197	
90	–	0.0	0	fully closed

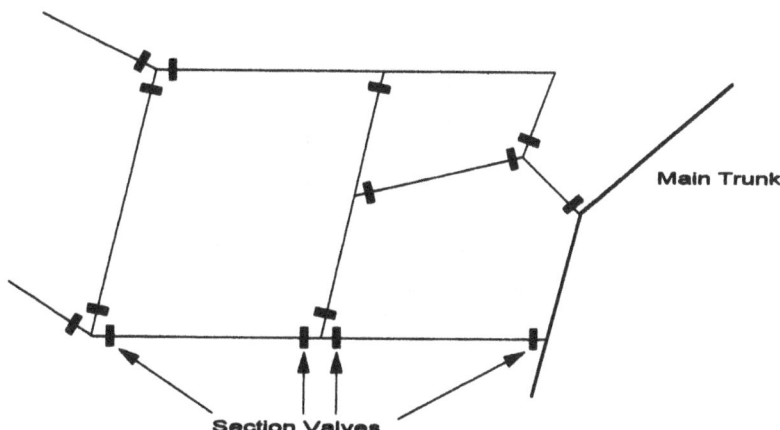

Main Trunk

Section Valves

Figure 3.73 Section valves in a network.

valves are simply ignored in modelling. Attention should be given only to those which are closed, for one reason or another. One should always bear in mind that in every *real* system there are some closed pipes, for very good reasons! Staff close them to isolate one pressure/supply zone from another, or to redirect flow from one part to another, or to prevent too rapid filling of a low-placed reservoir, or ... the list is endless!

In the model, section valves are not included. Normal policy is to allow the user to 'close' any pipe he/she chooses. One practical problem with valves is that sometimes, having closed a valve for a very good reason, an operator forgets to open it again. Locating such 'forgotten' valves is not easy, but model calibration (see Chapter 6) often helps to uncover some old mysteries of this sort.

Float valves

A FLV closes (or opens) as water in its reservoir approaches its upper limit (or falls away from it). Older types of FLVs had a buoyant body and a system of levers to operate the valve itself. New designs are more sophisticated, but the principle remains much the same. The function of a FLV is to control inflow to a reservoir and avoid overflows and consequent loss of water.

A FLV is a valve similar to the others, but it is difficult to obtain its characteristics in most cases, and a typical curve is given in Fig. 3.74 to facilitate modelling. Note that:

- a FLV does not control outflow from a reservoir and cannot prevent a reservoir emptying;
- the effect of a FLV is felt only within its regulating range (usually 0.5 m measured below an overflow weir or some lower level).

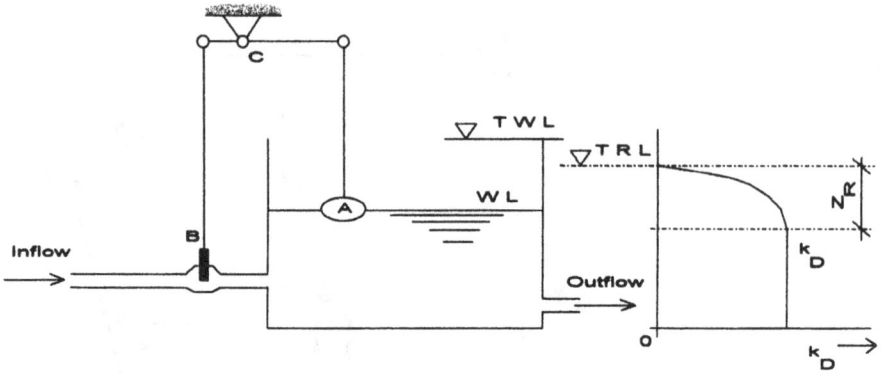

Figure 3.74 Protection against overspills using a FLV.

Another important question is where the FLV is located – whether in a pipe that enters the reservoir below water level or in a pipe that just pours water into the top of the reservoir. Typical arrangements are shown in Fig. 3.75.

Figure 3.75 FLV arrangements in which the FLV regulates inflow: (a) from below, with no bypass pipe; (b) from below, with a valve in a bypass pipe; (c) from above; (d) from above, with NRV that permits rapid backflow.

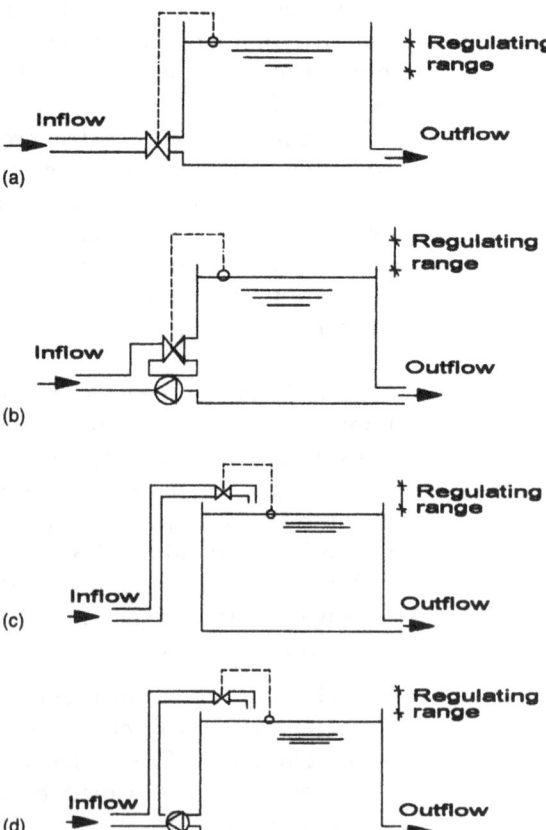

Float valves are very common devices in every water distribution system. If a reservoir is fed through a water main by gravity (i.e. no pumping), then a float valve is almost certainly fitted; otherwise, it will be difficult to prevent loss of water whenever the reservoir is full. During the build-up of a model, one can safely assume that there is a float valve in each pipe that fits the description given above.

Non-return valves

Non-return valves (NRV, check valve) are used to prevent a backflow where it might cause harm. A NRV may be placed downstream of a pump or a pumping station (very frequently) but are rarely needed within a distribution network. A NRV in normal operation can only be fully open or closed (1 or 0). If open, the loss of pressure is relatively small and could be treated as other minor losses. The other state is 'closed' – then the flow is stopped completely. Which state the NRV is in, 'open' or 'closed', depends only on the pressure difference across it: if the upstream pressure is higher, NRV is open, and vice versa. No particular data are needed for a non-return valve.

Non-return valves are sometimes used in distribution networks to prevent undesired flow. One case is shown in Fig. 3.76. The pumping station supplies both the main reservoir and a local demand – but only when its pumps are operating. The NRV is closed and water cannot enter the smaller service reservoir, which has its own source of water. However, when all pumps are stopped, the NRV will open and feed the local demand. Situations like this one must be spotted and modelled accordingly.

Figure 3.76 NRV feeding a local demand.

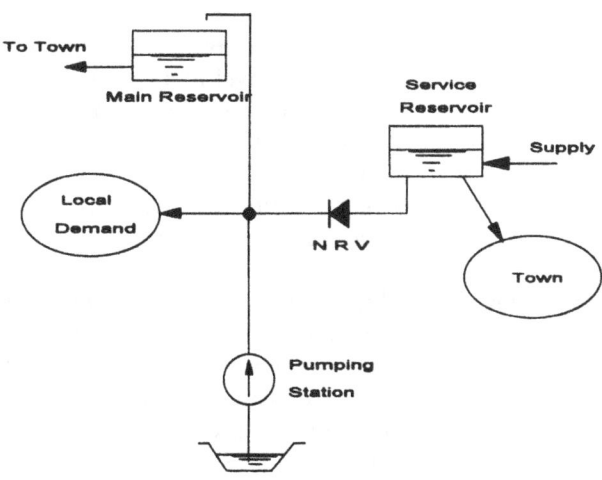

Control valves

These valves are the most important in modelling, because they often perform vital roles in a water supply system. A control valve is normally equipped with an actuator (pneumatic, hydraulic, electric, electric–hydraulic, self-regulating or manual), a control device ('magic box'), a sensor to get information from the control point in the system, a link between the two (electric, hydraulic, mechanical) and accessories. In selecting the actual valve to be operated under some form of control, it is often useful to select a valve that has as nearly linear a characteristic as possible.

The principle of operation is as follows.

1 The sensor picks up the value of the control variable: pressure, water level, flow, or time, and converts it to a signal.
2 The signal is transferred to the control device.
3 The control device analyses this signal and issues a command ('open', 'close', 'stall', etc.).
4 The actuator obeys the command and changes the opening of the valve accordingly.

The process is sometimes simpler (valves under local control) and sometimes more complicated (involving the use of process computers), but the general idea is the same.

Regarding their mode of operation, control valves may be classified as follows.

• *Throttled valves* (THV) work by 'throttling' and have a constant opening (from 0 to 100%, i.e. from closed to fully open).
• *Pressure-reducing valves* (PRV) or *pressure-sustaining valves* (PSV) try to keep pressure in a certain point at the desired level.
• *Pressure-controlled valves* (PCV) are controlled by variations of water level in a reservoir or variations of pressure in a far-away location, following pre-set rules.
• *Time-controlled valves* (TCV) operate according to pre-set timing.
• *Flow-controlled valves* (FCV) are controlled by variations of flow in a certain pipe.

Throttled valves

This is the simplest case, when the opening of the valve remains constant during simulation runs, acting as a local obstruction to flow. However, the same valve may easily be converted to a more sophisticated device (a PRV, a TCV, a PCV or a FCV) later in its life, as the distribution system becomes more complex!

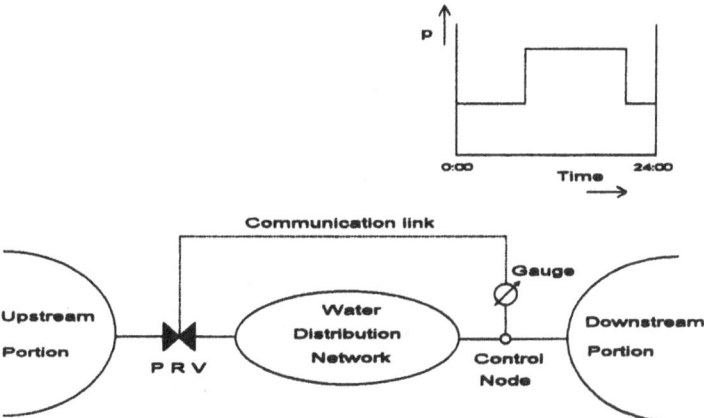

Figure 3.77 Pressure-reducing valve.

Pressure-reducing valves

Originally, this was a device to keep the immediate downstream pressure at a constant, pre-set level (Fig. 3.77). It operates according to these simple rules:

- If $H_i > H_{cont} > H_j$, PRV throttles the flow until the pressure immediately downstream of it is equal to H_{cont}.
- If $H_{cont} > H_i > H_j$, PRV opens fully – it tries to reach (unsuccessfully) the set pressure H_{cont}.
- If $H_j > H_i$, PRV closes down, to prevent the backflow.

This is rather a tricky element in a model (see Davis and Jeppson, 1979) and, if several PRVs are fitted in the loops of a complex water distribution system, there is no unique solution for flows. Fortunately, such arrangements are rare in real systems (if they do exist in a real system, it might well be worth while investigating the reasons for the arrangement). PRVs are normally used to control pressure in discrete areas of the network, as shown in Fig. 3.78.

Note that the control point is far from the valve itself, at a pressure-sensitive point, normally the highest point in the area, where pressure must be maintained at a desired level. The

Figure 3.78 A PRV in the branches of a network.

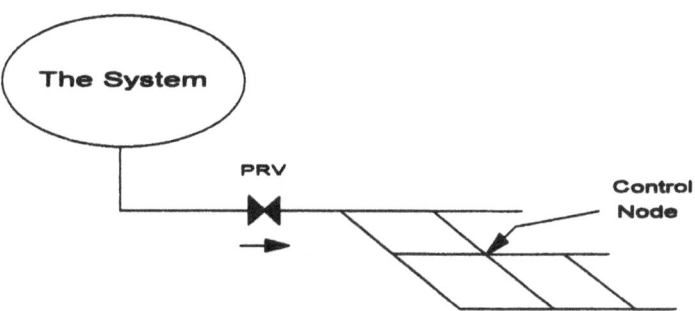

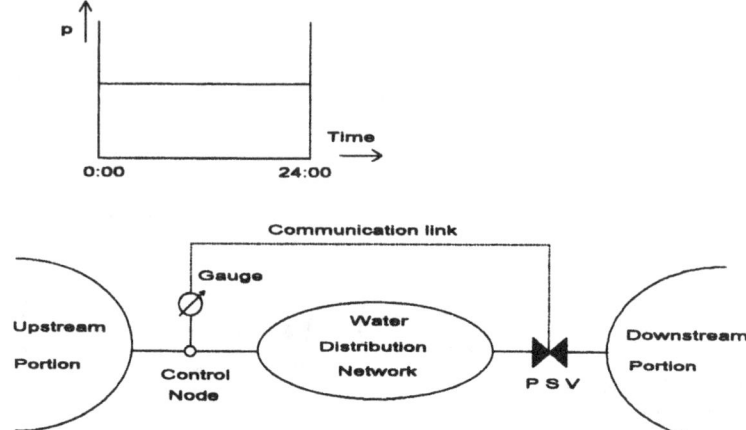

Figure 3.79 A pressure-sustaining valve.

pressure is measured by a sensor and the value transmitted to the control device on the PRV.

Traditionally the PRV controls pressure *downstream* of itself, but it can be used to maintain pressure *upstream*, in a similar manner, and then it is called a pressure-sustaining valve (PSV). An example is shown in Fig. 3.79. Note that the control 'logic' of a PSV is the reverse of that for a PRV: when the pressure falls too low on its upstream side, a PSV starts to close down, and vice versa.

Level- and pressure-controlled valves

These devices (PCV) are a recent development of well established conventional valves. As before, a pressure gauge is situated at a key point in the distribution network and measures pressure constantly. The signal is transmitted to the control unit of the valve via a communication link (Fig. 3.80).

Figure 3.80 Pressure-controlled valve.

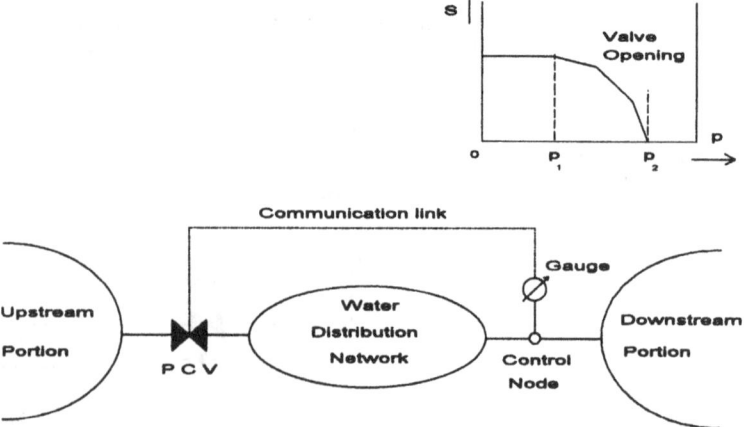

The control unit is a microprocessor with some 'logic' built in. The relationship between the control pressure (H_k) and the amount by which the valve is open (S) must be programmed in advance to suit local needs. In the case shown here, any increase of pressure in node 'k' will close the PCV a little, and vice versa. What this will mean to the system as a whole also depends on other factors.

The effects upon the control node are not immediate, as in the case of a PRV (see above). The slope of the 'control curve' is, naturally, an important factor in bringing the system rapidly to a stable state.

Three different pressure control systems are shown in Fig. 3.81 (from Huntington, 1979). These systems are:

(a) local sampling (i.e. measurement of pressure) and local control;
(b) remote sampling and local control;
(c) remote sampling and remote control.

Figure 3.81 Pressure control systems: (a) local sensing and control; (b) remote sensing and local control; (c) remote sensing and control.

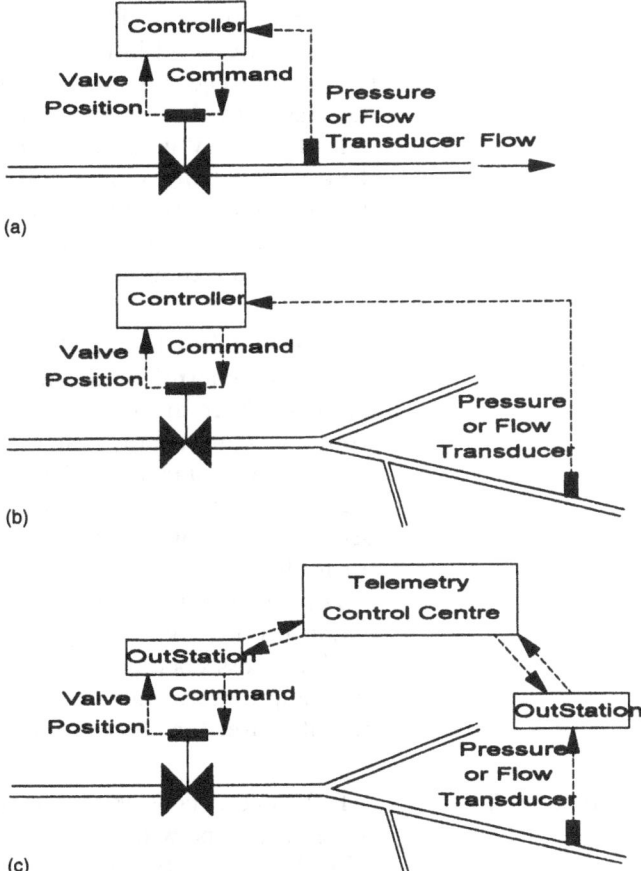

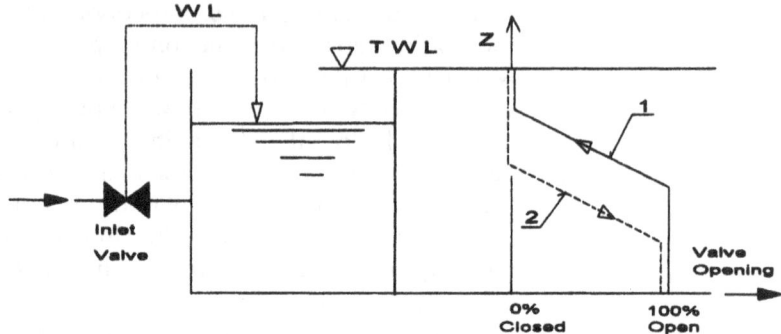

Figure 3.82 Two-speed
control valve.

The first system – local control (Fig. 3.81a) – is the most widely used, recently with quite sophisticated programmable logic controllers. The other two can give better control over pressures in the network, but may be less reliable because of vulnerable communication lines.

In these examples the control variable is pressure in the network, but both PRVs and PCVs are frequently used to control water level in reservoirs and water towers. An example is shown in Fig. 3.82. The valve is controlled by water level in the reservoir. An additional sophistication is that the control unit of the PCV is sensitive not only to the precise level but to the trend of changes in level: if WL rises, it keeps the valve open up to a point near the reservoir's top; if WL falls, the valve remains closed until it reaches a pre-set low level. The effect of responses to a trend is seen in Fig. 3.82.

Time-controlled valves

A TCV is a less frequently applied device in WSS. It is used mostly for control of the transfer of water from one area to another. There are cases when a TCV regulates the deliveries from one water supply system to another at the opportune time of day. For instance, the TCV might be fully open during the night time, but almost closed (never completely!) during peak demand hours, when the supplying area needs all its resources for itself.

The behaviour of a TCV is described by a function showing variations of TCV opening with time, $S = f(t)$. This means that the local loss coefficient, ζ, is also a known function of time, assuming that the valve characteristic, $\zeta = f(S)$, is given by the manufacturer (or taken from literature data).

Flow-controlled valves

A FCV reacts according to changes of flow through itself – or in some other pipe within the system – in much the same way as a PCV (Fig. 3.83). It is a rather sophisticated device, consisting of

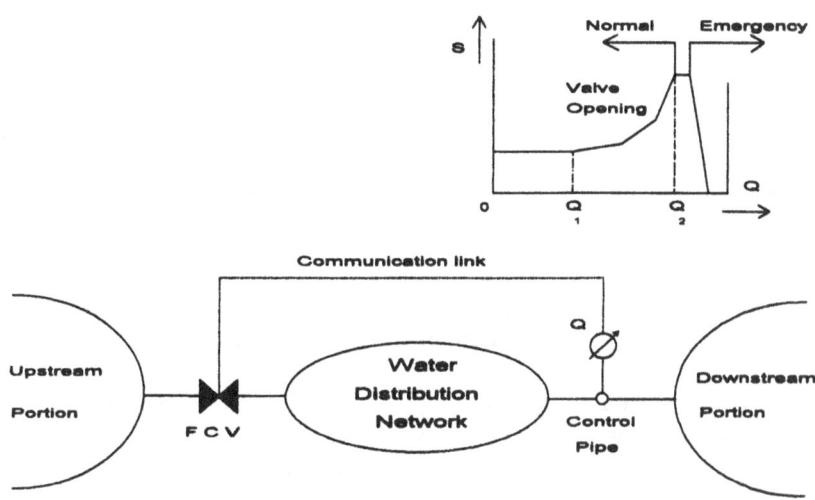

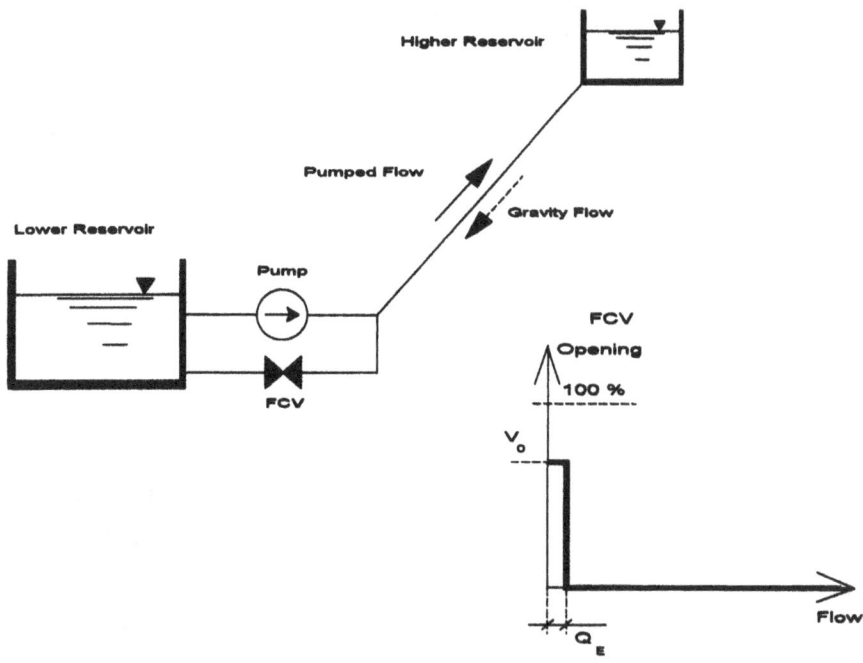

Figure 3.83 Flow-con-
trolled valve.

Figure 3.84 FCV opens
in emergencies.

a reliable flowmeter and a microprocessor that can be programmed by the user (see Lonsdale, 1984).

A case from real life is shown in Fig. 3.84. In this case, the
valve opens up with the flow – but only in the normal operating
range. If the actual flow exceeds a set value, the control unit

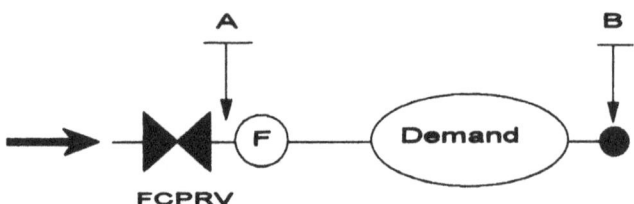

Figure 3.85 Flow-compensated valve.

automatically closes down the valve – because the increased flow was certainly caused by an accident, a burst pipe in the downstream region, for instance.

Flow-compensated pressure-reducing valves

A relatively recent innovation is the flow-compensated pressure-reducing valve (FCPRV) (Bessey, 1985, 1990), which has become very popular with water companies for pressure management schemes. The standard valve is now equipped with a flowmeter (Venturi type or other), which sends the signal to the PRV to increase the set pressure proportionally to the flow. The idea is to compensate for head losses between this point ('A' in Fig. 3.85) and a far-off point within the network ('B' in Fig. 3.85), so the service pressure *there* will be almost constant (Fig. 3.86).

The PRV has to be tuned by a person with local knowledge; however, if this is well done, the night consumption can fall dramatically, as reported by several companies (Lonsdale, 1984; Bessey, 1990).

Figure 3.86 FCV in operation.

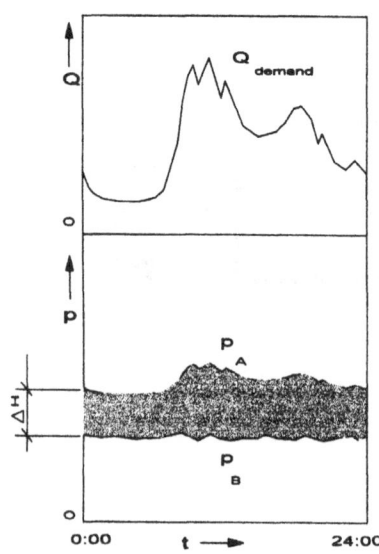

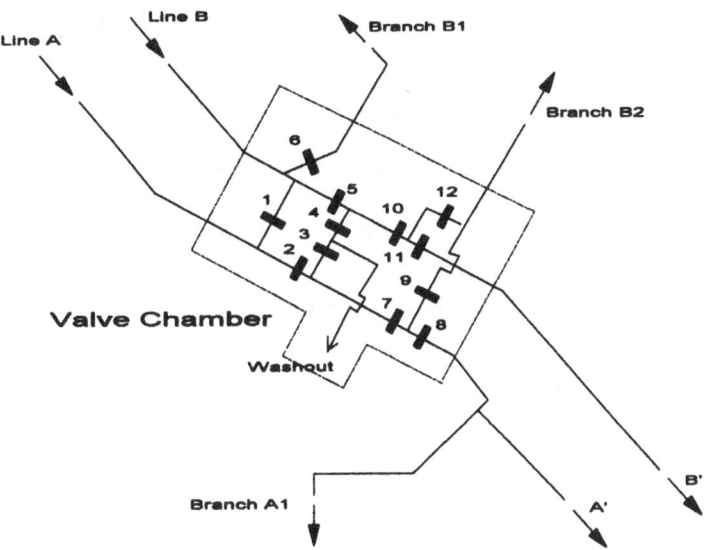

Figure 3.87 Modelling a valve chamber.

Modelling of multiple valve chambers

In real life the valves are not alone but bunched in clusters as shown in Fig. 3.87. There will be many valves, mostly of simple gate type for on–off operation and perhaps one sophisticated and expensive PRV. To make a full model – with all these pipes and valves – would be very wasteful, so the first task is to acquire reliable information on which valves are permanently closed and which are used from time to time for redirection of

Figure 3.88 Different options.

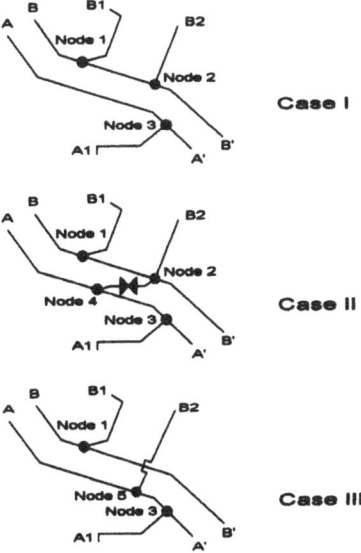

flow. Do not forget that such changes are few and far between, and view any contrary statement with a healthy dose of scepticism! The model should retain only the permanently open pipes. Some examples are shown in Fig. 3.88 and listed below:

- Case I – valves 1, 3, 4 and 9 closed, others open.
- Case II – valves 1, 3 and 4 closed, valve 4 could be open in emergencies, others open.
- Case III – valves 1, 3, 4 and 12 closed, others open.

Note how many other combinations are possible!

4 Modelling of demand

4.1 Introduction

A water company abstracts water from rivers, lakes, impounding reservoirs or aquifers, treats it to a particular standard, normally based on World Health Organisation (WHO) recommendations, and delivers it to individual users. The company makes strenuous efforts to do this without interruption, although in special circumstances they may place temporary restrictions on their customers.

This apparently simple task is really quite complex and incurs significant risk and cost. First, the company must have access to sources that are adequate to meet demand in all circumstances. Secondly, the distribution system must have sufficient capacity to supply the system in normal and abnormal conditions. Thirdly, the staff must be competent, well trained and adequately equipped. Last, the management should be providing support through planning, design and all the other activities necessary to maintain the service.

Fundamental to any planning, design or management activity is a sound procedure for recording historic demand data and forecasting short-term and long-term demand. The time horizon for planning and design may be several years (or even decades), whilst operational management deals with everyday problems – a few months at most. The following text is oriented towards *operational* problems. Further information on analysis and forecasting for the long term may be found in papers by Barrufet (1985), Prasifka (1988), Quevedo and Cembrano (1986) and Bessey and Garrett (1994); also in textbooks like Twort *et al.* (1994).

4.2 Definitions and terminology

Demand in any particular area and period comprises consumption, public use and losses.

Consumption is water delivered to meet customers' needs. It is usually metered to individual customers to provide data for billing (although other bases of charging exist). Generally, a water company seldom enquires how the water is being used, although the details of plumbing installations within customers' premises may be subject to technical regulations, one object of which is usually the minimisation of waste. The regulatory framework may also allow the water company to take corrective measures if cases of waste or abuse of water are identified. Concerns about the environmental impact of source development or the high cost of new source works (e.g. desalination plants) may compel water companies to take an increasingly active role in reducing losses through misuse or profligate use of water by individual customers.

Public use includes street washing, watering of parks, fountains, public taps, flushing of mains, etc., plus emergencies like fire fighting. These quantities are not normally measured. In some areas, water for some public uses is delivered through an entirely separate system, and a notable example of such systems is in Paris.

Losses in this context are those quantities of water which were delivered to the distribution system, but lost through seepage and evaporation from reservoirs, leakage from mains and service pipes, etc. Losses may be caused by inadequate control over the system, e.g. spilling of water from reservoirs or leaking valves, or by unauthorised use of fire hydrants, etc.

Water companies use the following terminology:

- *Water production* – quantities of water treated and delivered to the distribution system.
- *Water sales* – quantities billable to the users (either metered or on some other basis).
- *Unaccounted-for water* (UAW) – the difference between those two quantities.

UAW consists of 'authorised' and 'unauthorised' water use (after Prasifka, 1988). Authorised use includes:

- unmetered use from fire hydrants (fire fighting, mains flushing, construction works, street washing, sewer cleaning, etc.),
- unmetered connections (public buildings, schools, nurseries, cemeteries, parks, etc.),
- unavoidable losses (evaporation, for instance).

Unauthorised use comprises:

- leakage from water mains and pipes,
- illegal use (theft, either directly or by meter tampering, etc.),
- inadequate system control (overflowing reservoirs or malfunctioning valves),
- inaccurate water meters (which under-register with age or fail to register low flows),
- incorrect meter reading or billing.

For mathematical modelling purposes, it is easier to use the initial division because it splits total demand into just three groups, each with special characteristics:

- *Consumption* (measured or not) is a predictable variable, based upon people's habits, living standards and technology, which do not change quickly or erratically, and is influenced by known factors such as weather conditions, population growth, tourism, etc.
- *Public use* is just a fraction of total demand and can also be predicted to a great extent, even if it follows a different pattern to consumption.
- *Losses* remain an unpredictable quantity, and are estimated on a few facts and some hypotheses.

The analysis of consumption, public use and losses is a task that should be completed before one starts to model the whole water supply system. It has an important intrinsic value as input for almost all planning, design and management activities in the industry – besides providing basic data for the mathematical model.

Prediction of water demand is partly a *deterministic* and partly a *stochastic* process. Some factors keep demand within a certain range, and others cause random or chance fluctuations. The tasks of analysis are therefore:

- To *identify* deterministic factors (e.g. population changes or seasonal weather changes).
- To *quantify* deterministic factors – creating simple formulae to compute demand once the value of these factors is known (e.g. monthly factors over a 12 month cycle).
- To *estimate* the probable deviations from the forecast, to make appropriate allowances for the effects of random factors.

Deterministic factors are quite numerous (De Moyer *et al.*, 1973; Rao and Don, 1977), for instance:

- long-term, general trend (usually rising, as a result of increasing population, improving living standards, introduction of water-borne sewerage systems, etc.);
- difference between working days and weekend days;
- seasonal effects;
- diurnal variations;
- weather (air temperature, rainfall, humidity, cloudiness);
- service pressure;
- national holidays, major sporting events, etc.

Local circumstances are sometimes important (e.g. watering of gardens during the growing season, which may coincide with low rainfall – important in rural and suburban areas but perhaps less important in an urban area).

The probable effect of some of these factors is illustrated below, using cases from practice.

4.3 Analysis of demand

Demand needs to be analysed at several levels, from that of individual customers to total demand of the company area. The analyses needed for demand forecasting are used for both 'operational management' and 'planning and design'. The simplest statistical tools are adequate for the task, given the usual accuracy of such data. The problems lie elsewhere, primarily in establishing the mechanisms of collecting and retrieving reliable data and in designing reporting structures. It is not enough to make forecasts for the system as a whole, and demand data of all the 'sinks' generate a plethora of data. Some order needs to be introduced by separating the whole into hierarchical units, taking account of existing distribution systems, their zones, appropriate reporting intervals, communal and political boundaries, etc. It is not possible to establish general rules on how this should be done, given the diversity of water supply systems. Figure 4.1 depicts a fairly general case that might provide some guidance.

Figure 4.1 Hierarchy in water supply.

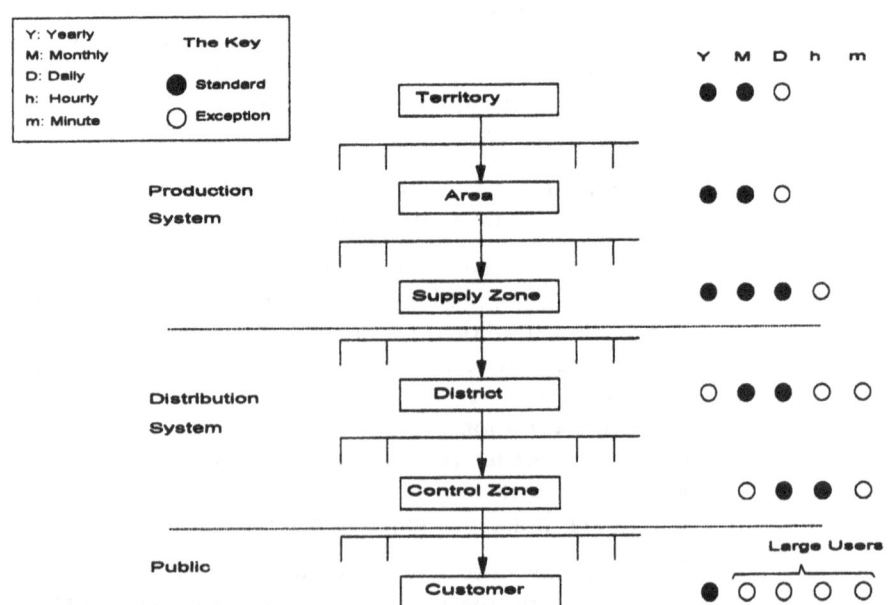

In the general case depicted in Fig. 4.1, the whole region or territory served by the company is the top 'layer' of the system. Data are archived at annual and monthly intervals (full circles) but daily demands aggregated for the whole territory are only held at these 'high' levels of the data system on a short-term archive. Similar records are held for each of the areas (say three or four) into which the territory is divided at the next level in the hierarchy. At the next level, the supply zone, it is suggested that records of demand are held on a permanent archive for annual, monthly and daily intervals, and hourly records on a short-term archive. The short- and long-term archive records and suggested intervals are shown for smaller and smaller groups down to individual users. At that level, it will be useful for a number of forecasting and planning purposes to take sample records for intervals as frequent as one minute, to build up a library or archive of demand patterns for a range of consumer types, various levels and types of domestic user, various industry types and so on. Only annual figures may need to be on a permanent archive to secure evidence of individual customers' demands and perhaps to develop information on long-term trends in consumption of the various user types.

The number of levels on which the data collection and archiving structure is based will be determined by a number of factors, not the least of which is the size of the whole company. However, the type and number of sources and/or treatment plants, the deployment of the customers (Is population concentrated in large towns or evenly spread through the area?), etc., all affect the number of levels that are appropriate. Those factors will also affect the frequency of data held in short- and long-term archives.

Note that all important values of flows and storage changes should be monitored, so one can always compute the actual demand and discover leaks, at least the big ones. In modern water supply systems, all these instruments are usually incorporated into telemetry.

The lowest level in the hierarchy of data collection is the demand of individual customers. It would perhaps be sensible to invert the hierarchy – the customer is clearly the most important element in the whole system and ultimately has the dubious pleasure of paying for it all! There is, though, an important reason why customer records should be regarded as the top of the heap of data. The reason is that other levels of data are, in many ways, derived from the customer demand data.

The customer is connected to the distribution system through one or more service connections. Good practice (and accurate billing) requires a meter on each connection. The water company then regularly reads actual consumption of water and charges accordingly. The majority of supply systems, worldwide, aim to meter all their customers, but a number of systems are only partially metered, notably in the United Kingdom, where, although most industrial, commercial and agricultural supplies are metered, only a minority of domestic properties are charged by volume. The majority of domestic customers are charged on the basis of a property valuation system, which itself is swiftly becoming redundant. The proportion of metered to unmetered customers has a definite influence on overall demand – which needs to be studied in each individual system. It is believed that metered customers use significantly less water than unmetered ones, but the difference cannot be quantified in advance, and should be investigated in each individual case.

The subdivision of any given area can only be done by the staff of the water company – they know the local circumstances best. It should not be forgotten that it is not enough just to *create* the system, it is also necessary to *update* and *maintain* this system over time. Note that for each territorial unit the following data are needed to compute water demand: inflow, storage and transfers. Demand can be computed from this information as:

demand $=$ inflow $-$ transfer $\pm$ storage variations

Note also that all data have to be collected continuously and with similar accuracy – or the whole effort is pointless.

There are two basic approaches in constructing the demand data structure:

- From the bottom – building demand by summing individual needs and adding likely losses.
- From the top – by monitoring the total demand at all levels.

The first approach has, historically, been favoured by planners and designers, the second by operational management. In practice, a combined approach should be adopted by planners and designers. They are usually looking at forecasting in the longer term, and need to absorb information from a wider perspective than operations staff, who are concerned with the short term. For example, planners need to monitor trends to determine the probable growth points in the 'territory'. The redevelopment of a low-water-using industrial area into a residential area might well be of great importance in 10 or 20 years. Similarly, developments in water usage might be of long-term interest but not of immediate concern. In our view, planners and designers should continue to build up their longer-term forecasts on the basis of 'bottom-up' component building, but modulate their results by keeping a note of current demand trends as revealed in 'top-down' demand monitoring.

Data about demand — In an ideal situation and using the example in Fig. 4.1, water demand in one supply zone will equal the sum of demands in its districts, the value for an area equals the sum of demands in its supply zones, etc. Naturally, in real life this cannot always be achieved – some areas are not entirely 'covered' by the instruments, while other devices become redundant (for example, by rezoning within the system), so the picture is not complete. Nevertheless, it can still give reasonably accurate information about the level of total demand – and the continuing development of new technology should produce increasing accuracy, hopefully with decreasing real cost. Figure 4.2 shows variations of monthly demand in an area in southern England. Note:

- Steady rise of demand from 1981 to 1988/89, then a temporary decline until 1993 and a new rising trend afterwards. The changes coincide with a recession in the UK from 1989 to 1993.
- Seasonal variations of demand. Summer peaks are clearly visible every year, although in some cases are less pronounced than in others (e.g. compare 1983 peak with 1987).
- Stochastic component – local deviations from the main trend.

The same pattern is repeated at lower levels, but less clearly because local conditions mar the picture. Figure 4.3 shows data

Figure 4.2 Historical data for one area: total water production, 1981–1995.

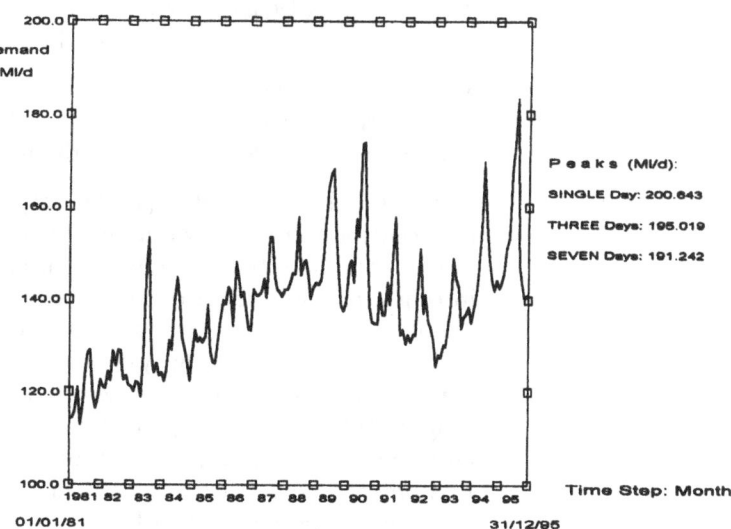

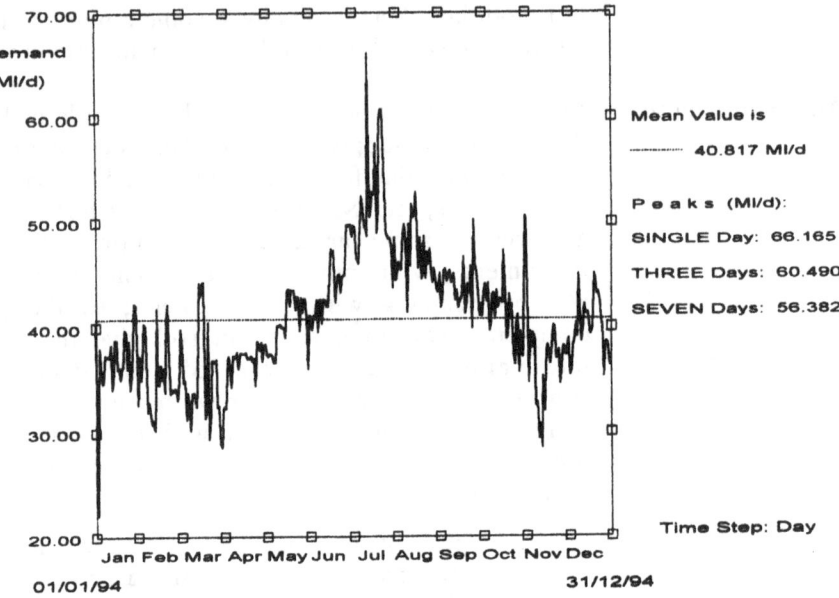

Figure 4.3 Historical data for one supply zone: seasonal changes of demand, 1994.

for a supply zone of the same area throughout 1994. Note seasonal effects – higher-than-average demand during summer months (July–August), but also local peaks during winter and late autumn.

With few exceptions, demand at higher levels of the hierarchy is rarely monitored in real time; the data are often collected at the end of the day or later. The time step is a day or even a week.

The demand in smaller units – such as a district – should be monitored more closely. Figure 4.4 shows data for an industrial district. This district is covered by two electromagnetic flow-meters and data are taken at 15 min intervals. Note:

- regularity of daily patterns;
- similarity between working and weekend days;
- relatively high demand during the nights.

Figure 4.5 shows one day (11 January 1995) in greater detail; the contribution of individual flowmeters is shown in Fig. 4.6.

The sheer number of individual users makes constant monitoring a prohibitively expensive exercise. However, some data about all customers are already available in every water company from the billing system.

Customers are charged for the water that they use; water is no longer cheap, so the company must explain the bill, not only to customers, but to higher authorities if bills are disputed. Customers' meters are installed in the pipe connecting their premises to the water distribution network. This instrument is

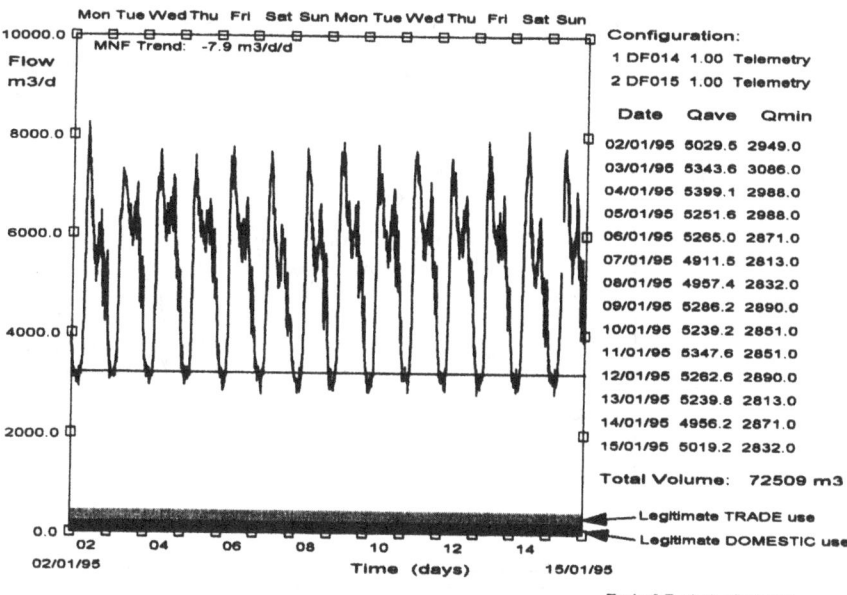

Figure 4.4 Structure of demand in one industrial district.

Figure 4.5 Daily demand in a district.

usually an inexpensive device, showing the quantity of water only locally (this situation may change dramatically with the introduction of remote reading systems). The company reads the meter at regular intervals, usually once a month, or quarterly or twice a year, and readings are transferred to the billing system for computation of charges. These data are processed on a

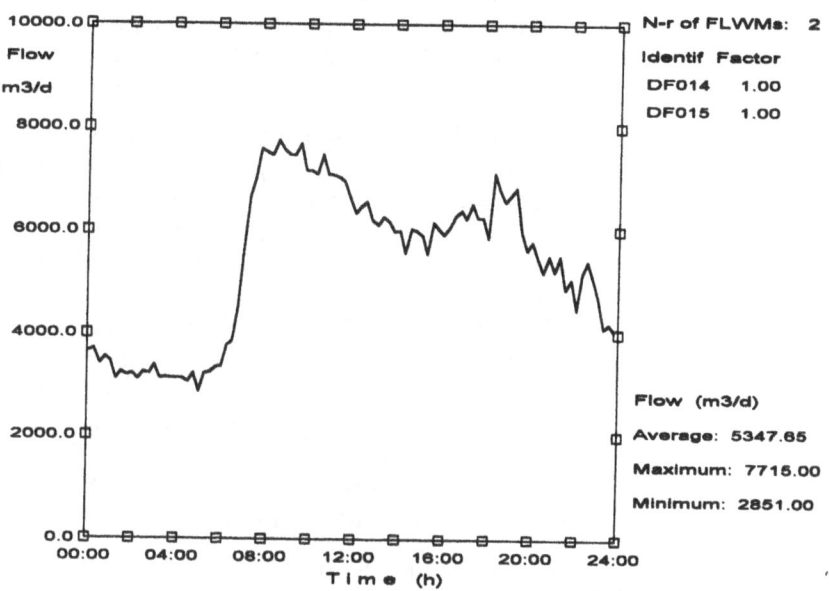

Modelling of demand

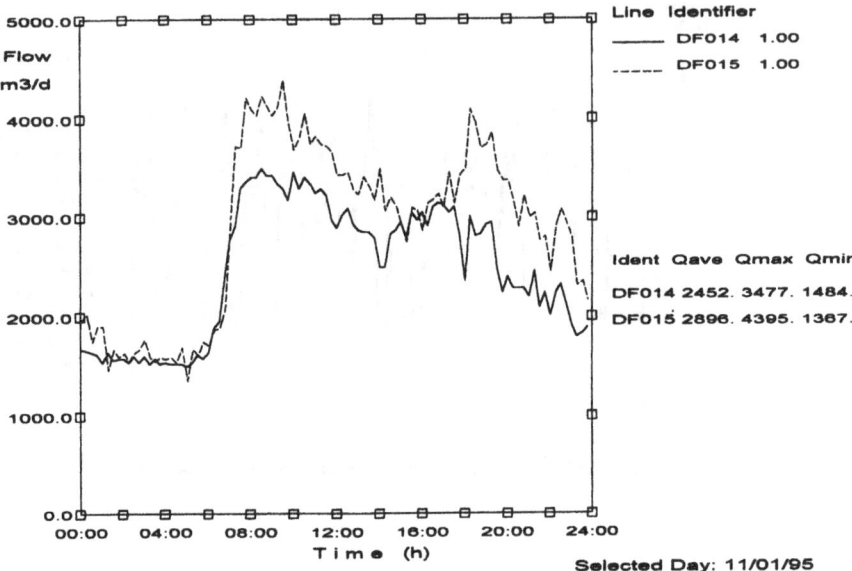

Figure 4.6 Individual contributions.

computer and could easily be used in other applications – such as modelling.

One example – for a large engineering company – is shown in Fig. 4.7. Note:

- Irregular variations – perhaps genuine, indicating changes in activity within the company, but could be due to poor readings or errors in data processing.
- Seasonal effects – increased activity (presumably) just before the summer period, or before the end of year.

One should be careful not to take such deliberations too far on the basis that the data may be of poor quality. However, the value of this information is enormous. The average consumption of all customers can be established with a high degree of confidence; also the larger customers could be identified and singled out for detailed analysis.

The problem of how to relate individual customers to the distribution network was discussed in the previous chapter. The conclusion is that a separate system must be devised for this task, the requirements being low cost, permanence, easy amendment to match changes in the distribution network and straightforward checking. This system should (preferably) be based on a public addressing system such as:

- Postcodes (in UK) or ZIP Code (in USA),
- street names and house numbers,

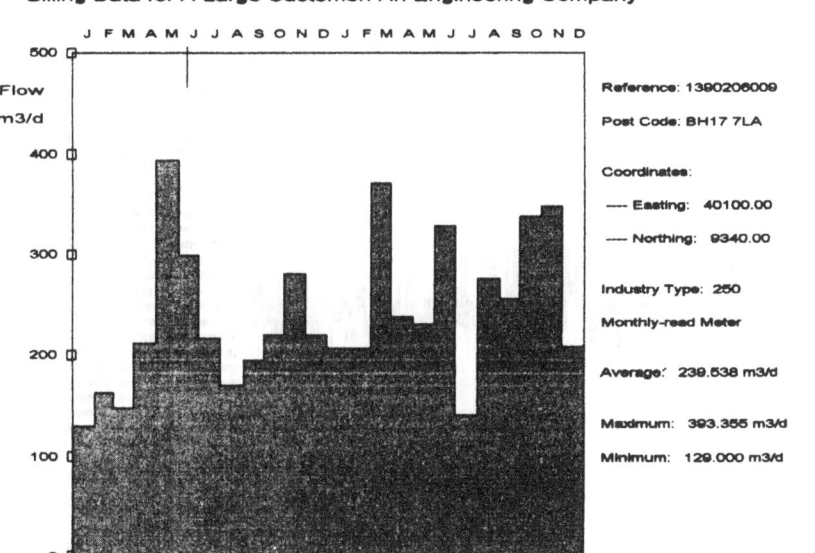

Figure 4.7 Billing a large customer.

• coordinates of individual houses – if available.

None of the systems is ideal or cheap, but some link to the system is essential.

As the cost of instruments and communications falls, many water companies can afford to monitor selected large users more closely – either by installing data-loggers on their water meters or even by extending the telemetry system. One case is shown in Fig. 4.8.

This diagram shows the variation of water consumption of a large fossil-fuel power plant. Note the almost constant demand, oscillating within a relatively narrow band. Closer examination reveals that changes occur at random during the day, without any recognisable pattern.

The data for any other month are similar to Fig. 4.8, except when important maintenance works are carried out.

Other large customers may use water in different ways. For instance, a large food-processing factory reduces its activity over weekends, with consequent reduction of water consumption, as shown in Fig. 4.9. Note the similarity between working days (Monday to Friday), with occasional high demands. In other industrial enterprises, the activity may stop altogether at weekends and no water is taken (Fig. 4.10).

This fact raises doubts about the weekend consumption in Fig. 4.9. Is it genuine consumption or leakage within the plant?

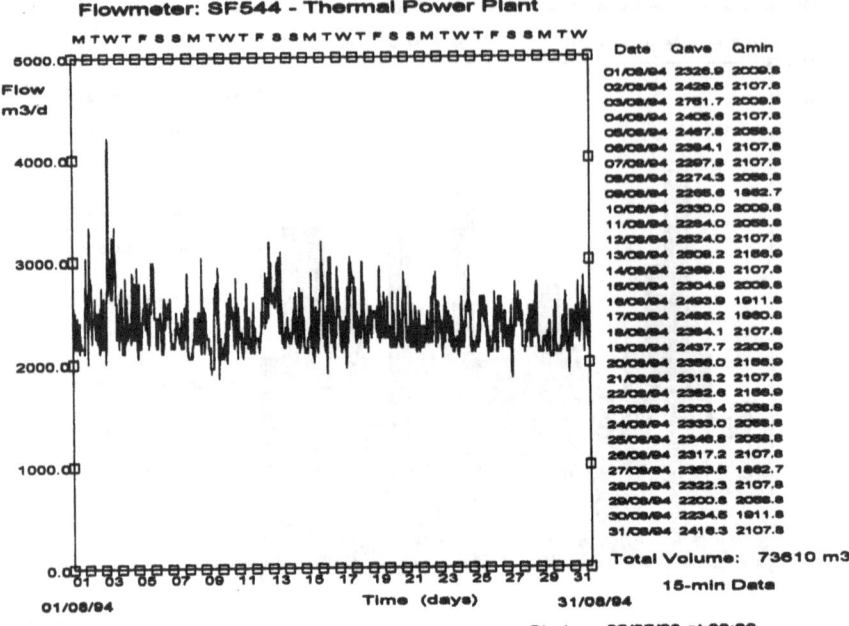

Figure 4.8 Thermal
power plant.

Figure 4.9 Chocolate
factory.

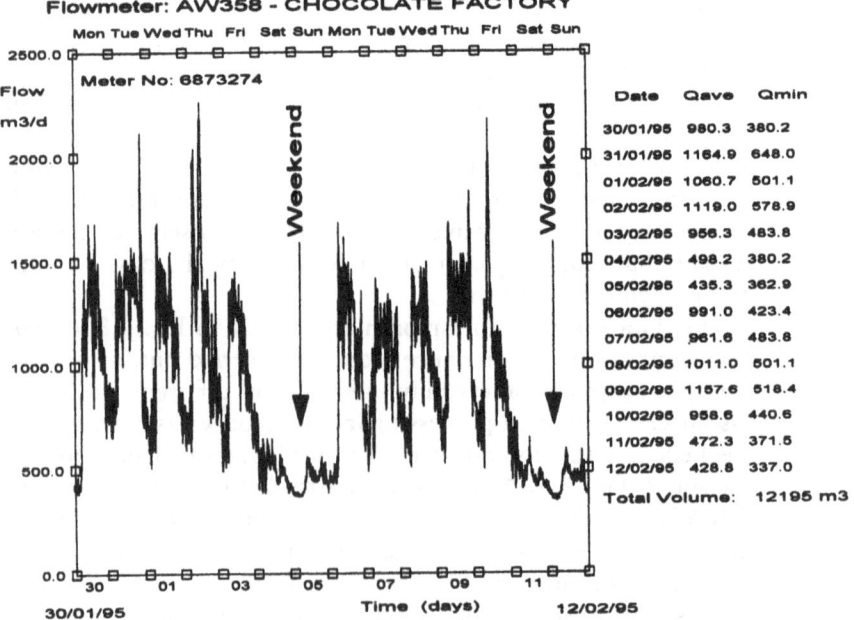

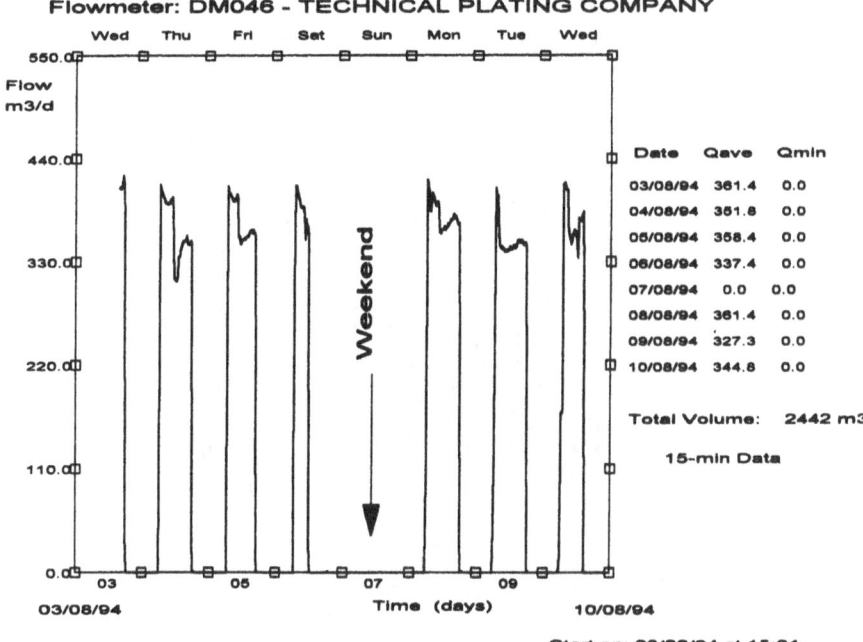

Figure 4.10 Industrial plant.

The answer to this intriguing question cannot be found without a closer examination of the distribution system in the food-processing plant – which may not be an easy task.

Other customers may need more water over weekends – for example hotels, holiday camps, the leisure industry. An example is shown in Fig. 4.11. Note the remarkable similarity between, say, Saturday and other days.

Finally, there are customers with regular consumption patterns interrupted by sudden bursts of high demand, dictated by their activity – and quite beyond the knowledge and control of any water company. An example is shown in Fig. 4.12.

Note the randomness of high demand and that it can occur during the weekend as well. Note also the lowest level of consumption – it might show the leakage level within the premises, hence the extent of possible savings if the internal pipework were mended. Another point, related to the high and random nature of demand, is that, if at some time increases in mains capacity are needed in the area, a cheap option might be to provide the customer with an individual small service reservoir rather than install very expensive mains.

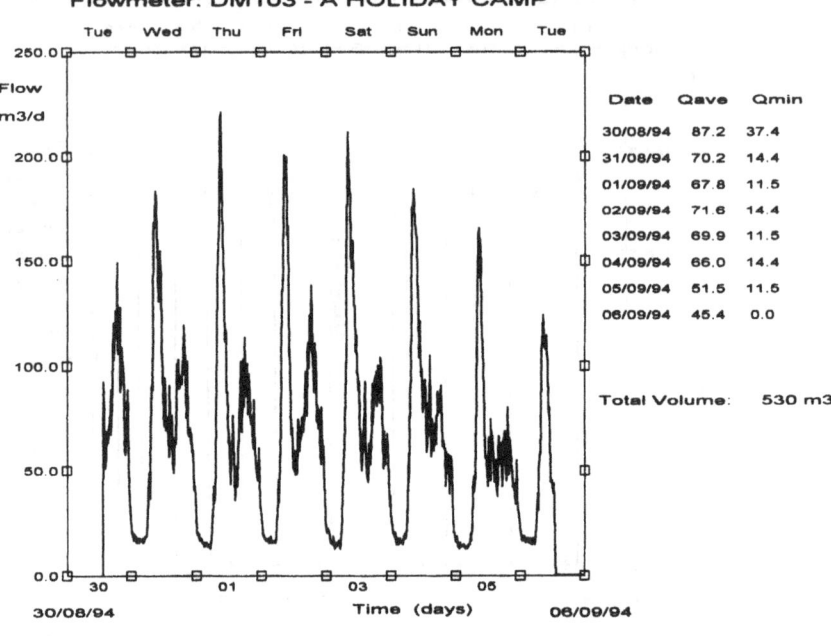

Flowmeter: DM103 - A HOLIDAY CAMP

Date	Qave	Qmin
30/08/94	87.2	37.4
31/08/94	70.2	14.4
01/09/94	67.8	11.5
02/09/94	71.6	14.4
03/09/94	69.9	11.5
04/09/94	66.0	14.4
05/09/94	51.5	11.5
06/09/94	45.4	0.0

Total Volume: 530 m3

Start on: 30/08/94 at 00:00

Figure 4.11 Leisure camp.

4.4 Capturing, holding and converting data to information

The examples shown earlier show that careful analysis of demand is worth the effort. The procedure is shown – schematically and much simplified – in Fig. 4.13.

Once data are available, analysis can begin, but looking for what? Briefly, one should try to spot anomalies – similarities and differences in the endless flow of data. Basic training for this job is necessary to extract the full information. Remember that there may be several alternative explanations for everything. For instance, a sudden jump in demand might be caused by:

- genuine increase in demand, perhaps unusual but real (and legitimate);
- rezoning of the distribution network;
- false data due to the malfunctioning of an instrument or transmission line.

A trained person can usually draw the correct conclusion as to cause (some examples were discussed in Chapter 2) and act accordingly, while an untrained person can overlook valuable information or draw incorrect conclusions. The training need is obvious, but should be blended with local knowledge. Some

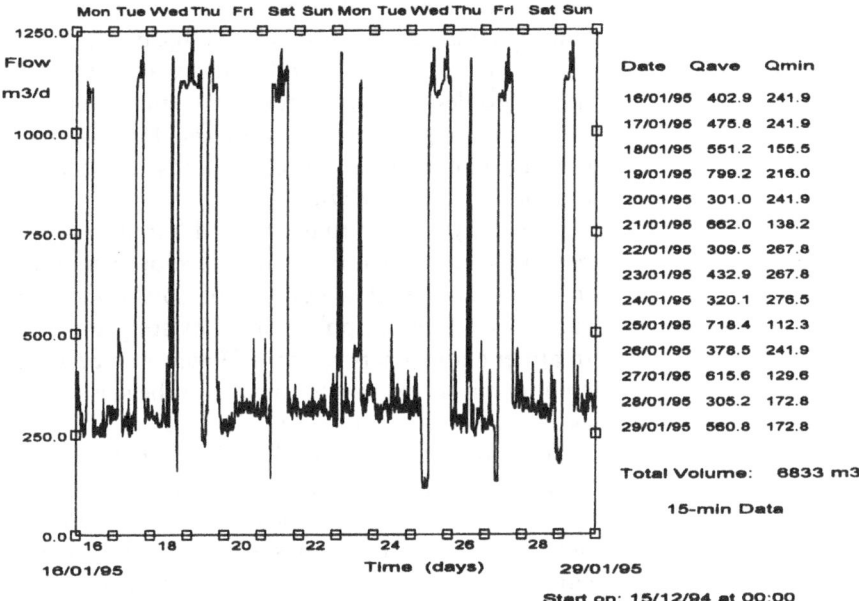

Flowmeter: SW176 - A Large Industrial Customer

Date	Qave	Qmin
16/01/95	402.9	241.9
17/01/95	475.8	241.9
18/01/95	551.2	155.5
19/01/95	799.2	216.0
20/01/95	301.0	241.9
21/01/95	662.0	138.2
22/01/95	309.5	267.8
23/01/95	432.9	267.8
24/01/95	320.1	276.5
25/01/95	718.4	112.3
26/01/95	378.5	241.9
27/01/95	615.6	129.6
28/01/95	305.2	172.8
29/01/95	560.8	172.8

Total Volume: 6833 m3

15-min Data

Start on: 15/12/94 at 00:00

Figure 4.12 Irregular daily patterns.

Figure 4.13 Analysing the data.

general guidance on how to analyse data and interpret anomalies can be provided as shown below; see also Prasifka (1988).

A final remark should be made: demand forecasting can be compared to weather forecasting – occasionally it may be badly wrong, but on the whole it provides a valuable guide for planning and operational management.

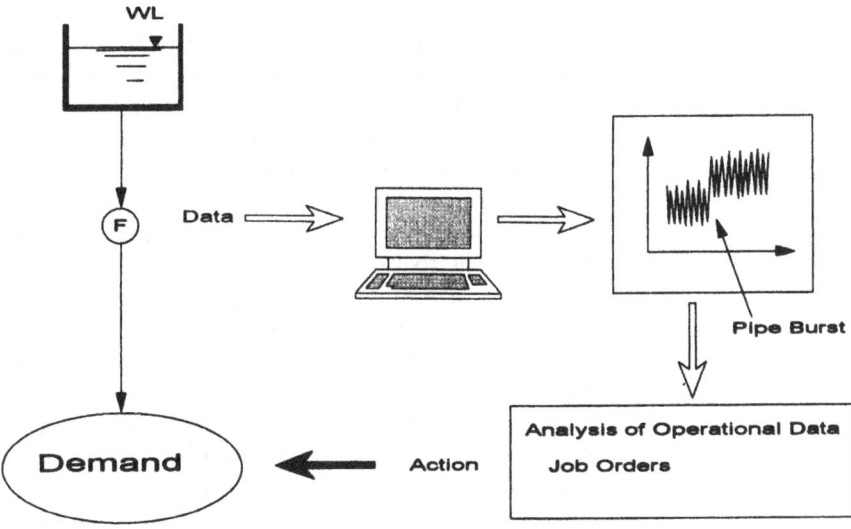

Long-term trends

Most communities have plans for physical and social development. The master plan provides a basis for specific plans on employment, traffic, housing, education, tourism and other activities. Just a few decades ago water supply was, generally, not a problem. It was a relatively small charge compared with other municipal services, and the local water company was usually able to meet its obligations and maintain quite large margins of capacity with ease.

Nowadays the situation is different – water supply has become a factor that may limit development of certain areas due to one or more of a number of constraints (scarcity of clean sources to develop, pollution of existing sources, high prices of treatment and energy, etc.). Water has become an important component in any master plan – on the same level of importance as, say, public transport. The price of new infrastructure projects is very high, and there is a natural tendency to optimise both timing and size of these expensive additions to existing systems. There are two possible traps: either to postpone the new projects until serious shortage become unacceptable for the community, or to build oversized capacities to meet a demand that may never materialise.

Planning usually starts from historical records, trying to identify and quantify the important factors, and from there, extrapolating a forecast of needs. All the planners' hard work might well come to nought if some economic or political crisis or technological change occurs, causing a decline in the demand for water. The planner then has the difficult task of predicting when the decline will stop, when demand will rise again and at what rate, etc.

Computerised databases and statistical packages make the task of ordering historical data and extrapolation easier and at least better documented. However, the reliability of any plan is not much better than before, unfortunately. For a planner, the only certain thing about the future is that it is uncertain (see, for instance, Castensson, 1989).

Total demand forecasting, based on extrapolated historical records, modified by overall social, economic and development plans, needs to be modified also by another important parameter: the consumption of water per capita per day, sometimes called *specific consumption* (in some countries consumption per property may be used). It can be either total consumption of water (including industrial and other uses) – per capita consumption for all purposes – or just domestic use. We believe the latter form is more useful, used in conjunction with industry-specific measures of non-domestic consumption (e.g. per head of livestock in agricultural consumption or per bed space for

Figure 4.14 Specific consumption.

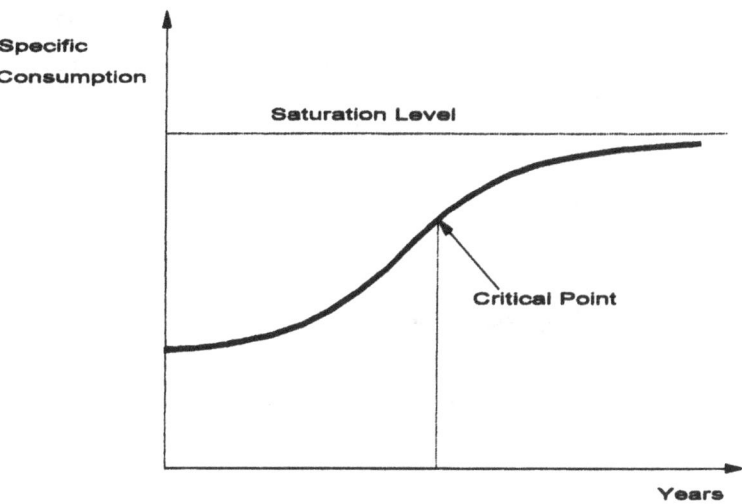

hotels). The value of domestic specific consumption is determined by local conditions such as:

- occupants per residential unit (house, flat, etc.),
- social status (rich/poor) and living standard,
- seasonal effects (always interrelated with other effects),
- quality of plumbing and leakage inside premises,
- metering and charging mechanisms.

Specific consumption has a tendency to rise over time, reflecting improving living standards. This is sometimes used to produce alarmist predictions of shortage of water resources, but in reality there must be a saturation level. There must be a limit to the number of washing machines or dishwashers used in the community – see Fig. 4.14.

The saturation level may also depend on local circumstances, but is a reality, and some effort should be made to determine the probable saturation level in each area under study, to enable better and more realistic planning of resources. An example is shown in Fig. 4.15 – the variation of specific demand in the United Kingdom since 1974.

The introduction of universal metering is certainly an important factor in reducing consumption and waste, although research has indicated that after a period of about 10 years increase in demand, possibly at a slower rate than before metering, resumes. Another benefit from full metering is tighter overall control – how else to spot excessive use in individual houses. The results of an investigation are shown in Fig. 4.16 (UK, 75 locations, six months). Note that:

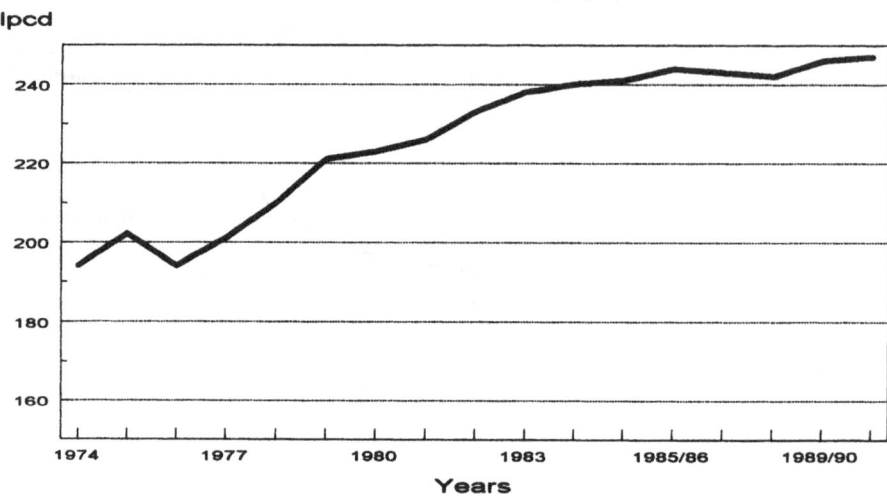

Specific Consumption in UK 1974 to 1990/91

Source: Waterfacts 1991

Figure 4.15 Situation in UK.

- The most likely consumption is about 156 litres per head per day.
- The average consumption is significantly higher at 220 litres per head per day.
- Some cases of high demand (up to 1000 litres per head per day) lift the average use to 220 litres per head per day.

Figure 4.16 Metered domestic demand.

The high users are of particular interest. Do they really use so much? It is possible, if, for example, they constantly top up a

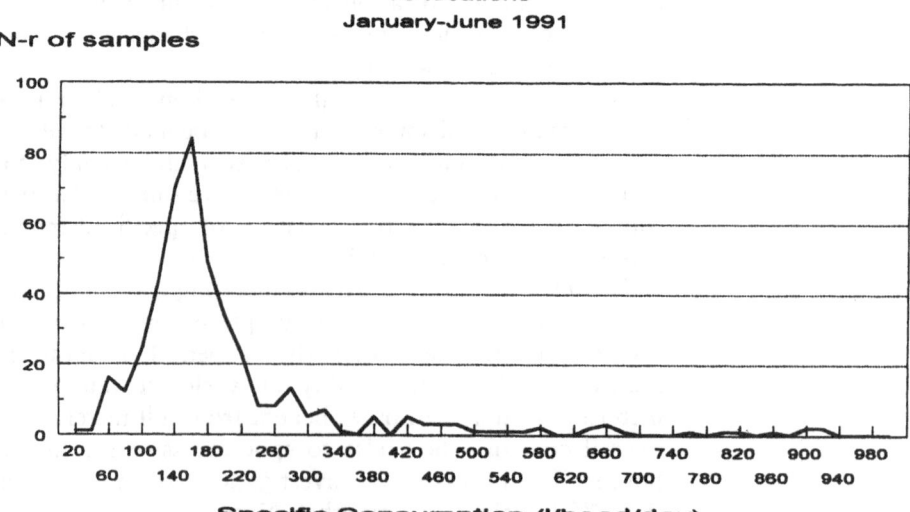

75 locations
January–June 1991

swimming pool rather than recirculate water through a treatment plant, but the excessive users might be persuaded to adopt more public-spirited habits, to the benefit of all. Perhaps the piping is in a poor state? Even if water is paid for, the water company still has an interest in suppressing wastage and misuse of water.

Seasonal effects

The seasons definitely influence water demand in both urban and rural areas. Demand normally rises during summer: more frequent bathing, watering of gardens, etc., all increase consumption. However, this assumption – and others of a similar kind – should be made with care. There may be other factors that influence seasonal demand, so each case needs to be examined. This is illustrated by a case (much simplified) shown in Fig. 4.17.

Curve 1 is for a large inland town. Note that there is no 'summer peak' but a 'summer trough' below the annual average; critical months are June and September – if the weather is hot and dry, the population is reduced because people leave to refresh somewhere else. Curve 2 is for a tourist area with many hotels, holiday camps and resorts of every kind. Note the summer peak (June–August) when the tourist season is at its highest – a normal graph!

Rainfall – or lack of it – may also have an influence on the actual demand, as illustrated in the next example for an area in southern England. Figure 4.18 shows monthly demand for 1983 when the summer was dry and hot (in relative terms); note the peak in July–August. The same data for 1991 are displayed in

Figure 4.17 Seasonal trends.

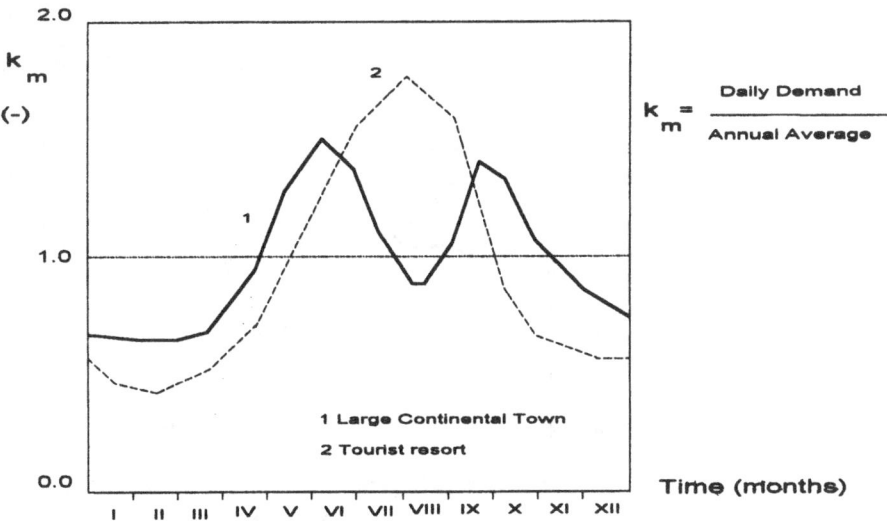

k_m = Daily Demand / Annual Average

1 Large Continental Town

2 Tourist resort

Time (months)

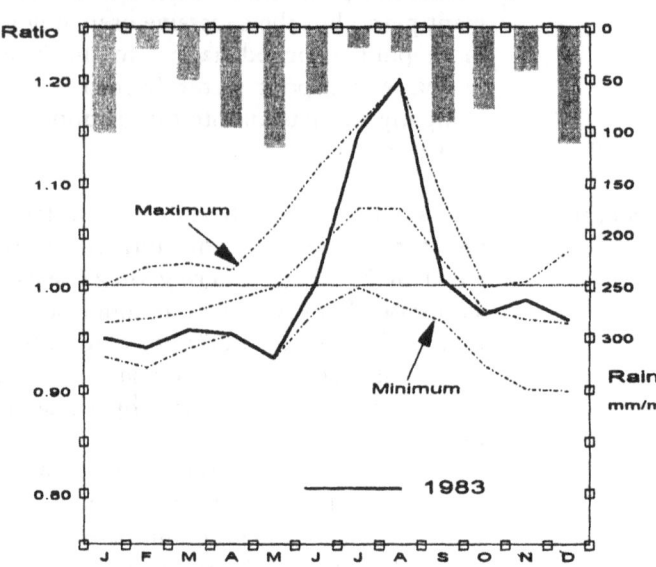

Figure 4.18 Dry and
hot summer.

Fig. 4.19, when the summer was different: colder and with more rainfall in the second half of the summer. The drop in demand is clearly visible.

The effect of rainfall is important during the growing season of horticultural and/or agricultural crops (roughly May to

Figure 4.19 Wet and
cool summer.

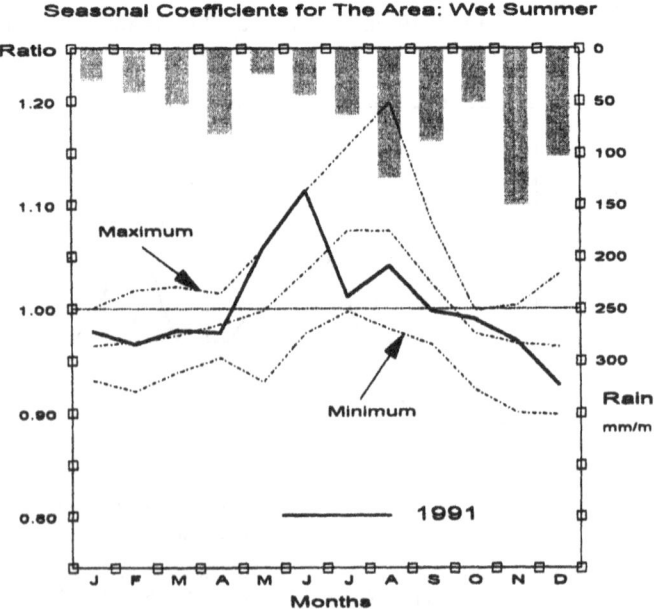

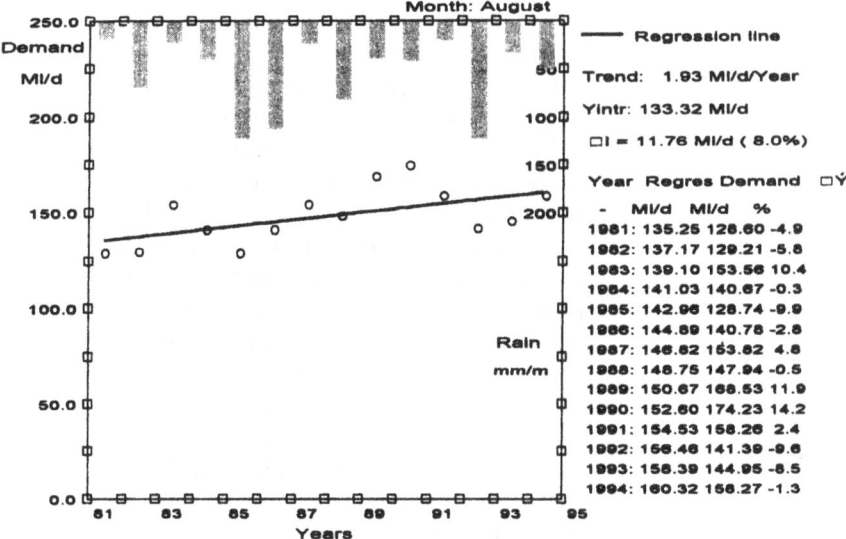

Figure 4.20 Monthly demand – August.

September in the UK). Data for August and February, respectively, support the point in Figs 4.20 and 4.21. There is a relationship between demand and rainfall in August (Fig. 4.20): during relatively dry weather (1983, 1989, 1990) demand exceeds the regression line, and is below it during years with a wet August (1985, 1986, 1991). No similar relationships exist in Fig. 4.21 – for February – as one can easily see.

Figure 4.21 Monthly demand – February.

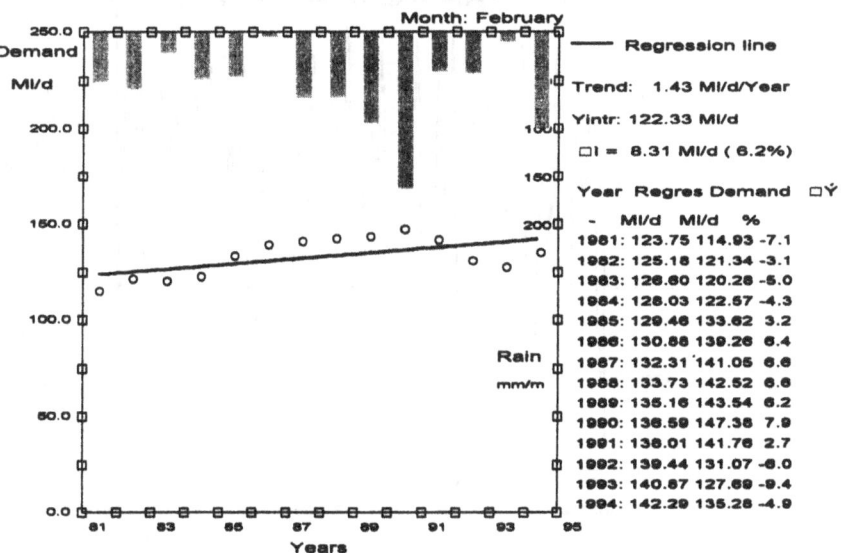

Quantifying the effect is difficult, as other factors distort the picture; in this case the overall trend in demand rose to 1990, then fell for three years, when the demand started to rise again. Therefore, 'mathematisation' of the problem could be dangerous – one should examine the evidence and then decide the probable range due to the effect.

Seasonal effects can be seen within a smaller area, at district level. Figures 4.22 (summer) and 4.23 (winter) are for a small town in the area. These are data collected at 15 min intervals. Consumption is obviously higher in summer, but the shapes of the daily diagrams are similar. Note the morning and evening peaks of demand, high night use (leakage?) and similarity between working and weekend days.

These effects are even more noticeable over a single year (Fig. 4.24). Note the seasonal rise and fall of maximum and average demand; also changes of minimum demand reflecting the success in reducing leakage in the area. There are indications that the minimum demand is also somewhat higher during summer months.

Working days versus weekend days

Figure 4.22 District demand in summer.

This factor may be important, but not as much as the other factors – at least, that is the authors' experience. The five working days of the week, from Monday to Friday in most Western countries, are slightly different from Saturday and Sunday, normally non-working days. Again, the difference is not

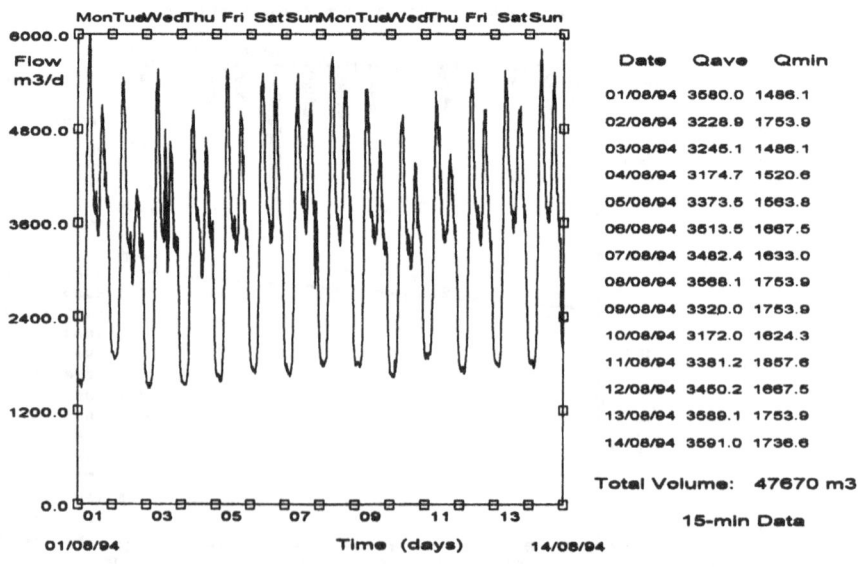

Data for District Meter: DW012

Date	Qave	Qmin
01/08/94	3580.0	1486.1
02/08/94	3228.9	1753.9
03/08/94	3245.1	1486.1
04/08/94	3174.7	1520.6
05/08/94	3373.5	1563.8
06/08/94	3513.5	1667.5
07/08/94	3482.4	1633.0
08/08/94	3568.1	1753.9
09/08/94	3320.0	1753.9
10/08/94	3172.0	1624.3
11/08/94	3381.2	1857.6
12/08/94	3450.2	1667.5
13/08/94	3589.1	1753.9
14/08/94	3591.0	1736.6

Total Volume: 47670 m3

15-min Data

01/08/94   Time (days)   14/08/94

Start on: 20/10/93 at 00:00

the same in a large inland town as in a resort area (Fig. 4.25). The demand during the weekend falls somewhat in inland urban areas, but increases in tourist areas. The reason is obvious, but the significance needs to be examined in each particular case.

In many towns, especially large ones, the difference between working and weekend days has almost disappeared. However, when analysing demand at the district level, this factor might still be important because large individual customers are relatively more important – and their activities could be affected by the weekly cycle, as the following examples demonstrate. The cases – all taken from practice – are:

- a food-processing industry (Fig. 4.26) – other data for the same customer were given as Fig. 4.09;
- a large hospital (Fig. 4.27);   ,
- a large bank (Fig. 4.28) – note the similarity between the week in January and the week in July – no seasonal effects here;
- a college (Fig. 4.29).

In all these cases, demand is drastically reduced over the weekend, to be increased again on the first working day. Note also that the patterns for working days are remarkably similar.

Figure 4.23 District demand in winter.

The demand on Saturdays and Sundays in all these cases is almost constant, albeit at different levels. This might be a

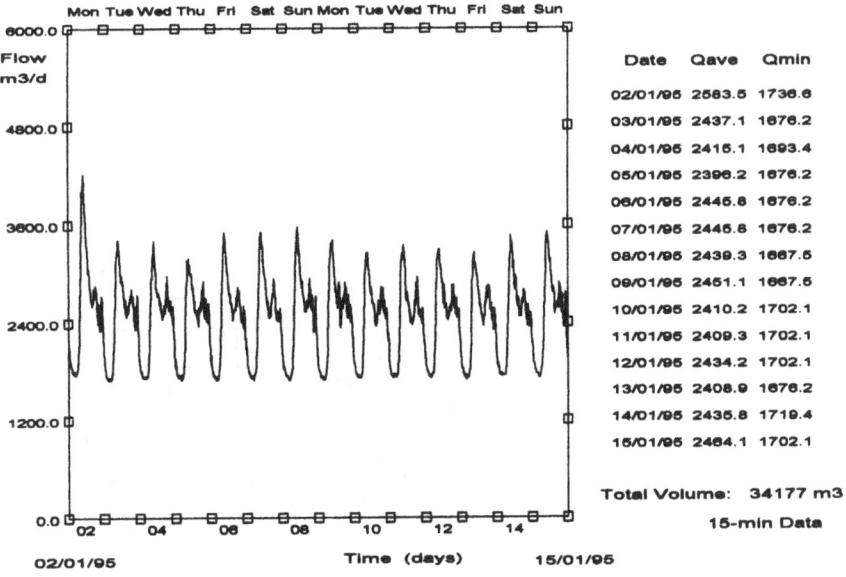

Data for District Meter: DW012

Date	Qave	Qmin
02/01/95	2583.5	1736.6
03/01/95	2437.1	1676.2
04/01/95	2416.1	1693.4
05/01/95	2396.2	1676.2
06/01/95	2445.8	1676.2
07/01/95	2445.8	1676.2
08/01/95	2439.3	1667.5
09/01/95	2451.1	1667.5
10/01/95	2410.2	1702.1
11/01/95	2409.3	1702.1
12/01/95	2434.2	1702.1
13/01/95	2408.9	1676.2
14/01/95	2435.8	1719.4
15/01/95	2464.1	1702.1

Total Volume: 34177 m3

15-min Data

Start on: 20/10/93 at 00:00

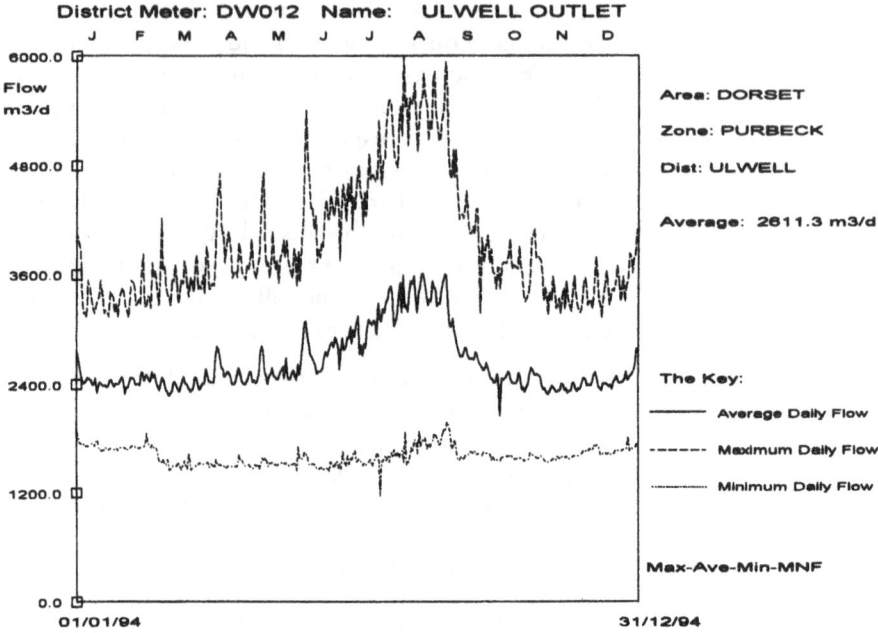

Figure 4.24 Average, maximum and minimum flows.

genuine demand, but it is more likely to be leakage in the customers' piping. This is worthy of further investigation by both water company and the customer.

It is interesting to note that the differences within the weekly cycle tend to disappear as the area under consideration

Figure 4.25 Effects of the weekend.

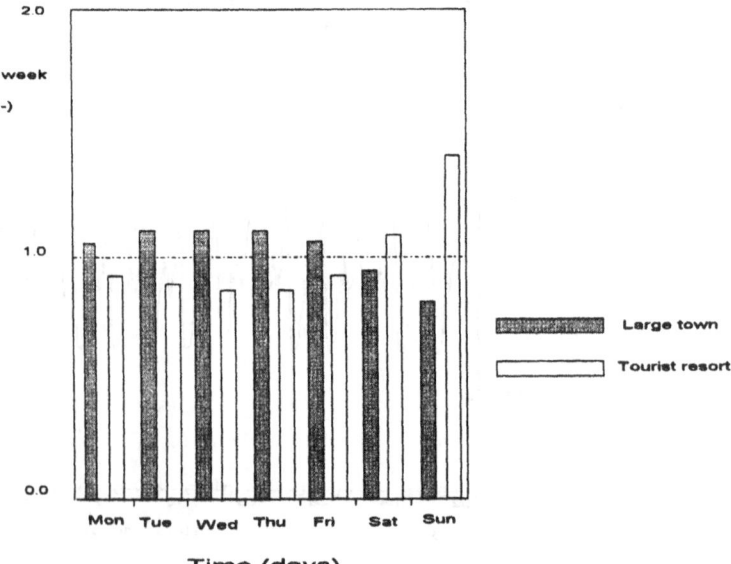

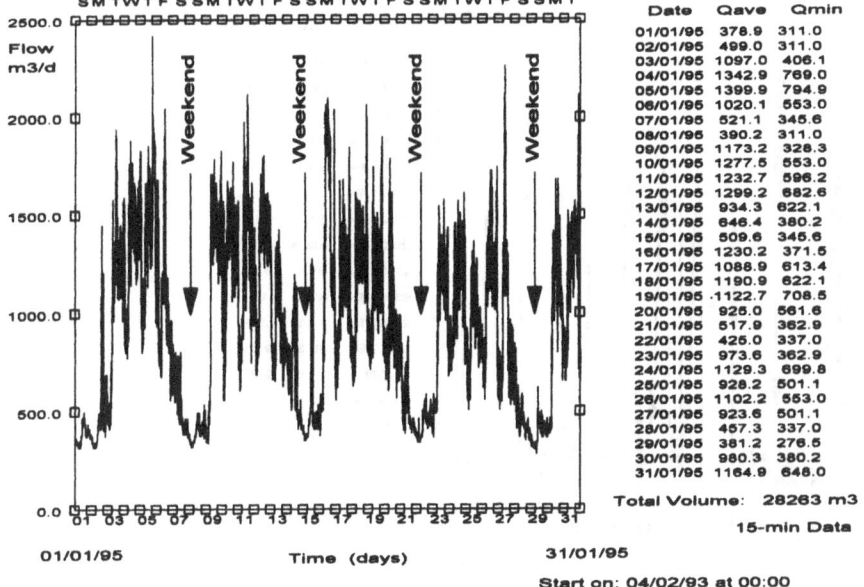

Figure 4.26 Weekend effects – a chocolate factory.

Figure 4.27 Weekend effects – a hospital.

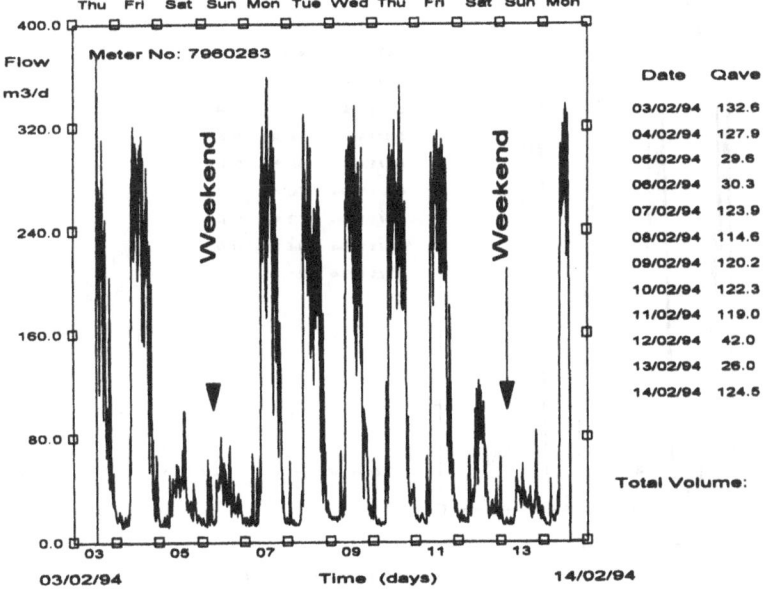

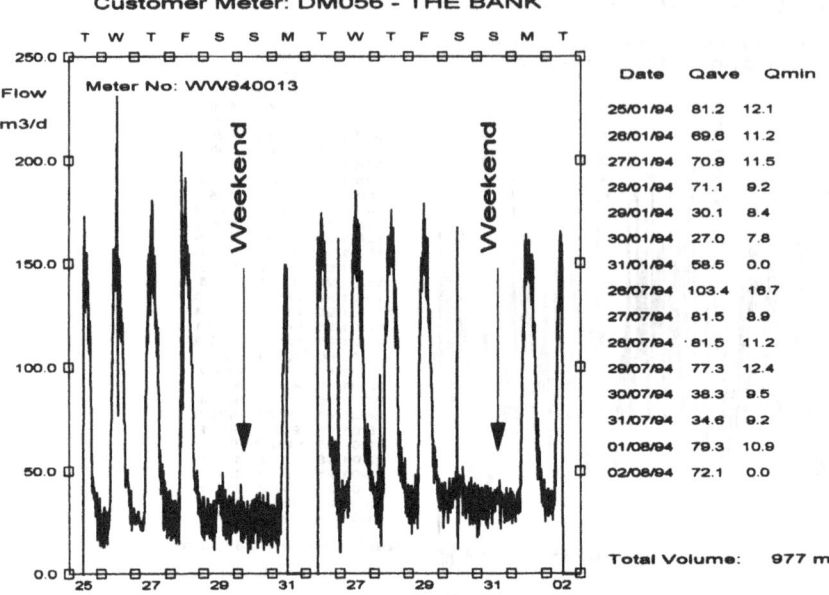

Figure 4.28 Weekend effects – a bank.

Figure 4.29 Weekend effect – a college.

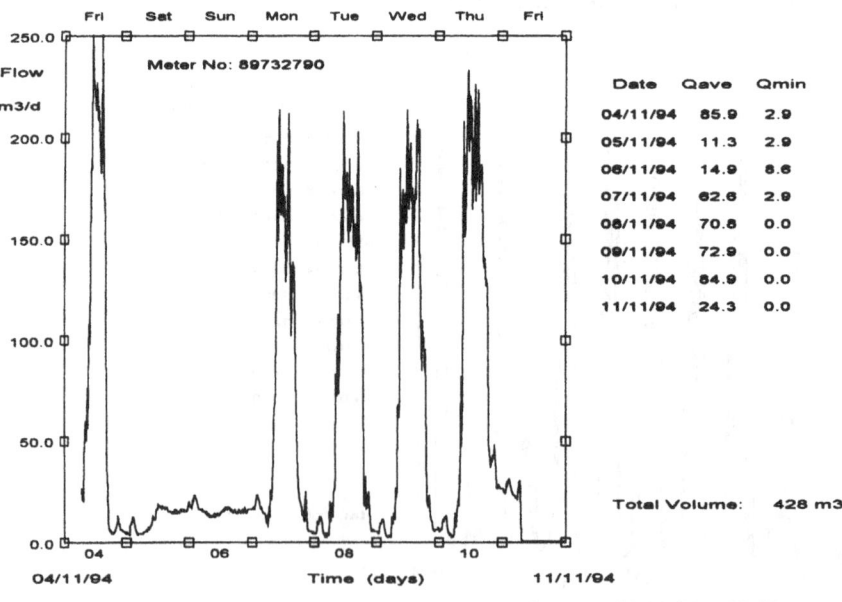

increases, partly because other large customers use more water over weekends (hotels, restaurants, places of leisure) and partly because the domestic demand is almost insensitive to the weekly cycle, as will be demonstrated below. The latter is probably the largest component of overall demand in a large area.

Note of caution   Readers should not forget that all the data in the figures above are taken from southern England; and the conclusions drawn from these data may not be valid elsewhere. Readers are, however, encouraged to collect data for their own systems and carry out a similar analysis.

Diurnal variation at district level

Human activities are geared to the 24 hour cycle, so demand patterns follow a daily cycle, clearly visible in the previous pages – see, for instance, the regular patterns in Figs 4.22 and 4.23.

A typical daily diagram could be constructed by use of statistical analysis simply by overlapping the data for several days on one graph. The data could be selected for a certain period with a distinction made between working and weekend days, or between dry and wet days, and so on. Figure 4.30 has been constructed in just such a fashion by computing the following three items:

- average daily demand (full line),
- maximum daily demand (above the average),
- minimum daily demand (below the average).

Figure 4.30 Domestic daily patterns.

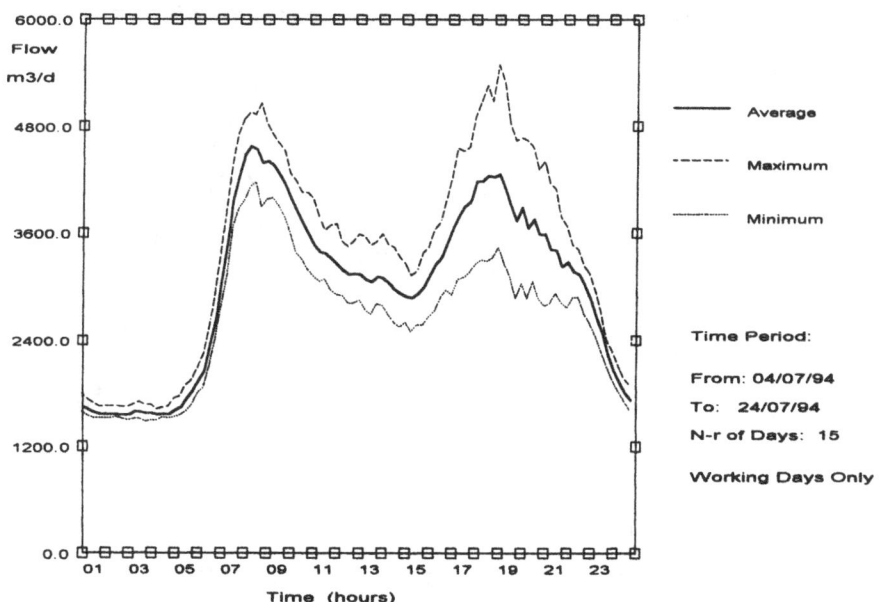

Modelling of demand

There are two distinct peaks: in the morning and in the evening. Note the low night-time demand and the steep rise of demand in the early morning hours. The diagram is fairly typical for many European towns – only the parameters (maximum/minimum) change from one place to another. The night demand provides a possible indication of leakage losses; in a residential area, with no industry, it should be as low as 30–40%, if not lower (see section 4.5).

Of course, individual days have slightly different patterns. For the example in Fig. 4.31, the solid line depicts variations of demand on Wednesday, 20 July 1994. The other lines represent other working days from the same period. They are all remarkably similar to each other. A similar analysis but for weekend days of the same period is shown in Fig. 4.32.

All the patterns in the residential area fall within a rather narrow band, despite occasional changes of weather during that time (July 1994) – and further investigation of possible effects of weather on demand does not seem to be justified in this case. The band is particularly narrow during night hours, so there were no sudden bursts during the period investigated here. However, the night consumption is somewhat high, probably due to leakage through several minor cracks somewhere in the system (in connection pipes, perhaps?).

Figure 4.31 Domestic demand patterns in a residential area – four working days.

It is interesting to note that a similar exercise for a number of winter days produced similar results to those in the previous two

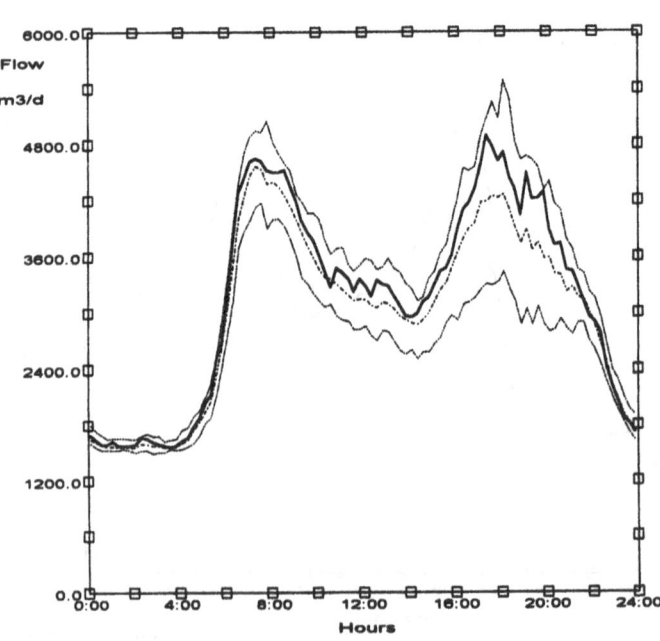

Working Days

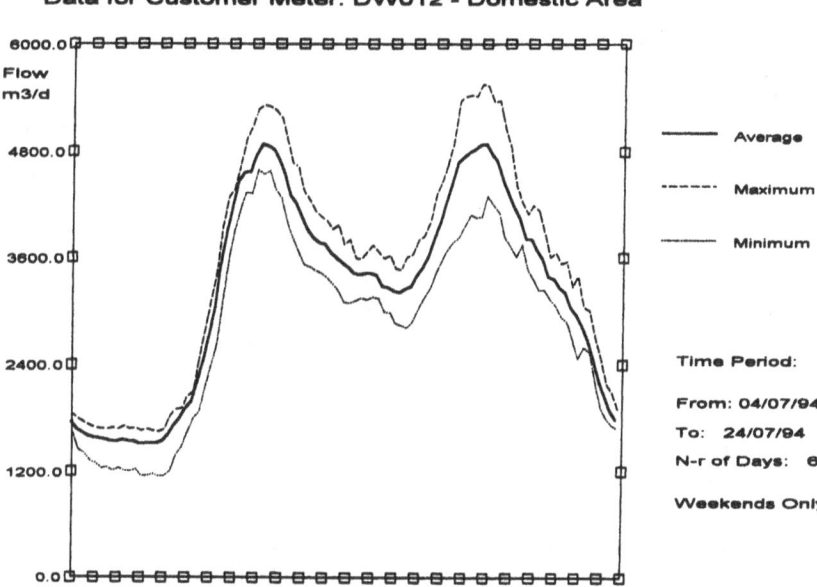

Data for Customer Meter: DW012 - Domestic Area

Figure 4.32 Domestic demand patterns in a residential area – four weekend days.

figures. The only significant difference is a marked reduction of the early evening peak in winter, when there was no call for gardening water.

Diurnal variation at individual domestic user level

Valuable data on domestic use of water in England were published by Bailey *et al.* (1986). Their results did show some scatter – as one would expect – but still some general conclusions could be drawn.

Figures 4.33 and 4.34 show average daily demand patterns for a small residential area in the summer and winter, respectively. Working and weekend days are separated. Weekend peaks always lag behind peaks for working days by a couple of hours; note also how the first peak is more prominent. There is no obvious explanation, however, why the winter peaks are as high as the summer ones – one should expect that they must be considerably lower. The peaks are shifted by one hour from winter to summer because of the adoption of 'British Summer Time' in the United Kingdom.

Flooded by the huge quantity of fresh operational data, one can be tempted to go into very fine detail and produce hundreds of different demand patterns, differing only in small matters. This will not only be a waste of effort but very confusing. To prevent this, one needs to develop the ability to recognise general patterns; it will be seen that most of them could be

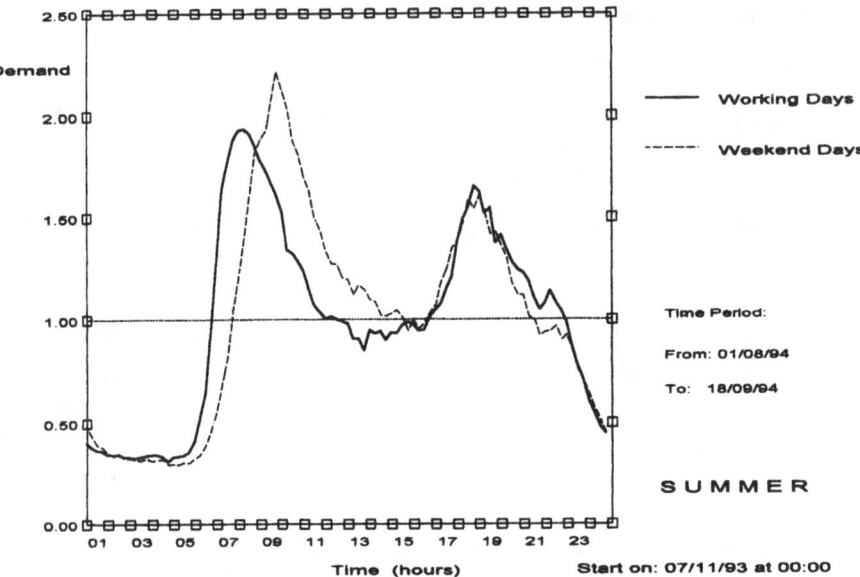

Figure 4.33 Domestic use – summer data.

Figure 4.34 Domestic use – winter data.

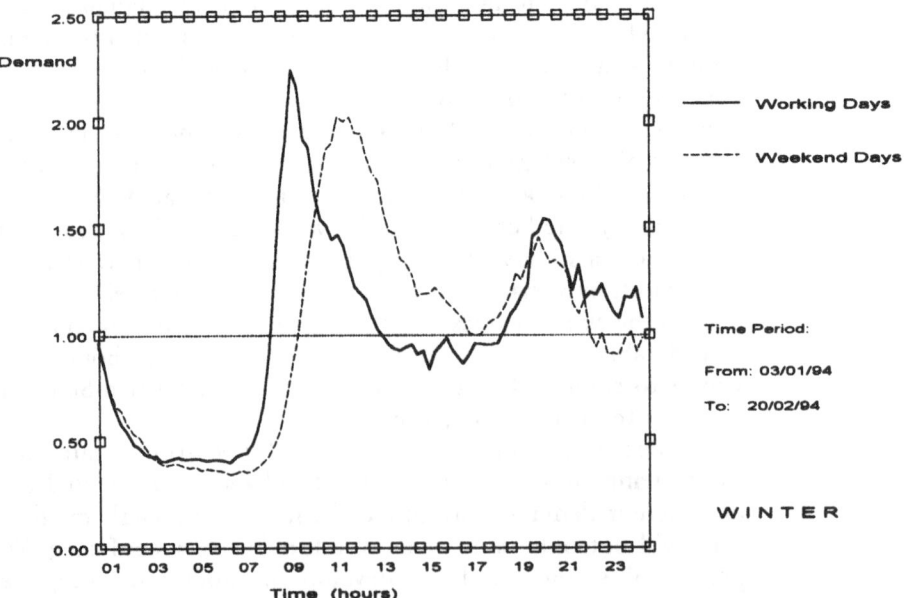

classified into one of three or four basic shapes. The important fact about different users is what they need water for and when, and the pattern is much more important than the actual consumption. To that end, the data should be made unitless by dividing actual values by the average annual consumption; the resulting unitless diagrams could then be grouped together according to their shapes.

Three basic types emerge as the result:

- *one-peak shape* – describes the demand of one-shift industries, banks, offices, hospitals and the like (Fig. 4.35);
- *two-peak shape* – domestic use, hotels, leisure camps and resorts (Fig. 4.36);
- *three-peak shape* – boarding schools, guest houses, prisons, etc. (Fig. 4.37).

Note how these diagrams – taken for very different users – are remarkably similar!

The three basic shapes are compared in Fig. 4.38. Note that leakage within customers' premises can shift the whole diagram upwards without changing its shape beyond recognition. The key parameters are:

Figure 4.35 One-peak demand patterns.

- night consumption – points a_1, b_1 and c_1;
- first (morning) peak – points a_2, b_2 and c_2;

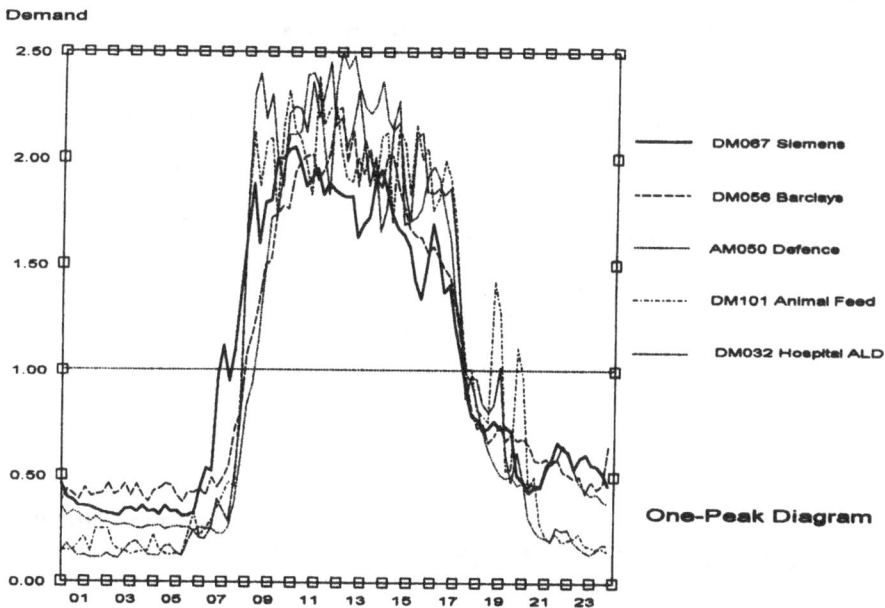

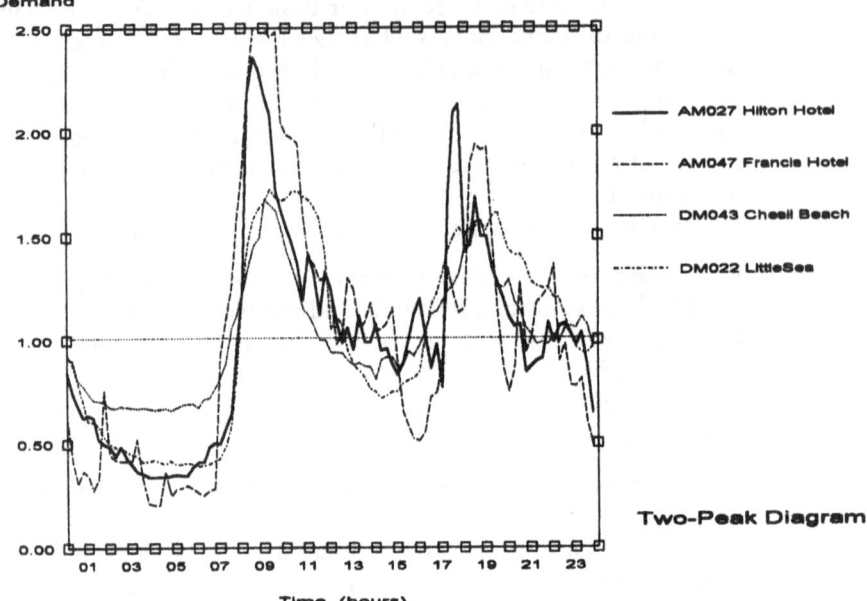

Figure 4.36 Two-peak
demand patterns.

Figure 4.37 Three-peak
demand patterns.

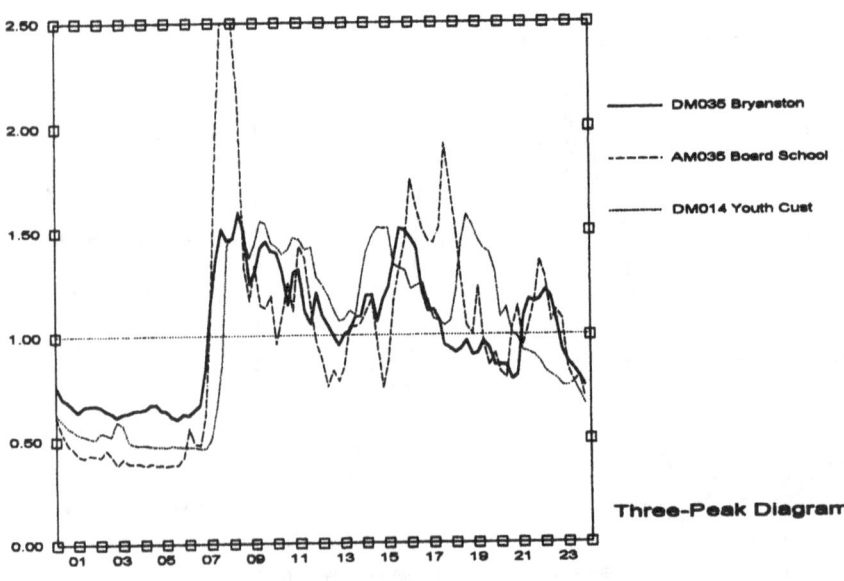

- second (afternoon) peak – points a_3, b_3 and c_3;
- third (evening) peak – point c_4.

Night consumption is usually low, but it does exist in most cases. A genuine zero night consumption is very rare indeed.

A sudden rise in consumption could be genuine (not a flowmeter fault) or a burst – see Fig. 4.39 for a hospital in England. The basic shape (in this case a 'one-peak' pattern) can be recognised by focusing on the lower part of Fig. 4.39a and screening out 'spikes' (Fig. 4.39b). Note that the exceptional demands happen only in the mornings, and none are after 15:00; also note the low night line. The causes are still under investigation at the time of writing; but similar phenomena have been seen on other individual supplies – perhaps it is caused by refilling of local water tanks. (Further diagrams of demand patterns are included in Appendix F.)

Figure 4.38 Different demand patterns: (a) one peak; (b) two peaks; (c) three peaks.

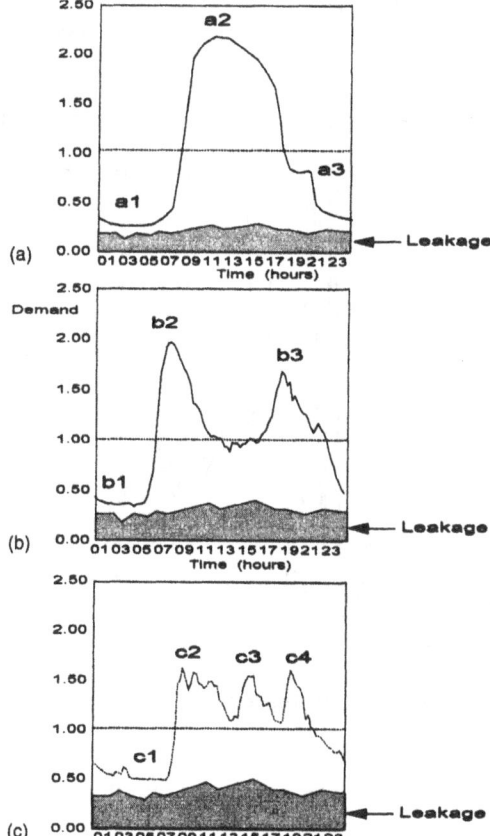

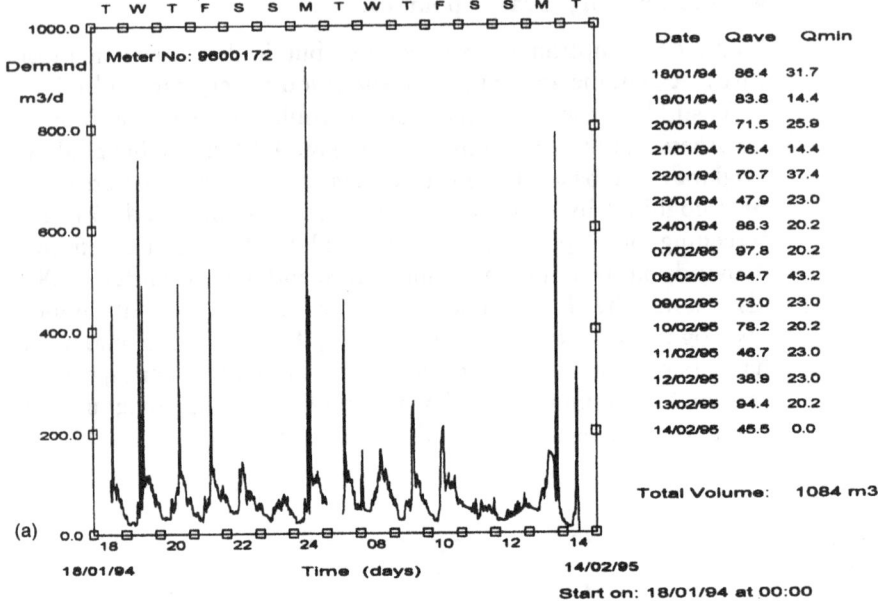

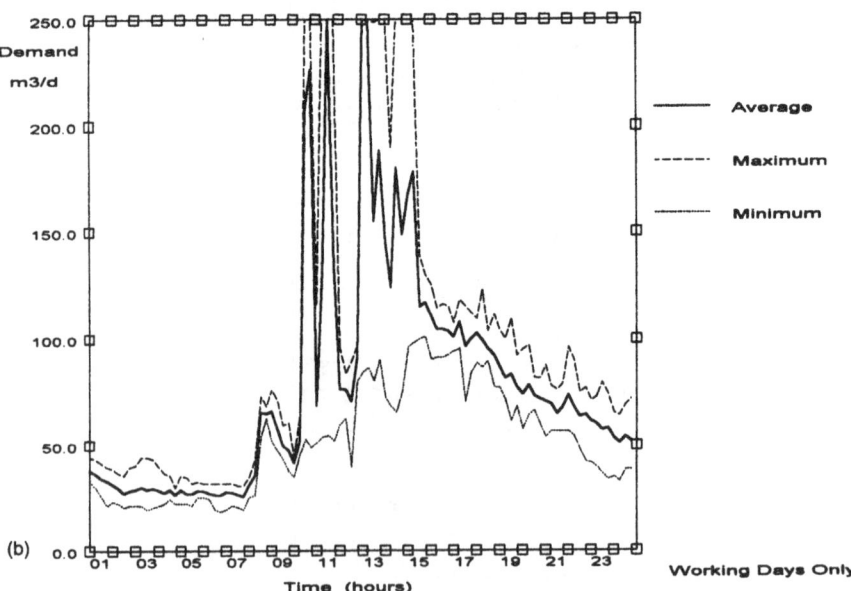

Figure 4.39 (a) Genuine increase of demand? (b) A closer look at the data.

Effects of weather conditions

Weather conditions can have an almost immediate impact on demand, especially during the growing season (from the late spring to the early autumn). In winter, this factor is less important, although freezing temperatures may cause pipe bursts and considerable losses in northern countries. The important parameters are: maximum air temperature, rainfall, humidity, cloud cover, etc. Unfortunately, general rules are hard to establish, so each particular case is a new problem. Figure 4.40 illustrates the point.

Comparing these two diagrams, one can see that the second peak (in the evening) is dramatically reduced on a rainy day – there is no need to water the garden. Consequently, total demand was also smaller than on a hot and dry day. To quantify this effect is, however, very difficult because of other parameters that come into play (e.g. rain in the morning produces a different effect to rain in the evening, weather over the preceding days also has an effect on watering regime of any specific day, etc.). Figure 4.41 shows the effect of a rainstorm on subsequent evening peaks. Note the high evening consumption before the storm; note also how the flow has exceeded the range of the meter.

Service pressure

It is known that actual consumption of water is related to pressure in the distribution system: the higher the pressure, the more water escapes through the tap during, say, washing fruits in a free water stream. However, no reliable or complete data can be traced in the literature, to the best of the authors' knowledge. Some measurements have been carried out in more advanced water companies, but the results have so far remained unpublished. The relationship between service pressure and leakage is better known, as will be discussed later in this chapter.

National holidays, sporting events, etc.

National holidays, sporting events and other exceptional days may affect overall consumption, according to local customs and circumstances. These events usually result in significant alteration to the normal demand pattern, but such relatively short (one/two day) events can be absorbed by operations staff without altering their basic routines.

General conclusions

The experience from different countries and different places has identified the factors that have (or may have) an influence on water demand in a system or an area. It is clear that each case is different, reflecting local conditions: climate, living standards, habits, level of technology, etc. However, a few general conclusions are common to most locations. These are as follows.

- There is certainly some repeatability in water use patterns, especially if the users belong to the same general category (for instance, domestic use).
- It may be misleading to use the patterns obtained elsewhere, in different climate and social environment.

Figure 4.40 Specific consumption (a) on a hot day and (b) on a rainy day.

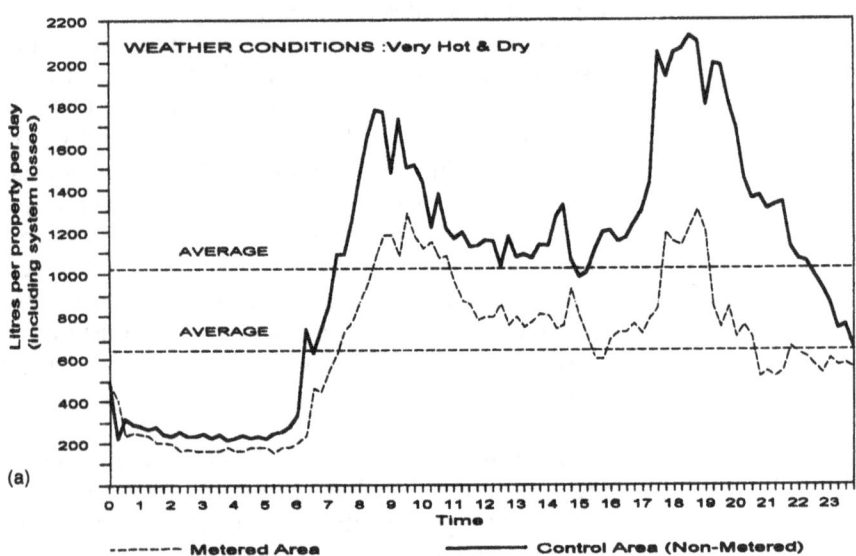

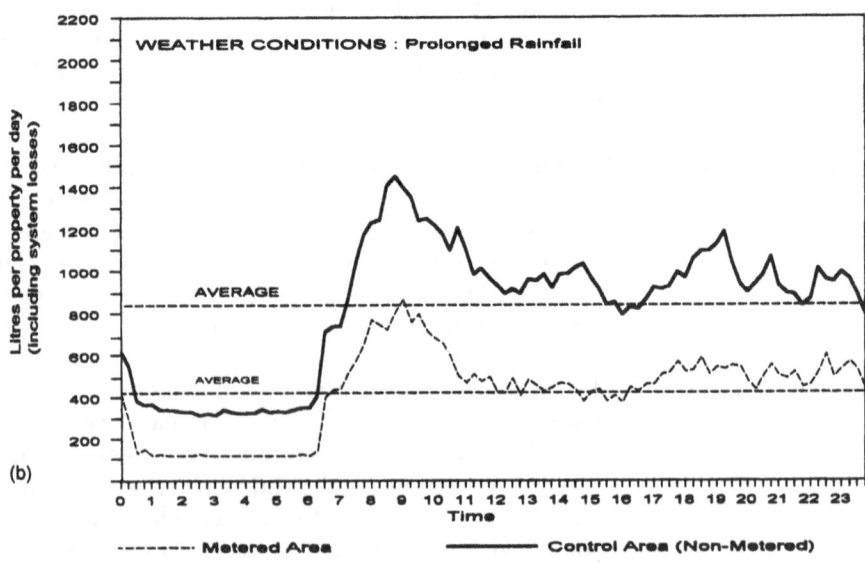

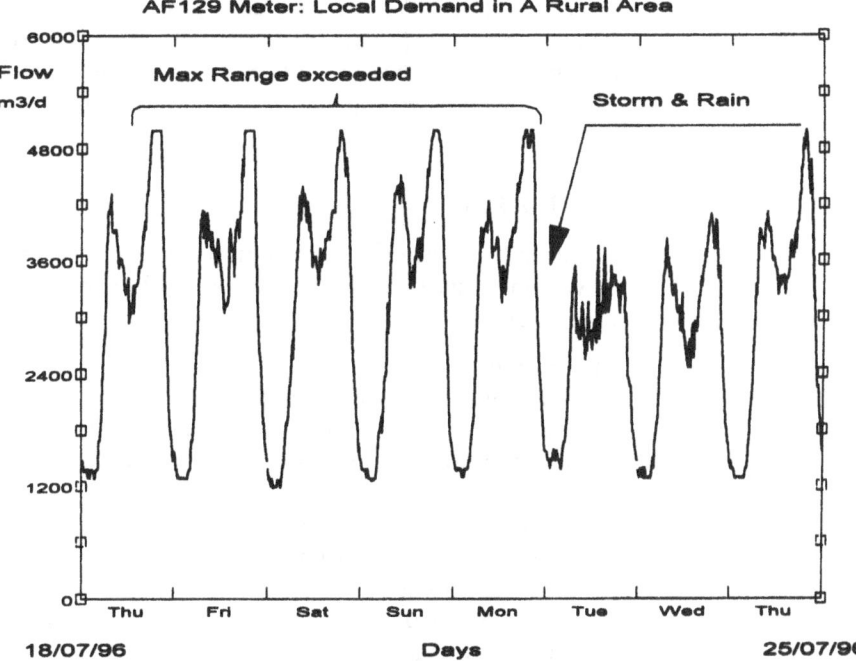

AF129 Meter: Local Demand in A Rural Area

Figure 4.41 Effects of a storm in summer.

- Each water company should from time to time conduct investigations of this kind within its own territory, then analyse the data and save the results for future analysis (design, planning, operational management).
- A water company should divide its territory into smaller territorial units (pressure and supply zones and blocks), identify the most important users and systematically start to collect data on water use.

All customers should be classified into a few categories, for example:

- domestic (residential) use,
- industrial use,
- commercial use,
- tourism,
- hospitals, schools and other public institutions,
- other uses (military, transport, crafts, etc.).

The analysis may be based on a more elaborate classification, such as SIC (Standard Industrial Code in USA), but the additional effort is rarely justified.

It is tacitly assumed that an individual user will use water in a similar way to others belonging to the same category. The total daily quantities may differ from one user to another, but the

patterns – daily diagram – are usually similar and can be reduced
to a few basic shapes as demonstrated above. The reader should
be aware that all these results and conclusions cannot simply be
used elsewhere without confirmation through local investigations.

4.5 Losses

Water is lost from the distribution system in several places due
to various causes, as shown in Fig. 4.42. The losses are usually
caused by:

1 spilling from full reservoirs,
2 leakage from water mains,
3 leakage through imperfectly closed valves,
4 leakage through open washout installations,
5 unrecorded water use – otherwise legal,
6 unauthorised use of street hydrants,
7 leakage in the distribution network,
8 leakage on service connections,
9 water meter under-reading,
10 theft of water,
11 vandalism – damaged hydrants, valves, pumping stations.

Losses are very expensive: every drop was treated, chemicals
were added, energy was used to get it into the system, etc. It is
understandable that the company tries to minimise losses.
Special teams are organised, trained and equipped with expensive equipment to find and repair leaks in the shortest time
(Swamy, 1986; Bessey *et al.*, 1994). The costs of this effort
should be balanced against the gains (Tustin, 1994, 1996).

Figure 4.42 Various
causes of water losses.

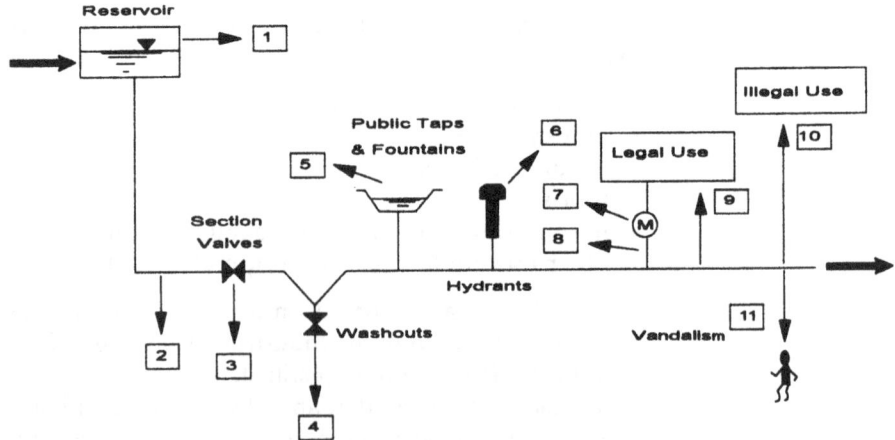

The modern data management system can offer invaluable help by capturing and processing flow data from several places almost in real time. A few examples are included here to illustrate this point. Figure 4.43 shows the variations of demand in a district. Note the two sudden bursts – revealed through sudden jumps in the night line in Fig. 4.43a. The bursts – once discovered – were repaired within a few days, bringing the night line back to the previous level (Fig. 4.43b).

Losses can be detected as a steady rise of night line (Fig. 4.44), so-called 'progressive leaks'. Leaks may also be small and numerous – but not obvious – indicated by a high night line (Fig. 4.45). This fact alone does not prove high leakage losses – perhaps this is genuine use if there are important customers using a lot of water round the clock (food-processing factories are a good example of such usage), but it is worth investigating.

The losses are sometimes more visible by taking a wide view – see the two examples in Figs 4.46 and 4.47. The first shows a progressive leak, creeping slowly to a high level and really only detectable by the daily patterns of several months compressed onto a single diagram. When eventually noticed and repaired, the night line fell to a respectably low level. The losses could be calculated by integrating the flows below the minimum flows. The other case, in Fig. 4.47, shows how the existence of larger leaks is detected earlier and they are repaired quickly, while minor leaks – sometimes as important in total – can last a very long time.

The successes in this endless struggle can be dramatic, as in the case shown in Fig. 4.48, where an important leak was discovered and repaired in a few days. In some cases the best answer is to replace the pipe because the costs of incessant repairs become prohibitive (Fig. 4.49).

Leakage and service pressure

Numerous investigations have been conducted to discover the factors that affect leakage, but many questions remain unanswered. Pressure is certainly an important factor. The flow through an orifice is proportional to the square root of pressure, but in the context of flows from points in a distribution system there is considerable evidence that the square-root relationship does not apply. Increased pressure causes a much higher increase of flow rate than the theory would suggest. One important report (Ridley, 1980) states that leakage is roughly proportional to the local service pressure instead of its square root. This assertion is repeated in more recent reports (Bessey et al., 1994), and disproportionately large reductions in discharges have been observed when service pressure has been reduced.

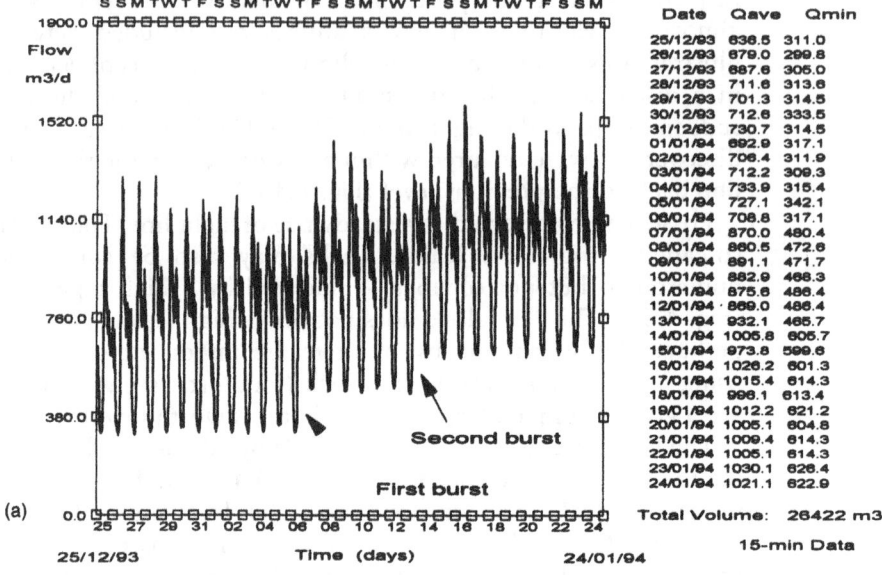

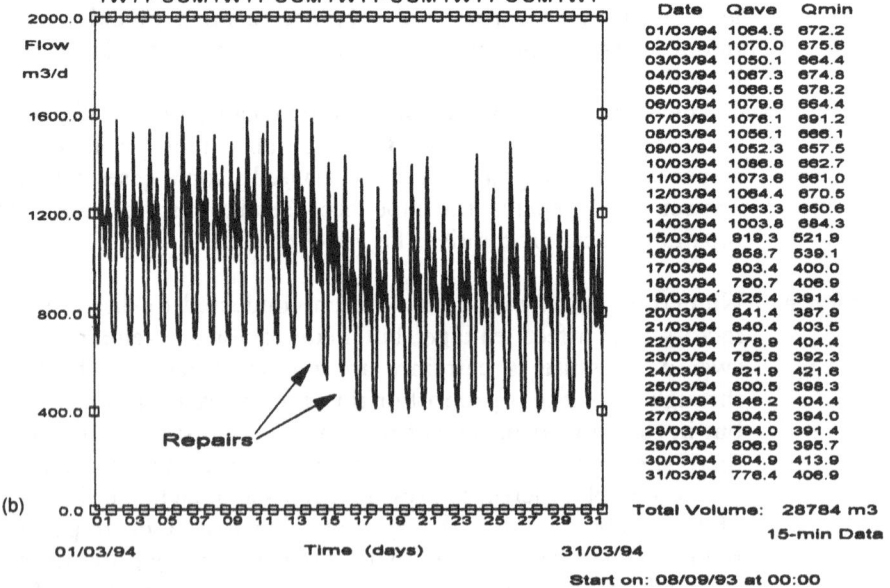

Figure 4.43 (a) Two
consecutive bursts
(b) later repaired.

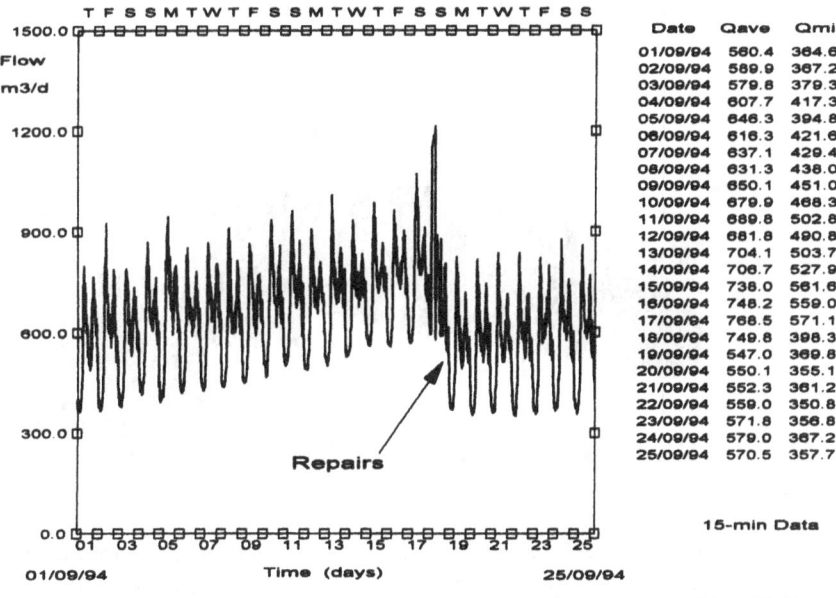

District Meter: DW111 - A Rural Area

Date	Qave	Qmin
01/09/94	560.4	364.6
02/09/94	589.9	367.2
03/09/94	579.8	379.3
04/09/94	607.7	417.3
05/09/94	646.3	394.8
06/09/94	616.3	421.6
07/09/94	637.1	429.4
08/09/94	631.3	438.0
09/09/94	650.1	451.0
10/09/94	679.9	468.3
11/09/94	689.8	502.8
12/09/94	681.8	490.8
13/09/94	704.1	503.7
14/09/94	706.7	527.9
15/09/94	738.0	561.6
16/09/94	748.2	559.0
17/09/94	766.5	571.1
18/09/94	749.8	398.3
19/09/94	547.0	369.8
20/09/94	550.1	355.1
21/09/94	552.3	361.2
22/09/94	559.0	350.8
23/09/94	571.8	356.8
24/09/94	579.0	367.2
25/09/94	570.5	357.7

15-min Data

Start on: 07/11/93 at 00:00

Figure 4.44 Progressive leak – repaired.

Figure 4.45 Genuine high night demand?

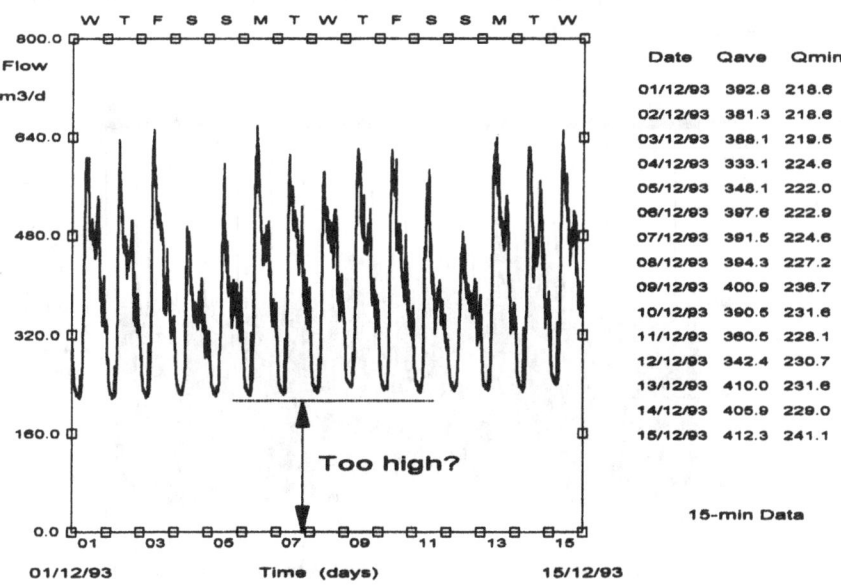

Date	Qave	Qmin
01/12/93	392.8	218.6
02/12/93	381.3	218.6
03/12/93	388.1	219.5
04/12/93	333.1	224.6
05/12/93	348.1	222.0
06/12/93	397.6	222.9
07/12/93	391.5	224.6
08/12/93	394.3	227.2
09/12/93	400.9	236.7
10/12/93	390.5	231.6
11/12/93	360.5	228.1
12/12/93	342.4	230.7
13/12/93	410.0	231.6
14/12/93	405.9	229.0
15/12/93	412.3	241.1

15-min Data

Start on: 06/11/93 at 00:00

Modelling of demand

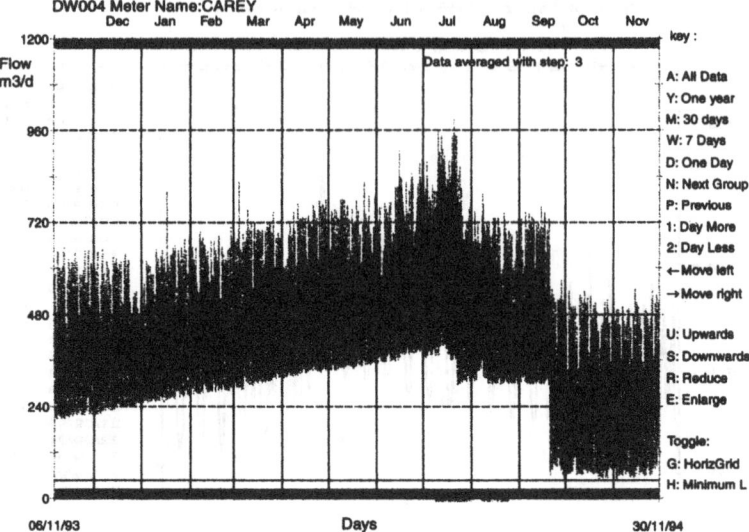

Figure 4.46 History of a progressive leak and its repair.

The effect is shown in Fig. 4.50. Note that the measured curve is steeper than the square-root curve. Why? It seems that high service pressure makes cracks and holes in the network progressively larger, thus increasing leakage losses almost linearly with pressure.

The variations of service pressure and leakage losses (hypothetical) over a day are displayed in Fig. 4.51. The leakage increases with the pressure, and falls with decreasing pressure. Of course, this graph shows only the nature of this relationship, without quantifying it.

Figure 4.47 The struggle against leaks.

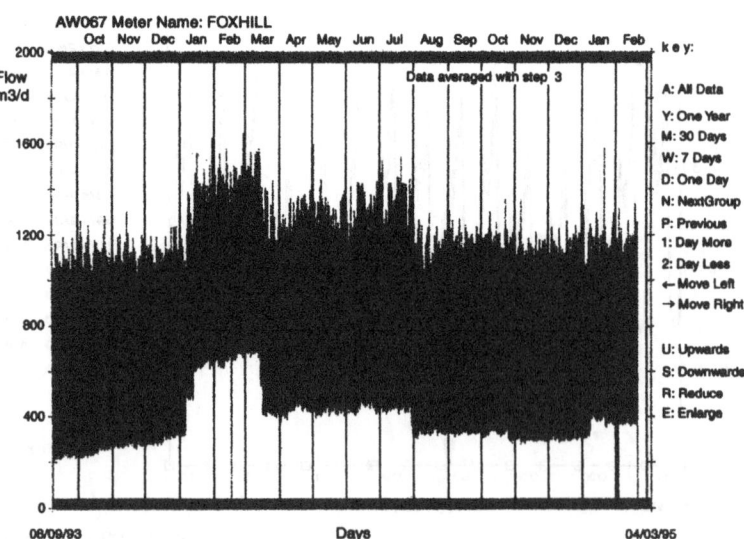

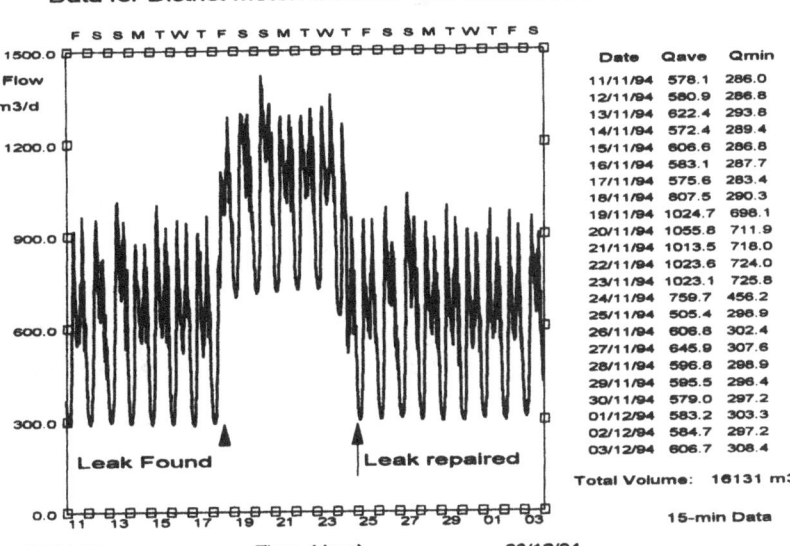

Figure 4.48 Leak found
and repaired in 5 days.

Losses were not treated explicitly in modelling until recently.
The usual practice was to compute consumption as well as
possible, and then add 20–30% to account for losses to give
total demand. This meant that losses 'travelled' all the way from
sources to service connections – inflating flows in mains,
sometimes considerably. A common argument for the practice
is that a safety margin is introduced in this way, especially useful

Figure 4.49 A case for
replacement of a pipe?

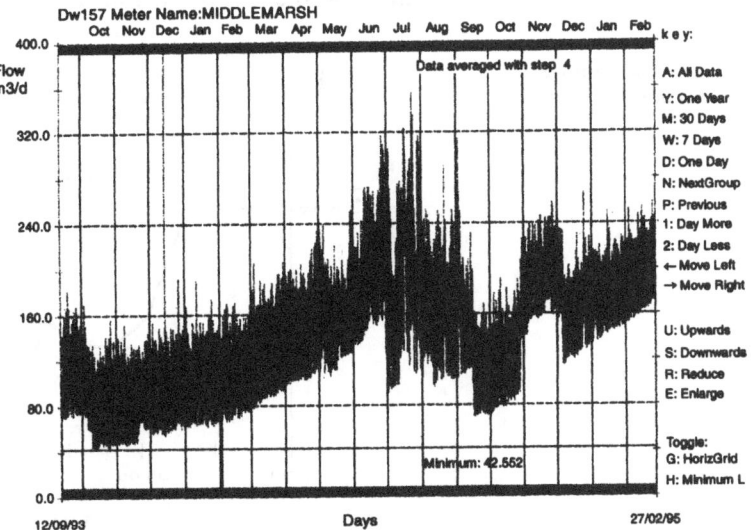

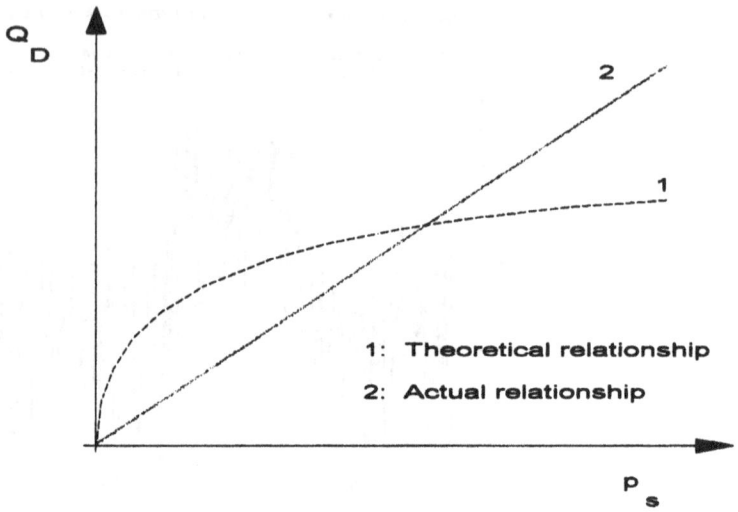

Figure 4.50 Leakage index.

when designing a new main. However, the practice is a serious distortion of reality. First, water escapes from the distribution system at many points, well before it reaches consumers, so the real, average flow is smaller than the 'inflated' one. Secondly, variations of service pressure in the network have a serious effect on leakage losses – and a much smaller effect on real consumption, which is taken through orifices (taps or faucets, etc.) that do not distort as pressure changes. The consumption by customers will tend to follow the square-root rule. This is illustrated in Fig. 4.52, which indicates that the night line of individual consumers is a much lower proportion of the average consumption than in the whole area.

Figure 4.51 Leakage versus pressure.

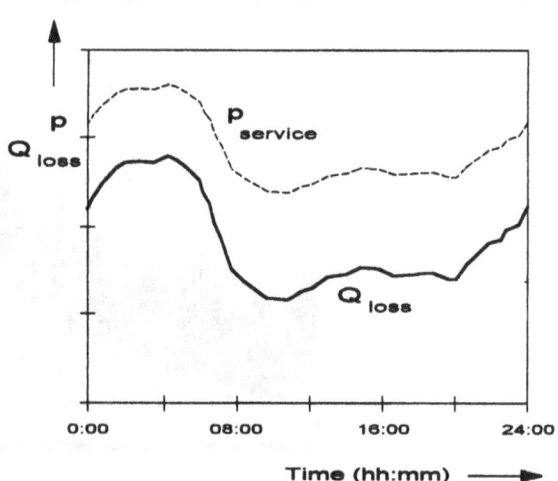

Note that night flow at a service connection is rather low and that leakage losses are much higher during the night, which is due to higher pressures in the network. The resulting diagram for total demand – as seen from the control room – is flattened; the night flow is increased and both peaks considerably reduced. In a system with a very high percentage of losses (40–50%), the night flow is very close to the average daily demand.

A 'flat' diagram of daily demand is a reason for water company staff to worry because it can signal:

- high leakage in water mains and the distribution network, or
- high leakage in house installations (poor plumbing?), or
- both evils at the same time, unfortunately.

Effort should be devoted to analysing the night line thoroughly and explaining anomalies. A case when strange variations of the night line were spotted is included here in Fig. 4.53. Note that minimum night flows (MNFs) increase over working days and drop to a lower level over weekends. The event repeats every week, so it cannot be just a freak incident. By examining water bills in that district, the large customers were identified and their consumption was monitored for a few weeks. It was discovered that a factory producing chemicals uses water five days a week, continuously, and then stops for the weekend (Fig. 4.54).

Figure 4.52 A view from the control room: (a) the whole area; (b) an individual user.

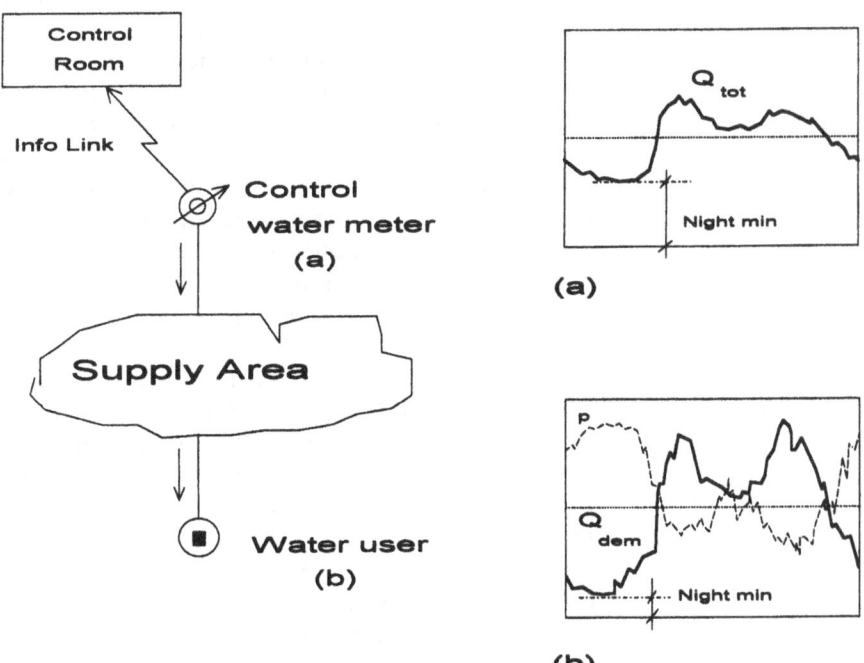

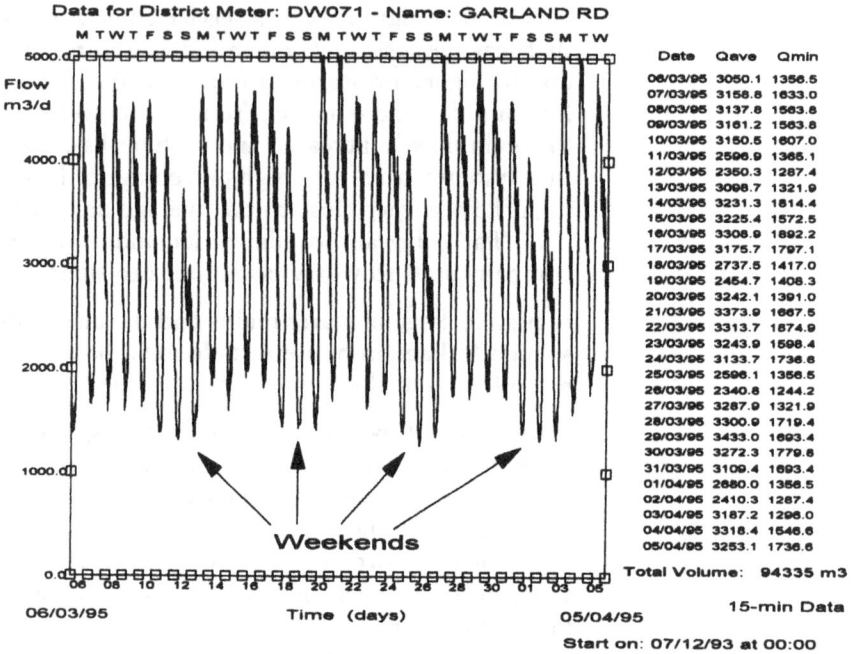

Data for District Meter: DW071 - Name: GARLAND RD

Date	Qave	Qmin
06/03/95	3050.1	1356.5
07/03/95	3158.8	1633.0
08/03/95	3137.8	1563.8
09/03/95	3161.2	1563.8
10/03/95	3150.5	1607.0
11/03/95	2596.9	1365.1
12/03/95	2350.3	1287.4
13/03/95	3098.7	1321.9
14/03/95	3231.3	1614.4
15/03/95	3225.4	1572.5
16/03/95	3308.9	1892.2
17/03/95	3175.7	1797.1
18/03/95	2737.5	1417.0
19/03/95	2454.7	1408.3
20/03/95	3242.1	1391.0
21/03/95	3373.9	1667.5
22/03/95	3313.7	1874.9
23/03/95	3243.9	1598.4
24/03/95	3133.7	1736.6
25/03/95	2596.1	1356.5
26/03/95	2340.8	1244.2
27/03/95	3287.9	1321.9
28/03/95	3300.9	1719.4
29/03/95	3433.0	1693.4
30/03/95	3272.3	1779.8
31/03/95	3109.4	1693.4
01/04/95	2680.0	1356.5
02/04/95	2410.3	1287.4
03/04/95	3187.2	1296.0
04/04/95	3318.4	1546.6
05/04/95	3253.1	1736.6

Total Volume: 94335 m3

15-min Data

Start on: 07/12/93 at 00:00

Figure 4.53 Changes in MNF – why?

A water company should not tolerate high losses in house installations, even if the water is paid for through volumetric charge (water meter), because this is an intrinsic waste of water, which effectively penalises all parties. Very soon sources will be used up and the company will be forced to find new ones – which may be very costly. The company should follow a wiser policy, investigate how the water is used, offer advice on the maintenance of plumbing installations, promote programmes of water conservation, influence manufacturers of water-using appliances to make efficient products, etc. The policy may decrease income a little, but it will also postpone important investment and save scarce resources for later generations.

Pressure management

High pressures in the network, especially during the night when demand is at its lowest, lead to higher leakage. Figure 4.55 shows variations of both demand and pressure in a district. Note how the pressure increases during night hours and drops during peak demand time. The water company has to make sure that the pressure will never drop below a prescribed limit; the fact that pressure exceeds this limit during the night is of no consequence – except that the losses are worse.

This well known fact has induced several water companies to try to manage service pressure, usually by reducing it overnight

Data for District Meter: DW071 - Name: GARLAND RD

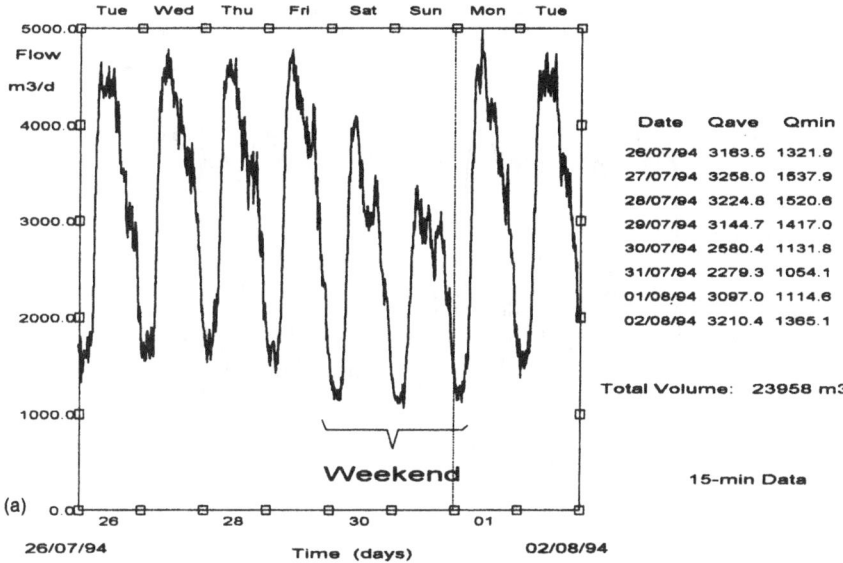

Start on: 07/12/93 at 00:00

Data for District Meter: DM018 - Chemicals

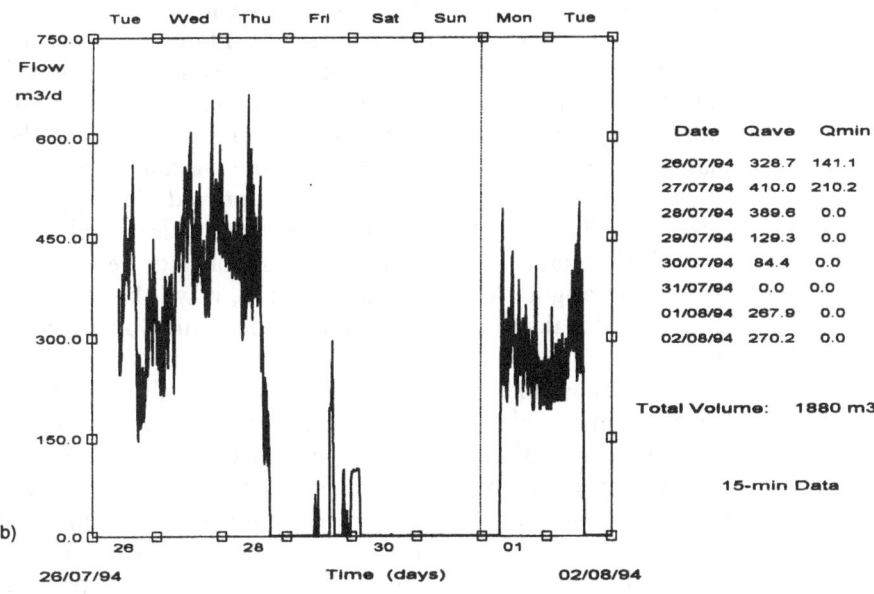

Start on: 26/07/94 at 09:03

Figure 4.54 (a) Close-up analysis of anomalous night line and (b) the explanation.

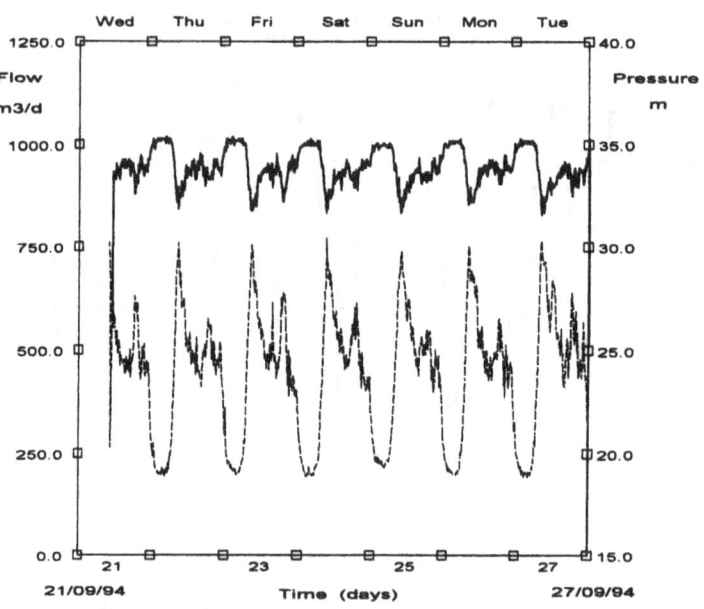

Figure 4.55 Flow and
pressure variations.

and increasing it during peak hours (Lonsdale, 1984; Bessey, 1985; Bessey *et al.*, 1994). The results could be very good, as in the case shown in Fig. 4.56. Note that, by bringing pressure down from very high values (between 60 and 80 m in this case), the night flows are dramatically reduced and typical daily diagrams reappear. The benefits are many: wastage is cut down, bursts are less frequent and service improved (Bessey, 1990; Jeffery and Taylor, 1993). There is, however, a drawback – once started, the management of pressures should not be discontinued. If the original pressures are restored, it will be found that the original leaks will reappear but accompanied by many new friends!

This point is more important than is usually realised, and it can cause serious problems in managing the water supply system after major reconstruction works. The increased service pressure will lead to higher demands and losses, so the additional quantities of water – supplied by new facilities – could become insufficient in just a few years, not in several decades as confidently predicted on the basis of previous consumption data, taken when pressures were being controlled and considerably lower.

Where to direct the leak detection and repair effort?

Modern urban distribution systems have been developed over many years, and in a country with almost all of its population connected to the system, such as the United Kingdom (where many water companies supply more than 95% of the residents within their area of supply), even some rural systems are of significant age.

Over the decades and in some cases more than a century of system development, many changes have taken place in the materials used in the construction of mains and service pipes. Some of the developments have left significant parts of the networks particularly vulnerable to leakage. In conjunction with the suggestions for identifying leaks, either singly or *en masse*, described above, it may be advisable to take account of some general points about pipe materials and ground conditions.

Cast iron

This was a very common pipe material up to the late 1960s. Early CI pipes had minimal external and internal protection, and are prone to corrosive internal attack by soft, acidic, upland waters and external attack by aggressive, heavy, wet, clay soils. Conversely, CI pipes have given good service when carrying 'hard', alkaline waters in well drained and/or alkaline soils such as chalk or limestone areas, sands or gravel. Later CI pipes are often cement-lined, and pipes that are lined under a rehabilita-tion programme give good service. CI is mechanically inferior to

Figure 4.56 Effects of pressure management.

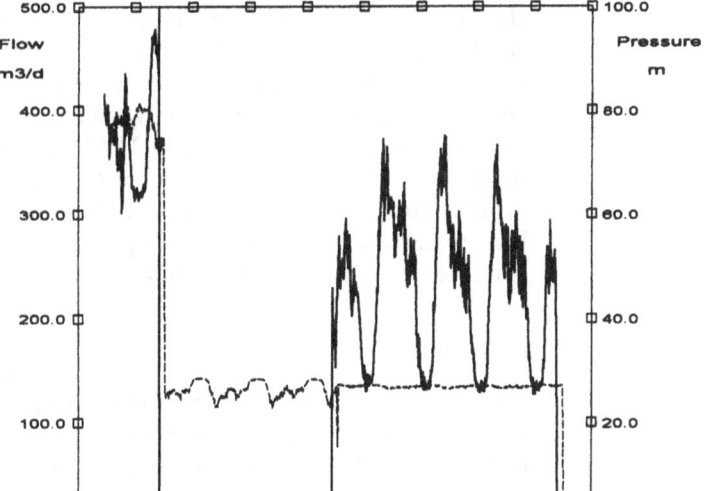

ductile iron and prone to cracking or splitting if badly laid or in ground prone to movement. CI pipes are often subject to seasonal breakage with temperature changes in late autumn and early spring.

Ductile iron

DI pipes were used commonly from the early 1970s and were thought, on their introduction, to be almost indestructible. They are mechanically very strong but prone to attack in aggressive soils and from aggressive waters if unprotected. Cement-lined pipes with external protection should give good service.

Asbestos cement

This was a commonly used material from pre-war days to the late 1960s in areas where CI was unsuitable due to ground conditions. It is an excellent pipe material but there are significant hazards from asbestos when cutting or turning pipe. It is prone to cracking in a similar way to CI, and mechanical damage can induce delamination of the pipe material over time. Asbestos cement is not usually prone to attack by aggressive soils.

Unplasticised PVC

This was a common pipe material from the 1950s to the 1980s. It is an excellent material but subject to some specific problems. Unplasticised PVC is not normally affected by chemical ground conditions or water quality, but needs to have been laid on appropriate bedding material to prevent mechanical damage. A common mode of failure is longitudinal cracking. Early production was prone to damage by UV light and exhibited jointing problems. Early pipe was jointed with solvents that were difficult to make in wet weather. Later jointing was mechanical. Larger sizes, above 200 mm, are prone to joint problems, as the spigot end tends to an oval shape.

Polyethylene

This is an excellent pipe material in most ground conditions and water qualities. A number of methods of joint welding have been developed, and after a few teething problems appear to be very successful.

The above are just a few of the 'pros' and 'cons' of the pipe materials commonly used in water supply. Other materials can be found, but in relatively smaller lengths (e.g. steel, concrete, etc.), and a whole range of materials are used in service pipes. It is useful to gain some knowledge of the history of pipe material development, and to use that knowledge in conjunction with the other methods of fighting the universal problem of leakage and waste in supply systems.

4.6 Modelling demand

It is not an exaggeration to state that the success – or failure – of the whole modelling project depends largely on a good analysis of water demand, both in time and in space. Other information, regarding the network, reservoirs, pumping stations, valves, etc., can be found and checked relatively easily – but not the demand. Moreover, the procedures to determine which data are needed, and how to gather and analyse them, and having confidence in the results, etc., are still largely an art rather than a science. See De Moyer *et al.* (1973), Gilman *et al.* (1971), Barrufet (1985), Maidment and Miaou (1985) and Quevedo and Cembrano (1986) for various approaches to this problem. The lines that follow are intended as a pointer towards formulating a general procedure of modelling demand.

The analysis of demand starts from the data gathered by telemetry, loggers or direct observations. The result should be a model of demand, described by typical seasonal, weekly and daily diagrams, and a list of factors influencing the demand and a measure of probable noise – likely deviations from the expected values. This can be done through the relatively simple procedure shown in Fig. 4.57 using statistical tools.

The first task is to read data available in various files. Potential sources are:

- *Telemetry system* – data on flows through water mains and on variations of storage in reservoirs.
- *Data-loggers* – measurements of flow to and from a discrete supply zone, a block of dwelling houses or an industrial estate.
- *Water meters* – regular readings of consumption (1–2 h intervals) at important users' service connections.
- *Special investigations* of water consumption variations in factories, hospitals, schools, hotels, etc.

These data may be of indifferent accuracy and reliability, so a thorough examination is absolutely necessary before any analysis. Those values which are clearly erroneous (due to a false reading, malfunction of an instrument, temporary loss of power supply, typing error, etc.) should be deleted. In some cases, rejected data may be replaced by more reasonable estimates; in

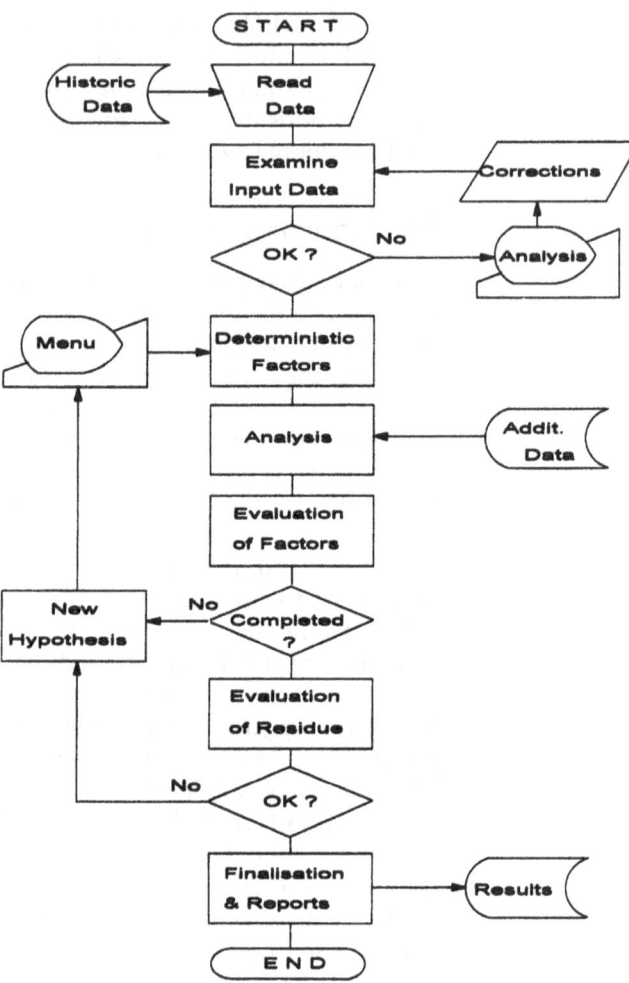

Figure 4.57 The procedure.

others, they are just ignored. The usual checks of data (besides control of type) are:

- *Control of range* – lower and upper limits.
- *Control of increment* – any large, sudden and unexpected change is suspicious and may signal an error.
- *Control of relationships* between data from different sources (if applicable).

The importance of this task cannot be overestimated. If the 'freaks' are not rejected, further distillation and subsequent analysis may easily become totally senseless. Unfortunately, in the later stages of data processing, it is very difficult to spot erroneous data that have evaded controls during the data capturing process.

The data that have passed the validation controls are then used for real analysis. The next step is computation of deterministic factors: general trends, seasonal effects, working/ weekend days and others. This is done in a *filter*: a program that extracts this information by using simple statistical tools (e.g. linear regression). In some cases more sophisticated methods are applied (see Barrufet, 1985; Cosgriff *et al.*, 1985; De Moyer *et al.*, 1973; Gilman *et al.*, 1971; Maidment and Miaou, 1985; Quevedo and Cembrano, 1986; and others).

The design of this filter is a serious task, which starts with a list of possible deterministic factors. Each one has to be evaluated on the basis of available information, and an adequate mathematical model (usually quite a simple one) is created that can give a measure of the factor in question. The analysis then picks up the next one, and so on. The net result is a *deterministic* model of expected demand, where independent variables only are needed to be input, like date, time, maximum temperature of air, humidity, etc.

The next step is computation of residue, i.e. difference of actual demand and the value computed by using this model. It is the only verification of the proposed model. The necessary condition is that the differences follow a *normal distribution* pattern (the characteristic 'bell shape'). Note that this does not mean that the differences are small!

If this is so, the model of demand can be accepted (until the next check) as valid. If not, then the whole process should be repeated, sometimes from the very start (if there are grounds to suspect that erroneous data have evaded all controls and 'polluted' working files). The usual procedure is to revise the list of deterministic factors and establish new hypotheses.

The *stochastic* residue that remains after all deterministic factors have been extracted normally amounts to 4–6% of the total demand (De Moyer *et al.*, 1973; and the authors' own experience).

The analysis is not as straightforward as it seems. First, the selection of deterministic factors is a 'hit-or-miss' affair, with few solid rules. A model may yield excellent results for a few months, and then just fail, due to facts unknown to the water company. The task is never fully completed.

It is not enough to make a model of total demand of the whole system. Models will be needed for particular zones and even for a few big consumers. This is another field where an expert system will be very useful – it is expected that these tools will be available soon. Tasks of this kind are too difficult for a water engineer who has a lot of other duties to perform and cannot devote a lot of time to such tasks like the one just described.

Note of caution A statistical analysis of this kind is always very sensitive to data. If the quality of data is poor due to inaccurate instrumentation and/or incompetent maintenance, the results may be worthless. Problems like this one can be solved only within the water company, by the staff.

Finally a few key pieces of advice:

- Base the model on available *facts*, not on *fiction*.
- Avoid 'black-box' concepts, i.e. methods and techniques not known to the user.
- Do not use sophisticated statistical tools, as the accuracy of data is too poor for that.
- Involve local staff from the very beginning and give them an active role throughout.
- Make the most of local knowledge – it is irreplaceable.
- Experience from other places should be treated with utmost caution, although it can provide valuable guidelines.
- National averages have to be compared with local values, not taken blindly.

5 Mathematical methods applied in modelling

5.1 The principles of modelling

The elements of a modern water system model were described in Chapter 3, particularly the general structure of models and the individual features of real systems that need to be represented in models. In this chapter, we look at the mathematical basis of water supply network modelling, but first, for convenience, we recapitulate some of the main points made in section 3.1.

With computing power now available at a relatively low price, it would be possible to include in a model every single pipe, pump and reservoir, even each customer connection. However, for a number of reasons, this is currently impracticable and some simplification is called for.

A supply system may be represented quite adequately by a set of *nodes* and *links* representing all the significant water conduits of the real system (the links) and their junctions (the nodes).

Many researchers, notably Bhave (1991), Coulbeck and Orr (1988), Davis and Jeppson (1979), Donachie (1974), Epp and Fowler (1970), Fallside (1977), Gessler (1981), Gilman *et al.* (1971), Hamam and Brameller (1971), Isaacs and Miles (1980), Jeppson and Davis (1976), Rao and Don (1977), van der Zwan (1988), Walski (1984), Wood and Charles (1972) and Zarghamee (1971), have contributed to solving the problem of creating useful models.

All nodes of a system fall into one of three groups:

1 *fixed-head nodes*, where water level remains unchanged during the simulation;
2 *variable-head nodes*, where water level rises if inflow exceeds outflow and vice versa;
3 *ordinary nodes*, usually the largest numerical group in a model, and where all the inflows and outflows are known and the head needs to be computed.

In cases 1 and 2, water level is a known quantity, while outflows and inflows need to be determined through computation. Examples that fall into case 3 are:

• a transfer point, where a known quantity of water is delivered to or from the system,

- a junction (branching point) of two or more pipes, with or without local consumption,
- a service connection to an important user, where local consumption of water is known,
- any other point within the system that is important for any other reason.

All links connect two nodes and may be one of three types:

1 *pipes*, which simply transport water from one node to another;
2 *pumping stations*, which add energy to water;
3 *control valves*, which regulate flow of water by reducing its hydraulic energy ('throttling').

These are mnemonic descriptions: they describe the function of elements of the model by using the name of a similar, physical entity. This does not mean that each valve in a water supply system will be represented by a 'valve' in the model. Only a few important physical valves (control valves, for instance) will be shown as such in the model – and the majority simply included as minor losses in their pipes.

'Pipe' in the model represents any conduit, tunnel, water main, large or small pipe – anything used to convey water under pressure (i.e. not through open channels). In contrast, a real pumping station, housing several pumps, may need to be represented in the model as several 'pumping station links' if the suction or the delivery node is not common to all the units.

It can be easily shown that all elements of any water supply system could be described by these items – either directly or in combination.

Each of the various elements of a network model, the nodes and links, need to be described by a number of key physical parameters and control loops or policies. Although acquiring and tabulating these data is quite a large task, it is relatively straightforward. Much more difficult to establish are the volumes and timing of water leaving the system, and some of these difficulties were dealt with in Chapter 4. Again we shall recall the main points of this difficult area.

Dealing with demand

The most difficult task in model making is to estimate the real demand and allocate it to individual nodes. It is clearly impossible to represent each customer individually in the model, so several customers are aggregated as 'sinks'. A relatively simple case is shown in Fig. 5.1 to illustrate the difficulties (taken from Figs 3.22 to 3.24). Note that all individual houses are shown. The customers could be identified by their address

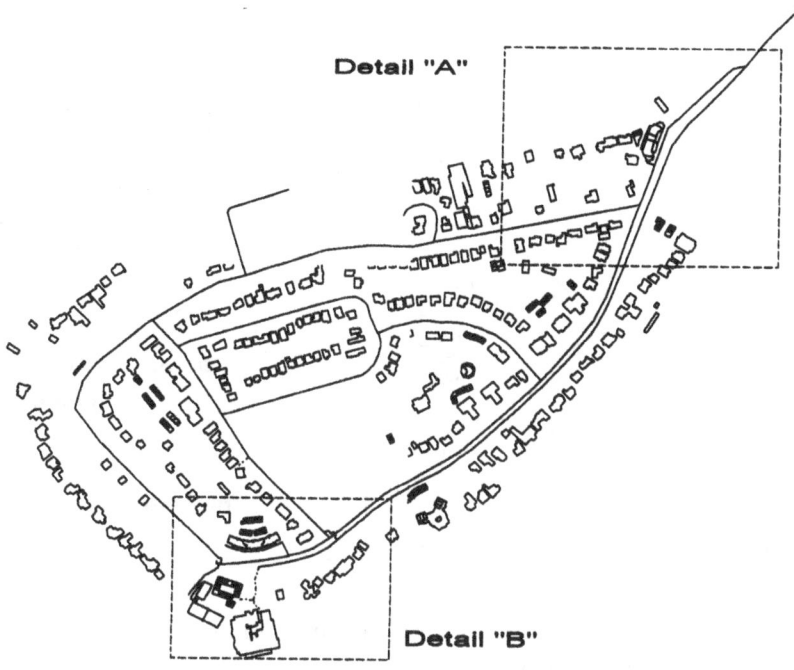

Figure 5.1 Distribution network.

Figure 5.2 Detail 'A' of the network.

and consumption data taken from billing records. By examining these records, customers could be classified and large ones singled out, as in Figs 5.2 and 5.3. The interesting ones are the small hotel, whose main trade is as a pub, a few shops nearby (Fig. 5.2) and one larger hotel (Fig. 5.3); there is also a row of

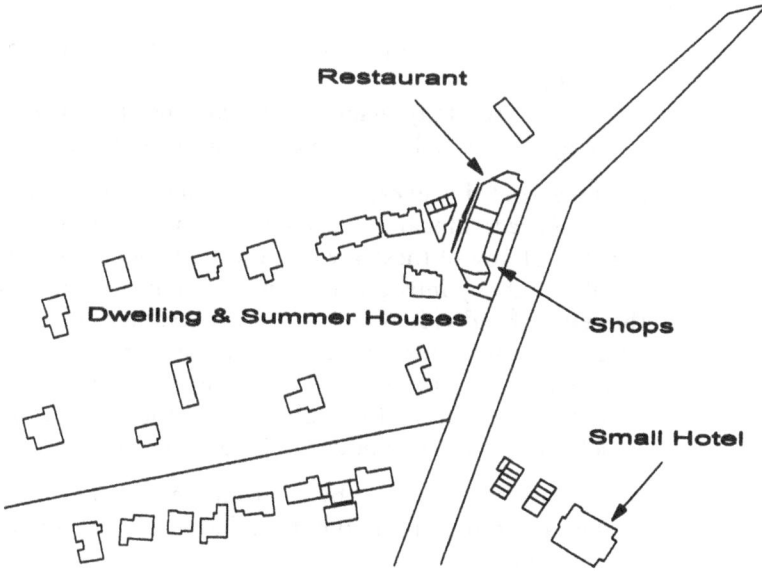

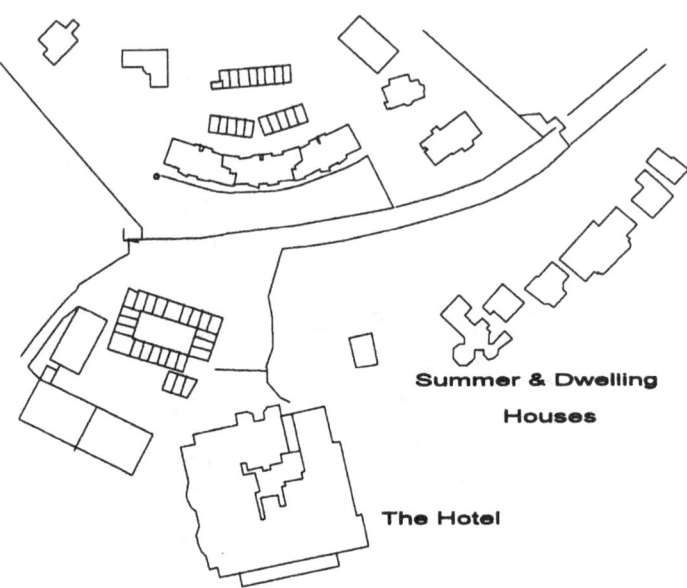

Summer & Dwelling

Houses

The Hotel

Figure 5.3 Detail 'B' of the network.

summer houses where consumption could be very low outside the summer season.

The analysis of billing records is certainly useful for the water company and may give precious data for planning and design work. From the modelling point of view, it should not be carried too far (because resources and time are always limited) and a simplified picture is quite acceptable:

- All customers should be allocated to the nearest node – in the hydraulic sense.
- Average consumption should be computed as precisely as possible.
- Appropriate daily diagrams or patterns should be developed using all available data and sound judgement.

At the end of this exercise, the whole territory will be split into *demand areas* allocated to individual nodes – see the example in Fig. 5.4. If any of these areas is too large, it should be split into smaller units by introducing new nodes into the model, even to the level of individual customers if necessary. This is, unfortunately, a procedure that cannot be formalised given the discrepancy in size between real systems and mathematical models. An additional complication is the allocation of losses, which cannot be overlooked. The options are:

- Increase consumption by a certain percentage (which implies the rather dubious assumption that losses are proportional to consumption).

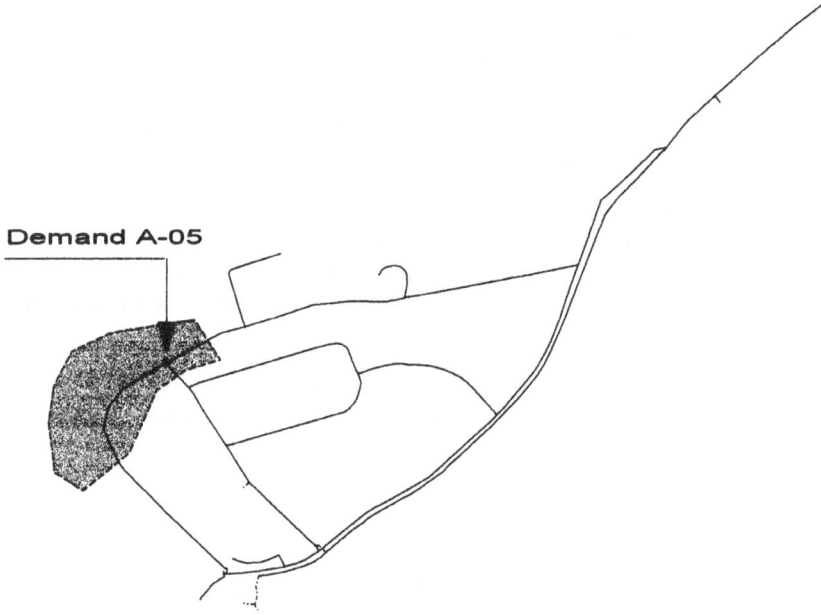

Demand A-05

Figure 5.4 Area supplied by node A-05.

- Allocate losses to particular nodes where, for instance, service pressure is higher than average.
- Distribute losses equally to all nodes – if there is no evidence to justify anything else.

Assembling the model elements

At present, the modeller has to make choices as to what will and will not be included in the distribution network part of a model. The extent to which pipes should be included was discussed in detail in Chapter 3, but to recapitulate briefly, the model should include most pipes that are in a loop and, depending on the use to which the model will be put, most closed pipes and small branches can be discarded. A policy of discarding all pipes below a certain size should be avoided and the inclusion of small pipes determined by their function in the network.

A degree of good hydraulic judgement is always necessary. The final solution is therefore highly personalised (which, incidentally, makes auditing of models rather difficult). With better and more powerful computers, GIS and user-friendly programs based on graphic interfaces, the task is now less difficult, but it is still formidable. It is expected that new procedures for making models from GIS data will be developed and standardised in the near future.

The final result of our assembly of data – the physical data on links and nodes and demand data – is the creation of a model schematic diagram. In the case of the network represented in

Fig. 5.1, the model schematic shows a rather simple network of nodes and links (Fig. 5.5).

The computation could start now, but in many cases it is worth while to examine the model and simplify further where it is feasible to do so without significant loss of accuracy. One should look for:

- closed pipes – these could be left out completely;
- short pipes with small hydraulic resistance – these might be dropped and their sinks merged into one node;
- branches – these could be removed from the model, at least temporarily.

In the case of the model in Fig. 5.5, it is assumed that all pipes are open, but the size of the model could be reduced by:

- merging together nodes A-08, A-09 and B-05;
- allocating consumption in branches to their base nodes (e.g. demand in node B-07 added to demand in node A-06, and pipe 'A-06 to B-07' removed);
- deleting nodes with zero demand (e.g. B-02).

Figure 5.5 First model – with all nodes.

The simplified model is shown in Fig. 5.6, with all nodes denoted by numbers and pipes denoted by letters. Note that the

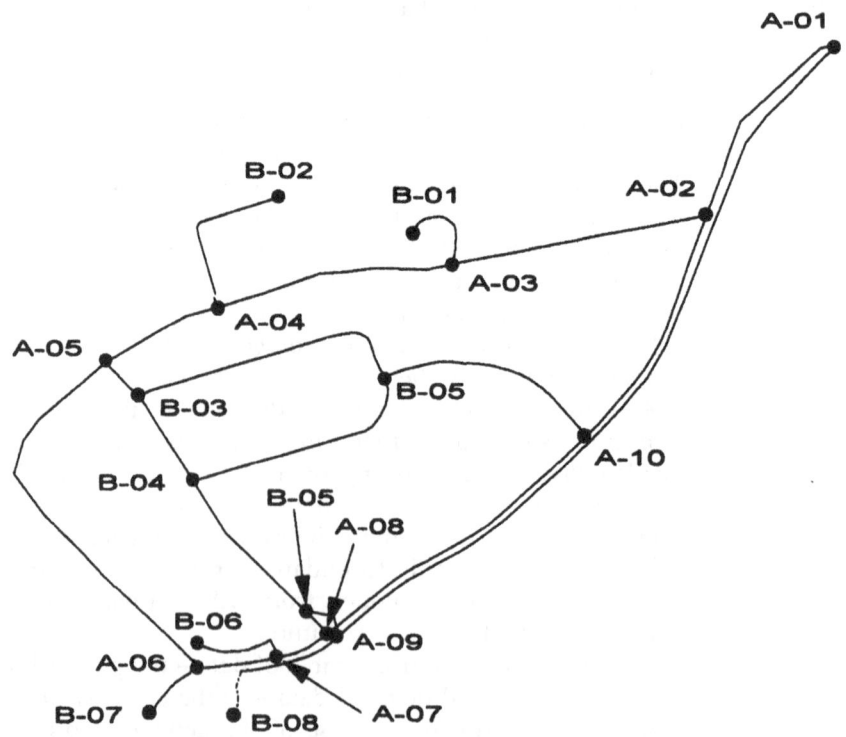

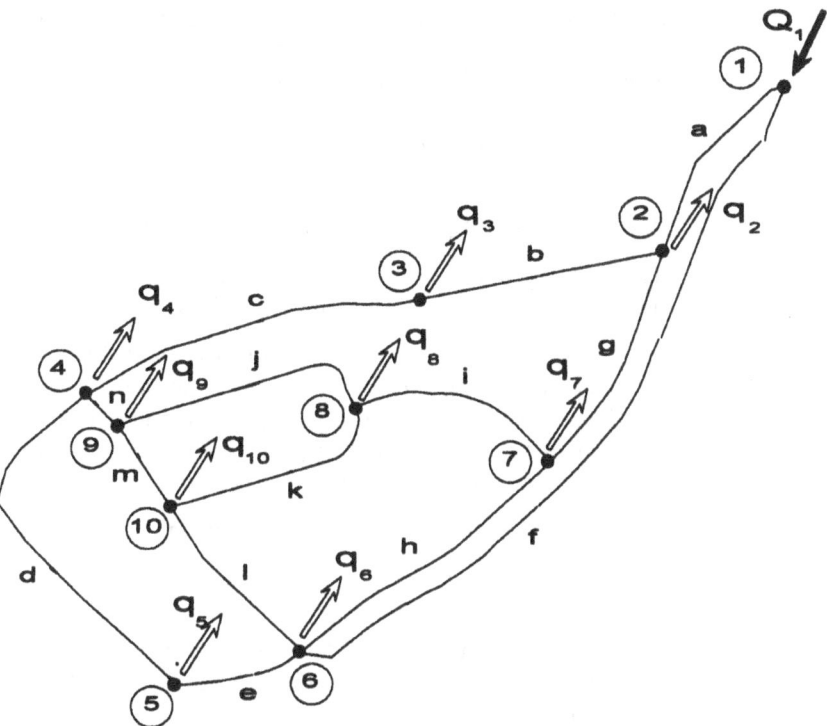

Figure 5.6 Second model – simplified.

sum of all demands, denoted by q_2 to q_{10}, has to be equal to the inflow Q_1. This model consists of:

- 10 nodes, denoted by numbers 1 to 10;
- 14 pipes, denoted by letters a to n.

Note also five loops such as the one formed by pipes b–c–n–j–i–g (others are j–m–k, d–e–l–m–n, i–k–l–h and a–g–h–f).

A simple relationship exists between the numbers of nodes, pipes and loops:

$$\text{loops} = \text{pipes} - \text{nodes} + 1$$

A 'loop' here means a closed circuit with the minimum number of pipes, called 'natural loops' (see Epp and Fowler, 1970). The reader may easily check this rule by starting from a branched network and adding one pipe at a time to close another loop.

Once the model is assembled, the next question is, 'What should be simulated?' The options are hydraulics, transient states, water quality changes, even economics (investments, revenue, costs). Here only the hydraulics of the system will be analysed: the behaviour of a water supply system over a given period of time (from several hours to several days). Assuming that all changes are slow – as is usually the case – the problem can be reduced to solving a series of discrete steady states of the network (called 'snapshots') linked by an integration scheme. The first task is therefore to solve one steady state of the system given the boundary conditions: demands, water levels in reservoirs, control policy for pumping stations and valves.

5.2 Computation of one steady state

The basic rules

The flow through any network, branched or looped or mixed, must satisfy two basic rules in a steady state.

1 *Conservation of mass*: all inflows must be equal to all outflows at every node (i.e. water is neither lost nor gained).
2 *Conservation of energy*: the algebraic sum of all increments of head around a closed contour (called 'loop') must be equal to zero (cf. Kirchhoff's voltage law in electrical circuits).

This is illustrated in Figs 5.7 and 5.8. From Fig. 5.7 it is obvious that:

$$Q_x = Q_y + Q_z + q_{demand} \tag{5.1}$$

This equation satisfies rule 1: conservation of mass. Applied to the three-pipe loop shown in Fig. 5.8, this principle yields three linear equations:

$$Q_i - Q_j - Q_k' - q_8 = 0 \tag{5.2}$$

$$Q_j - Q_m - Q_n - q_9 = 0 \tag{5.3}$$

$$Q_k + Q_m - Q_l - q_{10} = 0 \tag{5.4}$$

Here all inflows are positive ($+$) and all outflows negative ($-$). Local demand is an outflow and so is negative. The unknown variables are Q_j, Q_m and Q_k – three in all – while the other flows and local demands must be either given or computed through other equations. By summing up these three equations, the result is

$$Q_i - Q_n - Q_l - q_8 - q_9 - q_{10} = 0 \tag{5.5}$$

Figure 5.7 Node balance for a general node.

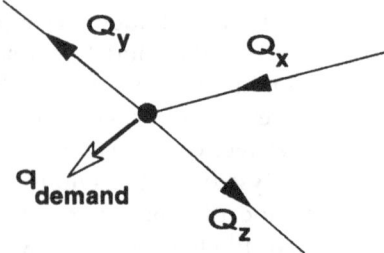

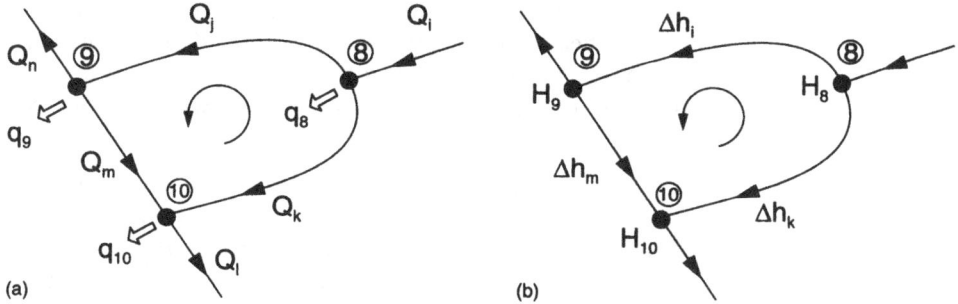

Figure 5.8 Balance around a loop: (a) flows and (b) head losses.

Note that three unknown variables have disappeared from this equation, which proves that they are not independent; only two of them could be used in computation, and therefore another equation must be provided. This arises from the conservation of energy – rule 2 above.

Hydraulic losses in the pipe j (Fig. 5.8b) could be computed as

$$H_8 - H_9 = \Delta h_j = K_j |Q_j|^r Q_j \qquad (5.6)$$

where the coefficient K_j is defined in Chapter 3 and the exponent r is given by:

$r = 1.0$ in the Darcy–Weisbach equation

$r = 0.85$ in the Hazen–Williams equation

for computation of head loss in a pipe. Note that the flow Q_j is positive in the direction from node 8 to node 9; then head H_8 is higher than H_9 and consequently Δh_j is positive too. In the opposite case, flow Q_j will change sign, but equation (5.6) remains valid.

Summing up these losses around the loop (in this case from node 8 to 9 then to 10 and back to 8), the following relationship can be established:

$$\Delta h_j + \Delta h_m - \Delta h_k = (H_8 - H_9) + (H_9 - H_{10}) - (H_8 - H_{10}) = 0 \qquad (5.7)$$

which satisfies rule 2. Note the negative sign for h_k because the assumed direction of flow (from node 8 to 10) is against the direction of summation.

By introducing flows in equation (5.7) the result is:

$$K_j |Q_j|^r Q_j + K_m |Q_m|^r Q_m - K_k |Q_k|^r Q_k = 0 \qquad (5.8)$$

which is a non-linear equation linking three unknown flows: Q_j, Q_m and Q_k. Such a system cannot be solved directly, and some iterative procedure must be applied instead.

Hardy Cross method

There are several solutions to this problem, but no particular one has been universally accepted to date. Most of the algorithms are based on Hardy Cross's ideas, published in the mid-1930s (Cross, 1936). He proposed two solutions to the problem.

1 *Loop-oriented method*: Assume initial values of flow in all pipes of the network in such a way that conservation of mass (rule 1 above) is satisfied, and compute the sum of head losses around each loop – if any value exceeds permissible error, start to correct flows in loops in such a way that the conservation of mass is not destroyed, and then repeat the procedure until an acceptable balance is achieved.
2 *Node-oriented method*: Assume initial heads in all nodes to satisfy rule 2 above (because at each node there is only one value for head), then compute flows in each pipe from an assumed difference of pressure at the end nodes, compute a balance for each node and see if the error is within acceptable limits – if not, correct the initial heads and repeat the procedure.

These concepts were used for analysis of flow in pipe networks by other scientists, like Cornish (1940) and others. However, in precomputer times it was difficult to apply either concept easily. The first one 'looked more promising for manual calculations, and was further developed and widely applied until recent times.

The first step is to make a reasonable initial assumption about the flows, taking into account local demands – as noted in Fig. 5.9. The next step is to compute the sum of all head losses

Figure 5.9 A single loop.

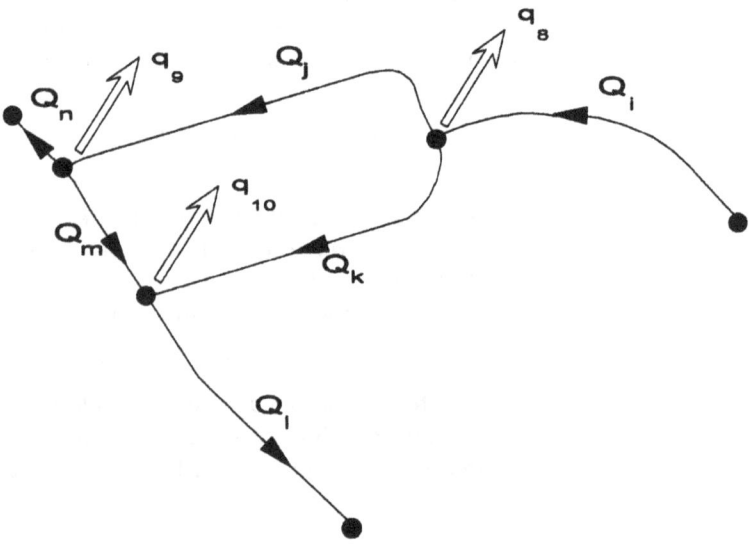

around each loop (equation (5.8)). The sum of head losses round the loop will not equal zero (except in very rare cases when one hits upon the solution by pure luck), so the initial flows must be corrected. The correction could be computed by the classical *Newton method* (or Newton–Raphson method).

The zero of a non-linear function $F(x)$ is found by expanding the function in a series around a known point (x_0) as

$$F(x) = F(x_0) + F'(x_0)\Delta x + O(x_0) = 0 \qquad (5.9)$$

where F' is the first derivative at the given point x_0 and $O(x_0)$ is the residue, presumed to be small enough to be ignored. The unknown correction (Δx) is approximately equal to

$$\Delta x = -\frac{F(x_0)}{F'(x_0)} \qquad (5.10)$$

By applying the same technique here, one easily finds (see equation (5.6))

$$F = \sum K|Q|^r Q \qquad (5.11)$$

and

$$F' = \frac{\partial \sum\limits_{\text{loop}} \Delta H}{\partial(\Delta Q)} = \sum_{\text{loop}} \frac{\partial(\Delta H)}{\partial(\Delta Q)} = \sum_{\text{loop}} \frac{\partial[K(Q + \Delta Q)^r(Q + \Delta Q)]}{\partial(\Delta Q)}$$

$$= 2\sum_{\text{loop}} K|Q|^r \qquad (5.12)$$

The correction of flow is then equal to

$$\Delta Q = \frac{\sum\limits_{\text{loop}} K|Q|^r Q}{2\sum\limits_{\text{loop}} K|Q|^r} = \frac{K_j|Q_j|^r Q_j + K_m|Q_m|^r Q_m - K_k|Q_k|^r Q_k}{2(K_j|Q_j|^r + K_m|Q_m|^r + K_k|Q_k|^r)} \qquad (5.13)$$

The new values for flow (denoted by asterisk) are:

$$Q_j^* = Q_j + \Delta Q \qquad (5.14)$$

$$Q_m^* = Q_m + \Delta Q \qquad (5.15)$$

$$Q_k^* = Q_k - \Delta Q \qquad (5.16)$$

The next step is to replace the originally assumed flows with the newly computed values. The absolute value of the error, computed through equation (5.8) again, should be smaller than before – and always is! By repeating the procedure a few times, one can find the solution to the desired accuracy. The reader may find more details on this procedure in van der Zwan (1990), with an elaborate case.

This simple technique is very effective if the network has just one loop. If there are more loops, as in the example used here

(Fig. 5.10), the same procedure might be used to solve the first loop, then the second loop and so on, until the last loop is reached. Unfortunately, if there are pipes that belong to two loops (and normally such pipes *do* exist – pipes g, h, i, j, k, l, m and n in Fig. 5.10), then correcting flows in the subsequent loop introduces a fresh error in all the loops, previously solved, that have some pipes in common. In Fig. 5.10, pipe j, for instance, belongs to both loop I and loop II. The flow through this pipe will be determined first when solving the first loop. In the next step, when dealing with loop II, the flow through pipe j will be corrected – which simply means that the balance in loop I, just found, is disturbed – maybe significantly. The same problem will reappear later with all pipes that are common to two loops.

In precomputer times, engineers developed skill in solving networks in the way just described. They had various tricks: compute the error for all loops and start with the one where the maximum error appears; introduce a few 'macro loops' to get a near solution, and then proceed with small fry; reduce the error in one loop a little and then move to the next one; etc.

Despite the skill and tricks that developed, all these expedients were not good enough for general use or for networks of any complexity. It is clear that a better algorithm was needed, which will attack the whole network simultaneously. The reader may find several different solutions to this problem in the references. The most popular methods are outlined below.

Figure 5.10 A system with several loops.

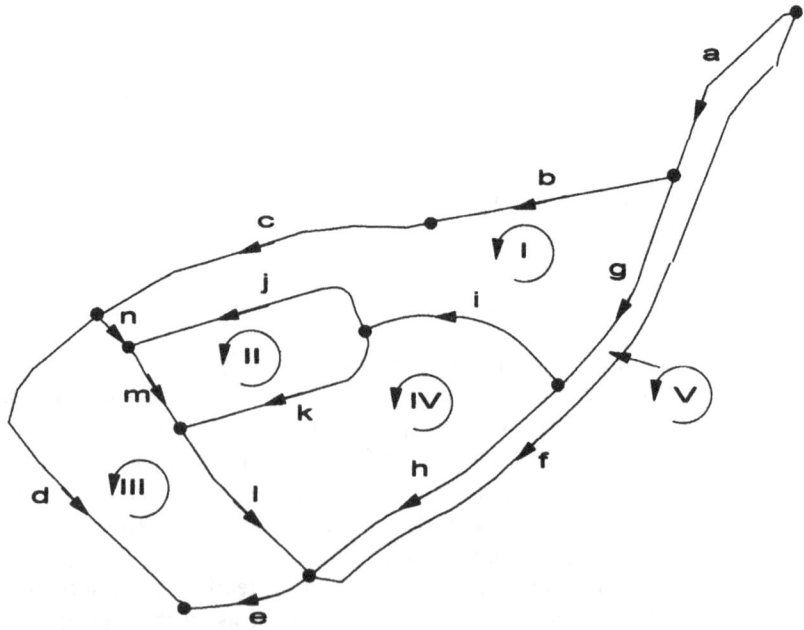

Loop (ΔQ) equations method

This method is the logical extension of the 'one loop at a time' procedure described above (Epp and Fowler, 1970). A system of N_k non-linear equations is formed, where N_k is the number of loops in the system. The initial activities are:

- identification and numbering of all loops,
- initialisation of all flows in such a way that continuity is satisfied at all nodes of the system (i.e. the law of conservation of mass is satisfied).

These tasks could be performed manually or (better) by using a computer.

The next step is to check the proposed solution – by determining how nearly the law of conservation of energy is satisfied in all the loops. For a loop k, this is measured by the algebraic sum of all increments of head (frictional losses or pump heads) along its pipes:

$$f_k = \sum_{i=1}^{N_k}(\Delta H_i) = \sum_{i=1}^{N_k}(K_i|Q_i|^r Q_i) \tag{5.17}$$

The summation is carried out over all pipes that belong to the loop k. It should be close to zero, but almost certainly will not be so at the beginning of the calculation. To find a better approximation, different methods might be used; one of them is the Newton method, very efficient in most cases.

The value of f_k is a function of estimated flows in all its pipes, so it can be expanded in series – taking these flows as independent variables. The result is:

$$f_k = f_k^{(0)} + \frac{\partial f_k^{(0)}}{\partial(\Delta Q_k)}\Delta Q_k + \sum_{j=m}^{N_k}\frac{\partial f_k^{(0)}}{\partial(\Delta Q_j)}\Delta Q_j + O(\Delta Q_k^2) \tag{5.18}$$

Only the linear terms are shown – the higher ones are in the residue and are discarded (denoted by $O(\Delta Q_k^2)$ above). Here ΔQ_k denotes flow corrections for this loop (k), while ΔQ_j is the correction for an adjacent loop (j) that shares the same pipe(s) with this loop k.

The partial derivatives are calculated as follows. For the loop k:

$$\frac{\partial f_k}{\partial(\Delta Q_k)} = \frac{\partial\left[\sum_{i=1}^{N_k}K_i|Q_i + \Delta Q_k - \Delta Q_m|^r(Q_i + \Delta Q_k - \Delta Q_m)\right]}{\partial(\Delta Q_k)}$$

$$= 2\sum_{i=1}^{N_k}K_i|Q_i|^r \qquad \text{for } \Delta Q_k \to 0 \tag{5.19}$$

and for the adjacent loop m:

$$\frac{\partial f_k}{\partial(\Delta Q_m)} = -2\sum_{n=1}^{N_m}K_n|Q_n|^r \qquad \text{for } \Delta Q_m \to 0 \tag{5.20}$$

The value of f_k should be as close to zero as possible. New corrections of flow are again found by the Newton method. This means that:

- the residue O is rejected,
- equations similar to equation (5.18) are compiled for each loop in the system,
- the system of linear equations is formed in the shape of a matrix equation

$$[A_{km}]\Delta q_m = -f_k^{(0)} \tag{5.21}$$

where the term in the kth row and mth column of the matrix is

$$A_{km} = \frac{\partial f_k^{(0)}}{\partial(\Delta Q_m)} \tag{5.22}$$

The unknowns are flow corrections q. Note that *all* corrections are computed simultaneously, thus greatly improving the stability and rapidity of the computation. The matrix for the network shown in Fig. 5.10 is included here as Fig. 5.11. Note its small size and few zeros.

The system (5.21) can be solved by any standard routine. Note that this matrix has the following characteristics:

- It is symmetrical.
- It contains a lot of zeros (whenever the kth and mth loops have no common pipes).
- Diagonal members are always the largest in their row and positive, while off-diagonal members are smaller and negative.

All these characteristics should be used to very good effect when preparing the program for the computer.

Figure 5.11 The matrix for the loop method.

A_{11}	A_{12}	A_{13}	A_{14}	A_{15}
A_{21}	A_{22}	A_{23}	A_{24}	0
A_{31}	A_{32}	A_{33}	A_{34}	0
A_{41}	A_{42}	A_{43}	A_{44}	A_{45}
A_{51}	0	0	A_{54}	A_{55}

Reservoirs, pumping stations and valves

The network considered in the previous section was of the simplest kind – no reservoirs, pumping stations or valves. These facilities need special attention in modelling.

Reservoirs One reservoir at least is needed per network – just to give a datum for the system. As the whole method concentrates on head increments (ΔH), one can easily see that head must be known at one point. Pressures at other points are found just by following pipes and adding or subtracting corresponding values of ΔH.

In systems with two or more reservoirs, imaginary links must be introduced as shown in Fig. 5.12. The fictitious link between reservoirs A and B closes the corresponding loop. It differs from ordinary pipes because no flow goes through it, but its ΔH value is real:

$$\Delta H_{\text{fict}} = Z_A - Z_B \tag{5.23}$$

A fictitious link contributes to the sum of head increments around its loop, f_k (equation (5.17)), and nothing else. The rest of the computation is the same as above.

If there are several reservoirs, say M_r, the number of such links is $M_r - 1$ (one less). For a system with, say, seven reservoirs, there are six fictitious links (Fig. 5.13).

Pumping stations A pumping station may either supply the system from one end or boost the pressure within the distribution network – as in Fig. 5.14. The pumping station is a link with special properties. It needs to be modelled in one of two ways, as appropriate. Those are:

- The state of the pumping station is known and with it also the overall flow–head characteristics – the operating regime has to be determined through computation.

Figure 5.12 Link between two reservoirs.

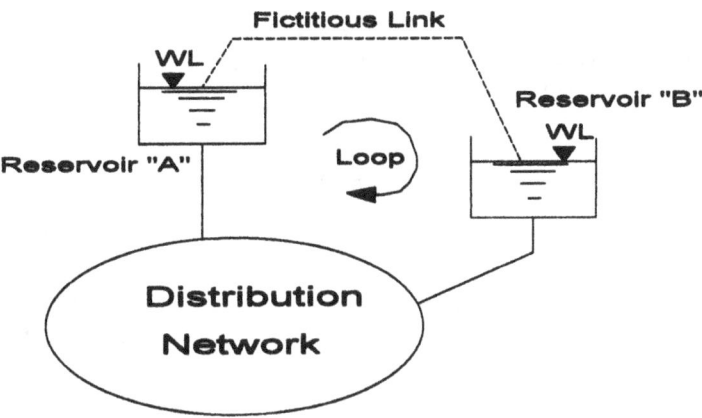

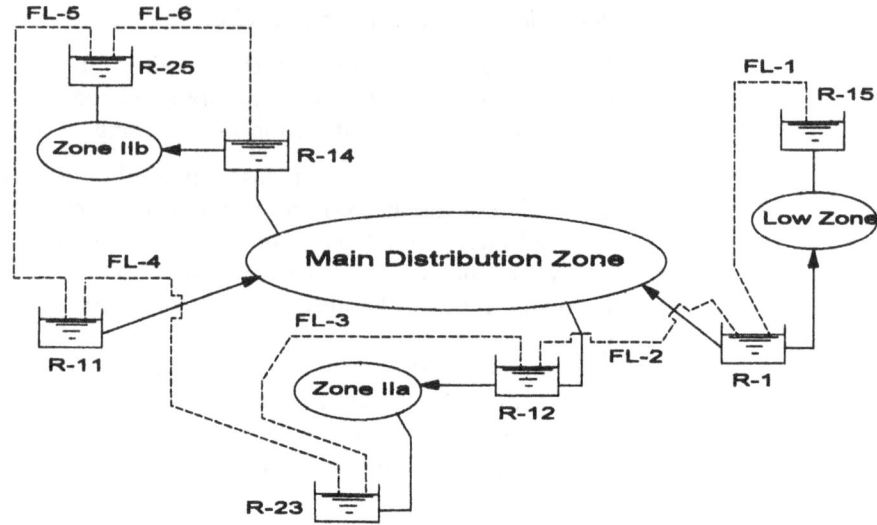

Figure 5.13 Seven
reservoirs in one system.

• Flow through the pumping station is given in advance (called *target* flow), and the head has to be computed.

Note that in both cases the power used by the pumps is not mentioned. Strictly speaking, it is not needed for hydraulic analysis of this kind – but it is wise to compute this variable too, and save it for subsequent analysis and evaluation of results.

In the first case the operating characteristics of the whole pumping station must be computed first, taking into account the number of active units, pump types and characteristics, rotational speed, throttling of individual units and all the other relevant data described in Chapter 3. The net result is a curve as displayed in Fig. 5.15. Note that this curve may be smooth or not – depending on the current state of active pumps and on the similarity or not of the pumps in a multi-pump station.

Figure 5.14 Modelling
of a pumping station: (a)
source PST; (b) booster
PST.

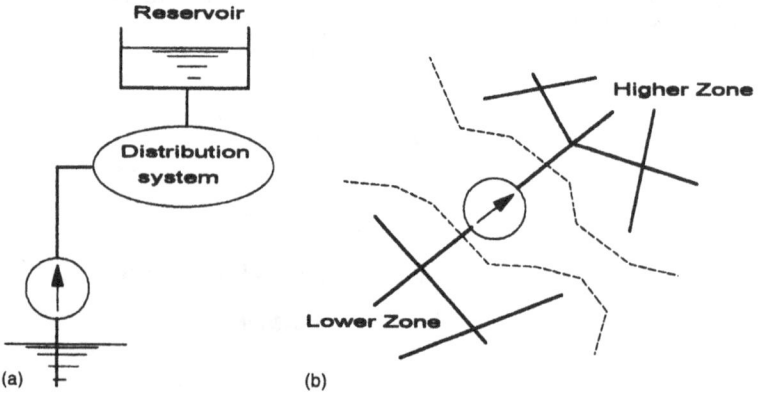

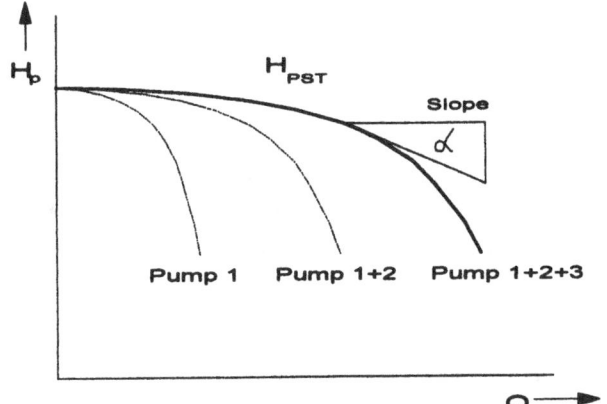

Figure 5.15 Operating characteristics of a PST.

The pumping station is now a link that produces an increment of head in relation to the flow:

$$\Delta H = H_p(Q_p) = F(Q_{ij}) \tag{5.24}$$

This increment should be added to the f_k value (in equation (5.17)).

Moreover, the first derivative, needed for equation (5.18), is now equal to

$$\frac{\delta(\Delta H)}{\delta(\Delta Q_k)} = F'(Q_{ij}) \tag{5.25}$$

where F' represents the slope of the overall operating characteristic (Fig. 5.15). This value enters equation (5.19) or (5.20) – as required by the position of the pumping station in the network. Computation then proceeds as before.

The operating characteristics of a pumping station may be represented by a parabola (or in some rare cases by some other analytical function) to facilitate computation, but this is not obligatory (and may lead to considerable differences between real data and this approximation). The other method is to apply interpolation – linear or higher – with a procedure for data 'smoothing'.

The other case, when target flow is given, is slightly more complicated. The loop where this station is located must be cut at that place (Fig. 5.16). The link is temporarily removed; target flow is now an *outflow* for the suction-side node, and an *inflow* for the pressure-side node. The loop does not exist as such, and the number of equations is reduced.

The resulting system is solved (target flow included as local inflow/outflow as described above) and all heads computed. Now the pumping head is just the difference

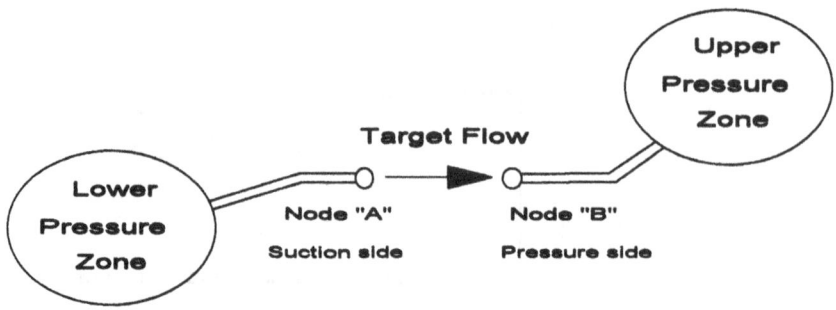

Figure 5.16 PST with a known discharge.

$$H_p = H_B - H_A \qquad\qquad (5.26)$$

Whether this can be achieved with the pumps available in the pumping station – and how – is a different problem. One possible, simple algorithm is given in Fig. 5.17.

The pumps are 'started' (in the model) one by one until the *target* flow is reached. The contribution of each unit is the flow increment ΔQ, which corresponds to the computed head H_p (from equation (5.24)).

A distinction must be made between pumping stations that have a variable-speed unit and those equipped solely with fixed-speed pumps. In the first case, the variable-speed unit will

Figure 5.17 Selection of the proper combination of pumps.

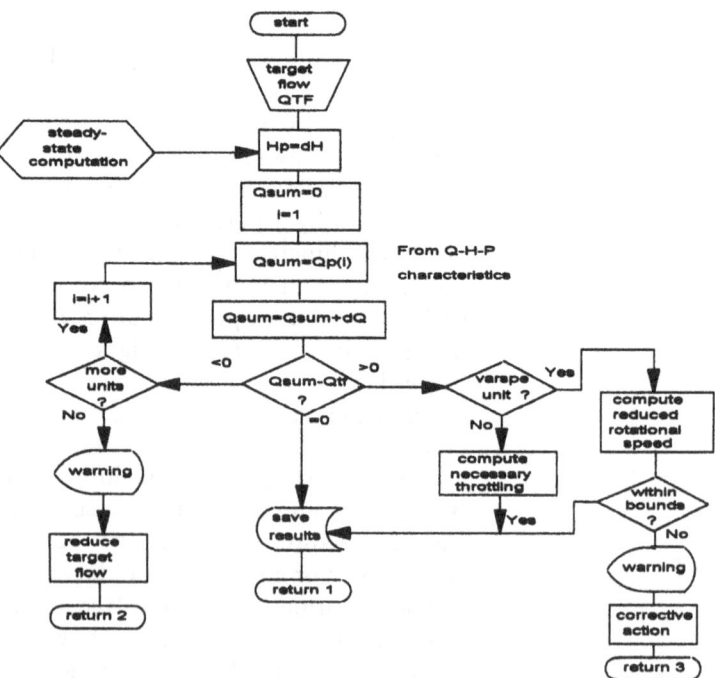

always be active – with or without other (fixed-speed) pumps. The eventual excess flow (over target flow) will be eliminated by adequate decrease of rotational speed of this pump (Fig. 5.18).

The appropriate rotational speed should be computed by applying the affinity laws (Chapter 3). The particular mechanism controlling the pump speed will normally limit the range of variation to between 70% and 100% of the rated speed. A frequency converter can give an even greater range, either down to zero or upwards as far as the capacity of electrical installation would permit.

Valves The majority of valves in a system are gate valves, used as simple open or shut devices to create zones or to allow work on a particular pipe in the network. They may be treated, in most cases, as simple minor losses, included in the equation for the coefficient $K(\zeta_{reg})$:

$$K = \frac{\lambda(L/d) + \sum \zeta + \zeta_{reg}}{2gA^2} \tag{5.27}$$

The value of ζ_{reg} is a function of the valve's opening, and in cases when this opening is predetermined – as for throttled (THV) or time-controlled (TCV) valves – the solution is a simple computation of the value of the coefficient K, based on the valve characteristic (see Fig. 3.72).

In cases where the valve opening cannot be determined in advance because it is a function of variables that have to be computed (e.g. pressure- or flow-controlled valves), the problem becomes serious. One simple expedient is to supply the missing information (e.g. pressure or flow) from the previous full model

Figure 5.18 Variable-speed pump operating in parallel with several fixed-speed pumps.

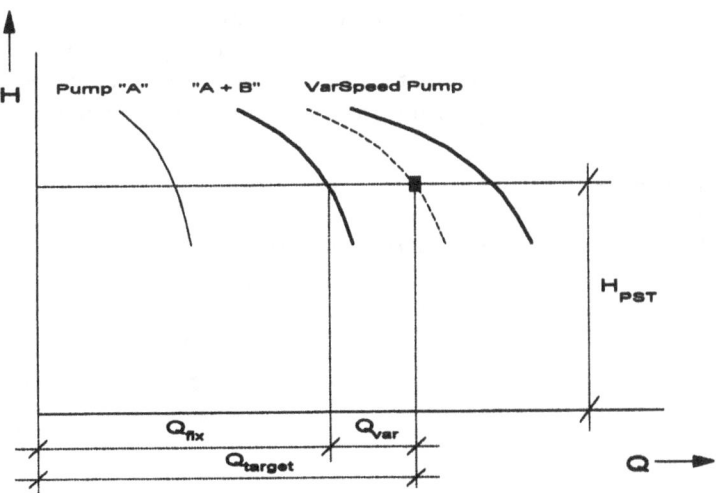

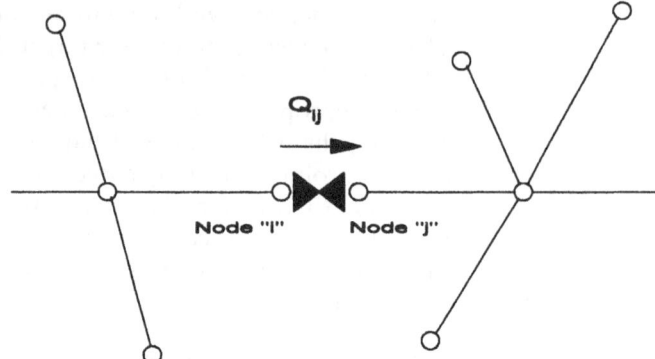

Figure 5.19 A control valve in the distribution network.

computation, but then the problem of instability may appear. One way or the other, the opening of a control valve has to be estimated prior to proceeding to a new steady-state computation. The link containing a control valve (see Fig. 5.19) is then treated as any other ordinary link.

Pressure reducing valves (PRV) are a special case altogether. Their function is to keep pressure in a chosen node as close as possible to a given value (as explained in Chapter 3). This is difficult to simulate in the general case (Jeppson and Davis, 1976).

However, there are cases where a PRV can be simulated in a straightforward manner – when the device is situated upstream of a branched or looped portion of the network and there is no reservoir in the part of the system affected by the PRV (Fig. 5.20). The procedure is clear: computation is done in the usual way, assuming that the PRV is fully open. The computed value of head at the control node is compared with the desired value: if there is any excess, it is reduced by throttling the PRV by an adequate amount.

A PRV that is fully incorporated in the loops of a network can be treated as a pressure-controlled valve, using the value from the previous iteration. It may be a problem in modelling, but a more

Figure 5.20 Case when a PRV can be easily modelled.

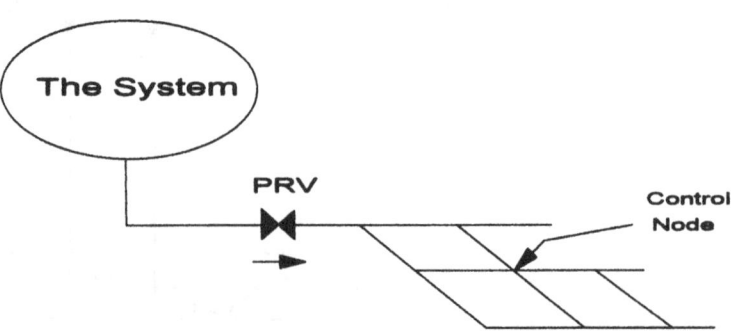

important point is that such a solution is not recommended in practice – it may cause instability in real systems as well as in the model, as changes to the pressure on the downstream side of the valve may, through the loop(s), affect the upstream pressure at the valve. There are certainly situations, in real systems, in which pressure control to a looped system appears attractive to the operator, but the installation should only be carried out if a discrete or semi-discrete supply-zone is created, downstream of the PRV, using zone valves and/or non-return valves.

Note of caution The Newton method (also called the Newton–Raphson method) described above has a tendency to overshoot the solution, with consequent oscillations around it. To reduce this tendency, it is advisable to apply a reduced correction of flow. Then the new value of flow through pipe i–j, belonging to the nth loop, will be

$$Q_{ij} = Q_{ij}^{(0)} + \kappa \Delta Q_k \tag{5.28}$$

where κ is less than 1 (normally 0.6 to 0.9).

Node (H) equations method

The other possibility – to start from assumed heads and check the continuity of flow – is somewhat simpler and easier for programming (Shamir and Howard, 1968; Zarghamee, 1971).

A part of the network, around node i, is shown in Fig. 5.21. All heads are initially estimated, so the flow through all pipes can be computed. Then the balance for node i will show whether the estimate was correct or not.

The flow through one pipe – for example, from node i to node j – can be computed as:

$$Q_{ij} = C_{ij}|H_i - H_j|^u \, \text{sgn}(H_i - H_j) \tag{5.29}$$

where C_{ij} is the unit capacity of the pipe (i.e. the flow that corresponds to 1 m of pressure difference), exponent u is given by

Figure 5.21 Part of a distribution network around the node i.

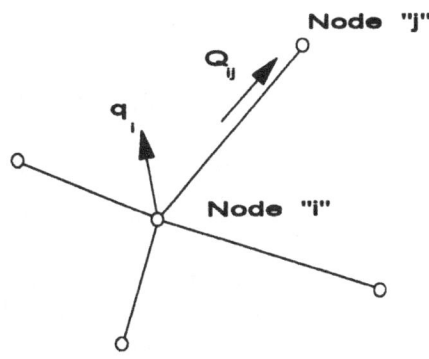

$$u = \begin{cases} 1/(r+1) = 0.50 & \text{for the Darcy–Weissbach formula} \\ 0.54 & \text{for the Hazen–Williams formula} \end{cases}$$

and the function 'signum' is defined as

$$\text{sgn}(x) = \begin{cases} +1 & \text{for } x > 0 \\ 0 & \text{for } x = 0 \\ -1 & \text{for } x < 0 \end{cases} \tag{5.30}$$

By computing flows through all pipes connected to node i and deducting the local demand, one finally arrives at:

$$\sum_{j=1}^{N_i} Q_{ij} - q_i^{(0)} = \delta_i(H_i; H_1, H_2, \ldots H_{N_i}) \tag{5.31}$$

Here δ_i represents the magnitude of error – it should be equal to zero. The value of δ_i is clearly a non-linear function of assumed values of head in all neighbouring nodes. By varying these heads, one should find the solution – which is indicated by all δ_i values close to zero. This can be achieved by different methods. Here the Newton method is used again – see other methods in van der Zwan (1988) or Gessler (1981).

The problem is to find corrections of head, denoted by h_i:

$$h_i = H_i - H_i^{(0)} \tag{5.32}$$

in such a way that non-linear functions d are reduced to zero. By expanding the function d in a series and retaining the linear terms only, one gets finally

$$\frac{\partial \delta_i}{\partial h_i} h_i + \sum_{j=1}^{N_j} \frac{\partial \delta_i}{\partial h_j} h_j = -\delta_i^{(0)} \tag{5.33}$$

where h_i is the unknown correction of head in node i and h_j is the same for other neighbouring nodes.

The partial derivative for h_i correction is equal to

$$\frac{\partial \delta_i}{\partial H_i} = \frac{\partial \delta_i}{\partial h_i} = \frac{\partial}{\partial h_i}\left(\sum_{j=1}^{N_i} Q_{ij} - Q_i^{(0)}\right) = \sum_{j=1}^{N_i} \frac{\partial Q_{ij}}{\partial h_i}$$

$$= \frac{\partial}{\partial h_i}\left(\sum_{j=1}^{N_i} C_{ij}|H_i - H_j|^u \, \text{sgn}(H_i - H_j)\right)$$

$$= u\sum_{j=1}^{N_i} C_{ij}|H_i - H_j|^{u-1} \tag{5.34}$$

and for other correction, h_k, the derivative is

$$\frac{\partial \delta_i}{\partial H_k} = \frac{\partial \delta_i}{\partial h_k} = -uC_{ik}|H_i - H_k|^{u-1} \tag{5.35}$$

One equation of the type of equation (5.33) has to be prepared for each node in the network. After rearrangement, this results in a system of linear equations

$$[B_{im}]\Delta h_m = -\delta_i^{(0)} \qquad (5.36)$$

The similarity between this system and the one above (equation (5.21)) is then more apparent – they have the same characteristics, which is quite useful.

Reservoirs, pumping stations and valves

Reservoirs A reservoir is just a point where the corrections equal zero. Therefore, if it is situated in node k,

$$h_k = 0 \quad \text{and} \quad H_k = Z_k \qquad (5.37)$$

where Z_k is the current water level. The best way to handle this situation is to drop the equation for node k from the system (5.33) altogether. This calls for a slightly more complicated program, but it is worth the effort. An alternative is to change the matrix B (put 1 on the diagonal at place k and clear this column and this row by inserting zeros, also let $d_k = 0$), but this option is not recommended. A redundant equation is retained and programming difficulties are increased.

Pumping stations A pumping station should be modelled in much the same way as above. The operating characteristics of the pumping station should be determined prior to this point, as was done in the previous method. The station is then represented as a special link with the properties

$$Q_{ij} = \Phi(H_j - H_i) \qquad (5.38)$$

i.e. flow through this link is a function of the actual pump head (difference of pressure downstream and upstream). The partial derivatives – needed for equations (5.19) and (5.20) – are easily determined as

$$\frac{\partial Q_{ij}}{\partial H_i} = -\frac{\partial Q_{ij}}{\partial H_j} = -\Phi = \frac{1}{F'} \qquad (5.39)$$

where F' is the slope of the operating characteristic (see Fig. 5.15). These relationships should be taken into account when matrix B is compiled.

Valves The valves are treated as before. The coefficient C ('unit capacity') for the pipe with a valve will be changed into

$$C = A\sqrt{\frac{2g}{\lambda(L/d) + \sum \zeta + \zeta_{\text{reg}}}} \qquad (5.40)$$

where flow area is $A = \pi d^2/4$.

Note of caution The Newton (or Newton–Raphson) method gives slightly overestimated corrections and the results may

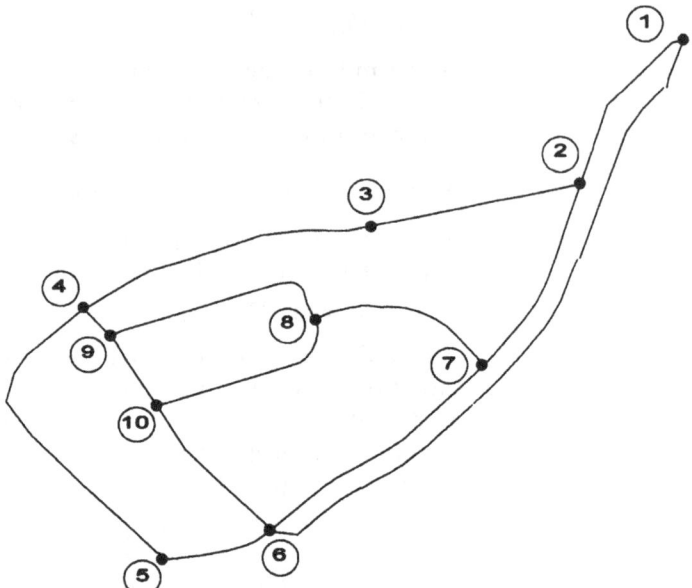

Figure 5.22 All nodes
for the selected network.

oscillate around the true solutions. Therefore, the corrections h should be reduced (again as in the previous method) before being applied. The new values of heads are then equal to

$$H_i = H_i^{(0)} + \kappa h_i \tag{5.41}$$

where κ is a factor less than 1 (usually between 0.6 and 0.9).

All nodes for the selected case (Fig. 5.6) are shown in Fig. 5.22 and the corresponding matrix is in Fig. 5.23. Note that the matrix is now larger (see Fig. 5.11) and sparse – several elements are equal to zero. The matrix is symmetrical in both cases.

Pressure-related demand

The node method has an advantage over other methods: equations describing the relationship between demand and pressure in the network can be easily included, as will be demonstrated below.

It is unfortunate that the dependence of actual demand upon local pressure has not been better highlighted in the literature. Most researchers have assumed that demands (including losses) are known – which is not the case. At the very best, one can have data on demands and losses captured by modern instruments and technology (see Chapter 4) *under normal circumstances*. If the same data were included in the model as demands, and the results – heads and flows – are close to the observed values, then everything is fine and acceptable. In the opposite case, the model

B_{11}	B_{12}					B_{16}			
B_{21}	B_{22}	B_{23}				B_{27}			
	B_{32}	B_{33}	B_{34}						
		B_{43}	B_{44}	B_{45}			B_{49}		
			B_{54}	B_{55}	B_{56}				
B_{61}				B_{65}	B_{66}	B_{67}		B_{610}	
	B_{72}				B_{76}	B_{77}	B_{78}		
						B_{87}	B_{88}	B_{89}	B_{810}
			B_{94}				B_{98}	B_{99}	B_{910}
					B_{106}		B_{108}	B_{109}	B_{1010}

Figure 5.23 The matrix for the node method.

should be calibrated by changing some parameters until a satisfactory agreement is reached. Note that *both flows and pressures* have to match the real value as closely as possible. But the model has to be applied to other situations as well, some very different from anything yet encountered in reality. The computed pressures can be outside the normal range, either higher or lower and sometimes even negative. Obviously such results must be rejected because pressures cannot be negative – the system will be emptied beforehand and air will enter the pipes, thus negating the initial assumptions of modelling: continuity of water mass. However, such events are extremely rare in practice – and much more frequent in computation. The problem is that demands (and losses) depend quite strongly on local pressures; as pressures start to fall, so will the demand, and the real system might survive an emergency that the model cannot.

This difficulty can be solved only by introducing a relationship between demand and local pressure; as hard data are extremely rare, assumptions must be made (Fig. 5.24). In one model (Fig. 5.24a) the relationship is linear (Jones, 1984), while in the other (Fig. 5.24b) it is proportional to the square root of pressure (Wessex Water, 1993a, b). Of course, these assumptions

Figure 5.24 Demand–pressure relationship: (a) linear; (b) square root.

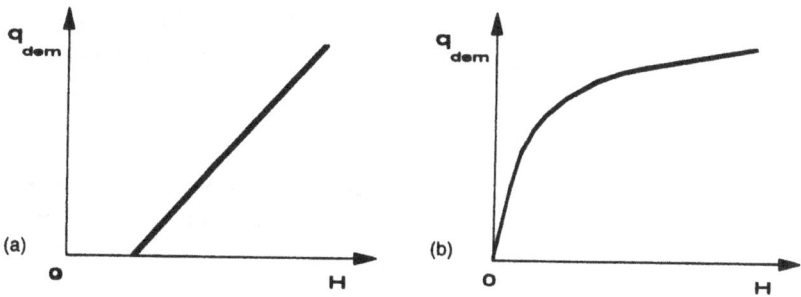

(a)   (b)

are too simplified, but might serve well until better data are made available through practice.

Note that this relationship – if known and described by mathematical means – could be easily introduced in the model. Node-oriented methods are particularly well suited for this. Introducing in equation (5.31) the new equation

$$q_i = F(H_i) \tag{5.42}$$

the local demand q_i can be computed from the current value of pressure, H_i. By taking the partial derivative of this function – as part of equation (5.33) – one obtains the new equation (5.34) as

$$\frac{\partial \delta_i}{\partial H_i} = u \sum_{j=1}^{N_i} C_{ij} |H_i - H_j|^{u-1} - F'(H_i) \tag{5.43}$$

This means that only the term on the diagonal is changed, while the rest of the system remains the same. Note that all nodes must be retained in the model – at least those with some demand.

The difficulty is, how to define the unknown function F in real life. Above all, more data about demands and pressures in buildings and control zones are needed, which could then be analysed by statistical means and general conclusions formulated. It seems that a lot of spadework is needed before any practical solution is found.

Other numerical methods

The Newton (or Newton–Raphson) method is very efficient in most cases, but there are other methods that are acceptable (Wood and Charles, 1972; Isaacs and Miles, 1980). The proposal is to linearise the basic equations and solve the problem by an iterative procedure. In a sense it is a combination of the node and loop methods, and can be used to compute either heads or flows.

The basic equation of the node method is

$$\sum_{i=1}^{N_i} Q_{ij} - q_i^{(0)} = 0 \tag{5.44}$$

i.e. the sum of all inflows and outflows should equal the local demand q. The unknown flows are found by successive approximations. After n steps, the flow through pipe i–j is equal to

$$Q_{ij}^{(n)} = k_{ij}^{(n)} (H_i - H_j) \tag{5.45}$$

where the new coefficient k is not a constant, but related to the previous value of flow through the same pipe as:

$$k_{ij}^{(n)} = C_{ij}^2 / |Q_{ij}^{(n)}| \tag{5.46}$$

The coefficient C is the same as in the node method (see equation (5.29)).

Introducing this result in equation (5.44), one obtains the linear equation with unknown heads H in the shape of:

$$\left[\sum_{j=1}^{N_i} k_{ij}^{(n)}\right] H_i - \sum_{m=1}^{N_i} k_{mi}^{(n)} H_m = q_i^{(0)} \qquad (5.47)$$

An equation like this one can be composed for each node in the system. Together, these equations permit the unknown heads H to be computed. The process is then repeated using equation (5.44) to estimate whether the solution is reached or not. Note that the number of equations equals the number of nodes. Convergence might present a problem, so some safeguards must be included in the mathematical procedure.

Flow (Q) equations method

A similar technique could be used to compute flows directly, by using the loop method combined with the node method as explained below.

In a network with N_k nodes, there are $(N_k - 1)$ continuity equations like

$$\sum_{i=1}^{N_k} Q_{ik} - q_k^{(0)} = 0 \qquad (5.48)$$

which are linear with respect to flows Q_{ik}. For a system without loops, nothing else is needed. However, when loops are present, for each loop an equation – which sums up all increments of head around its contour – should be created in the shape of

$$\sum_{j=1}^{N_k} K_j^* Q_j = 0 \qquad (5.49)$$

where new variable coefficients are introduced as

$$K_j^* = K_j Q_j^r \qquad (5.50)$$

Here Q^r denotes the flow found in the previous iteration.

There are as many additional equations of this type as there are loops within the network, so the total number of equations is equal to

equations = loops + nodes − 1

The set of linear equations is composed of equations (5.48) and (5.49). The solution is found by standard routines. The model is shown in Fig. 5.25 and the corresponding matrix in Fig. 5.26. Note that this system does not have the useful properties of A and B matrices (loop and node method, respectively), because the matrix is the largest of the three and unsymmetrical

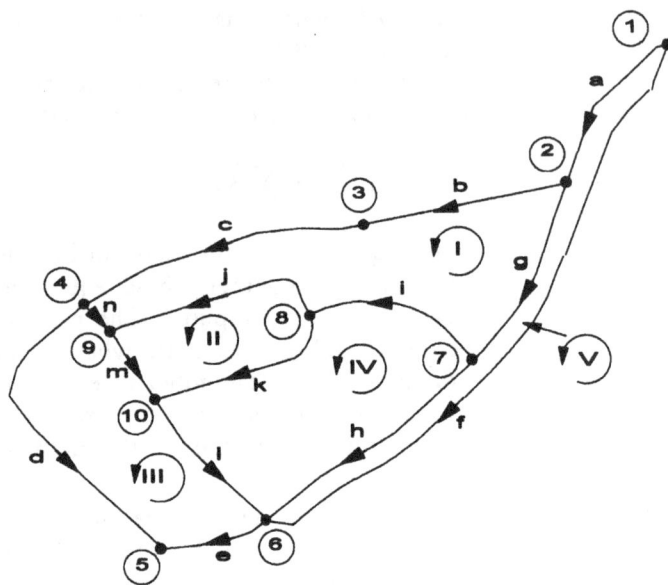

Figure 5.25 The flow method.

(although again it is sparse, permitting the use of special techniques developed elsewhere).

The results are unknown flows. If the new values agree well with the assumed ones, the solution is reached – if not, the whole procedure has to be repeated. Convergence could be accelerated by judicious choice of flows in each step. One possibility is to use the previous values to anticipate the new one as:

$$Q_j^* = \frac{\left[Q_j^{(r-1)} + Q_j^{(r-2)}\right]}{2} \tag{5.51}$$

Special features, such as reservoirs, pumping stations and valves, are handled in a similar way.

Figure 5.26 Matrix for the flow method.

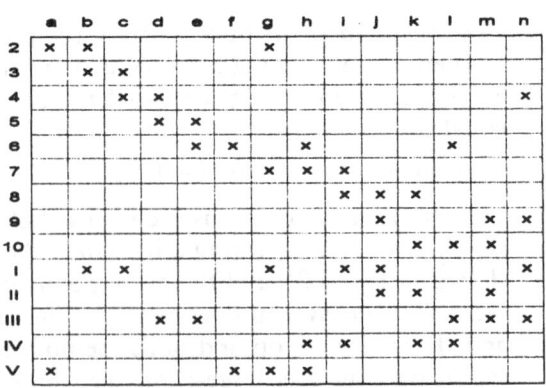

**Possible
improvements**

Handling large networks, with hundreds of pipes and nodes, is not easy, absorbing a significant number of work-hours and computer time. So effort to find a more efficient procedure and/or program is justified.

Branched network

In any normal distribution system, there are many branches – pipes that are not part of a loop. They are sometimes quite numerous in a network. With a little additional effort, all these pipes and nodes could be stored away to reduce the size of the problem (Fig. 5.27). This is particularly important in node-oriented methods.

Flow through each branch is readily found – by summing all downstream demands (easily seen on Fig. 5.27). The total for the whole branched part is attached to the root node (node F in Fig. 5.27) as additional local demand. The procedure then continues as normal, minus all the branches.

The solution will yield exact value of head in the root node (node F here). Other values are found easily starting from the root node downstream, towards the ends.

Reducing bandwidth

The matrix in both main methods – loop and node method – is symmetrical and sparse (i.e. it contains a lot of zero elements). By better numbering of elements (loops in the first case, nodes in the second), one can group non-zero elements close to the diagonal. The benefits are obvious: a banded matrix needs less space in computer memory and less time for solving (the reduction goes only as far as the band goes). In complex systems, with hundreds of elements, the savings can be tremendous.

Figure 5.27 Reduce the matrix size – store away all branches: (a) before and (b) after computation.

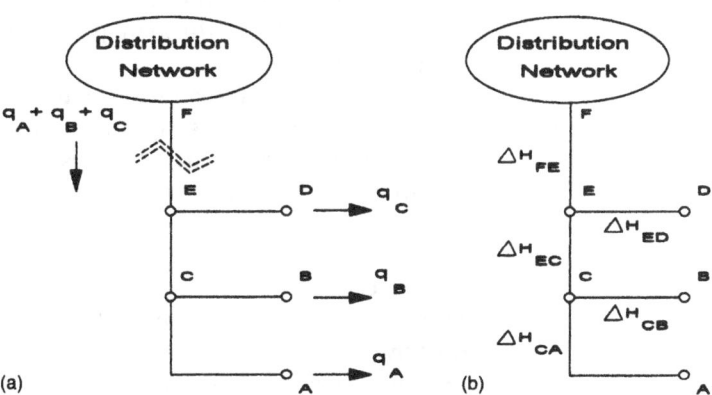

Comments on the different methods

It is always useful for a beginner to solve a few cases by simple means, just to get a 'feeling' for the problem, even if a sophisticated computer program is available. After all, people have to evaluate the results of any model and simulation – not the computer (as yet).

Novices can solve these exercises manually or prepare a simple program of their own. For the former case, an example of a manual exercise is given in Appendix G. In the latter case, beginners should be aware of the relative merits of the methods described above.

Loop method This method has a smaller number of equations (only as many as there are loops in the network), and is more suitable for manual calculation, aided by a simple calculator.

Node method This method is easier for programming, because there is no need to identify the loops, and is less sensitive to poor initial estimates.

As for handling of reservoirs, pumping stations and valves, both the above methods can be equally effective. We believe that most of the differences listed in other papers (Gessler, 1981) could be alleviated by better programming.

Flow equations method This method does not look very promising, as it leads to the largest matrix and to an unsymmetrical matrix; however, it is still used in practice.

Hybrid method

As none of the earlier methods were completely satisfactory, other proposals have been suggested. One of the most interesting was made by Hamam and Brameller (1971), and combines the node and loop methods – a hybrid method. The procedure is as follows.

1 Link up all nodes by using pipes with the least resistance and form a structure without loops (as in Fig. 5.28 for the selected case). Flow through pipes temporarily set aside is assumed to be zero.
2 Calculate flows in the reduced structure. There are no loops, so this is a relatively simple procedure. Then calculate the initial values of heads at all nodes.
3 Compute the flows through 'discarded' pipes using the heads computed in step 2. For instance, the flow between nodes 8 and 9 in Fig. 5.28, a pipe not in the structure, is now computed.
4 Flow in the 'set aside' pipes is now included as an inflow/outflow in appropriate nodes. In this case, flow Q_j is the outflow for node 8 and the inflow for node 9 (Fig. 5.29).

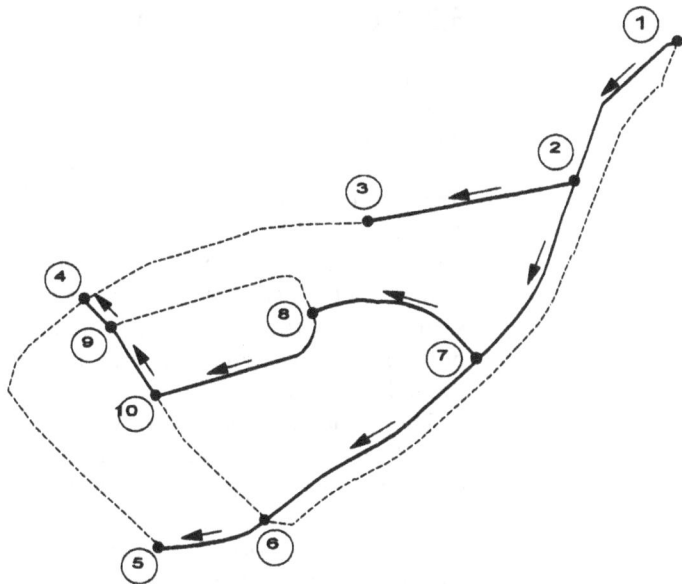

Figure 5.28 Branched structure.

5 If the changes are significant, repeat the procedure from step 2 until the solution is found.

Note that this method does not need a matrix, so its requirements for computer storage are very modest, and the size of the model is less of a problem than in other methods.

Figure 5.29 The hybrid method – the second step.

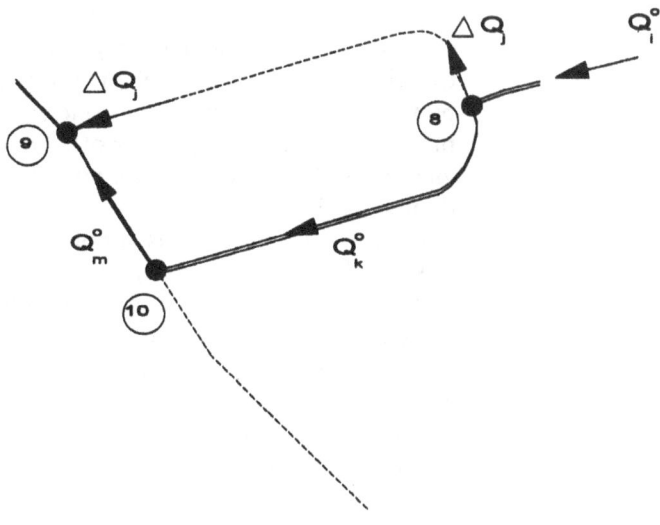

5.3 Continuous simulation

The previous section described methods used to create a single-state simulation of a network. Although the creation of a mathematical 'still photograph' was useful, and was used extensively during the 1960s and early 1970s, the practising operations or planning engineer really needs to understand the activity in the network over time. The 'snapshot' simulation needed to be developed into a 'movie' simulation. The initial and, as yet, only significant development from the single-state simulation is the creation of a quasi-dynamic simulation of the network. Movies have been invented, but we have not yet moved to video!

The basic idea of quasi-dynamic simulation is very simple: start from a steady state of the system, wait for a few minutes, compute all changes in boundary conditions, and use the new conditions as the basis of the next simulation. The process is repeated over a pre-selected period at appropriate intervals of simulated time. Several authors, independently, have come to the same idea (Gilman *et al.*, 1971; Obradovic, 1973; Rao and Don, 1977; and others). In the literature, it is also called 'extended time simulation' or 'dynamic simulation' (Walski, 1984; Hoogsten and van der Zwan, 1985; Bos and Jarrige, 1989; Cohen, 1990).

This approach means that all inertial effects are neglected, not to mention the elasticity of pipe walls and the compressibility of water. By implication, such models are limited only to those events in water supply characterised by slow change over time. Any sudden disturbance – the start of a pump, for instance – cannot be permitted to upset the overall balance.

This is certainly a drawback – but not a serious one. Most changes in a real water supply system are gradual, for example the changes in water level of a service reservoir. The start of pumps is just a change from one steady state to another, with pressure surges rapidly dying off. The effects are felt only in the vicinity of the station and are quite feeble, normally having been dampened by air vessels.

On the other hand, this simplification offers the great benefit of simplifying the program, allowing the various elements of real systems to be included in great numbers. The relatively simple programs also allow rapid simulation runs, so the user may try different cases in a reasonably short time and so get a better answer.

The changes in systems between two steady-state computations can be classified into four groups:

- changes in demand,
- changes of storage,
- changes that were scheduled or pre-planned,
- changes caused by the different state of the system after a steady-state computation (e.g. a pump start).

Changes in water demand

Total demand embraces both the real consumption of water and true losses (leakage and others). Consumption varies with time, as one would expect and as amply demonstrated in Chapter 4. A typical time diagram is shown in Fig. 5.30. At one moment, say t_1, the consumption has one value, but at another, t_2, it is changed. These changes must be taken into account before the next steady-state computation is started.

The majority of the losses are related not to time but to local pressure. As this pressure is higher during the night and lower in daytime, especially during the daytime peaks, one can construct a hypothetical diagram for losses too. Whether this effort is justified or not, one must decide in each particular case.

All this applies to the traditional simulation, in which the relationship between actual demand and service pressure is not explicitly given (or estimated). If this relationship is introduced into the model, it will be another change to be made between two steady-state computations.

Changes of storage

Water is stored in reservoirs, water towers, contact tanks and balancing tanks (all 'reservoirs' in hydraulic terms). The level of water (and therefore its volume) in each reservoir must be known initially. It remains unchanged during the steady-state computation ($H = Z$, where Z is water level).

The steady-state solution, when found, permits a balance of all inflows and outflows for each reservoir in turn. Assuming

Figure 5.30 Variations of water demand and losses (hypothetical).

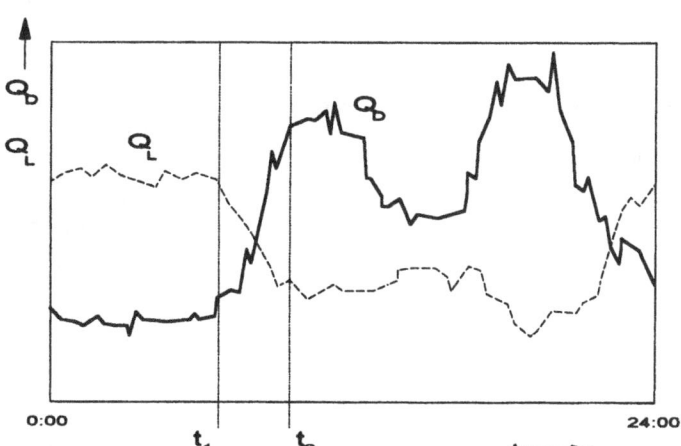

that this situation will last for the next Δt minutes, the storage will be changed for ΔV as

$$\Delta V = (Q_{inflow} - Q_{outflow}) = \int A_R(Z)\,dZ \qquad (5.52)$$

where $A_R(Z)$ is the area of the horizontal cross-section at the level Z, and integration is performed between level Z_0 (at time t) and level Z_1 (at time $t + \Delta t$). This is quite obvious (Fig. 5.31). Some difficulties may appear in abnormal situations, when a reservoir is either filled up or totally emptied. There is no straightforward answer as to how such emergencies should be treated in a model.

1 The first case – *overspilling* – is the easier. It can be assumed that water will flow over a weir (or reservoir's edge), keeping the level slightly higher than the top elevation. Quantities lost in the process can be computed and later shown in the balance sheet – as a loss.

2 The other case – *empty reservoir* – is more difficult. Water has vanished and air freely enters the network. The basic assumption of closed conduits theory is not valid any more. To avoid complicated algorithms, one can resort to crude simplifications:
   - assume that the reservoir has become an ordinary node without any storage and find the local value of head;
   - keep head artificially high at the elevation of the reservoir bottom and compute the deficit of storage needed to prevent the reservoir from emptying.

Neither of these assumptions is particularly good, but at least they permit completion of the run. Such an event is not normal and should be avoided in reality – therefore, it is not very important to model it very precisely.

Figure 5.31 Variations of water level in a reservoir.

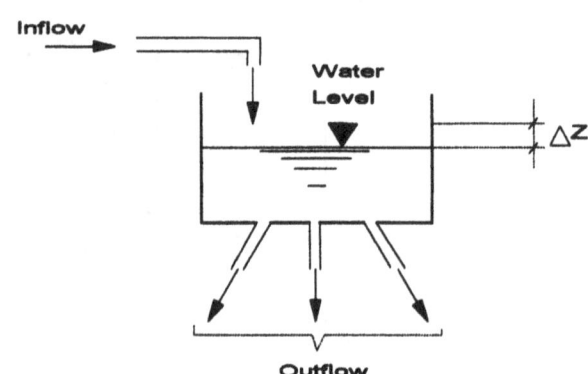

Scheduled changes   Typical examples of scheduled changes are:

(a) planned operation of pumps in a pumping station (switching units 'on' and 'off' at scheduled times);
(b) planned delivery of a pumping station ('target flow'), which varies in time (e.g. flow is higher over night – the low tariff period – and lower during the day);
(c) change of valve opening at a predetermined time (e.g. valve opens during night hours to let excess water refill reservoirs in a nearby system, but closes down in the morning, before peak demand time);
(d) control pressure of a pressure-reducing valve is raised during peak demand time (to raise pressure at distant points) and lowered during the night (to reduce leakage losses).

These cases are illustrated in Fig. 5.32. Time t_1 is the present state of the system, and time t_2 is the next one; the difference is

Figure 5.32 Scheduled changes in a water supply system: (a) PST provides flow as planned; (b) pump scheduling; (c) valve closes when needed; (d) control pressure changes over day.

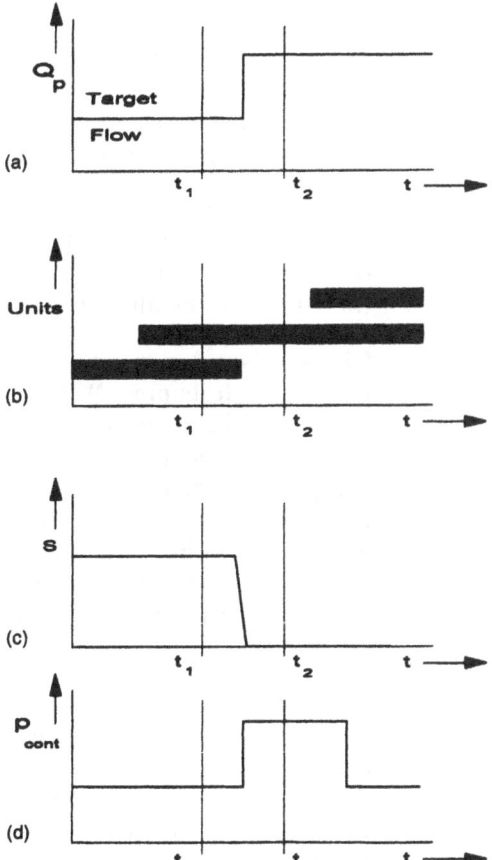

equal to the time step, Δt. The changes are somewhat exaggerated – to better illustrate the point. One can see that a significant change may happen within the time step (Δt), before time t_2 is reached. This small inconvenience may be handled either:

- by 'postponing' the change until time t_2, or
- by 'inserting' a new steady-state computation at the precise time of change.

Either method has its 'pros' and 'cons'.

Changes caused by the system

The changes of flow, water level and pressure in various nodes of the network may cause changes in the status of related facilities: pumping stations, float valves, check valves and control valves.

Pumps The pumps in a pumping station react to changed conditions: if inlet pressure drops below a certain level or outlet pressure exceeds a set value, all active pumps are automatically stopped. This applies to all pumping stations. To those operating in remote control mode (controlled by level, pressure or flow at some point in the system), further changes are also possible. If the level in the control reservoir or pressure in a control node transgresses a certain value, the status of the corresponding pump will be changed automatically.

Float valves A float valve closes as water level in its reservoir rises, and vice versa (see Chapter 2). Consequently, any change of water level will cause a change of float valve opening, and throttling will be altered.

Check valves A non-return valve (check valve) prevents reverse flow through its pipe. With a changed pattern of pressures, the NRV may be activated. It is either open or closed, with no intermediate state. As its initial state is not known, the computation starts with an assumption – which might be proved to be wrong later. The computation has to be repeated with the opposite state to that first assumed. In some cases, this may lead to unstable conditions in the network, especially if there are several NRVs present, influencing each other.

Control valves A pressure-controlled or flow-controlled valve reacts to changed conditions within the system as specified (see Chapter 3). Change of control pressure or flow will cause a corresponding change of valve opening and the degree of throttling. Modelling such behaviour is difficult because it is essentially an implicit problem, ill-suited to methods usually applied in this field.

Transient flows

So far it has not been defined whether the model will serve to analyse steady or transient regimes. Further development will be limited to relatively slow changes – where inertial forces and the compressibility of water can be ignored. This assumption is valid for all normal and most emergency situations in water supply systems, especially if longer time periods (up to several days) are being analysed. Of course, when analysing transient states – which last just a few minutes at most – inertial forces, the elasticity of pipe walls and the compressibility of water need to be taken into account, using more sophisticated analytical tools – see the classic textbook by Streeter and Wylie (1979). The use of these methods was proposed in 1967 by Streeter and several times since then (Stephenson, 1985; Karney and McInnis, 1992) but with little success in practice. The reasons are:

- additional, specialist training is needed;
- the models must be even more simplified and reduced to a skeleton network;
- a simulation takes much longer computer time.

These additional difficulties do not justify increased accuracy of simulation. What is the point of solving a crude model of a real system with great precision? The difference is not – most of the time – very significant, as the next simple case will illustrate.

Pressure surges in a pipe (Fig. 5.33) 500 m long, caused by a change of velocity from 0 to $0.5\,\mathrm{m\,s^{-1}}$ in the space of an hour (a relatively large change in a water supply pipeline), are of order

$$\Delta H = \left(\frac{L}{g}\right)\left(\frac{\Delta v}{\Delta t}\right) = \left(\frac{500}{9.81}\right)\left(\frac{0.5}{3600}\right) = 0.007\,\mathrm{m}$$

Figure 5.33 Inertial forces in pipe flow.

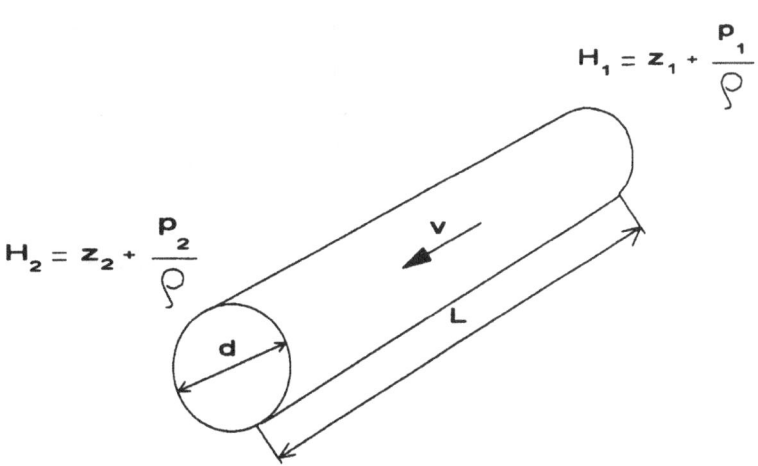

$$H_1 = z_1 + \frac{p_1}{\rho}$$

$$H_2 = z_2 + \frac{p_2}{\rho}$$

which can be safely neglected given the accuracy of other data and assumptions.

Solving the problem   The quasi-dynamic simulation might be a relatively crude image of the real process – but it is still very useful, not to say irreplaceable, in practice. The main features of the process are shown as a flow diagram in Fig. 5.34.

The first task is the acquisition of all necessary data: from model files or an operational database. Data are then cross-checked to eliminate, as far as possible, errors and inconsistencies – before the simulation.

After this stage has been successfully completed, the program formulates the initial state of the system, making all the

Figure 5.34 Flow-chart of the simulation program.

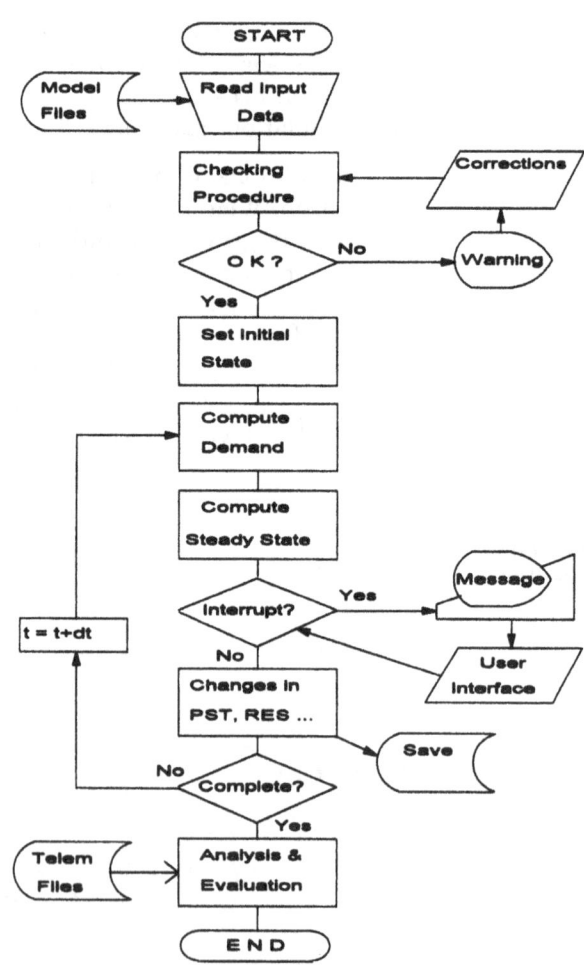

necessary preparations (arranges data, closes input files, opens runtime files, sets pointers, etc.). Then the 'clock' is set to 'on' and the simulation cycle starts.

The program solves the steady-state condition in the system for the given boundary conditions, using one of the available methods described in the next chapter. The results are then analysed in order to determine the changes in boundary conditions that have occurred during the simulation period:

- increase/decrease of water level in all reservoirs, including water towers and contact and balancing tanks;
- change of status of any pump, due to changed water level in the control reservoir or pressure in the control node;
- change of valve opening, for the same reasons, etc.

Relevant messages are displayed – so the user may intervene through the interface, e.g. to review current results. Then the procedure is resumed: the clock increases time for a certain increment (from 1 to 30 min, rarely more) and determines the water demand at this moment, before repeating the cycle.

The simulation lasts until the set time is reached – or the user stops it, for any reason. The main results are stored in data files for analysis. The user may get information on extreme values of flows, pressures or water levels in selected nodes, operating range and time for pumping stations and individual pumps, maximum/minimum openings and flows through control valves, various diagrams and snapshots at certain moments, etc. How the user will interpret this wealth of information is another matter. Some degree of experience and skill is needed before the user 'attacks' real-life cases. This might be achieved by analysing simple cases and models made by experienced colleagues.

Snapshots or continuous simulation?

Traditionally, the design of a water supply system will start with the computation of one state of the system – that of maximum demand at some point in the future, often described as the planning horizon. If the system copes, then a solution is near. The next snapshot might have been taken with a minimum (night) demand to see how the reservoirs will refill and if pressures will be within prescribed limits. A few other cases might follow – for average conditions, or weekend days, etc.

This approach is still valid in *design*, especially in the early phases of work. However, in real-life conditions, snapshots are of limited value and might even be misleading. Increasingly, design engineers, concerned with capital investment in the supply system, are using dynamic models to support their proposals. In the design of some elements of the system, a dynamic model is almost essential to avoid very laborious manual analysis on which to judge investment options.

For example, the determination of appropriate storage in a water supply system has often, until recently, been quite an arbitrary exercise, simply because software was not available to compute changes in stored water. The capacities of reservoirs were often based solely on a 'rule of thumb' proportion of the demand of the area. It was quite usual in the UK up to the mid-1980s to adopt a standard based on 24 hours of demand as a suitable volume of storage. This arbitrary standard might have been modified to some extent by individual water companies and for particular conditions, but very few companies carried out a detailed analysis of the real storage needs of the system. The dynamic model provides an immediate assessment of the diurnal fluctuation in an individual service reservoir. As a consequence, most water companies would now design reservoirs on the basis of daily fluctuation in stored water and add a relatively small volume (say 6 hours of daily demand) as emergency reserve. The result of the availability of dynamic models, aided by extended telemetry monitoring, has allowed service reservoir design to be much more precise, and in many areas has saved capital investment by allowing much smaller reservoirs to be constructed than would have been the case under the previous practice.

The availability of dynamic models has also allowed much greater precision in planning mains improvement programmes by running dynamic simulations in which as yet unconstructed elements such as service reservoirs, pumping stations and even source works can be fully tested.

The prototype network is a dynamic system and should be modelled as such. The validity (or not) of a mathematical model of this dynamic entity must be proved to be valid over time, not just at individual points in time.

5.4 Initial state

Prior to running the simulation, the initial state of the system must be known. The necessary data are:

- water level in reservoirs, wells, water towers and contact tanks;
- state of all pumps in the system ('on' or 'off'), also rotational speed for variable-speed units;
- opening of all control valves;

- all changes in the system (if any), like excessive demands, or closing down some pipes (maintenance works, for instance), etc.

This information is best provided by a modern telemetry system, which provides operational data practically on-line. However, the telemetry does not cover everything, and some data have to be estimated.

A good initial guess is that all reservoirs are half-full and that most pumps are operating – if there are no other clues. After the first simulation run, these assumptions should be examined and changed if found false. For example, if the inflow into a reservoir greatly exceeds the outflow, causing a rapid rise of water level, perhaps the initial value was too low? Better results could be obtained by raising the level to a zone where a float valve will throttle the inflow – as in reality. Never forget that the real system had ample time to 'settle' – start or stop all pumps as required and adjust the opening of all valves, etc.

5.5 Choosing adequate software

General requirements

The model of a particular water supply system has three components:

- a program to perform all computations,
- a database with operational data and other information,
- special knowledge about the system – control policy, etc.

A newcomer to the business of water supply system model making will quickly realise that the model cannot be bought 'off the shelf' – it must be made and preferably made by the company's own staff, in house. The only question is whether the model-making software package, the program, will be made 'in-house' or bought on the software market. A supplementary question is whether existing hardware is adequate or more powerful computers are necessary?

The potential user may be astonished by the wealth of software and hardware available, at prices from a few US dollars to many thousand US dollars – some software even being free of charge! However, there are a few key factors that should be considered when obtaining new software:

- No-one can buy a model of their water supply system, only the program on which to build it.
- No-one knows your system better than you. It is easier for a water operator to learn a new program than for a stranger to

the industry, however expert they may be in computing, to learn how a real water supply system operates.
- Whatever program you acquire, you will have to spend time and effort to learn and master it. The same applies to new hardware.
- The best program in the world cannot solve the problems in your real water supply system. It can only help you to do so.

These simple truths are often disregarded, because the user places too much faith in the new tool – with predictable results. However, this is just a warning for beginners and not advice to refrain from acquiring new software and hardware – quite the opposite. The text that follows may help in the task of deciding on modelling software.

The user should start by asking what the new tools will be used for. The main applications are:

- *Operational management* of existing water supply systems
- *Design and analysis* of existing and future water supply systems
- *Education, study and research* in hydraulics and water supply

It follows that the principal users are:

- engineers and operators in the control and information centres of water companies;
- engineers and technicians in design offices of water companies, consultants and contractors in the water industry;
- engineers and students in universities, high schools, research departments and institutes and the training departments of water companies and consultants.

The typical tasks arising from water supply system models are:

- analysis of how the system really operates,
- verification of control policies,
- evaluation of storage reserves,
- optimum pump scheduling,
- discovering 'bottlenecks' within the system,
- selection of post-chlorination sites,
- decision on which tariff should be used,
- completion of existing telemetry system,
- design of new facilities,
- design of enlargements and reconstruction of the system,
- design of a new control/telemetry system,
- 'war games' for staff training and retraining,
- training of novices on various levels, etc.

To deal easily with all these tasks, the new program must, ideally, have the capacity to:

- simulate any normal, emergency or catastrophic situation that can or has happened;
- accept live data from the telemetry system and provide on-line analysis of the information;
- simulate and assist the formulation of control policies best suited to local requirements;
- simulate changes of water quality in the distribution system, etc.

Program requirements

Modern software products are labelled 'user-friendly' because they do not ask for any special knowledge or training prior to use, but this is only partly true. The characteristic features of such a program are that it:

- is extremely easy to learn and use, because it is tailored to suit the user's demands;
- is completely menu-driven, so the user does not have to look very often in manuals;
- has a comprehensive system for capturing errors, with an explanation in plain language, so the user is protected against common blunders and mistakes;
- has on-line help available at any stage;
- can model a wide range of possible control policies for pumping stations and control valves, but keeps all computing 'technicalities' invisible to the user;
- has the ability to keep the user informed about the state of the system during a simulation run;
- always provides an answer to the user regardless of the size or complexity of the case – and never loses control;
- has a good graphical user interface (GUI) available to the user for both input and output jobs;
- can compare telemetry data with simulation results on the same diagram;
- permits the user to create the output according to particular needs and wishes in a very versatile way;
- is a very efficient tool.

As for the user, it is assumed that the user:

- is well informed about the water supply system that he/she is modelling, but does not have to be an expert in mathematical modelling or computers;
- should describe the real system as it really is, without 'inventions' – and the program must be versatile enough to cope with real-life complexities.

There are also other, less obvious, requirements, like:

- good documentation,
- well-organised database.

A widely used program should be accompanied by manuals, and supported by a wealth of references to textbooks, papers describing interesting cases, reports and tutorials for beginners.

As for the database, it should be organised bearing in mind the internal structure of the water company and the responsibilities of its various sections. This is shown schematically in Fig. 5.35. In this case each file is under the sole surveillance of an appropriate person in the water company. For example, depending on the staffing structure of the company, these people might be:

- Network geometry: Design Engineer
- Pumps and valves: Distribution Engineer
- Sources and WTPs: Production Engineer
- Water demand: Customer Services (water bills)
- Control policy: Operations Manager

Updating of model files should only be done through the person authorised for this job and preferably only by one person.

Environment

Figure 5.35 Database structure for a mathematical model support.

The modelling software or program could be used alone in a water company, design office, research institute or college, for any one or more of the tasks listed above. However, perhaps its most outstanding role is as the 'heart' of a real water distribution system, integrated with many of the other information systems installed by the company. Ideally, if a telemetry

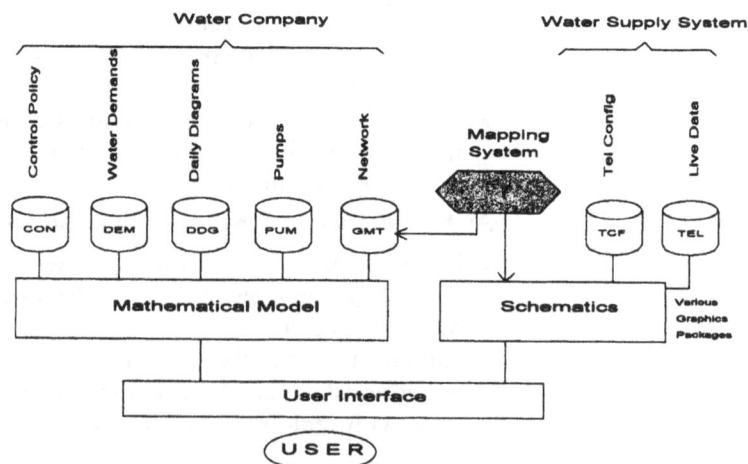

system is installed, it is that system which will be used to feed fresh and reliable data to the model and ensure that it is possible to recalibrate the model regularly. In such conditions, the model should then develop its full potential and give sterling service for various design, planning and operational management tasks.

The next chapter describes the way in which telemetry is used in the development of models integrated with the ongoing use of the model as an operational tool.

6 Making a model

6.1 The general approach to model making

Mathematical modelling was treated, until recently, as an academic exercise that could only be done by specialists. Even the basic skills of modelling were considered quite sophisticated, and few models were used in operational practice. Those which were produced were used by planning and design engineers.

However, as more powerful, more compact and cheaper computers became available, as well as more user-friendly software, the barriers were broken down. The only knowledge required by the user is of the real water supply system – their own system – and the rest is relatively easy and straightforward.

A simplified procedure for making a model is shown in Fig. 6.1. Some major tasks may be organised in parallel:

- acquisition and training in the use of necessary hardware and software;
- collection of data and compilation or 'construction' of the model;
- organisation of a database of operational data that will support the model later.

After completion of these tasks, which will produce an initial model, the model has to be calibrated, as described in section 6.6.

Differences between the model and the real system will appear and have to be ironed out (by changing the model) or accepted as the limitations of the method. This is a critical phase of the project. If the model fails to live up to expectations and produce benefits, the whole project might founder.

However, if the effort is sustained, it is almost always worth while because valuable knowledge about the *real system* will have been gathered, many myths will be dispersed, and new efficiencies introduced to the company. Certainly, significant effort will be needed to succeed, but success will be found to be rewarding both for the company, in terms of tangible and intangible benefits, and for the participants, in terms of their professional development.

Experienced members of staff, at all levels of the company, should be involved in the project from the early stages, and the circle of people with a genuine interest in the results of modelling should widen as the models develop.

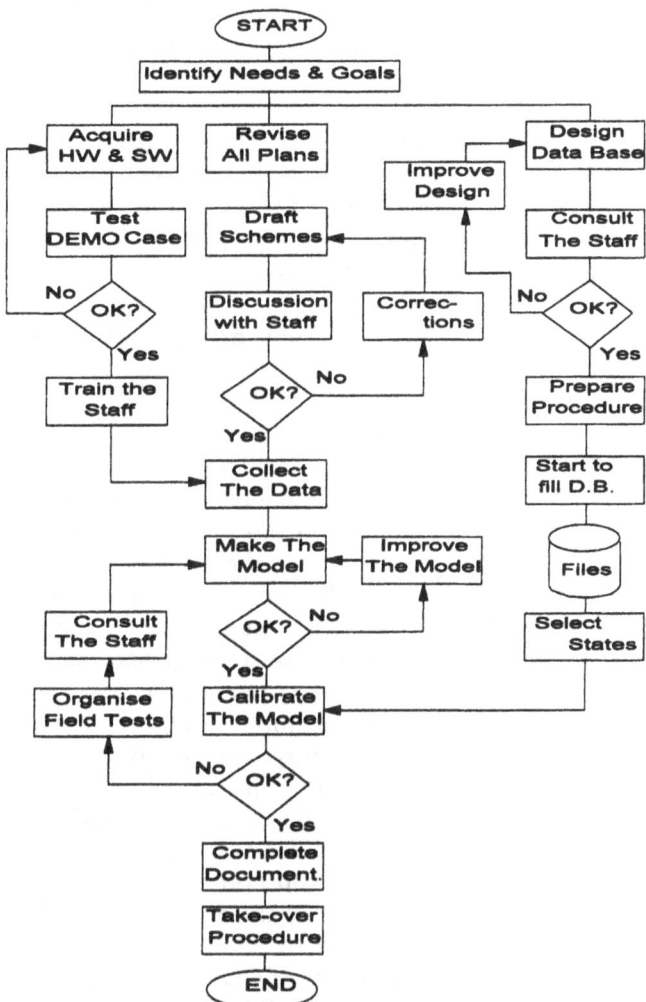

Figure 6.1 Organising the modelling effort.

The calibrated model should be available where it will be most usefully employed – the control room, information room, planning and design office, management. Someone must be in charge of the model in the capacity of a coordinator to ensure that all changes to the basic model(s) are correctly carried out and documented and to provide help and support to the users, *but* access to the models should be possible at all levels of the company to derive the maximum benefit from the development.

The database of *operational* data is an asset in its own right and should be supported all the time, not just occasionally. Ideally, a permanent link to other large systems – telemetry, data-loggers, billing system, GIS – should be organised on the basis described in Chapter 1.

The importance of fieldwork should be stressed once again; one of the benefits of modelling is the review and completion of the data capturing system, both the permanent one and special, task-oriented ones. Staff should not only participate in these exercises, but should make up most of the workforce and provide local knowledge. The beneficial results of fieldwork often exceed those of the modelling itself and provide valuable information for operational management and other sections of the company.

6.2 Practical advice on model building

There are a few guidelines for the novice model builder:

1 *Try the DEMO case first.* Good water system modelling software will include some form of demonstration model with guidance on how to use and change a previously prepared model. Before trying to build and use your own model, try the DEMO case. This is the best way to learn quickly how the program works and is *not* a waste of valuable time – quite the opposite.

2 *Model size.* Having had 'hands-on' experience of the demonstration model, the next step will be to build your own. For this initial 'trial', why not attempt to model a real or part of a real system. Do not try to make a huge model needing an enormous quantity of data, as you will be swamped by them. It is better to start by modelling a small part of the system or a simplified version of the whole system, with no more than 70 to 100 nodes. Following either course will allow subsequent expansion of the model by extending its geographical area or by adding detail.

3 *Purpose of model.* Whenever you make a model do not forget what it will be used for. If, for instance, you want to have a tool for pump scheduling, then the model should not go too far into distribution network detail – just the trunk mains would normally suffice.

4 *How to model various peculiarities.* Every water supply system has some unique features and sometimes one is at a loss how to make a good model. The advice is – avoid artificial and complicated solutions.

A few tips for dealing with the individual elements of models are given below.

Water distribution system

The key points are as follows.

- The first job is to make an overall schematic that is acceptable to all concerned.

- The major features (e.g. pump stations) should be included in the schematic.
- Name all nodes, *do not* number them – the names should be mnemonic.
- Do not make 'provisional' sketches – it will be very hard to change them later! Better spend a few more days on this job until everybody is happy with the schematics!
- Use simple and clearly defined symbols.

Nodes

The identification of nodes is not as easy as one may think. Remember that the nodes are:

- sources of water,
- all reservoirs and water towers,
- junctions and dead ends,
- important water users (hotels, hospitals, etc.),
- any other point that *you* regard as significant.

The simple rule is: 'Minimise the number of nodes'.

Reservoirs (including water towers and sources)

The factors to consider are as follows.

- Each of these items may be treated either as a *variable-head* node (a reservoir or tank with variable flows in and out and therefore a changing water level) or as a *fixed-head* node (where water level effectively remains constant, e.g. a well source or impounding reservoir). Be aware of the difference.
- For a variable-head reservoir, one needs data about the reservoir shape and the top, bottom and regulating range elevations, plus a description of inflow/outflow arrangements.
- Does the reservoir have one or several chambers? If there are several chambers, are they identical? Are the chambers interconnected, or should each chamber be treated as a separate reservoir in the model?
- Does the inflow pipe enter the reservoir from below or from above? Check, by sending a reliable observer to visit the site.
- Is the reservoir protected by a float valve (FLV)? If there is a FLV, the data needed for the model are: regulating range (default 0.5 m); and height of the upper regulating range limit, measured from the reservoir bottom (m). Is there a bypass with a check valve (to permit rapid emptying of the reservoir)?

Pumping stations

Pumping stations (PST) are links, just as a pipe is a 'link' in the model, and its defining nodes may be either ordinary nodes or reservoirs. However, the pumping station is a more sophisticated 'link' than an ordinary pipe, and the data required to make it acceptable to the model are rather more than those for a pipe.

The initial step is to make a clear schematic of the PST, showing pumps, valves, pipes, etc. Show the result to the operators and make all the changes necessary until they agree that the diagram is a true representation of how the station is configured.

The key points to check in assembling the pumping station data are:

- Is there a bypass pipe around the PST (with a check valve, of course)? If the answer is 'yes', collect the data for this pipe – the data may be important.
- Investigate whether the PST takes water from a suction tank (i.e. a reservoir) or from an ordinary node. If there is a suction tank, it should be modelled either as a variable- or as a fixed-head node, regardless of size.
- Try to get information about all the pumps. If there are no Q–H–P curves (flow, head, power), find out for each unit: rated values of discharge, head and rotational speed; and rotational speed and rated power of electric motor. Is this a variable- or a fixed-speed unit?
- Is this PST supplied over the high- or low-voltage system?
- Which tariff is applied? Get information about the tariff structure.
- Protection against low suction pressure – find the level that trips off each pump. It is possible that units have different 'off' levels. Over-pressure protection – is outlet pressure at this station limited to protect the pumps?
- Variable-speed pumps – most VS devices can change the speed in the range between 70% and 100% of the rated speed, but the latest devices may have a broader range.
- Number of stages – some pumps are made of several stages put together in series (well pumps, for instance). This must be checked with the staff.
- Rotational speed – the characteristic curves (Q–H–P) are provided at the rated speed (2950, 1450, 960 rpm, etc.). Check if the actual speed differs from the nominal value – this is very important! If the difference is greater than, say, 1%, then new characteristics should be computed using affinity laws.

Finally, one needs to consider the control mode. The pumps may be controlled in one of two basic ways in an automatic control loop.

1 *AUTO* mode of operation (pressure- or level-controlled pumps) should be used in a model if:

- there cannot be more than one control node for one pumping station – if individual units in the pumping station are not controlled in the same way or by the same node, the station must be 'split' into several smaller ones, each modelled as a separate pumping station, controlled in the model by a single node;
- the control node may be either a reservoir (control value is depth of water) or a node (pressure);
- PST control may be established to refill a reservoir or increase service pressure at the control node, or to empty it or reduce service pressure at the control node (the latter case is called 'reverse logic').

2 *FLOW* mode of operation (where PST discharge is a 'target flow') should be used in a model if:
- the PST has a variable-speed unit;
- the discharge of the PST is known (monitored?), but not the extent of throttling of pumps.

Pump characteristics The curve should start from $Q = 0$ (the ordinate axis): if the data are missing, extrapolate the curve to both axes. Pump head (H_p) should decrease with discharge (Q). If the actual data do not comply (an unstable operating characteristic!), it is better to adjust the data, rather than try to run the model with an unstable characteristic. This helps in computation, and the final results are not affected because the operating point is unlikely to be within the adjusted part of the curve. A few tips are the following

- Power (P) for $Q = 0$ is not equal to zero. Have this in mind when extrapolating this part of the curve $P = f(Q)$.
- Do not wait indefinitely for actual $Q–H–P$ curves. Start to work with the data from the literature (see Chapter 3).
- Compare the data from energy bills with the results of simulation. Is there any resemblance?

Pipes *Beware*: It is a fact that some pipes in a distribution network will be illogically *closed*! The closure was probably done for good reason, *at the time*! Try to discover if the closure is really needed.

- Try to get information about the pipe material and age, as well as the data essential to construct the model (length, internal diameter, roughness value, minor losses). Have a look on the scrap yard; one can get an idea about encrustation and quality of pipes by looking at discarded pipes.
- If you use the Colebrook–White formula, remember that it was derived under laboratory conditions, and consequently

computed values of the friction coefficient represent an 'optimistic' estimate. The real value may be considerably higher – unknown factors are all on the 'negative' side (accumulation of silt or debris, heavy encrustation).

- Do not attach too much importance to the 'friction coefficient question'. It is rarely crucial unless the pipe(s) are so badly encrusted by corrosion products or silt that the cross-sectional area is significantly affected.
- Avoid short pipes, for two reasons: they can make the computation unstable, and they unnecessarily increase the size of the model. A few examples are shown in Fig. 6.2.

If flow through a particular pipe is unquestionably *lower* than the computation shows, the reasons may be:

- a neglected 'minor' loss (forgotten valve or a partial blockage),
- heavy internal encrustation of the pipe.

It is easy to represent these facts in the model: in the first case, introduce an equivalent loss coefficient, ζ; in the second one, decrease the internal diameter. Do not insert an improbably inaccurate friction coefficient value – you may mislead your successors.

If actual flow through a particular pipe is unquestionably *higher* than the computation shows, check the internal diameter and all connections in the network. Once again, do *not* play with friction coefficients.

Figure 6.2 Short pipes in the model.

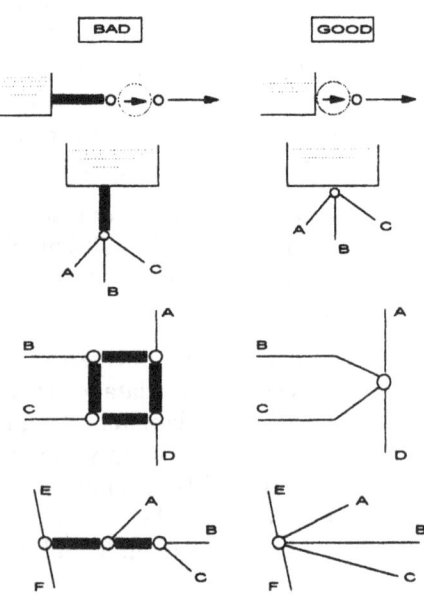

Float valves	A FLV can be positioned only on a reservoir inlet pipe. A reservoir may have several inflow pipes, each with its own FLV. The necessary information for a FLV is:

- nominal size (i.e. internal diameter),
- coefficient of local loss in the fully open position,
- regulating range (default 0.5 m).

Non-return valves	No particular data are needed for a NRV (check valve) except for the nominal diameter and the loss coefficient ζ when fully open.

Control valves	Control valves can be used for a variety of roles, which the model must represent adequately. The most usual cases are:

- Throttled valve (THV) – a valve with a fixed opening;
- Pressure-reducing or -sustaining valve (PRV/PSV);
- Time-controlled valve (TCV);
- A pressure- or level-controlled valve (PCV);
- A flow-controlled valve (FCV).

In each case the necessary data are:

- nominal diameter (i.e. internal diameter),
- loss coefficient (ζ_0) when fully open,
- mode of control, with additional data.

Beware: Ensure that a control valve is not placed in a loop. This is hydraulically unsound and may cause considerable problems, in both the model and the real system.

Water demand data	Some key factors are the following:

- Seasonal effects are usually important, so monthly data should be collected (bills and flowmeters).
- Check if the effect of the day of the week is important in local circumstances.
- Daily diagrams are the most important patterns to be determined. Data from telemetry, loggers, water meters, etc., should be gathered and analysed. If necessary, field tests should be organised (see Chapter 4). Patterns from other systems can be misleading and are only useful for initial trial runs.

Do not forget that demand patterns are the *weak link* in modelling because they are effectively forecasts of what will probably be the pattern of demand. They are also the weak link because of the arbitrary nature of individual customers' demands and of leakage distribution. Be prepared to readjust the patterns of demand and daily diagrams, as data accumulate in the database – again and again and again!

6.3 Solving difficulties and some common errors

Experience of constructing models has shown that a number of problems are quite regularly presented and if not resolved will prevent the simulation from running. Some of these 'regulars' are described below.

Float valves

It is certainly wrong, in practical and modelling terms, to bypass a float valve (Fig. 6.3); the valve cannot prevent overspilling of water delivered through the 'B' link. Either the float valve or the link through node 'B' is redundant. The float valve is a protection against overspilling and should not be needed if the pumping station feeds a reservoir directly, as in Fig. 6.4a. Overspilling should be avoided by controlling the pumps properly, but an excessively cautious operations engineer may require both protections against overtopping. In this case, the system may need to be further protected against excess pressure if the pumps do not stop when the reservoir fills. Altogether this would be a very cumbersome control system. A float valve might still be needed if part of the pump discharge is feeding the network directly (Fig. 6.4b).

Pumping stations

We have looked at the possible need, in untangling complex pumping stations, for creating several model stations covering a single site if the suction and delivery points are not common.

Figure 6.3 Redundant nodes and pipes.

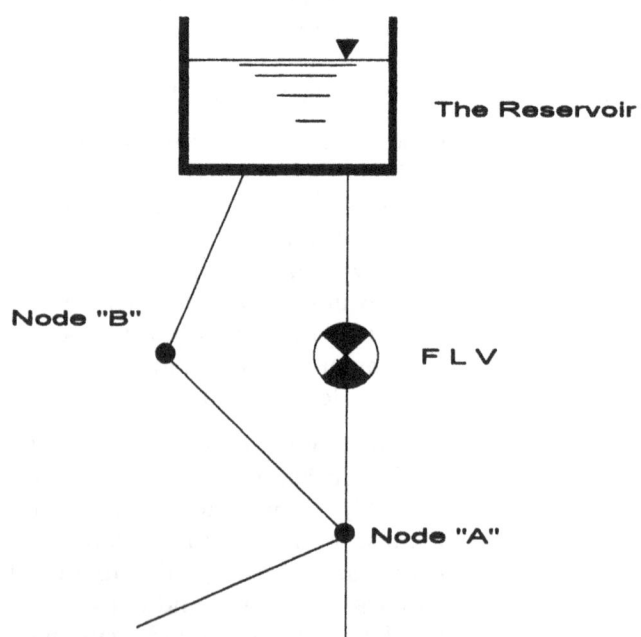

The Reservoir

Node "B"

F L V

Node "A"

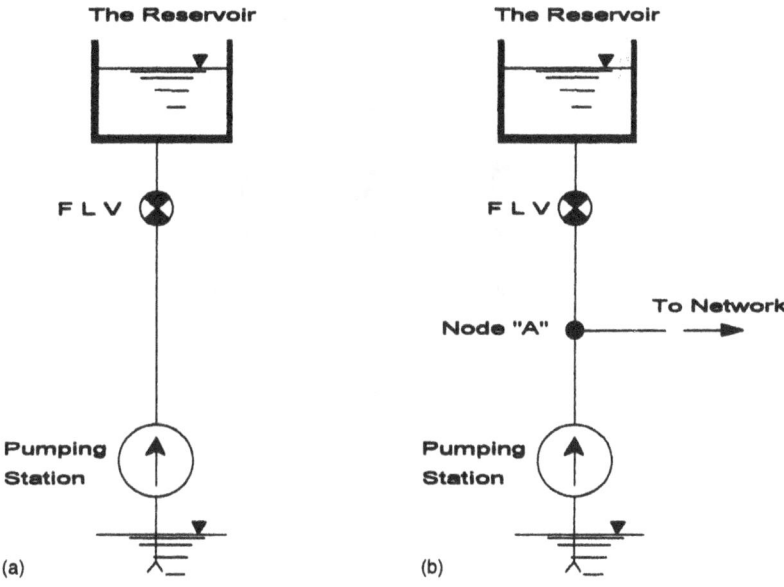

Figure 6.4 (a) Redundant and (b) non-redundant FLVs.

A further mistake in modelling stations with more than one pump is to try to model each item separately – as shown in Fig. 6.5a. There are three pumps, three valves and three intermediate nodes, each a separate item of the model. This is not an error but it would be more efficient to model the station as one element (Fig. 6.5b), reducing the model size and probably improving the speed of computation. An effective modelling package should be able to handle pumps arranged in parallel, quite adequately.

Burst protection valves

Another case of poor modelling (or practice?) is shown in Fig. 6.6. A valve controls the inflow of a small reservoir feeding an isolated demand. The inlet valve of a reservoir is almost always controlled by water level in the reservoir, and not by the outflow from it – at least in normal circumstances when the intention is to keep the reservoir full.

In advanced systems the outflow may be monitored, and if it exceeds a certain level – indicating a burst somewhere downstream – the valve is automatically shut. But that is only a secondary protection, and an attempt to model such features of a system is probably a wasted effort.

Breaks in the system

As indicated earlier, one should start with a model of a small part of the water supply system or a simplified model – not a huge model, which will be a problem even for an expert. The question is where to put the dividing line. In the case shown in

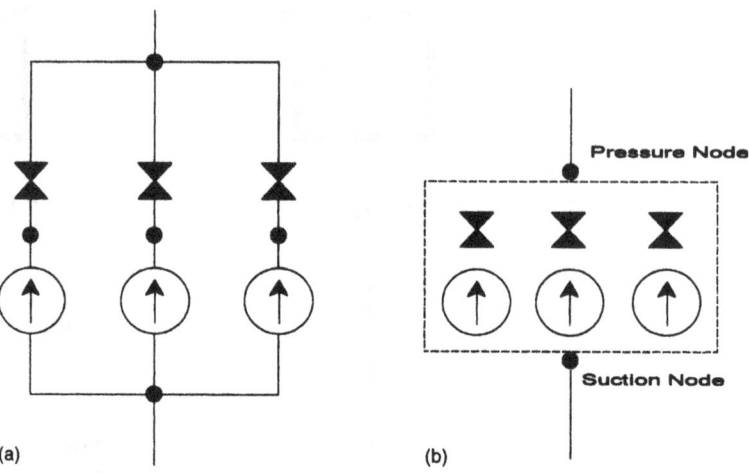

Figure 6.5 Avoid too many details: (a) too many nodes; (b) correct modelling.

Fig. 6.7 there are two pressure zones with a reservoir and a pumping station (for transfer) between them. Following this advice, one should start with one zone. But how are we to model the pumping station, which belongs to both zones?

The answer is simple. When modelling the lower zone, one has to take into account only the transfer of water (for the higher zone) – as an outflow from the reservoir. The pumping station is excluded (Fig. 6.7b). On the other hand, when modelling the higher zone, one should model the pumping station in detail, and assume that it takes water from a fixed-head node because levels in the reservoir cannot oscillate very much.

Figure 6.6 How the valve is controlled: (a) good; (b) bad.

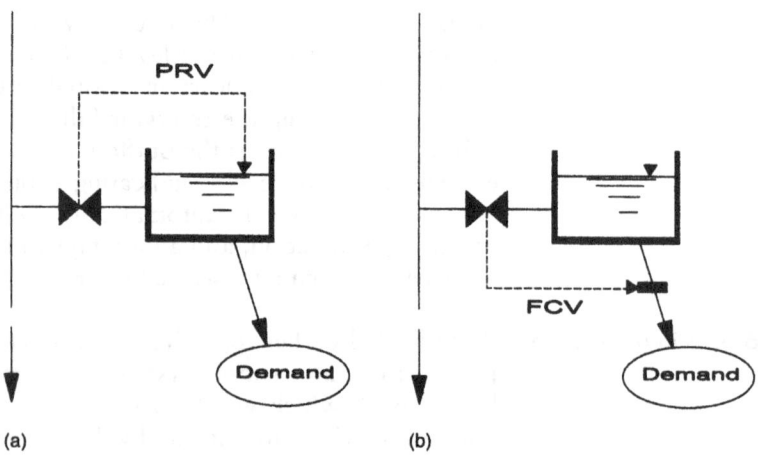

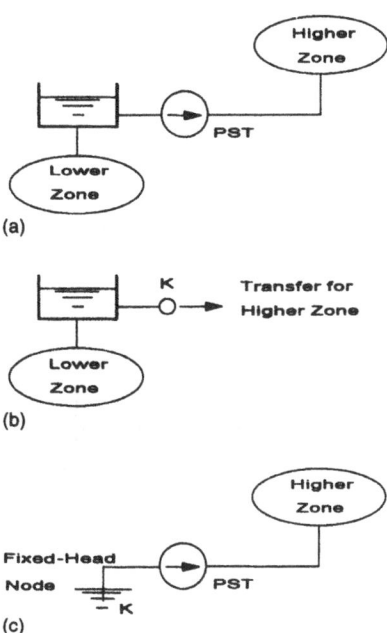

Figure 6.7 Splitting the model up: (a) the complete system; modelling (b) the lower zone and (c) the higher zone separately.

In other cases it may not be wise to separate two parts of one water supply system, ignoring the links that exist in reality; see the example in Fig. 6.8. A small town is supplied from two reservoirs, denoted by 'A' and 'B' in Fig. 6.8. The demand has been concentrated at two nodes (A and B) to simplify the model. This is perfectly safe, as long as the link between these two nodes exists – represented by a fictitious pipe. This pipe replaces several small-sized pipes that exist in the real system. Of course, this is just a stop-gap measure in the early stages of modelling; later the network should be described in greater detail.

'Tricks' are sometimes useful – like replacing a small reservoir by a PRV (Fig. 6.9). This will keep pressure at the real reservoir level and the continuity of flow will be preserved, while problems of simulating inflow and outflow in this reservoir can be avoided in the initial model making.

Another device is to replace a pumping station for which not enough data are given with a simple transfer node $Q = f(t)$. The value of Q, which can vary over time, represents the discharge of the pumping station, and the computation will yield the unknown head at that point. In the early stages of modelling, this approach can be quite useful, but no effort should be spared to get the real information about pumps and station – and check whether the real units are capable of providing the required head. Similar caution should be applied to other simplified boundary conditions.

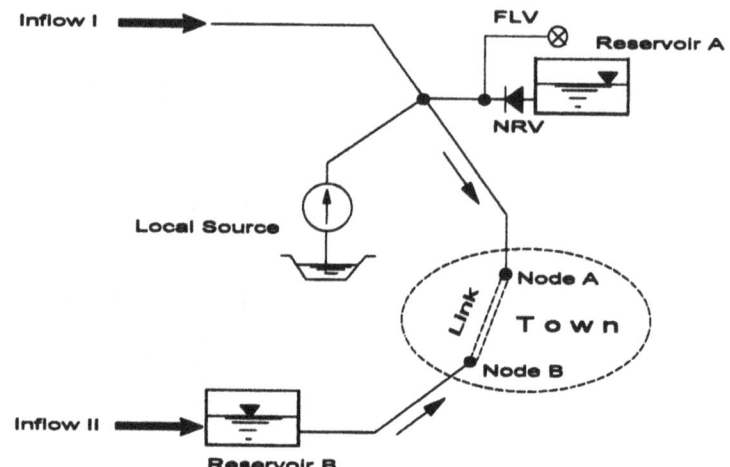

Figure 6.8 Links must be preserved.

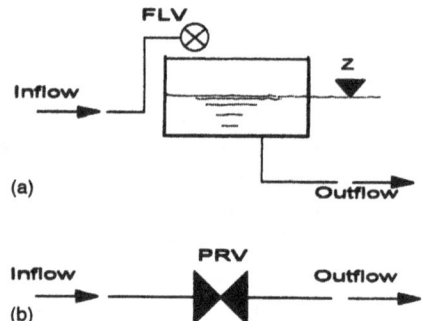

Figure 6.9 Replacing a tank with a PRV: (a) a *small* reservoir ... (b) modelled as a PRV.

Such tricks should be applied sparingly and judiciously, with adequate comments to explain the idea to other users of the model.

Isolated parts of the network

Pressure-reducing valves (PRVs) are used to control pressures in isolated parts of the network, as in Fig. 6.10. All is well as long as the PRV remains open. But what will happen if it closes because the upstream pressure falls below the PRV set pressure?

In real life, pressure in this isolated system will fall and eventually air will replace the retreating water. The model cannot simulate this situation because it (the model) assumes continuity of supply and will give poor results (negative pressures) or stop with a message. Therefore, some measures may need to be provided to prevent the PRV closing down completely.

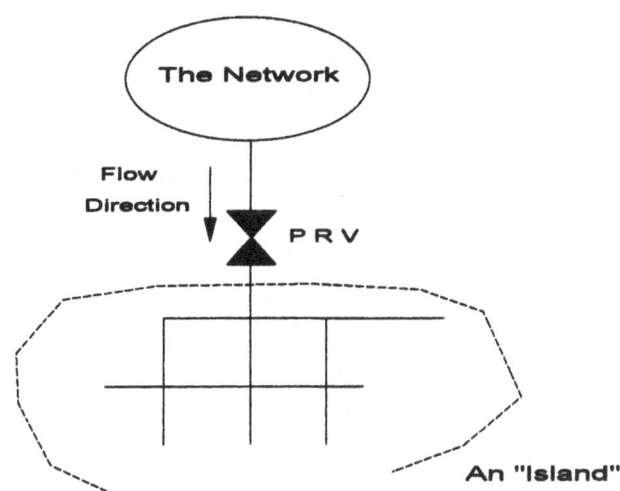

Figure 6.10 Closing the PRV will isolate a part of the network.

6.4 Recognising typical solutions

Conveyance systems

A gravitational system transporting water over great distances is represented in Fig. 6.11. Water is pumped from the source to the highest reservoir and from there it flows to a series of lower reservoirs by gravity. Note the following points:

- Float valves (FLV) prevent overspilling of downstream reservoirs.
- Level-controlled valves (PCV) prevent emptying of upstream reservoir.
- Both FLVs and PCVs are placed at the downstream end of the trunk main, so the pipe will always remain full whichever valve closes.

Figure 6.11 A gravitational water conveyance system.

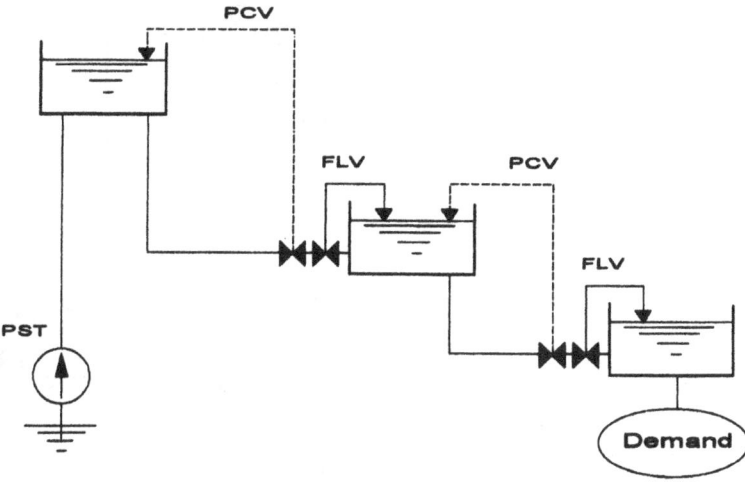

Emergencies – such as a serious burst – are dealt with by shut-off valves (not shown in Fig. 6.11). Shut-off valves could be remotely or locally controlled or manually operated.

Another common arrangement is reservoirs linked to a common trunk main (Fig. 6.12). Each of the reservoirs feeds a demand that may vary seasonally. When demand is low, reservoirs will remain full and their float valves will be partially closed. During high demand periods, demand may exceed the system capacity, or that of one branch, so the storage cannot be replenished. The reservoirs hydraulically 'nearer' to the source will then be favoured. Standard float valves are of no help, and remotely controlled valves should be installed to distribute water equitably between the reservoirs.

Pumping systems

In many cases water has to be transported by pumping; Fig. 6.13 is a typical case. Water is treated close to the source and pumped to a pressure-breaking tank on the top of a ridge, from where it flows to the main reservoir by gravity. Note the control valve at the main reservoir's inlet – it has to be a reliable float valve or, better still, a PRV controlled by reservoir level, permitting implementation of an active control policy.

Pumps are controlled by level in the pressure-breaking tank. In emergencies, water could be pumped straight to the town,

Figure 6.12 Reservoirs in parallel. Detail shows: (a) float valve; (b) remote-controlled valve.

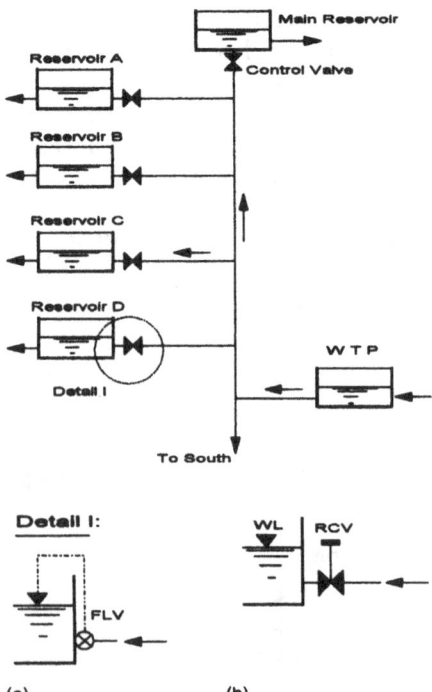

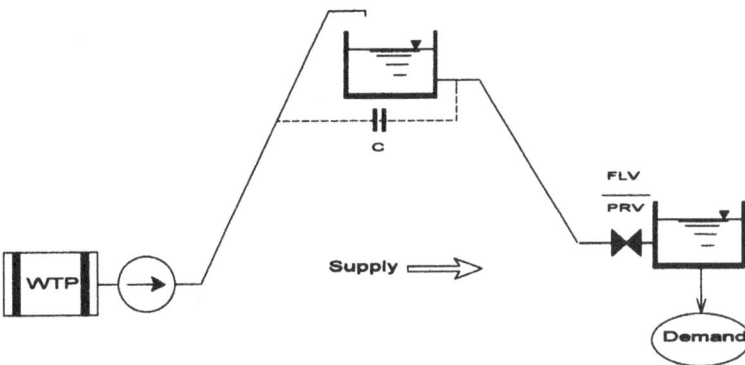

Figure 6.13 A pumped
water conveyance system.

bypassing the tank – which will then be isolated by shut-off
valves. But such operations are dangerous because of possible
pressure surges or air entrapment at the summit.

Contrary to popular belief the greatest danger from pressure
surges is occurrence of vacuum – not high pressures, which the
system can withstand in most cases. The vacuum can seldom
cause the collapse of a section of pipe. The real danger is that
polluted water around the pipe might be drawn into the system,
causing water quality problems (and therefore potential health
problems). This is the reason why vacuum is to be avoided in
supply pipes. In cases where the hydraulic grade line may fall
below ground level, special devices might be required to
maintain a positive pressure. An example is shown in
Fig. 6.14. A PRV is installed to operate as a pressure-sustaining
valve and 'lift' HGL above the ridge.

Real conveyance systems are typically more complex than the
few examples given here, with several branches supplying other
towns and large customers. One should never forget that these
systems grow with time, following changes in the area, and the

Figure 6.14 A pressure-
sustaining valve.

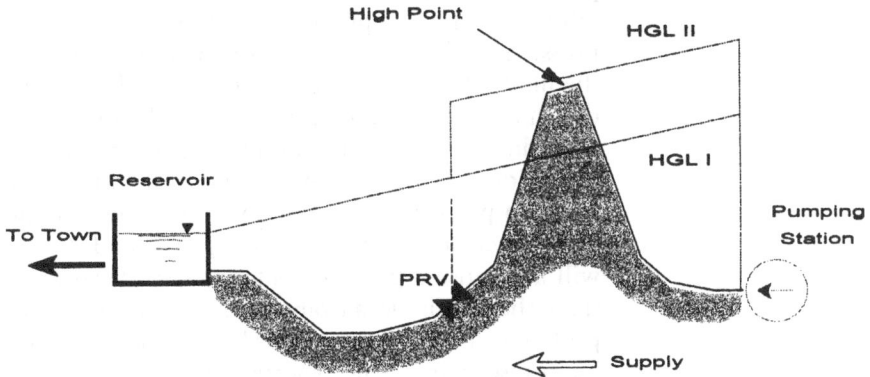

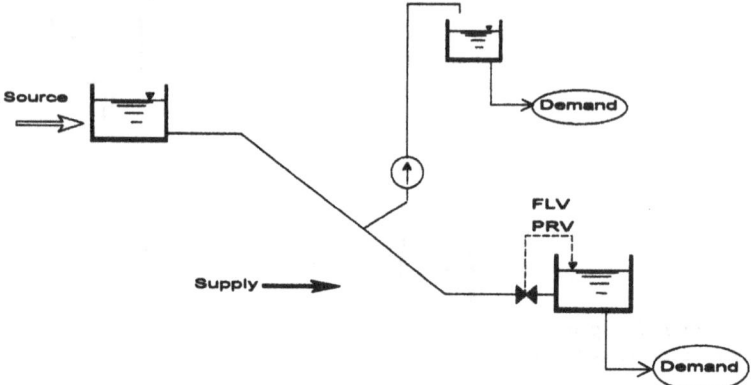

Figure 6.15 Pumping water from a gravitational pipeline.

new additions can cause serious problems by diverting water from the original destination and altering the originally conceived operating regime. A few cases are discussed below.

Water can be diverted to a new location by pumping out (Fig. 6.15). If the quantities are small in comparison with the capacity of the pipe (and the source), everything will be fine. A simple float valve at the reservoir's inlet is needed to prevent overflowing. Difficulties may start when the local demand grows. A model will then be quite helpful in devising an adequate control policy for the system.

Special attention should be given to the computation of the operating range of pumps in various regimes – because the pressure on the suction side can alter significantly with changes of demand. A similar system is shown in Fig. 6.16, but this time the water is transported by pumping from the source to its main demand area. The booster PST again takes water from the main trunk to meet a local demand.

This time the situation created by the new booster could be even more dangerous. When pumps in the main PST start, the pressure on the suction side of the booster will rise sharply, and fall back when the pumps stop. Note that the minimum suction pressure could be lower than the main reservoir – if all pumps in the main PST are stopped, then the local demand is taken from the higher reservoir and the local pressure will decrease accordingly. The model can be used to anticipate such potential operational problems before they start to plague everybody.

When pressure in a trunk main is sufficiently high, water can be diverted simply by laying a pipe with a PRV (Fig. 6.17). This will not cause a problem as long as the local demand is smaller than the available surplus. If not, this diversion might cause problems. Note that a serious leak in the local area can drain the main reservoir – if not spotted in time.

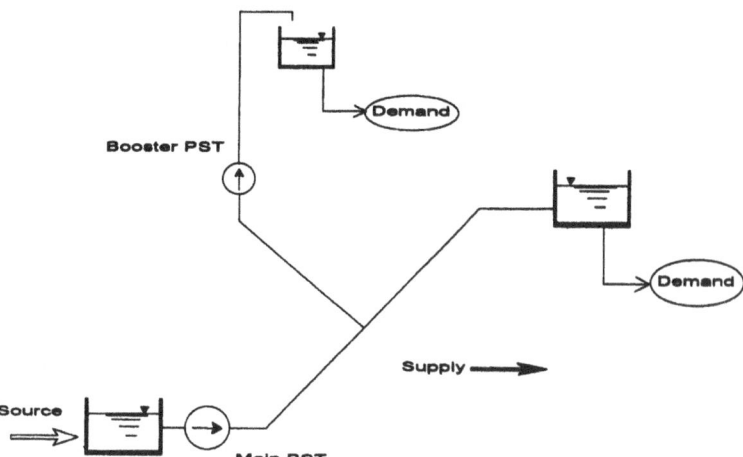

Figure 6.16 Taking water from a pumping main.

Water usually flows in one direction through the conveyance system; however, there are exceptions (Fig. 6.18). The local source can satisfy local demand most of the time and even have a small surplus. This surplus is pumped into the main pipeline by a booster station. The situation can change drastically during summer, if the weather is hot and dry for several days. The demand then outstrips the local source, and additional quantities have to be brought in – through the PRV (Fig. 6.18).

The problem of redistributing water between different areas may call for an additional booster station, used only occasionally, as in Fig. 6.19. Do not forget that the original system may not have been designed for such operation – and be aware of possible dangers elsewhere.

Fig. 6.17 Feeding the local demand.

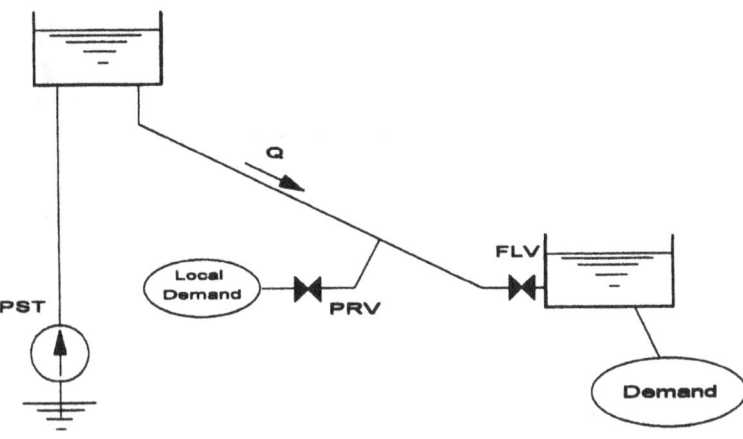

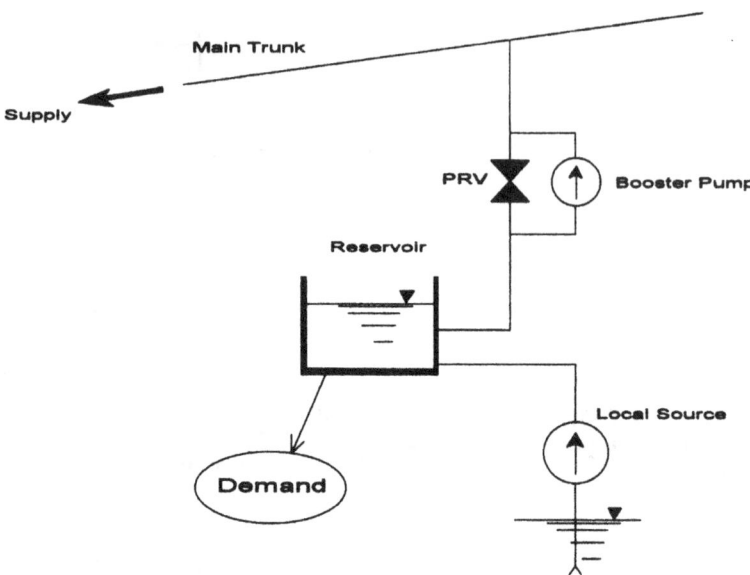

Figure 6.18 Two-way link.

Increasing demand often forces staff to operate the conveyance system to its maximum capacity, sometimes even beyond it. The original pumps are often replaced by larger units, out of necessity, and this may cause problems – as in Fig. 6.20.

A large tunnel supplies water to five major pumping stations, forming the backbone of a large water supply system. The design capacity of the system has been reached, causing problems in operation. The suction pressure of PSTs III and IV is already low and close to its limit, so restrictions have to be

Figure 6.19 Redirecting water where most needed.

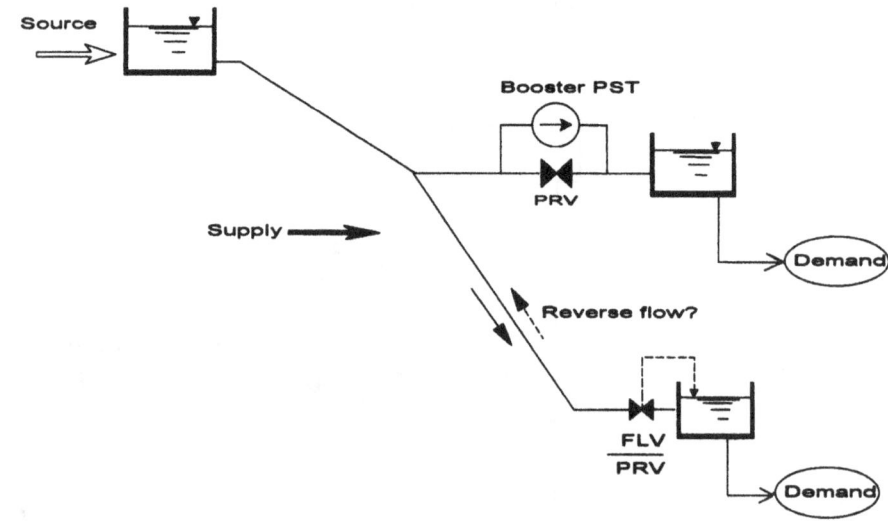

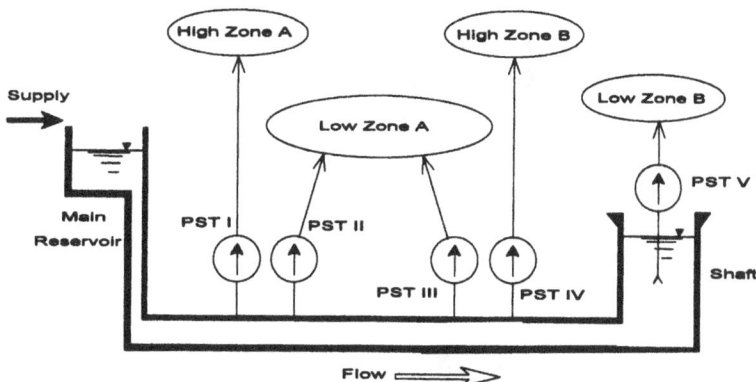

Figure 6.20 A tunnel linking five PSTs.

imposed on starting pumps in these PSTs, depending on how many pumps are operating elsewhere. Here modelling effort is well justified.

The model can be used to analyse unusual operating conditions. For instance, a trunk main is not used for supply during winter months – demand is too low – but it is kept full of water, ready to start in any emergency. To prevent a deterioration of water quality or even, in an extreme case, freezing, a small reverse flow is maintained from the higher reservoir, through a bypass with a PRV Fig. 6.21.

The model can also answer whether the capacity of an existing gravity pipeline could be increased by installing a booster station, for how much, and where to install it (Fig. 6.22).

Distribution system

Large distribution systems are usually divided into separate pressure (demand) zones, interlinked by pumping stations and reservoirs (Fig. 6.23). The water can be pumped from the low

Figure 6.21 Winter operation may differ from that in summer.

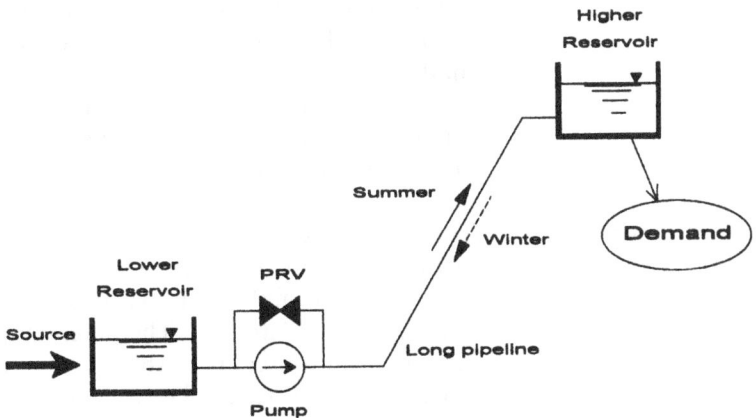

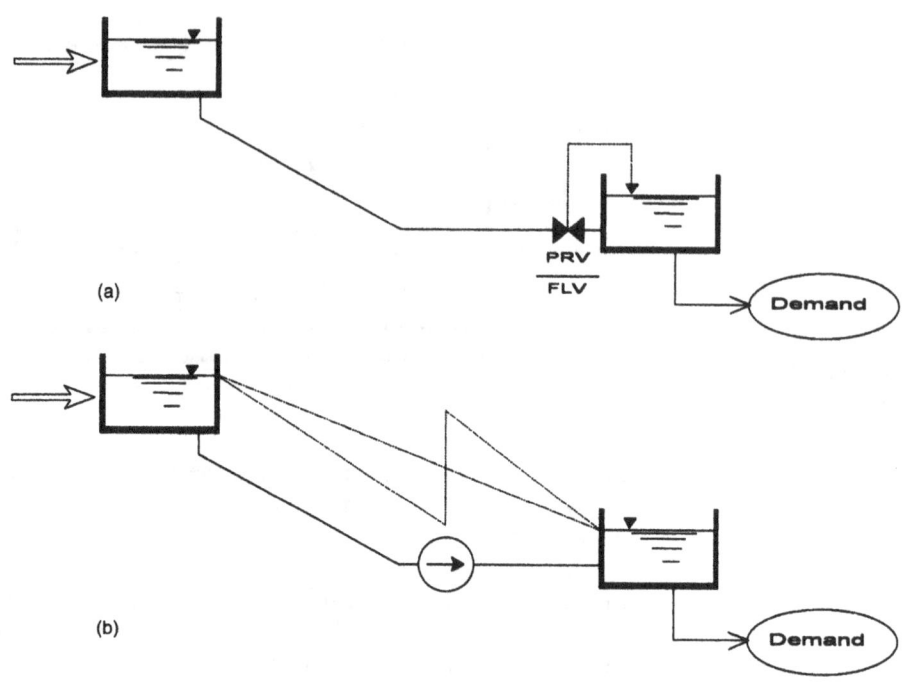

(a)

PRV
FLV

Demand

(b)

Demand

Figure 6.22 Where to place a new PST: (a) gravitational flow may not be sufficient; (b) it may be increased by installing a booster pump.

level through the network to the reservoir at the other end, and from there to a higher level, etc. (Fig. 6.23a). The other possibility is to pump all water to the highest reservoir and let it flow to lower zones by gravity (Fig. 6.23b).

Sometimes a pumping station has 'two masters', reservoirs at different sites (Fig. 6.24). The problem is clear: signals to the pumps from the two sites might conflict. If the distribution reservoir is almost full, the signal will be 'Stop pumps', while the water tower could be empty and calling to 'Start pumps'. Assuming that the tower is the higher, it could be given priority, and the reservoir protected from flooding by a float valve or a level-controlled valve. Optional policies should be tested on a model: summer demand, winter demand, various emergencies and so on. The situation might be complicated by a second pumping station supplying the tower (Fig. 6.25), making analysis even more necessary!

Dealing with emergencies

The model is an important tool for analysing emergencies, such as the sudden increase of demand or the failure of an important source. An example is shown in Fig. 6.26. A town is supplied from sources 'A' and 'B'. The potential problem is the long pipe leading from source 'B' to the town – it could easily be damaged because it is a fragile asbestos cement pipe. In such an event, a

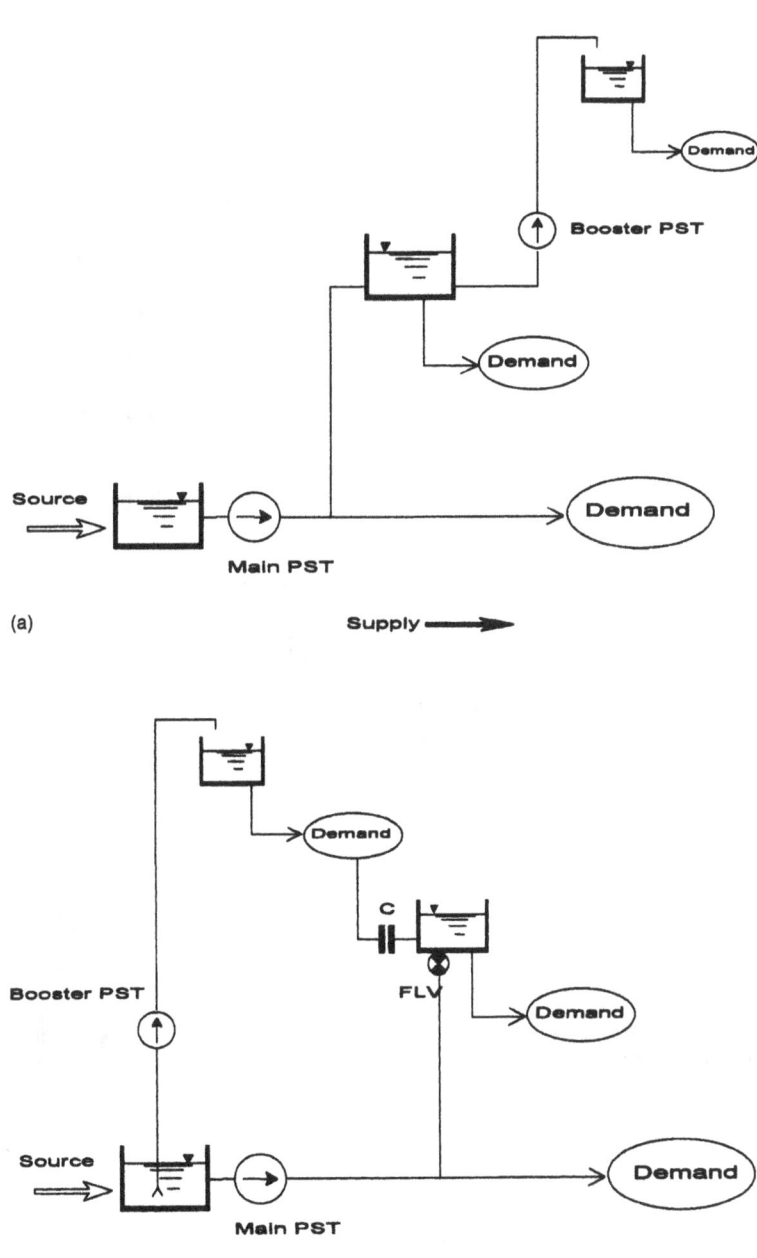

Figure 6.23 Supplying zones (a) from the bottom or (b) from the top.

(a)

(b)

shut-off valve will isolate the network from the leak, permitting repairs, but 'A' cannot provide enough water to meet the whole normal demand.

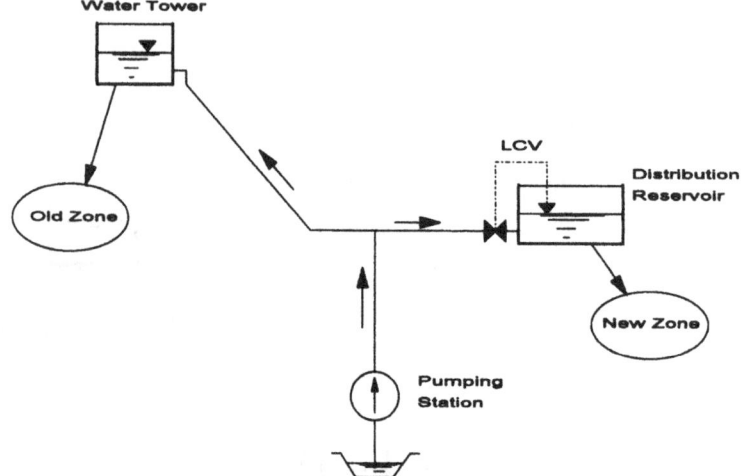

Figure 6.24 A PST feeding two reservoirs.

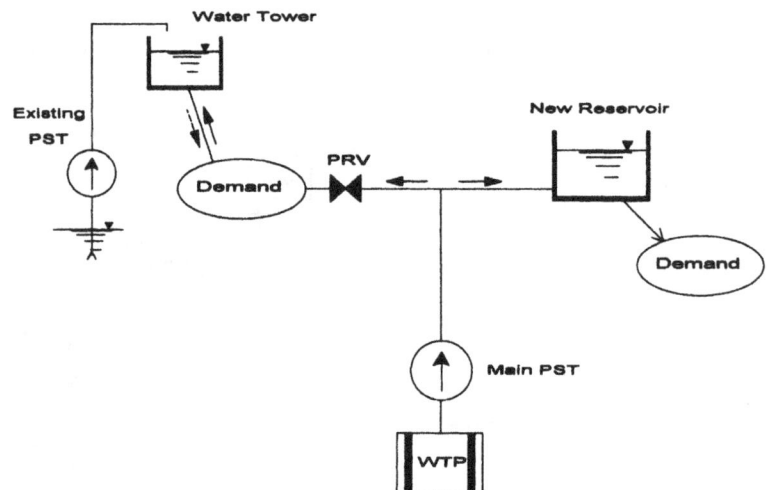

Figure 6.25 A real-life control problem.

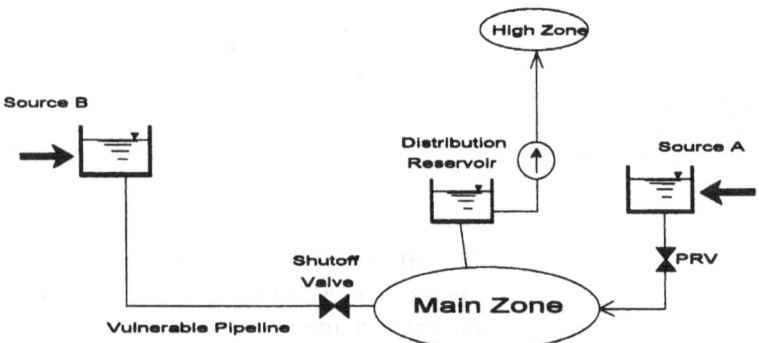

Figure 6.26 Dealing with an emergency.

One solution is to reduce pressure, and therefore demand, in the main zone, by using a PRV, reducing pressure to 1–1.5 bar (normally 4 bar). The high zone will not get enough water and will be left alone, to use the reserves of stored water as long as possible, although higher flats may be dry – but the mains will remain full and people will have at least some supply. The system remains under control and could be restarted immediately the damage has been repaired.

Devising a proper control policy

A new facility – a source, a reservoir or a pumping station – sometimes cannot fit into the existing system smoothly, because of either poor design or changes to the system over many years (a common case in practice). A few cases are discussed in the following text.

In the case shown in Fig. 6.27, the old PST and existing water tower served well until demand exceeded their capacity. A new source – with a modern treatment plant – has been built, but now the capacity of the water tower is too small. The pumps in *both* stations have to be switched too frequently, to 'track' demand variations. A larger reservoir is obviously needed – the question is how big it should be and where to locate it.

Another problem is described in Fig. 6.28. New customers are in awkward locations, where it will be difficult to maintain supplies. The new customers are on high ground, so their addition to the system will cause pressure to fall below the minimum pressure standard. An obvious solution is to build a

Figure 6.27 Putting in a new source.

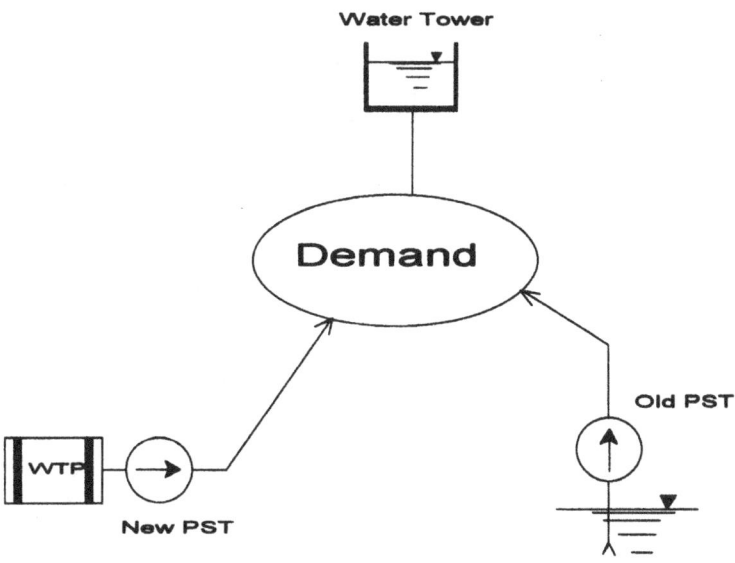

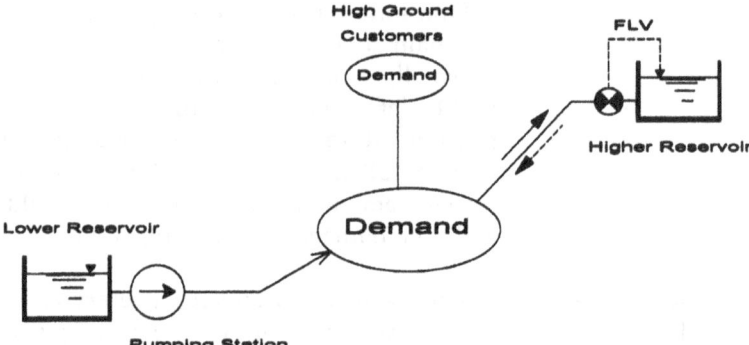

Figure 6.28 Customers on high ground.

small booster station, but in the meantime at least one pump has to operate in the existing pumping station to satisfy the new customers. The reservoir is protected against flooding by its float valve – closed for the better part of the day. This can be only a stop-gap measure, because the increased pressure, through pumps operating against a closed float valve, will certainly increase the number of bursts and losses through leakage.

Another example is how to operate a system with inadequate storage capacity (Fig. 6.29). There are several villages scattered over a large area, as well as a fossil-fuel power plant. All are supplied by a single pumping station. Pressure is maintained by three smallish water towers of different heights. The problem is how to prevent emptying of these small reservoirs without subjecting the fragile pipeline to excessive pressures.

Figure 6.29 Inadequate storage problem.

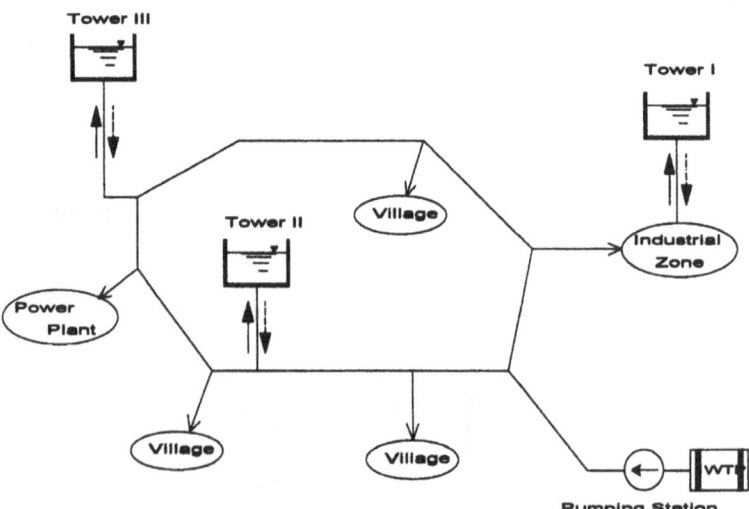

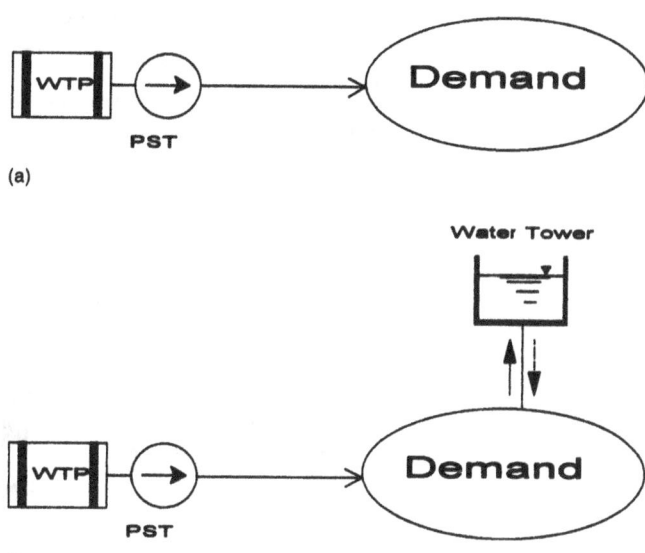

(a)

Figure 6.30 A water tower might help: (a) no storage in the system; (b) a water tower is added.

(b)

Maintaining the pressure in a distribution network by pumps alone is not a very efficient method, especially when all the pumps are fixed-speed units. Apart from frequent switching, the network is subjected to strong pressure surges, which result in increased leakage and bursts. The answer is either to introduce variable-speed pumps – which might be beyond local expertise and resources – or to build a water tower to equalise pressures (Fig. 6.30). Analysis on a model will help to find the right size and location for the new water tower – a rather expensive project that must be justified to the financial controllers of the company.

Figure 6.31 Combining sources.

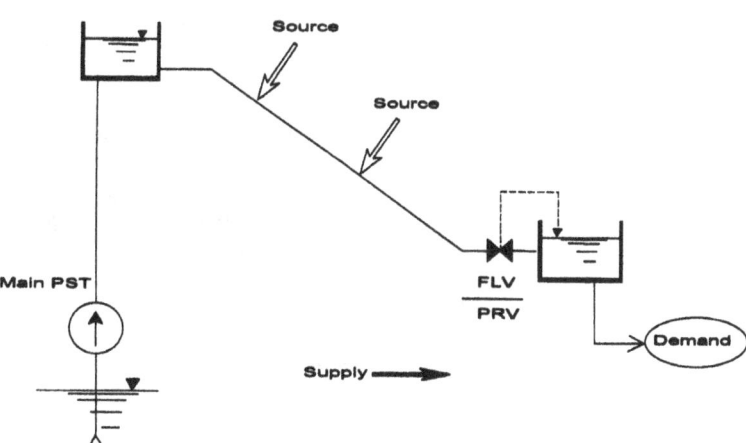

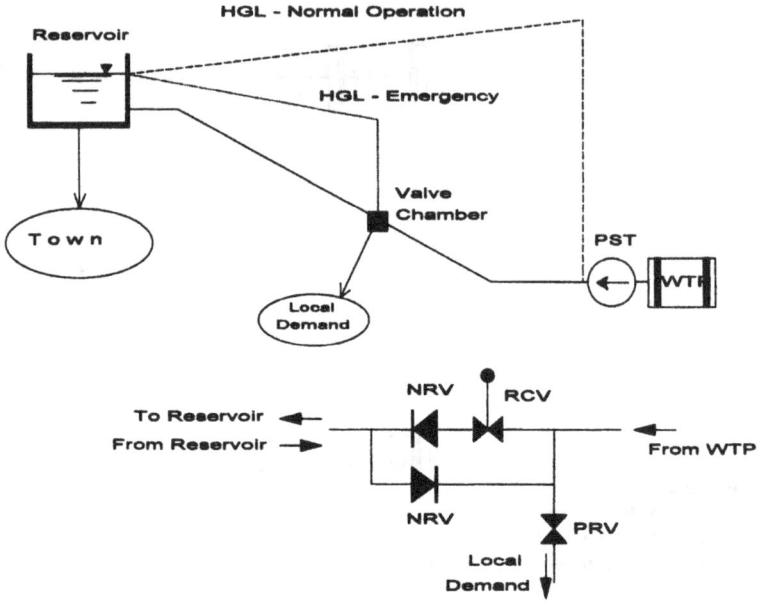

Figure 6.32 Emergency
supply.

In some cases, modelling is needed to strike a balance between
different sources (Fig. 6.31). The very long gravity pipeline takes
water, from a distant source, to the town. The water is of very
good quality, but has to be pumped to the higher reservoir – and
energy costs are high. There are two springs close to the pipe
that can provide additional quantities of clean water – but not
always; their capacity varies with hydrogeological conditions.

Figure 6.33 Alternative
supply routes.

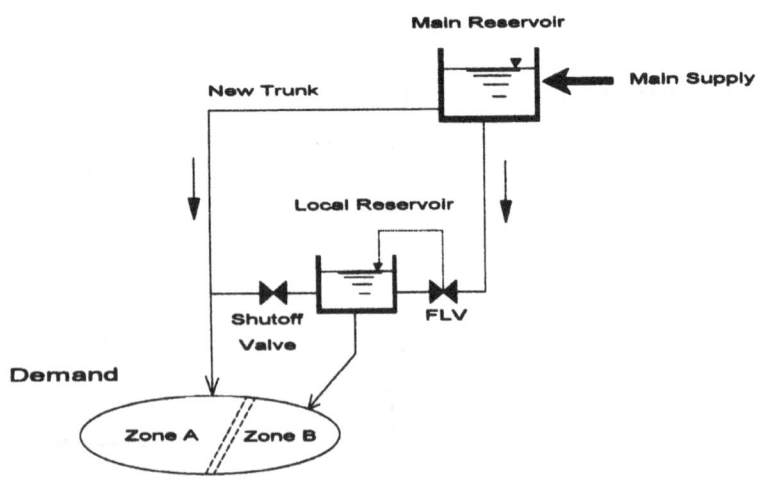

This is another case where modelling can help to find the best solution.

The last two cases show unusual solutions, made of necessity (Figs 6.32 and 6.33). In the first case, a treatment plant supplies a main reservoir and local demand alongside the pipeline. High local pressure is reduced by a PRV (Fig. 6.32). When the plant stops, local demand will be met by a reverse flow from the reservoir, controlled by check valves. In the other case (Fig. 6.33), the local reservoir could be supplied from either side; water from the main reservoir can flow through a float valve or a shut-off valve. Obviously the shut-off valve is closed in normal operation, to be manually opened in emergencies.

A similar solution was applied to control water level in the two reservoirs shown in Fig. 6.34. The larger main reservoir is supplied from the trunk main through a level-controlled valve. The smaller one (the service reservoir) is fed from the outlet side, where the pressure is lower, through a float valve. However, there is a link between the trunk main and the service reservoir that could be used in emergencies – but otherwise must remain closed.

Some arrangements might be unusual and even puzzling, but usually there were good reasons behind them. Usually, such arrangements are the result of changes in the system to add further customers or to improve the service. For instance, a small distribution reservoir is equipped with a level-sustaining valve at the outlet in Fig. 6.35. This is an unusual arrangement, but necessary in this case. It was introduced when new customers took most of the water coming from the source, especially a new brewery. The valve had to be closed to prevent emptying of the tank, which had become too small. Too great a fall of level will

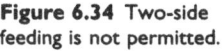

Figure 6.34 Two-side feeding is not permitted.

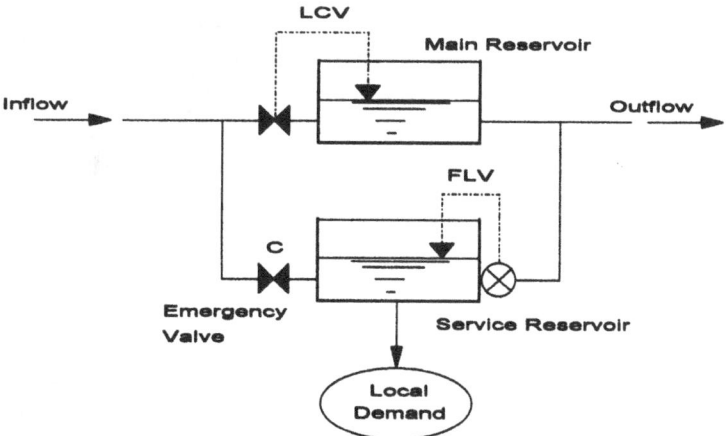

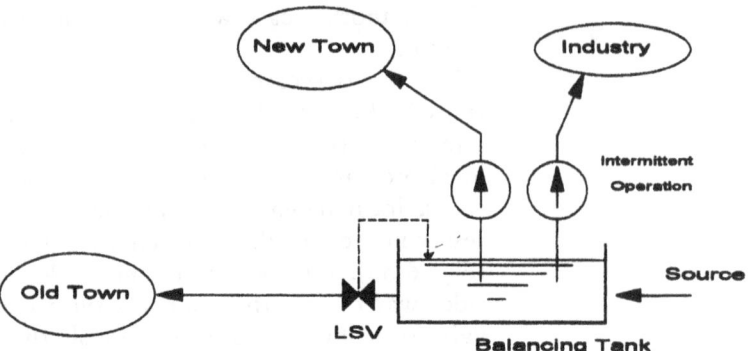

Figure 6.35 A balancing tank that is too small.

put pumps out of operation (activating suction protection) with dire consequences for industry and the local population.

Figure 6.36 shows a simplified diagram of part of a large urban distribution system. The pumping station can supply demand both directly, through a short pipe, and indirectly, via the reservoir. In reality, the direct link was closed, because otherwise the pressure in the network will be too high. This was found by monitoring pressure variations in the network while individual pumps were purposely started and stopped.

Valves can ,(and do) alter the operation of a distribution system, sometimes in an 'officially' planned way, but more often through the initiative of local staff. Consider, for instance, the

Figure 6.36 Is the bypass open?

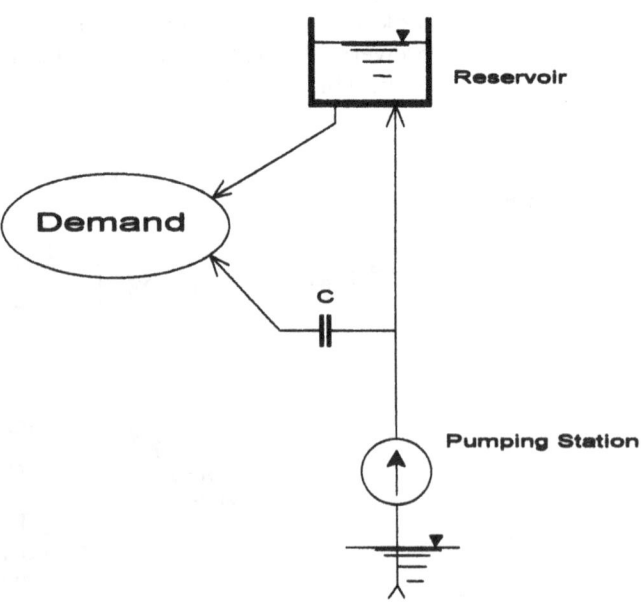

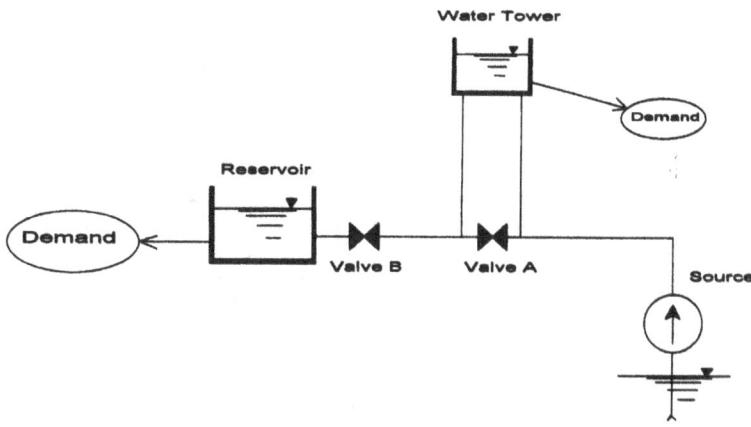

Figure 6.37 How does
the system work?

system shown in Fig. 6.37. The tower is higher than the main
reservoir, so how is it supplied? There are two feasible control
policies:

- Valve 'A' is closed, water is pumped to the tower and surplus
  flows by gravity to the reservoir; 'B' can act as a level-
  controlled or as an ordinary float valve, to prevent overspilling of water.
- Valve 'A' is open; 'B' sustains pressure to fill the water tower
  when needed.

Either policy is valid; but in the event the second one was
applied.

Another case is shown in Fig. 6.38: two pumps are linked
through a valve. How does this system operate? If the valve is

Figure 6.38 What do
the pumps supply?

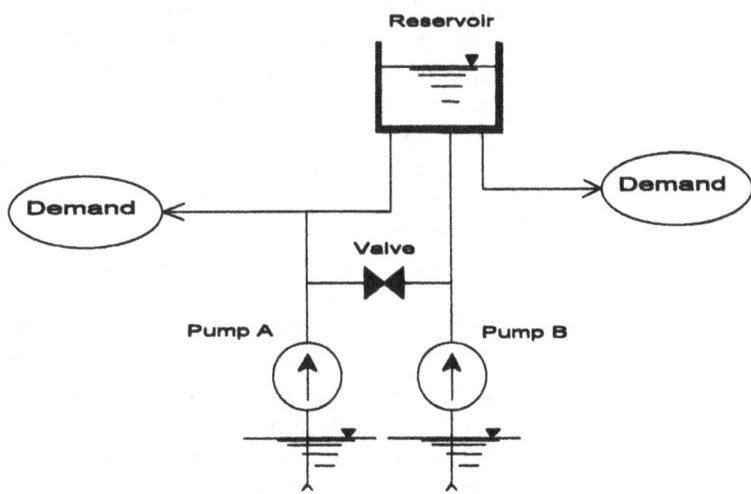

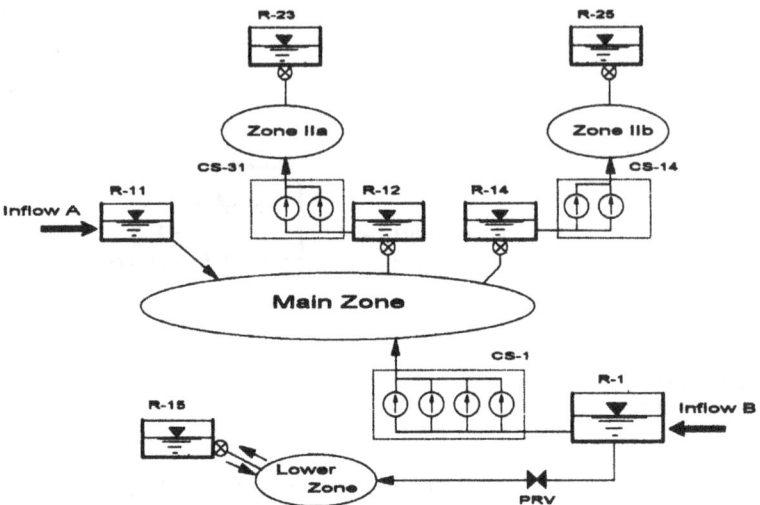

Figure 6.39 How is the system controlled?

closed, pump 'B' delivers water to the reservoir only, while pump 'A' can both feed the demand and send surplus water to the same reservoir. If the valve is open, then the pumps will interfere with one another, but either of them will supply demand directly. The odds are in favour of the first option, because staff prefer to have a clear and simple situation. However, in the real case the valve was actually open. This fact was discovered by careful analysis of operational data: whenever one pump started, the flow through the other was diminished, and vice versa.

The last case (Fig. 6.39) shows a real distribution system for a medium-sized town (population above 100 000). The town is supplied through two inflows, 'A' and 'B', through reservoirs situated at opposite ends of town (R-1 and R-11). The main pumping station, CS-1, delivers water from reservoir R-1 to the lower zone, while water from reservoir R-11 flows in by gravity. Any surplus is stored in reservoirs R-12 and R-14, to be used to supply two smaller pressure zones (IIa and IIb), each one with its own pumping station and a balancing reservoir at the far end.

The pumping stations in the higher zones, CS-31 and CS-14, are controlled by their reservoirs, R-23 and R-25. How about the main pumping station, CS-1? It could operate round the clock, providing the bulk supply, and minimise the water taken from the gravity source, or it could be time-controlled (to take advantage in changes of tariff), or it could keep one reservoir reasonably full, either R-12 or R-14 or (even) R-11, or ...? All these options are feasible, and should be discussed with the staff, perhaps opening new possibilities to them.

6.5 First simulation runs

The newly constructed model is best tested through simulation runs. In the beginning, everything goes wrong: all sorts of errors arise, each needing patient study and further investigation.

However, one should not hesitate to try the model as soon as possible. There is no point in postponing 'the moment of truth'. After a few initial difficulties, everybody can make the model 'work' reasonably well.

The next operation is to check if the results make sense or not. There might be some obvious errors and some less apparent ones. The program used for simulation will warn the user against the formal errors, which prevent analysis (e.g. an impossible Hazen–Williams C value of, say, 320). Other errors will be more difficult to detect (e.g. an incorrect pipe size, say 100 mm in the model for a real pipe of 150 mm) and may remain in the files for a long time, despite all the checks. The best way to clear most of the errors is to show the model to as many people as possible, who have knowledge of the real system – somebody will notice the error eventually.

Examples of *formal errors* are as follows.

- A node is mentioned in the list of nodes, but is not linked to the rest of the system.
- One (or both) end node(s) of a pipe was (were) not included in the list of nodes.
- Characteristics of a pump (or pumps) do not exist or contain impossible data (negative flow, etc.).
- The network is not fully interconnected – its nodes and links form not one network, but two (or more) separate networks.
- There is no reservoir or water tower or fixed-head node upstream of a pumping station, and, consequently, the head for this PST cannot be computed.

The novice may be exasperated when the program stubbornly refuses to accept erroneous data, but this is far better than to start the simulation with false data. The old slogan 'Rubbish in equals rubbish out' is not an excuse for a program with poor data checking abilities. The new slogan should be 'The user can make a mistake, but the program cannot', so the program must have a good error trapping procedure to spot and catch all mistakes.

The other question concerns what to do when an error is found. In some cases the program may stop, with an explanation; in others it will reject the erroneous data and give the user another chance to correct the error; in still others the program will just warn the user and continue with analysis.

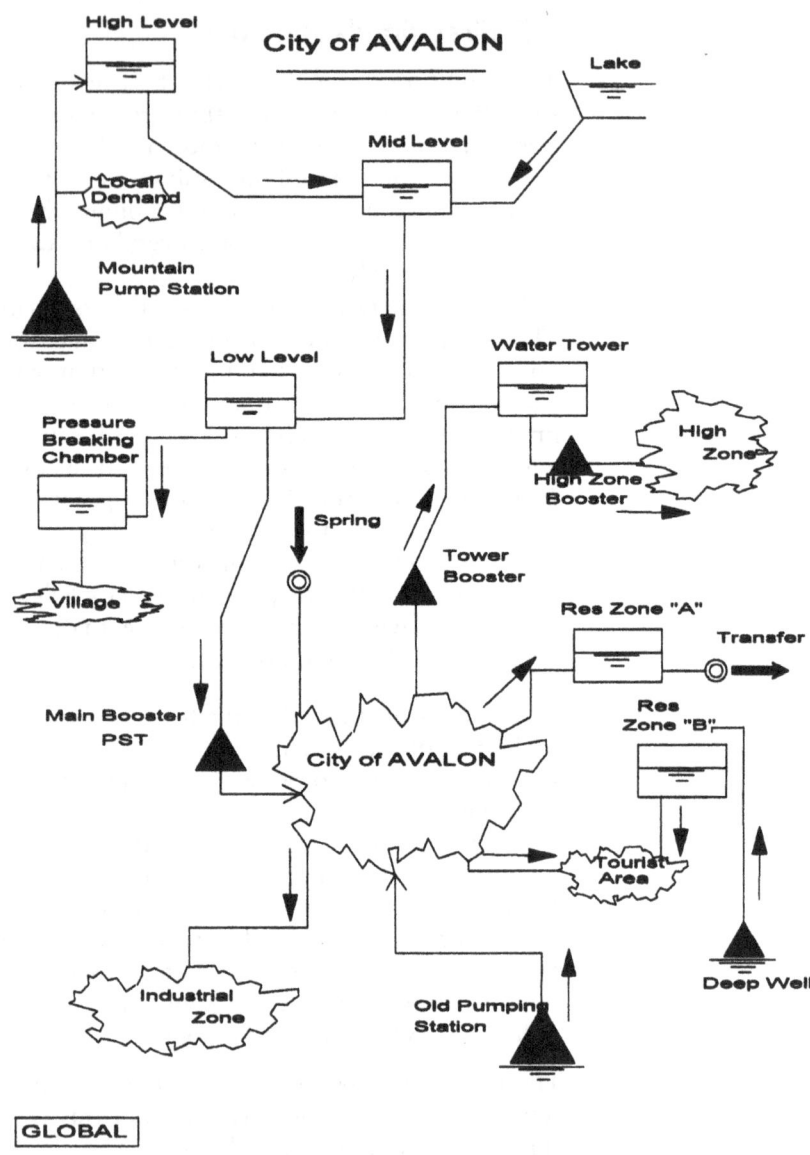

Figure 6.40 First information for the city of 'Avalon'.

Modelling is a process – it starts from a simple schematic such as the example shown in Fig. 6.40. The collection of data is usually a slow and tedious task, which must be completed systematically and thoroughly. A good schematic is very useful, but it must be revised and updated constantly. The first task is to identify all the major features – sources, pumping stations and reservoirs, as well as the major links of the network (Fig. 6.41). The town is supplied from the north (top of the diagram) via a

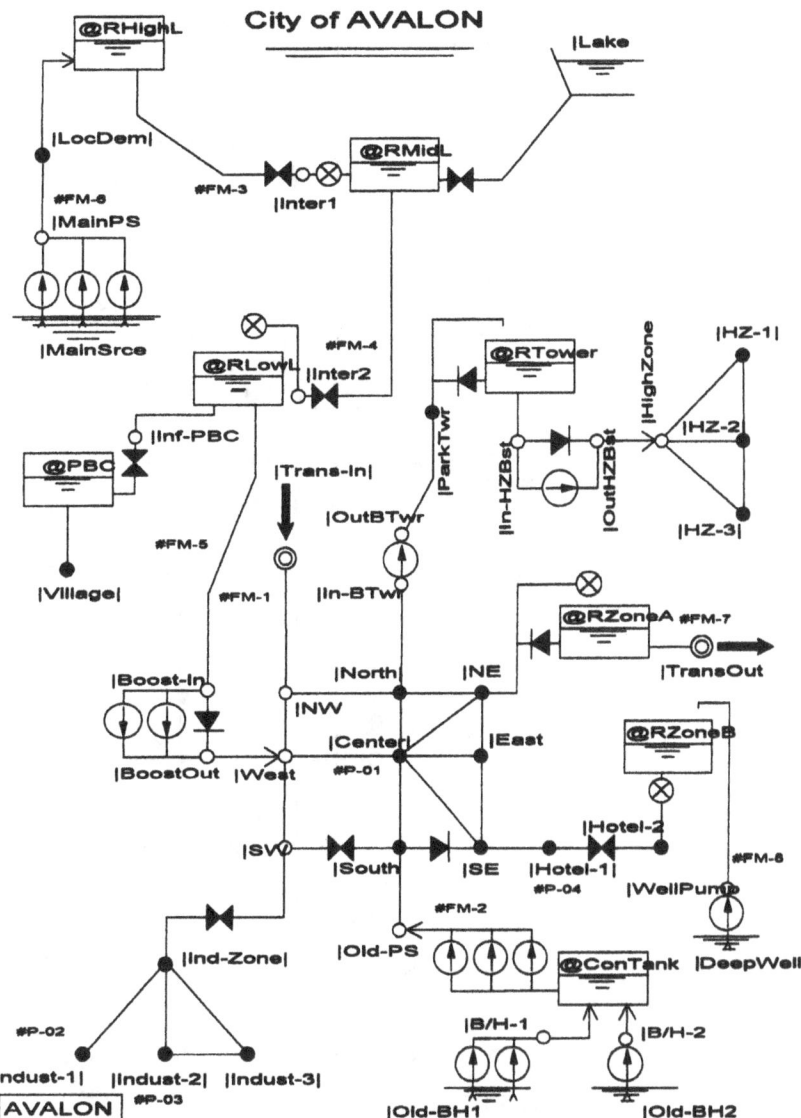

Figure 6.41 Details about PSTs and valves for 'Avalon'.

regional system, and from the south via an old source (Old Pumping Station). All pumping stations, reservoirs and some trunk mains are shown in this schematic.

Figure 6.41 shows much more detail: individual pumps in the pumping stations and control valves, for instance. The next step is to collect information about the control policy. The working hypothesis is shown in Fig. 6.42. It has to be verified on the spot – or in the control room, using operational data captured by telemetry.

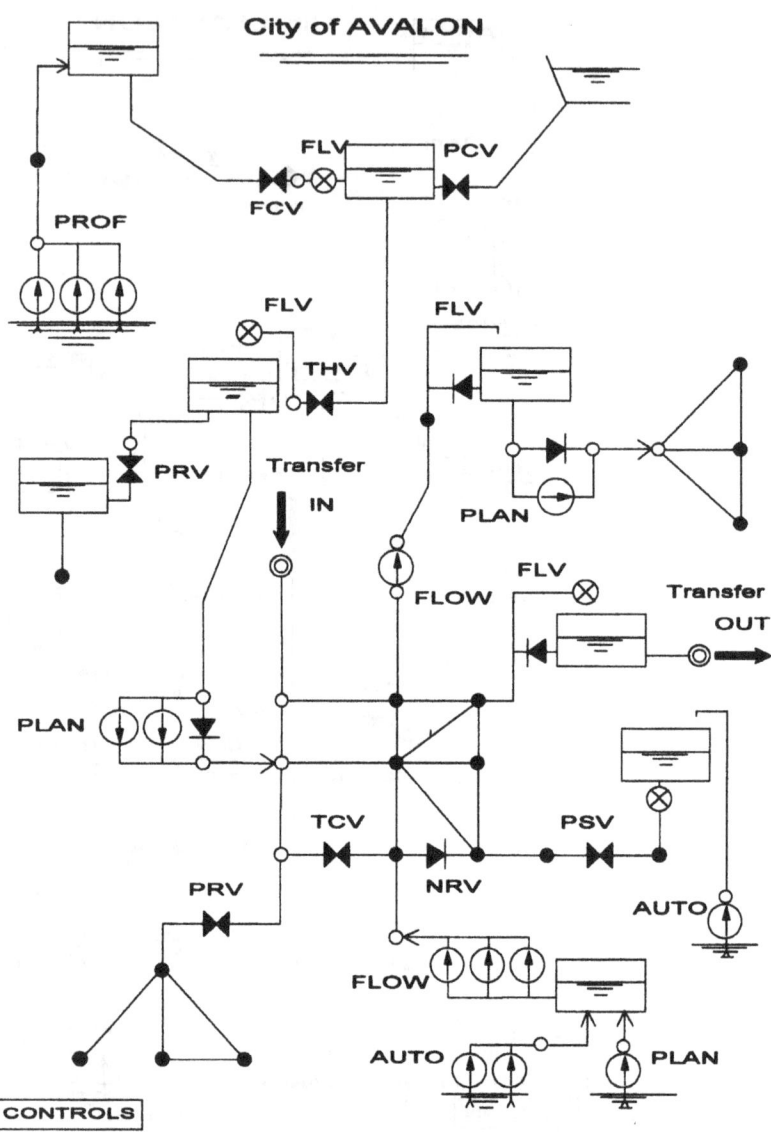

Figure 6.42 Hypothetical control policy for 'Avalon'.

The simulation results should be discussed with staff, not just once, but several times. Their cooperation and assistance are absolutely vital for the success of this task. It is important to master the program sufficiently well before starting the process of model calibration, which is discussed below in some detail.

6.6 Calibration of the model

Calibration of the model is the second stage of a three-stage process that needs to be followed in any engineering modelling project. The first stage is construction of the model, and that is what the previous sections of this chapter have discussed. The second stage (proving) is to satisfy oneself that the model represents the real world accurately enough to allow the user to move to the third stage of the modelling process. That third stage is actually using the model, and doing so with a reasonable degree of confidence. The proving or calibration stage is quite onerous and can be frustrating, but it is one that needs to be undertaken carefully.

During the 1970s, and even into the early 1980s, the modelling software, computer hardware and data collection equipment available to water system model makers were crude by modern standards. As has been pointed out, the work was carried out by specialists, and use of the models was largely confined to planning and design engineers. In all these circumstances, a model was usually calibrated using a small number of snapshot analyses and correlating a relatively small number of computed and measured flows, pressures and reservoir levels at some appropriate point in time, e.g. the time of peak demand or at a time of average demand.

Arbitrary, informal rules were developed by individual practitioners for the proving process. Typically, pressures were recorded (on clockwork-driven paper charts) at no more than 10% of the nodes forming the network. Major reservoirs would be monitored by temporary mechanical level recorders, and flow data captured by manually reading the integrators of meters at intervals for two or three hours across the target time.

In correlating computed and measured parameters, further arbitrary rules were developed. Pressures, for example, might be acceptable if they matched within 1 m head. Of the pressure points recorded, typically 10 or so for the proving tests, one or two might be discarded if outside the pressure correlation 'rule' on the basis that a mechanical problem might be the reason.

These approaches to the problem of calibrating or proving a model, with the benefit of hindsight, appear to be so crude as to be almost worthless. However, before the widespread availability of data-loggers and computers, and certainly when modelling software was only available to produce 'snapshot' analyses, the collection of data to calibrate the model was a major logistical exercise for most water companies. The cumbersome logistics and consequent delay in producing results was probably a further reason for the limited use of models in the water supply industry during the 1970s and early 1980s. Certainly the delays precluded any significant use by operations managers.

The development and application of loggers, primary instrumentation and, in particular, telemetry in the industry – together with the creation of 'dynamic' modelling software – has made the proving stage far less of a tedious operation and has boosted the confidence that users can place in their models. Nevertheless, careful calibration of models remains a key element of successful model making and use. Despite the accumulated fund of experience gained over the last two decades, no standard set of calibration rules have emerged. This fact of modelling life is quite understandable, given the range of uses to which models are put and the remaining gulf between the most and least sophisticated data gathering systems available to water supply organisations.

This gulf may mean that some organisations are still faced with the very real problem of mobilising sufficient resources to collect all the data desirable to complete a satisfactory calibration. It is hoped that the following notes will give some guidance on how to gain the most from the resources that are available. We would, however, make the point strongly that practitioners should not be deterred from creating models by a lack of data collection resources. Build the model, calibrate it to the best of your ability and use it, having in mind the level of calibration that you have been able to use. In many ways the whole process is iterative. A relatively crude model will allow cases to be made for the best location of monitoring equipment, and that in turn should lead to better calibration facilities.

The following text offers some suggestions on how to approach the task of calibration but we should first reiterate why we are calibrating in the first place. The reason is the following. *To establish the credibility of the model and allow decisions about physical and operational developments in the real system to be made with as high a degree of confidence as possible.*

If calibration is carried out systematically, a number of peripheral benefits should emerge, including:

- the creation of a 'benchmark' tool against which a wide range of investment and operational options can be tested;
- enhanced understanding by staff, at all levels, of the system and the way in which it works;
- the identification and elimination of problems and anomalies in the real system through the discipline of the calibration process itself.

The calibration process

The model of a WSS should embrace all the known facts (but some will be of dubious value, certainly during the initial stages of calibration) and a few hypotheses. The most common assumptions upon which these hypotheses are based and which are potential sources of error are as follows.

1 The modelling method itself is based upon assumptions such as:
   - all changes are slow;
   - pressure surges may be neglected;
   - consumption does not depend upon service pressure;
   - nothing significant happens within the time step.
2 The model of water consumption is accurate, *but* in fact it is wholly artificial and is no more than a very simplified picture of reality.
3 The database is complete, *but* inevitably it will not be complete. Many data will be missing or dubious – pump characteristics, exact locations and elevations of nodes, etc.
4 Information on losses is reasonably accurate, *but* information on the distribution of losses around the system is *not* likely to be accurate.

Calibration is normally done in two steps:

- Verification of the model
- Fine tuning of the model

Verification

Verification of the model comprises the following tasks:

- List information required to build the model and identify uncertain data (e.g. inflow/outflow arrangement at reservoirs, pump characteristics, which pipes are closed, valve size, friction coefficients, etc.) that might have a significant influence on the results of simulation.
- Assign a feasible range of values to the unknowns.
- Gather all available information, from written reports to telemetry and logger data, and carefully examine all this, trying to find hard facts and reduce the number of unknowns. It will be useful, for future model development, to identify and date the sources of information.
- Run a few situations on the model, varying the uncertain or unknown parameters. Find out the ranges of important variables, critical spots, likely reactions of the system, etc.
- Compare these results with observations. Try to locate the largest discrepancies – which may well be a blessing in

disguise - showing the most serious shortcomings of the model or imperfect knowledge about the real system. Discuss these findings with staff and try to find the reasons for disagreement.

Some typical errors and problems are listed below:

- The real network is different. The length or diameter of main pipes is wrongly stated in plans, there are some forgotten valves somewhere which isolate parts of the system or cut loops, etc.
- Reservoirs are not modelled well. Either data are wrong (volume, elevation, depth) or the structure is more complex than believed, with several chambers separated by (forgotten?) valves, leading to different water levels in chambers.
- Pumps have an incorrect characteristic curve due to speed different from standard value, or the ageing and/or poor maintenance of pumps, or throttling of one or other pump, or actual pump sets are not the ones from the documentation, or suction pressure is too high/too low, or etc.
- A 'black box' (a programmable controller) controls the operation of the pumping station, installed by another department and overlooked during the modelling exercise.
- Are all the large customers adequately identified and described in the model? How significant are seasonal effects?
- A poor assumption on losses was used in the model. Maybe the leaks were not estimated correctly or were distributed incorrectly in the model.
- Water consumption on this particular day was unusual (public holiday?, sports event?, bad burst?, unusually hot weather?, etc.).

The key advice for untangling these problems is always: *Visit the site and talk to the staff.* But the other advice is to collect additional, corroborative data from other sources, like the electricity and water bills, for instance.

The purpose of the first stage of calibration is to establish a reasonably good balance of the system. The details can be sorted out during the 'fine tuning stage' and as the model is expanded. Simple, manual calculations can help a lot in achieving a good balance.

After the obvious errors are discovered and corrected, there might be some anomalies that cannot be easily dismissed. These must be discussed with experienced staff in the water company – managers, operators, engineers. Some of the difficult problems may need to be resolved by carrying out field tests. Sometimes the differences may indicate some unusual characteristic of the

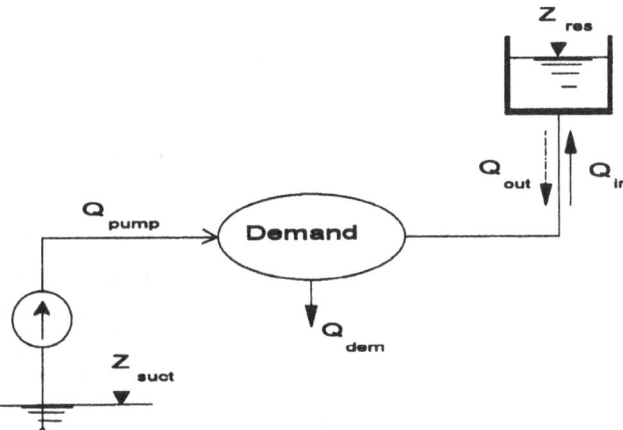

Figure 6.43 Calibration of a simple system.

particular system, improving everybody's knowledge about the real operation of the system.

If everything goes smoothly, do not be too complacent. The model cannot be perfected in one attempt (and should not be); usually several sweeps are needed before good results are achieved. The time is not wasted, however, especially if staff become involved in the process.

An example is included here to illustrate the point. There is a pumping station, demand area and a reservoir at the other end (Fig. 6.43). Assume that the simulation results show that the reservoir is overflowing, while in reality this is not the case. Possible causes are listed below – writing such lists can be a useful means of untangling the problems:

- The capacity of the pumps is overestimated; perhaps Q–H–P data are wrong.
- The actual speed of pumps is overestimated.
- Demand is underestimated.
- Losses are underestimated.
- The reservoir is modelled at too low an elevation, so the model pumps 'deliver' more water than is really possible.
- Water level in the suction tank is too high, having the same effect as in the previous case.
- The friction and minor losses are underestimated.

Perhaps the reader can add other possible causes of the discrepancy between the model and the real system. The list is already rather long, even in such a simple case, and it takes time and effort to discover the truth. This detective work repays well. After several computer runs, the model 'yields' and starts to behave more or less like the real system. The sensitivity of the

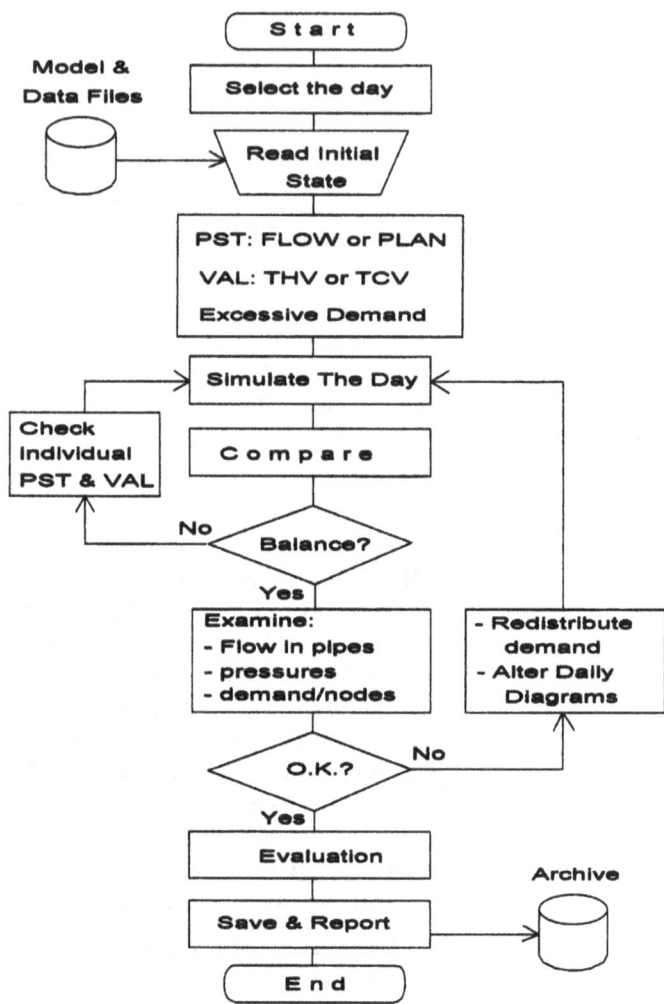

Figure 6.44 Fine tuning
of a model.

model to changes of parameters is determined in the process.
Then it is time for final calibration, the fine tuning of this
delicate entity.

Fine tuning

Fine tuning of the model is described in the flow diagram in Fig.
6.44. The first step is to select some historic and well
documented days and/or events. The first cases should correspond to normal situations, with all facilities in good order, and
the seasonal demand close to its maximum. The initial state of
the system should be read from the model database.

Pumping stations should be modelled through the FLOW or PLAN option (because the discharge and changes in operating regime are known for the chosen day). Control valves should be modelled as usual, but if true values of valve openings are known (through the telemetry system) then the TCV option should be used instead. Excessive demands – if there were any – should not be forgotten.

The simulation should be made for a significant period (a full or half day) and the results compared with observations (at reservoirs first) to see if the balances are right or not. If serious discrepancies are noted, then the causes must be found and errors corrected. The causes can be various: from errors in the model to bad timing of changes. Telemetry data are of great help in this analysis, as shown below in the examples.

After a while, balance should be achieved in most places. Then computed pressures and flows within the network should be compared with measured data, wherever possible. Differences may mean that either the demand pattern or the daily diagram are not correct (but not the totals – the quantities are already balanced at the level of the whole system). The patterns and/or the demand diagrams can be corrected relatively easily. Naturally, one should not change demand patterns and diagrams arbitrarily – this would not make any sense and will just confuse everybody.

Changes made to the model must be feasible. For instance, friction coefficients must be within the acceptable range – if the changes still leave unacceptably large differences between the model and the real system, the causes must lie elsewhere. Fictitious items should not be introduced (imaginary pumps or valves), nor should pipes be closed without an enquiry as to the effect in the real world of the action. Demands and losses are never precisely known, of course, but again they should not be distributed arbitrarily, or moved around each time as it comes handy. No 'unknown customer' should be allowed in the system, to cover the mismatch between the model and reality.

The results of simulation can never exactly match observed data – small random variations of water demand will prevent this – but the margin of error should not be too large. A few cases from practice will give an indication of how good this agreement can be (Figs 6.45–6.49). In all these examples, the solid line represents the results of simulation while the dashed line shows data recorded by the telemetry system. Sometimes the results are quite acceptable, even if the difference is far from negligible. One such case follows (Fig. 6.50).

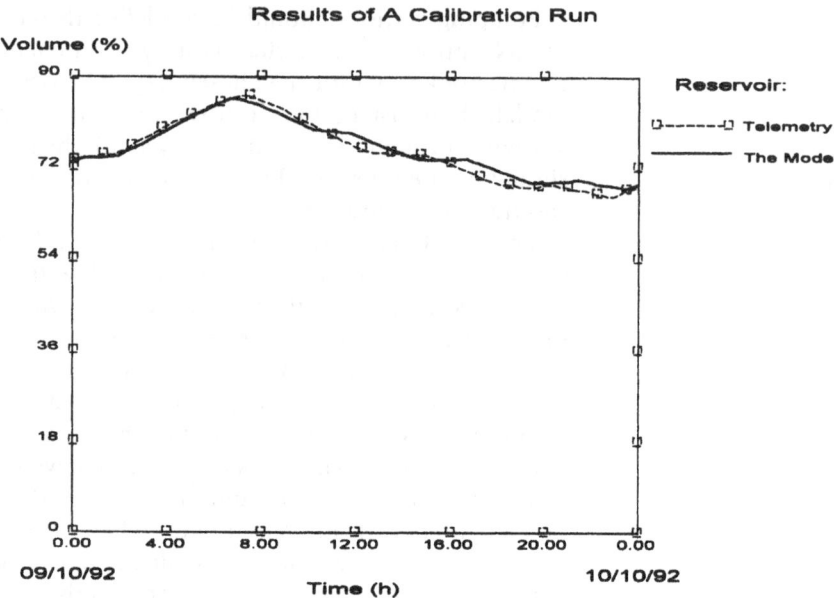

Figure 6.45 Changes of
WL in a reservoir.

Figure 6.46 Pumping
station output.

Figure 6.50a shows telemetry data: variations of water level in a reservoir and flow through its associated booster station. Pumps are obviously controlled by water level in this reservoir – note the perfect synchronisation between characteristic points on these two lines. Figure 6.50b shows corresponding model results

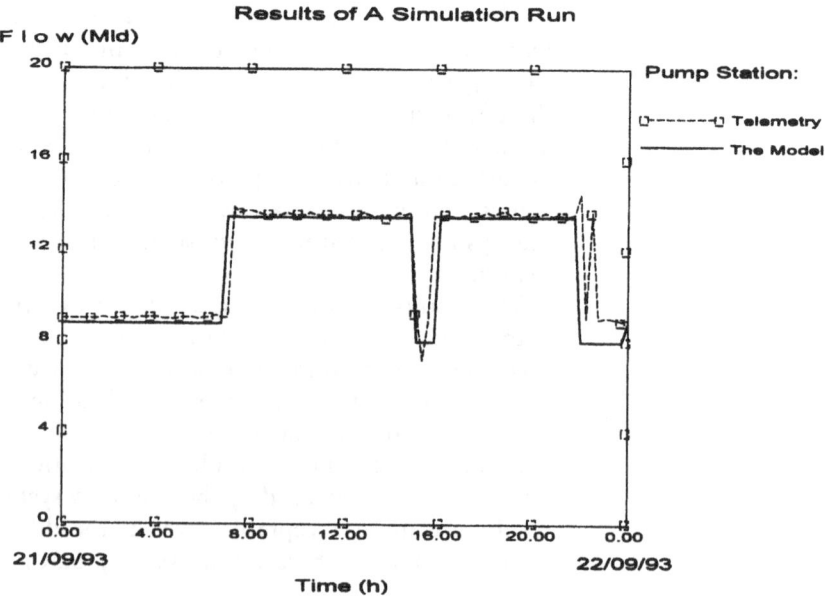

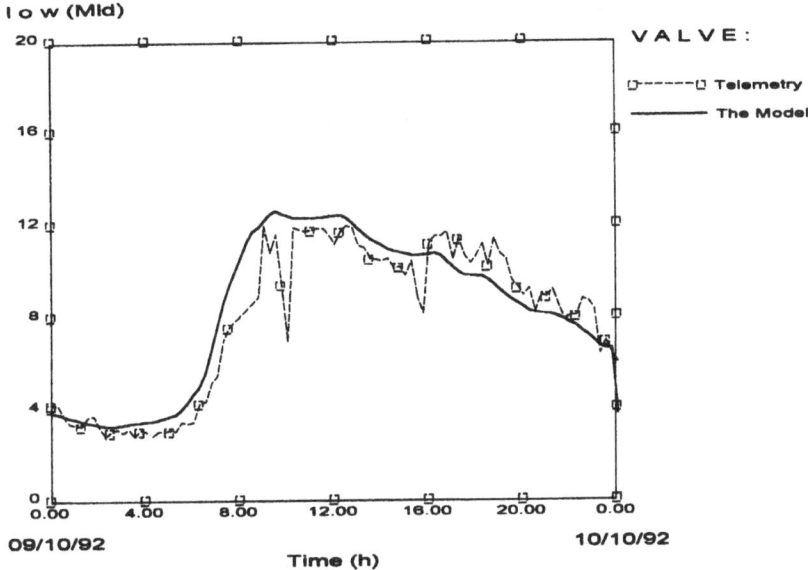

Figure 6.47 Flow into a demand area.

Figure 6.48 Pressure at a point in an area controlled by a remotely controlled PRV.

for the same system. The booster station again is controlled by water level in the reservoir – but not at similar times. Why? By comparing the timing of 'starts' and 'stops', one can see that they are different (four cycles in Fig. 6.50a and five in Fig. 6.50b). The answer is that the demand patterns and diagrams are not the same in both cases, i.e. in reality and in the model.

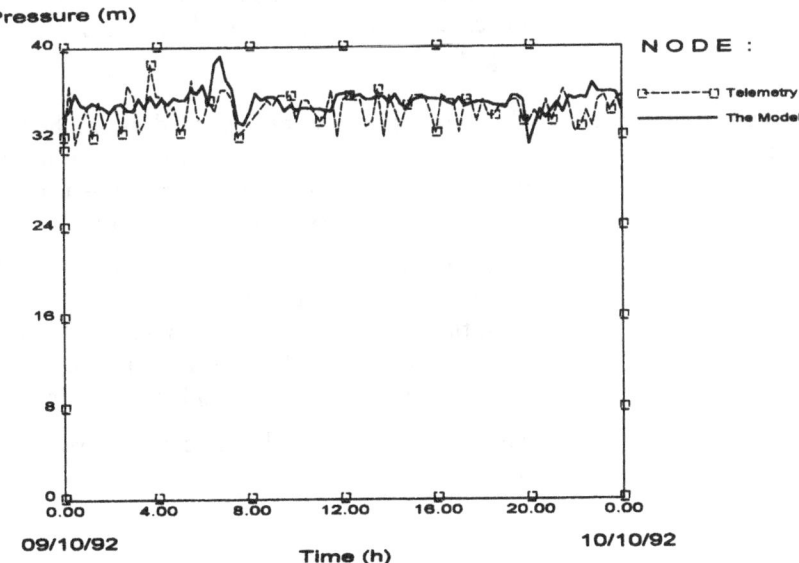

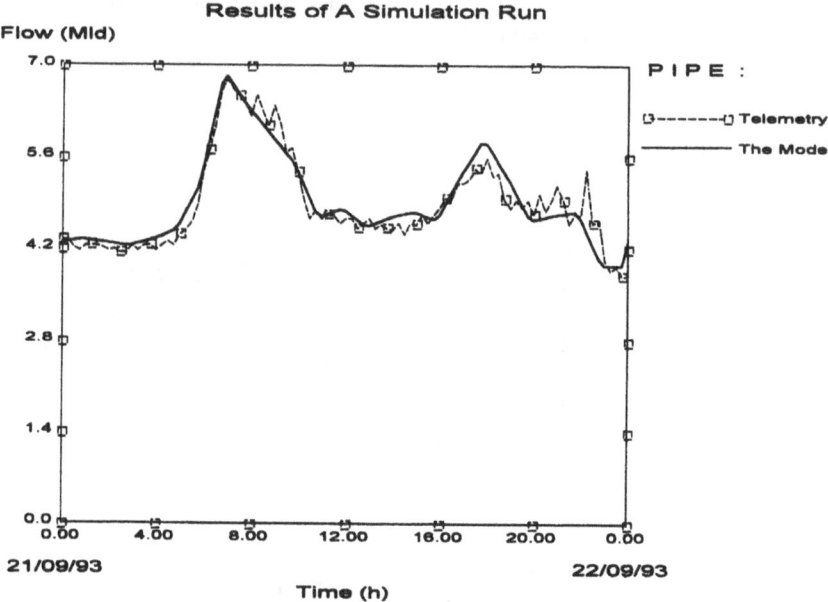

Figure 6.49 Outflow from a reservoir.

Going one step further, one can see that quantities supplied by the booster station are practically the same in both cases – so the problem is only in the daily diagram. Its peak is underestimated – although not by much – in the model. However, the results can be accepted – with more effort on the analysis of daily patterns in future.

Figures 6.51a and 6.51b show simulated and real data for a pumping station with Profiler control. Pressure control reduces losses *and* demand, so such systems are frequently modelled (Fig. 6.52).

Determining system capacity

The next task may be regarded as the first task of the model's useful life and the last task of the calibration process. This is to determine the capacity of the system – a sort of 'sea-trial' for the new vessel. This is a situation when:

- all pumping stations are operating with all units,
- all valves are fully open,
- initial water levels in reservoirs and water towers are half-way between the bottom and the top,
- water demand is at its highest expected peak.

This computation will indicate where the critical or 'pinch' points are:

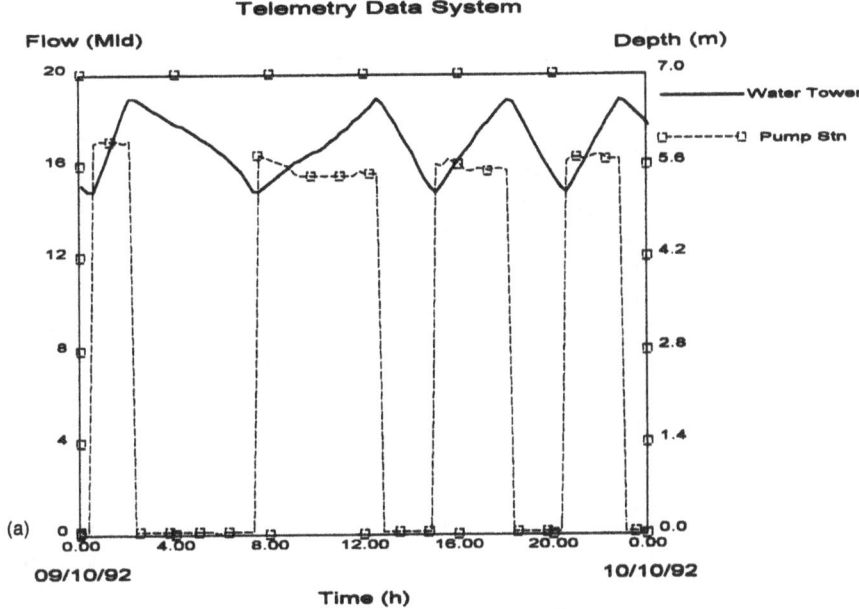

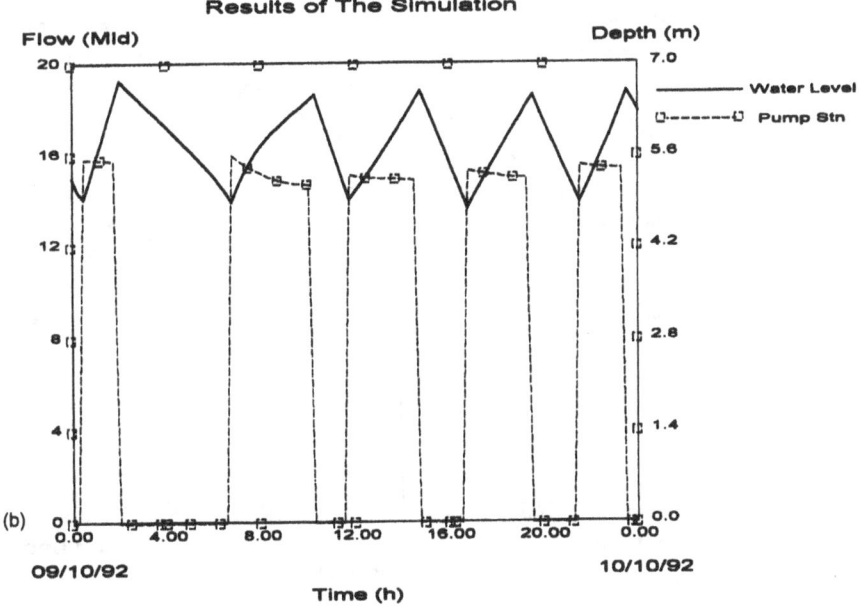

Figure 6.50 (a) Telemetry data and (b) results of simulation.

- water mains with insufficient capacity, or too great a capacity, which may cause water quality problems or accumulation of silt and debris;

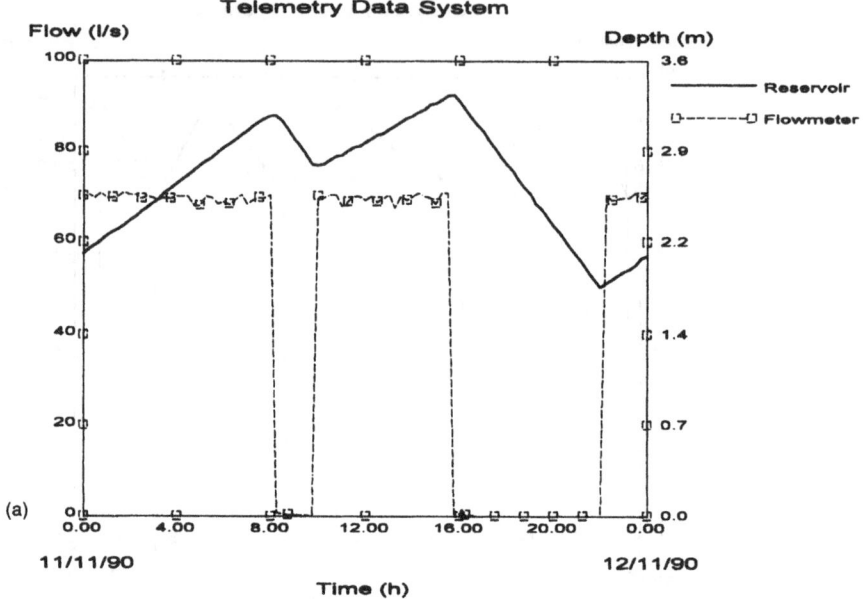

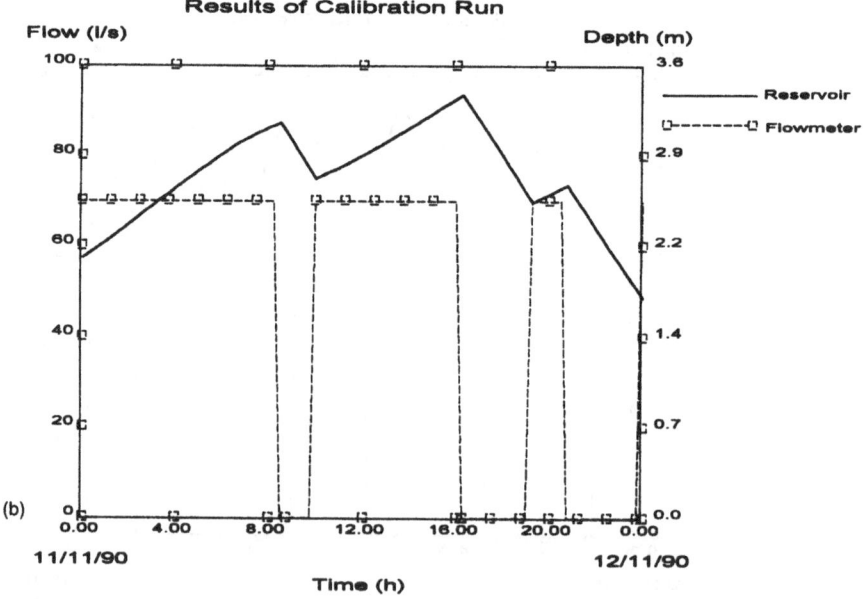

Figure 6.51 Profiler: (a) telemetry data and (b) results of simulation.

- badly placed reservoirs and water towers, which cannot be refilled sufficiently or which are always full and useless;
- inadequate pumping stations, which are either too small or too large;
- areas where service pressure is either too high (say, over 4–6 bar) or too low by existing criteria.

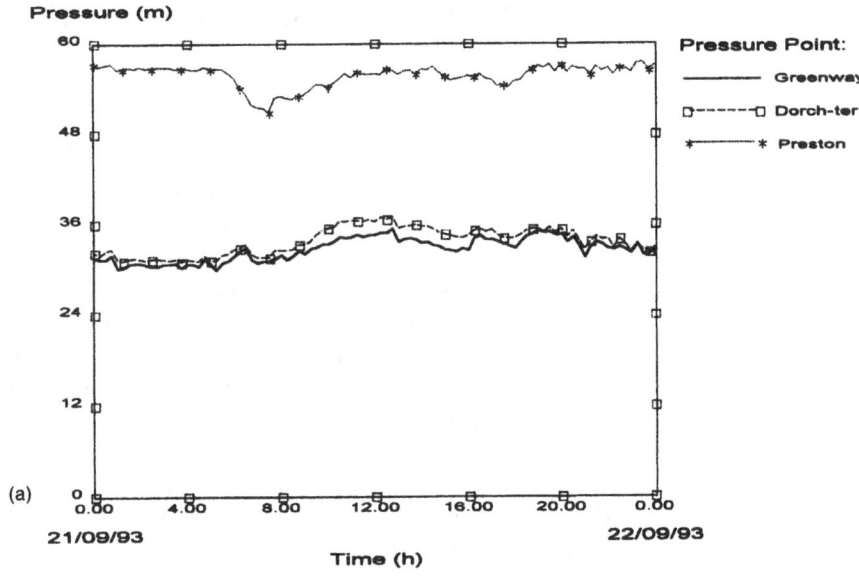

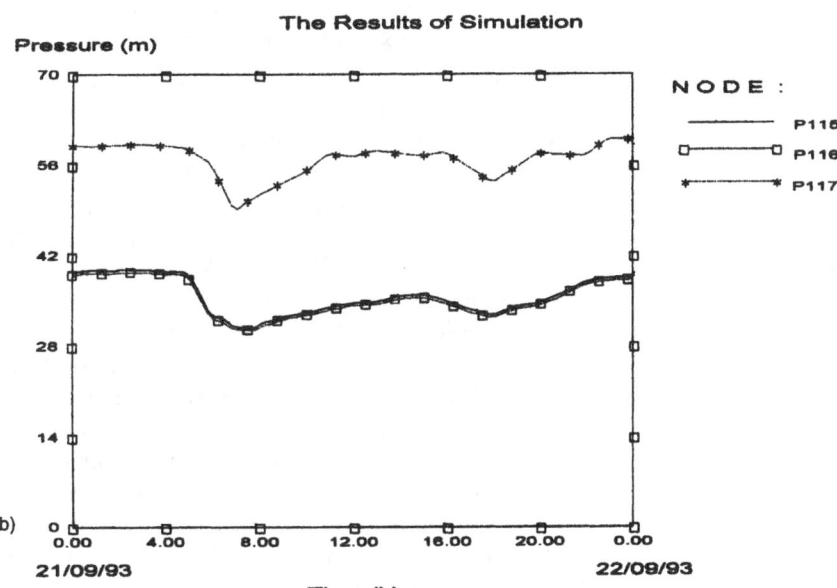

Figure 6.52 Pressure control: (a) telemetry data and (b) results of simulation.

The results of these runs should be discussed with the staff of the company. It is always very rewarding for all concerned when conclusions obtained by using the model are accepted by the staff as valid and useful.

7 Modelling and real-life problems

7.1 Simulation models as tools for design

The design process has its *creative phase*, when options are being formulated, and its *routine phase*, when accepted solutions are elaborated in greater detail. The phases cannot be completely separated, because every proposal has to be evaluated from various points of view: functionality, controllability, capital and operational cost, etc. This evaluation is based partly upon the designer's experience and judgement and partly upon hydraulic analysis. Here the models come into play, permitting the designer to examine quickly and efficiently different alternatives with precision and in detail. The idea is as old as designing itself, but computers have opened new possibilities, as amply demonstrated in the references (Alla, 1985; Bos and Jarrige, 1989; Brammer and Schulte, 1993; Clark, 1994; Cohen, 1990; Coulbeck and Orr, 1988; Hoogsten and van der Zwan, 1985; Jones, 1984; Obradovic, 1993; Obradovic and Kordic, 1986; Rao and Don, 1977; van der Zwan, 1988; Walski, 1984).

To illustrate this point, a simple case is included here (Fig. 7.1). The new system has to supply four communities – demands 'A' to 'D' – from a single source. One possible solution is to build a main pumping station, four large reservoirs (I to IV) and two booster stations (II and III), and to control the pressure at two places (A and D) by PRVs.

Obviously, this is not the only solution. Maybe reservoirs II and III could be replaced by a single reservoir, or demand 'C' could be fed by gravity flow from either reservoir II or IV (or from both). Clearly, many options are feasible, each with its advantages and shortcomings. Without modelling, one could only make arbitrary decisions (not necessarily the wrong ones!), which cannot be easily described to other parties; nor could an objective consensus be reached easily. With models, the situation is far better: every proposal can be examined on a model of the system in a relatively short time. All concerned can see for themselves why one solution is better than another. This consensus cannot be taken for granted; the owner of the future system must be consulted almost from the beginning of the design process and participate in decision-making.

Another simple case is included to illustrate the importance of good cooperation. Figure 7.2 shows the simplest possible

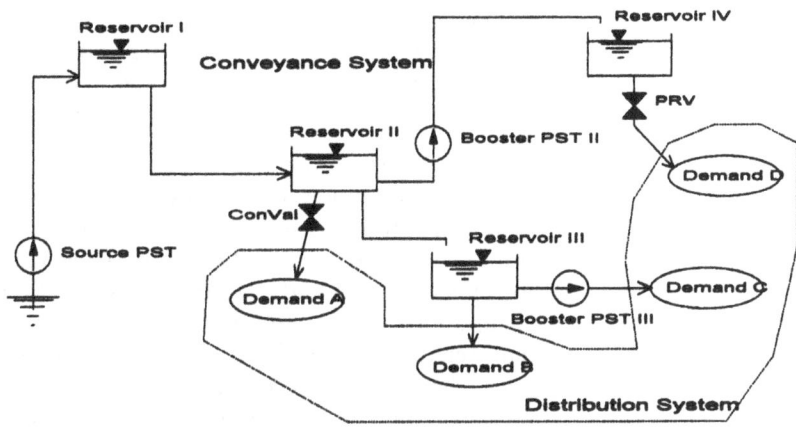

Figure 7.1 A water
supply system.

problem, satisfying a given demand from a source. Water must
be pumped into the area, but there are at least three possible
arrangements:

1 Pump all water through the demand area to a balancing
   reservoir at the opposite side.

Figure 7.2 Alternative
solutions.

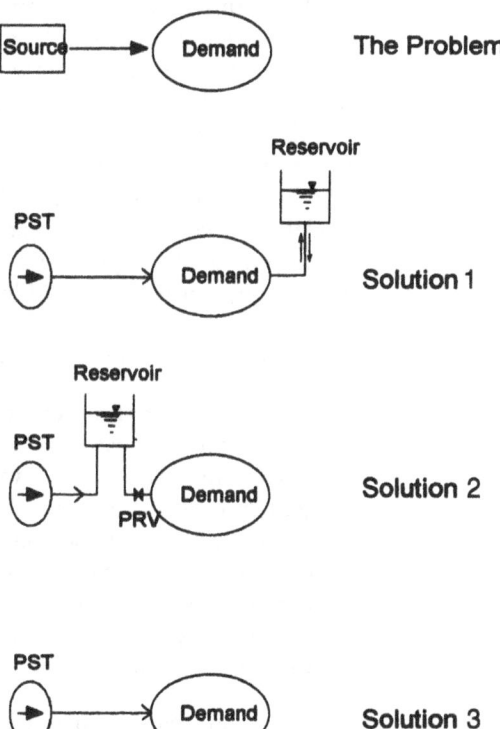

2 Pump all water to a reservoir from which it can gravitate to the demand area – the pressure can be controlled by a PRV.
3 Pump all water directly to the demand area by variable-speed pumps, without any storage capacity.

There are 'pros' and 'cons' for the hydraulics of each solution (without even mentioning widely different capital costs). For instance, solution 1 is simple, can be controlled easily and offers some security in emergencies – but the network will be submitted to relatively high and variable pressures. Solution 3 may be the cheapest to build, but operating costs could be high, and the maintenance of variable-speed pumps might be difficult. Moreover, there would be no reserve capacity if the power fails. (Is this probable? How often?) Next, pressure management would be almost impossible in solution 1 but quite simple in the other two. The list goes on. Obviously, water company staff should be consulted and their preferences taken into account from the very beginning of the design process. The model permits quick and systematic evaluation of options and avoids costly mistakes.

The designer is well advised to start the job by examining the existing system – 'green field' systems are extremely rare. The benefits are many: one can gather valuable information about local circumstances (e.g. specific consumption and seasonal effects), then contact the local staff and mobilise their support for the project. Local experience is usually irreplaceable, so the additional effort is well justified.

This exercise could be performed best on a mathematical model; in some cases this model will already be available, while in others it has to be made by the designer. In either case, a model of the existing system provides opportunities for the development of cooperation, collecting of information vital for the design, unravelling problems and anomalies in the existing system, and hopefully gaining converts to the procedure amongst the local staff. The list of tasks includes:

- Analysis of demand – current level, trends, specific consumption for various uses, patterns of use, local peculiarities.
- Unaccounted-for water levels – true losses, pressure management, billing system, non-billed use.
- Control policies in normal and exceptional circumstances – equipment, instrumentation, level of technology and expertise within the water company, etc.
- Data management – organisation, sensors, communications, computing facilities, etc.

The good and bad features of the existing system should emerge during this process, enabling the designer to offer sound advice with the maximum benefit/cost ratio. Even more important, the stages between the existing system and the future one will be clear to all concerned, not only to the designer.

The steps in the design procedure therefore should be:

1 Make or update a *mathematical model* of the existing system.
2 *Calibrate* the model in cooperation with the local staff using fresh operational data.
3 Analyse the *existing system* in great detail and find its good and bad features.
4 Define feasible *alternative solutions* for the future system.
5 *Estimate costs of each option* on a common basis using techniques such as internal rate of return, discounted cash flow, etc.
6 Determine *stages of development* from the present system to each alternative solution.
7 Analyse various *normal and emergency states* on the model and discuss the relative merits with the staff.
8 *Make a decision* by selecting the most promising alternative, possibly with one or two major subvariants.

What should the model include? There are two types of basic models:

• macro models, for the whole system;
• distribution models, detailing selected parts of the system.

The *macro model* must include all sources, reservoirs, pumping stations and control valves, but the distribution network can be reduced to a system of major pipes. Demand is allocated to relatively few nodes to keep the model to a manageable size. It is usually the macro model that allows planners to examine major developments of a system. However, before they start to use the model, there are a number of decisions that need to be made.

Probably the first decision is at what point in the future are they to plan for. In determining that point, a number of factors are at work and some of them are in conflict. For example, it is probable or even inevitable that demand will increase. If it is decided to plan for a situation in, say, 10 years time, then there will be some excess capacity in the system for 10 years. From an operational point of view, that is fine, but from a financial point of view, it is not so good. Some of the invested capital is not working fully for the company until the pipe capacity is matched by demand.

Other factors that need to be considered are briefly described below.

• The lifespan of the facilities being constructed needs to be addressed. Pipes, for example, might have a life of between 50 and 100 years. Should pipes be laid in the ground with a capacity to meet demand growth over that period, or should the planner recommend that duplicate pipes are laid at, say, 20 year intervals?

- A second factor to be considered is the lead time of the project – i.e. the time between deciding that an additional facility, like a new source, will be needed and its completion. In some cases such developments can take as long as 20 years to come to fruition.

- A common 'planning horizon' in the water industry in the UK is 20 years. In considering the development of facilities to meet increases in demand, therefore, the planner needs to forecast demand in 20 years' time.

Having determined a suitable horizon, the planner then stresses the model up to the demand levels at the planning horizon. Under that stress, the model will usually show a considerable number of shortfalls (if it does not, it means that some previous planner had provided excess capacity and probably a great deal of unnecessary investment!). The planner will then go on to consider a number of options to relieve the shortfalls in the system. All the successful options, i.e. the ones that do meet the planning horizon demand, can then be costed on some basis that takes account of *all* costs of the scheme, i.e. the immediate capital costs and future operating costs. To do such a comparative costing, the planner may need to use some technique that reduces all the costs of each proposal to a common base, such as discounted cash flow.

The *distribution model* goes into fine details of the network, so in some cases even the individual houses are presented. It is admirably suited for the analysis of water quality and/or inadequate pressure problems. The problem is to find a suitable starting point – usually a service reservoir or a pumping station – and how to model the boundaries between that part of the system and the rest.

The model must be verified and calibrated before the beginning of any serious analysis (see Chapter 6). The simulations and fine tuning or calibration of the model are of great help when designing changes to an existing system. The model of the existing system can be used both for design purposes and for evaluating short-term decisions, like rezoning the network, laying a new pipeline, replacing pumps, enlarging reservoir capacity, etc., which are usually made within the water company alone.

In conclusion, the final advice to designers is to involve local staff in the design process as early as possible, using the model of the existing system as a vehicle, adding the improvements on it as the work progresses. If the model of the system does not exist, it should be made (this is effort well spent). The model left after the completion of the design is a valuable asset in its own right and could be used as the backbone of a data acquisition and analysis system, not to mention opportunities for training and staff games.

7.2 Modelling and operational management

Advanced data management systems, with a centralised telemetry system and hundreds or even thousands of accurate and reliable instruments, seem to offer the ultimate answer to the problem of day-to-day control of complex supply systems. However, many questions remain unanswered, like:

- Can we predict future developments – and for how far into the future?
- What can happen in various circumstances?
- How far can the influence of an item – say a pumping station – extend?
- Is there a better control policy?
- What should staff do in an emergency?
- What are the effects of proposed operational changes (new control valve, variable-speed drive in a pumping station, enlargement of a reservoir)?

Clearly a tool needs to be added to a DMS, to make full use of the information it contains. Mathematical models and expert systems are clearly promising tools for the purpose, and have been proposed by many authors (e.g. Chase and Ormsbee, 1993; Coulbeck and Orr, 1989; Cosgriff et al., 1985; Cullen, 1987; Fallside, 1977; Halpern and Pascal, 1987; Huntington, 1990, 1993; Kado and Itoh, 1987; Nguyen, 1994; Santoni et al., 1987; Schulte and Malm, 1993; Snoxell et al., 1989; Takagi et al., 1983).

Mathematical modelling of water supply systems was first applied in design. The technique is superbly developed and widely used for analysing both existing and proposed water supply systems. This success led to the next step, the application of modelling in operational management. It seemed a very good idea, given the power of modern computers and on-line data capture systems. Why should the model not be 'the brains' of the water supply system, a template against which the real system could be constantly monitored, effectively controlling minute-by-minute operations, while humans could be relegated to a supervisory role?

In reality it is not easy to achieve. The main difficulty is that many existing models and techniques were developed for the designer's needs, but far less for the operational manager's needs. The following list summarises the differences between the two.

1 System to be dealt with
   - Design: a future system, with many alternative solutions, subject to many changes before realisation.
   - Operations: the existing system, which cannot be easily changed.
2 Timescale
   - Design: usually several years.
   - Operations: hours or days and certainly no more than a few months ahead.
3 Distribution network and reservoirs
   - Design: to be designed according to requirements and criteria.
   - Operations: pipes and reservoirs exist, although many parameters might not be known, e.g. exact location, size, encrustation and deposits within pipes, etc.
4 Pumps and control valves
   - Design: to be selected and placed as appropriate, sometimes with the manufacturer's help and expertise.
   - Operations: these items are already installed and may not be appropriate for their purpose; data may not be available on their characteristics.
5 Demand
   - Design: stated by the client – the starting point of analysis.
   - Operations: generally an unknown quantity – unless it is monitored by a well organised and supported data capturing system.
6 Leakage
   - Design: usually included in the demand through an accepted (or prescribed) UFW percentage, evenly distributed through the system
   - Operations: can appear anywhere and at any time, and is extremely difficult to spot and quantify.
7 Pressure management
   - Design: has to guarantee minimum pressures to be above the given level at all times; maximum pressures seldom analysed, still less reduced.
   - Operations: pressure should be kept in a narrow range, high enough to meet the standards but the maximum constrained – to reduce losses and pipe bursts.

8 Operational data
- Design: desirable, but only as a part of the information needed – the other data are forecasts, demographic studies, design standards and special requirements.
- Operations: the constant influx of fresh operational data is absolutely vital; the data must be complete and reliable – or the conclusions might be wrong!

9 Accountability
- Design: designer is accountable to the client only.
- Operations: operations manager is accountable to the Board of Directors and general public.

10 Final users
- Design: engineers, consultants – all experts with high qualifications and expertise.
- Operations: operators, technicians and managers.

The requirements of operational management are not fully met as yet, although progress is significant. The difficulties 'at grass roots level' are well described by Barker (1993), like faulty instrumentation and wrong data. However, the most difficult task is to model demand and losses in the real system. To the best of our knowledge, the problem is not yet fully formulated, let alone solved, despite its obvious importance.

Consider a simple case of a town supplied by two independent sources from opposite sides, as in Fig. 7.3 (from Obradovic and Kordic, 1986). If source 'A' is lost for a certain period of time, can source 'B' increase its output (assuming it has the capacity available) and make good the deficit? This begs the next question.

How are the additional quantities to be transported to the other side of town? If the network is to be used, this means that pressure will rise close to the operating plant, while at the other end the pressure might fall to zero.

That is all the *designer* needs to know to start looking for alternative solutions, such as laying another pipe between the two sources or securing additional sources of water, etc.

The *manager* needs to know exactly what will happen when the whole town is supplied from just one plant working at full capacity. One can expect that the demand will rise in areas close to the plant – due to the increased pressure – but some water will reach the other end, where the pressure (and demand) will be low. Whether the extreme end of the network will empty (a disastrous case) or the lower houses will still have some water (a far less dangerous condition) is highly conjectural – because the relationships between demand and pressure are not well known at present.

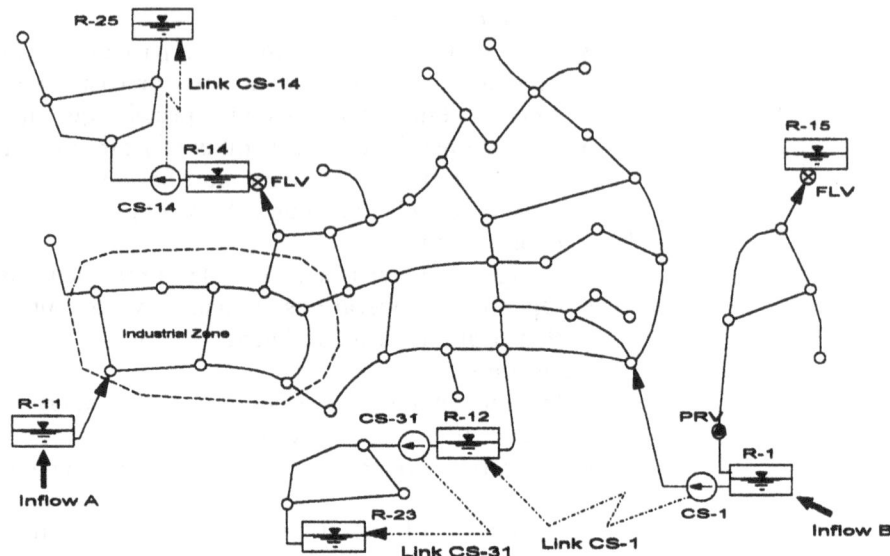

Figure 7.3 A town supplied from two sides.

In this particular case a decision was made to reinforce the network by laying an additional pipe between station CS-1 and the industrial zone, permitting quick transfer of additional quantities of water in the area supplied by reservoir R11 and thus providing adequate provision in the event of a failure (Fig. 7.4).

The best the user can do is to simulate several scenarios, applying different hypotheses and 'educated guesses'.

Staff in the control room are usually conservative in this regard, preferring well established routines rather than modelling, a largely untried tool – the difficulties are well described by Chase and Jones (1993). However, there are instances of successful breakthroughs – for instance in Paris (Nguyen, 1994) and the London system (Rance *et al.*, 1993a; Johnson *et al.*, 1993; Burnell *et al.*, 1993; Davies, 1993) – so one might expect that significant progress will be made within a few years. It seems that the principal difficulty is not in creating the model but rather in how to calibrate it initially and keep it well tuned for a longer time period. It should not be forgotten that, even after a successful calibration, questions about the model's credibility will constantly hang in the air. The notorious fact is that every water supply system is subjected to constant change (new consumers, new pipes, public works, pumps replaced, etc.). Is the model still valid? Again, the straightforward answer could be best obtained by comparing simulation results with observations at regular intervals and seeing whether the differences are significant.

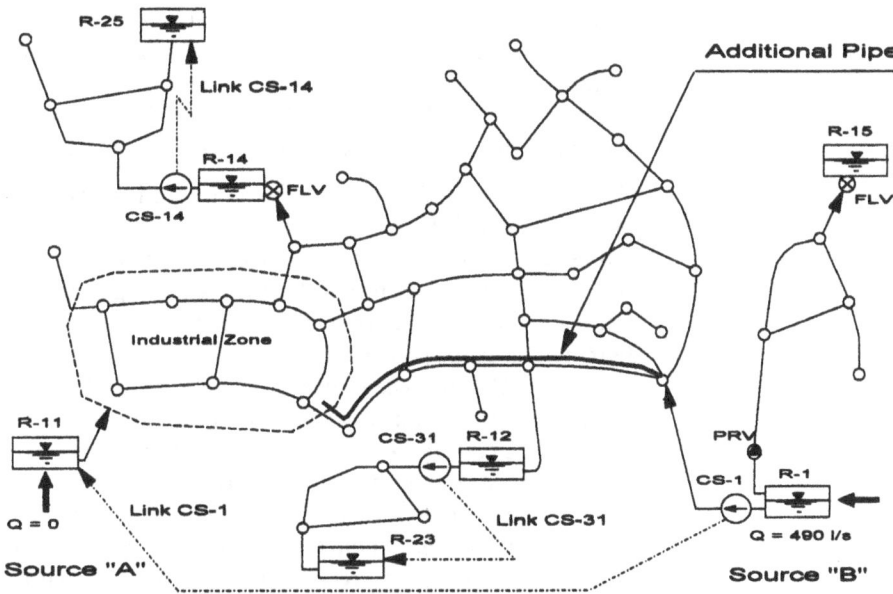

Figure 7.4 Operating without one source.

Both the calibration and later testing of a model are infinitely easier in remotely controlled systems, where data are readily available. In older systems, where data are logged manually on paper, attempts at regular re-calibration are usually less than successful.

7.3 The use of models in a water company

Training of staff

The first task is to train several members of staff in the use of the model. This can be done only through practice, providing that all have free access to the necessary facilities: data files, computers and documentation. We have stressed the need to involve local staff at all stages of the development of a model. This involvement should produce an indication of the aptitude of individual members of staff to the technology and offer an initial group of potential model operators. Special training exercises should be assigned to individuals, who have to solve them on the new tool: the model. A few typical tasks are:

- the analysis of past events, using either actual data (preferably gathered through telemetry) or results of simulations (old or new runs);
- forecast of future developments, starting from either known or estimated initial states of the system, under the existing control policy or some new one;
- selection of the 'best' control policy by analysing several alternative pump schedules, valve settings, etc.;

- analysis of various emergency situations (which may appear with a high probability or not) to devise action plans;
- study of possible catastrophic events (major power failures, loss of a major facility, earthquake, major fire) and evaluation of possible remedial actions and overall strategy;

The model must be under the care and surveillance of one person, who will be responsible for its maintenance. This person will be authorised to introduce necessary changes – anybody else can only place recommendations and suggestions, although there is no reason why they cannot develop their own models for their own purposes 'off-line'.

Figure 7.5 shows the structure of a model and a simulation run procedure. All files describing the model are protected as 'read-only' files. Fresh data are available through a telemetry system.

The user starts with a description of a 'scenario': selects an initial state and defines a control policy for pumping stations and control valves. Then the simulation begins. The user can stop the computation from time to time to 'have a look' and decide whether to let it continue or to stop it and examine the results up to that time.

After the simulation, the user starts to analyse the results. This is still a highly individual process, with only a few general rules. Whether expert systems will soon be used for this job remains to be seen (Alla, 1988; Cullen, 1987; Kado and Itoh, 1987; Nandy and Owen, 1993; Santoni *et al.*, 1987; Walker, 1988).

Control over a water supply system

Operators in the control room monitor the supply system continuously, round the clock, every day of the year. They are informed about the current state of the system via some form of data acquisition system, which may be quite crude (logbooks) or very sophisticated (with complete telemetry system, loggers and other modern equipment).

Important information – levels in sources and reservoirs, flows, status of pumps, etc. – is sent by some means or other to the control room. There, the information is analysed and evaluated. A forecast of future developments in the system follows, on the basis of past experience, simulation results or intuition. Control staff then decide what action – if any – is necessary, using established criteria and expertise. The result is a plan of operation. (Even 'do nothing' can be a valid plan!) The staff implement this plan by giving necessary orders and commands – and then wait to see the effects of these measures on the system.

The procedure briefly described above may be undertaken with or without a mathematical model of the real system. Staff

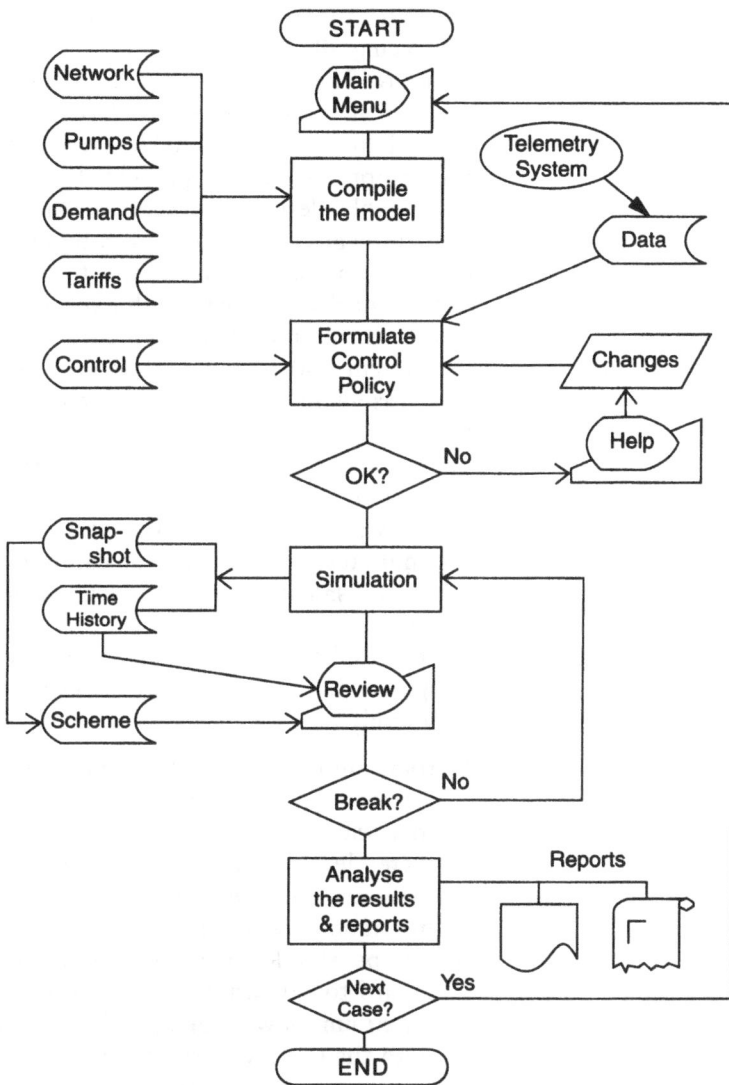

Figure 7.5 Simulation run.

usually rely entirely upon their experience and intuition. They may do excellent work, but some questions may remain:

- Is there a better way to run the system?
- What should be done in emergencies that nobody has had to deal with previously?
- How is it possible to 'introduce' novices to the system in a matter of days, not years?

The model of the system provides an efficient answer. The intention is not to discard the accumulated experience, quite the opposite: the model is an adaptable and flexible framework embracing all available knowledge and expertise.

Figure 7.6 shows how the model is used for operational decision making. The state of the system is monitored in the control room via SCADA (supervisory control and data acquisition). Data are processed in the usual way: displayed on suitably designed screens, alarms and messages flashed to attract the attention of the staff, reports are printed, archive files updated and saved, etc. If the fresh operational data show a significant disagreement with expectations (forecasts), staff will start a simulation of the model to analyse the developments in the system and possible consequences. The initial state of the model is available from archive data files, network data having been collected, checked and stored in appropriate files, so the operator in charge has only to describe the control policy – if there is any change at all – and start the simulation run. If the answers are not satisfactory, several runs could be made in a few minutes, so the operator can explore various options and select the safest one – or consult senior colleagues about the situation.

The results of simulation clearly show whether there is any danger or not. Examples of the criteria that a water company may apply are:

1 total inflows and outflows have to be balanced in the whole system and in all supply zones, taking into account changes in storage;
2 water levels in all reservoirs/water towers have to remain between the given limits – between top water level and minimum reserve level;
3 pumps should operate within the acceptable range with regard to cavitation and efficiency;
4 maximum power engaged should not violate the limit agreed with the electricity supplier;
5 service pressure in all zones should not be too low or too high;
6 reserves of water stored in reservoirs should always be adequate to satisfy emergency demands (fire fighting, for instance).

Each water company sets its own criteria according to its particular needs and customs.

The model – a tool for analysis

While operators/engineers in the control room use the model for supervision and control over the system 'on-line', other engineers can use the same tool for various analyses. The aims of these analyses may be:

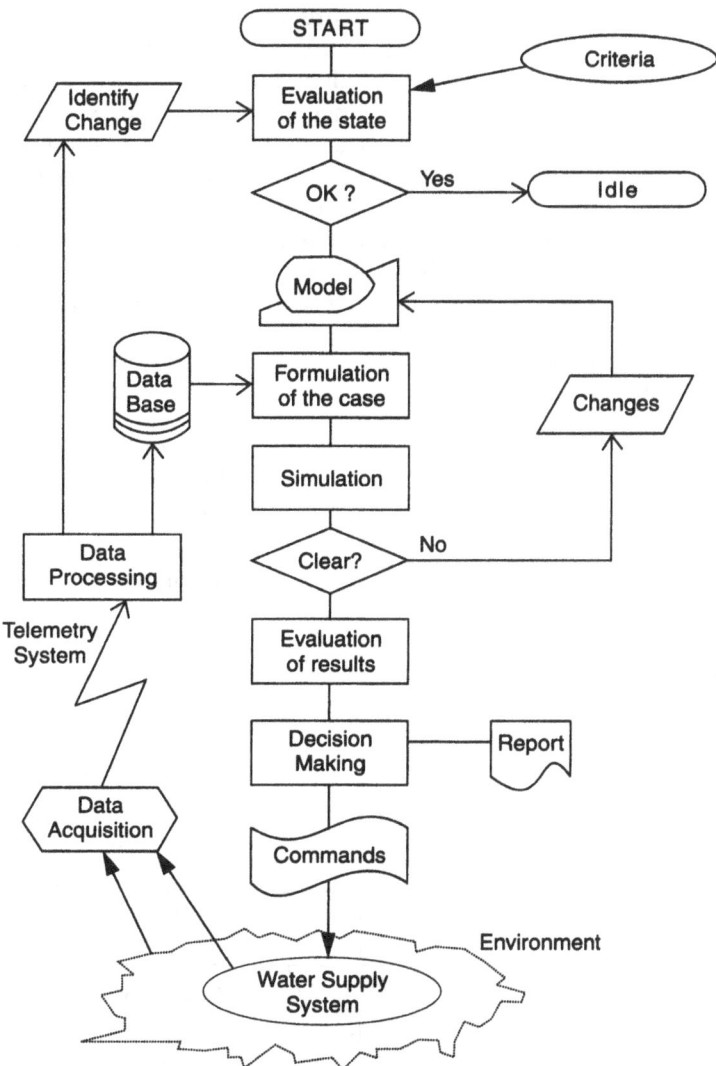

Figure 7.6 Operational decision making.

- to determine the maximum capacity of the whole system and its zones;
- to evaluate the role of each element within the system – pumping stations, reservoirs, water towers, control valves;
- to examine the adopted control policy from all points of view – efficiency, reliability, security;
- to establish and evaluate various alternative policies in all conceivable normal, emergency and catastrophic situations;
- to identify weak spots in the WSS and analyse feasible remedies, etc.

The procedure on using the model for these jobs is rather straightforward (Fig. 7.7). The procedure starts with an appreciation of demand in the area. Available sources should be capable of meeting the demand, which, of course, includes losses. The exercise should indicate if rationing of water is necessary or not in the near future, and if further infrastructure development (of sources, storage and distribution capacity) will be needed if rationing or inequitable distribution are to be avoided.

The next task is the definition of a suitable control policy for the system. It is always good to start from the existing policy and

Figure 7.7 Evaluation of control policy.

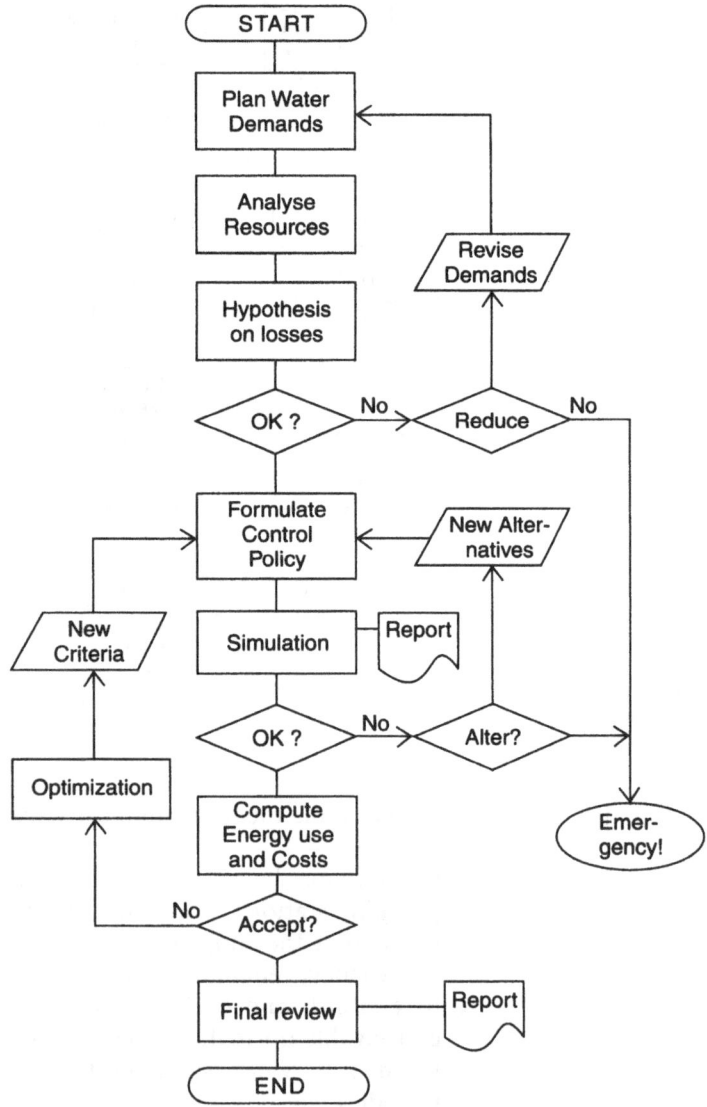

introduce changes one at a time, carefully pondering the effects on the whole system. The criteria are practically the same as above. An optimisation routine may also be included in the process to generate the proposals, but intuition could also lead to excellent solutions.

This is not an academic exercise, but a job where risks and gains should be carefully studied and the results debated by all levels of staff: for example, an 'optimised' control policy (minimising costs by restricting pumping to low tariff periods), superior to the existing one, may cause a certain depletion of storage that may or may not be acceptable to the management.

Credibility of the model

Even the best model is just an image of the real water supply system and is not the system itself. This simple fact is sometimes overlooked. It is quite possible that the model and the real system may, after some time, start to behave quite differently from each other, due to:

- inaccurate data used during model building, overlooked during calibration;
- unknown facts about the real system;
- incorrect hypotheses built into the model (e.g. about spatial distribution of demand);
- changes of the real system after calibration, etc.

Therefore, even the best model must be checked at regular intervals, more frequently (say, once per week) at the beginning of its life, but at least once per month. This is a tedious task, but much easier if a steady inflow of new data is being provided by telemetry. Staff should be trained to analyse this kind of information and to recognise important changes – if and when they appear.

Staff should briefly review operational data every day, even if everything is normal and according to their expectations. This daily review should include:

- computation of demand, source output, pumped volume;
- unusual variations of water level in main reservoirs;
- operating range of the main pumping stations;
- other points specific to the particular system.

If anything suspicious is noted, further analysis should be started immediately, helped by other members of staff.

First signs of trouble

The first sign that the model may need recalibration is when significant differences between simulation results and data obtained through the telemetry are observed, such as:

- water balance (total demand plus losses over 24 hours) is not within, say, 5%;
- variations of WL in main reservoirs differ considerably from observed data;
- daily diagrams for the whole system or for well-defined zones are quite dissimilar;
- pumps in main PSTs operate in a different manner;
- pressure profiles at key points disagree.

The reasons for the 'drift' may be numerous, but can be grouped into:

1 Shortcomings of the model
   - incorrect data on main facilities or network;
   - poor estimate of water demand (quantity, spatial distribution, variations in time);
   - poor estimate of true losses in the system;
   - other deficiencies.
2 Changes of the real system
   - demand patterns changing due to seasonal effects, population growth, new plant and buildings, etc.;
   - new control policy introduced on pumping stations, control valves and other devices;
   - some pipes in the network were closed (or open), without notifying the control room, often with potentially serious consequences for the system;
   - a distribution system was changed because some public works were executed;
   - existing facilities or the distribution network were reconstructed, etc.

Is a recalibration necessary?

One awkward day should not be taken too seriously. However, when the model clearly disagrees with reality, the staff from then on should pay more attention to it in the next few days. If the failure is repeated, staff should identify the sources of disagreement and recalibrate the model.

The procedure of recalibrating and recommissioning the model is the same as the commissioning of a new one. Two tasks are particularly important:

- analysis of energy consumption and, if possible, finding ways to cut down the costs without putting the safety of supply into jeopardy;
- analysis of water quality in the distribution network, in order to discover 'cul-de-sacs' and other sites of potential health hazard.

Dealing with emergencies:

- First analyse the system as it is, with no change in the control policy, just to see what is likely to happen.
- If circumstances dictate policy changes, alter the existing policy as little as possible.
- Keep an eye on developments and be ready to react.
- Warn others as soon as the situation gets significantly adrift from normal and involve them in observing progress.

A practical application of modelling is described in section 7.4 below: how to model energy and costs in a water supply system. The case is fictional, but based on real data from several existing water supply systems.

7.4 Modelling energy and costs

The task

Attempts to introduce 'optimum pump scheduling' to the water industry have been made frequently during the last two decades, but there has been little impact on everyday operational management. The reasons for this apparent failure are, in our opinion:

- The methods applied are too sophisticated for the average worker in the control room (just mention 'dynamic programming' and his/her eyes will glaze over!).
- Everything has been attempted in 'one go' – one model used for both hydraulic and energy calculations, so one aspect (or both) had to be oversimplified to get a solution at all.
- The existing organisational framework was disregarded. In the real world of operational supply, one person is usually responsible for the supply service and another for energy conservation.
- The expert – usually an outside consultant – wanted to formulate an 'optimum control policy', show where the money was wasted, and leave staff to implement the policy. In wrestling with everyday problems, staff did not have the time to experiment with the expert's ideas, even if they were quite good ones.

It is no wonder that the experts regarded the water companies as too conservative, while the staff returned the compliment by abandoning the new routines as soon as possible. However, the costs of energy continue to rise and so the problem should be revisited. This should be an 'in-house' job, because all the necessary expertise is available in the water companies.

All the circumstances dictate that optimising energy costs should be approached, not as a one-off campaign or even as a series of campaigns, but, like the approach to leakage control, as a continuous effort to save energy and costs, whenever and wherever possible. The policy should aim to accumulate small gains to keep total cost under control – without jeopardising the basic service: water supply. A big success might be obtained from time to time, but the real benefit is gained slowly by an everyday struggle. This point should be made clear to all concerned – and from the beginning.

Can one really cut down the costs? Before going into another expensive exercise, one should carefully evaluate the possible gains and compare them with the certain costs of maintaining the existing policies.

The case that follows will illustrate how the analysis could be done. The system – the city of 'Avalon' – was shown before in Figs 6.40–6.42. The simulation was made for 10 January 1991 (Thursday) on the model.

Estimation of possible savings

There are four possible ways to save energy and/or to cut down the operating costs:

1 by replacing low-efficiency pumps with better units;
2 by forcing the pump to operate around its best efficiency point in a narrow range;
3 by stopping the pump in peak rate hours;
4 by selecting a more suitable tariff.

Progress in pump design is not rapid, but steady. Modern units attain efficiencies of between 80% and 90%, significantly more than those of only 10 or 20 years ago. A yardstick is provided in the *Pumping Manual* (Dickenson, 1988, pp. 84–87). It states that the expected theoretical value of maximum efficiency $\eta_{0,\text{theor}}$ (–) is equal to:

$$\eta_{0,\text{theor}} = 0.94 - \frac{1.0}{(13.20 Q_0)^{0.32}} \tag{7.1}$$

where Q_0 (l s^{-1}) is the rated flow of the pump, at the maximum point of its efficiency curve.

The user can compute the effects of replacing an existing pump by a better one through some simple formulae.

Savings of power ΔP_1 (kW)

$$\Delta P_1 = P_0 \left(1.0 - \frac{\eta_0}{\eta_{0,\text{theor}}} \right) \tag{7.2}$$

where P_0 (kW) is the rated power of the pump and η_0 (–) is its best efficiency.

Savings of energy ΔE_1 (kWh)

$$\Delta E_1 = R_{E/H} \Delta P_1 \Delta t \qquad (7.3)$$

where $R_{E/H}$ is the ratio of electric/hydraulic power, usually from 1.20 to 1.40, and Δt (h) is the operating time of the pump in question. If the pump is not used (or rarely), there are no prospects of savings!

Savings of power ΔP_2 (kW)

$$\Delta P_2 = P_0 \left(1.0 - \frac{\eta_{ave}}{\eta_0} \right) \qquad (7.4)$$

where η_{ave} is the average efficiency of the pump and η_0 is its optimum efficiency (cf. equation (7.2)).

Savings of energy ΔE_2 (kWh)

$$\Delta E_2 = R_{E/H} \Delta P_2 \Delta t \qquad (7.5)$$

which is basically the same as equation (7.3) above; the savings of energy (ΔE) are easily converted into monetary savings by multiplying by the average price (p kWh^{-1}).

All this seems obvious, but in practice the computation of costs and possible gains is not easy or quick without modelling, which allows one to see, reasonably accurately, the points on the efficiency and power curves at which the pump operates and for how long. Even so, the results of modelling could blur the issue because too many details are made available.

In the case of our example, one can see the operating point of every pump in every pumping station – the data for Unit No. 1 in Mountain PST are given in Fig. 7.8. Note that the operating range is relatively narrow and close to the best efficiency point (BEP). The same data could be compared to the maximum, theoretical efficiency curve mentioned above, and this has been done in Fig. 7.9. Note that BEP is close to the theoretical curve (just a few per cent lower), and the line showing operating range is not very far either. The other two pumps – identical to the first one – behave in much the same way. The conclusion is that pumps in Mountain PST are used quite efficiently.

The situation is different in the Old Pumping Station (one larger and two smaller units). The smaller pumps operate in a broad range (Fig. 7.10), and therefore the average efficiency cannot be high. The larger pump operates in a narrower range (Fig. 7.11), but a long way from its BEP and on the wrong side (between BEP and zero flow), proving that this pump cannot fulfil its role and should be replaced at the earliest occasion. The comparison with the theoretical curve (Fig. 7.12) reveals that the

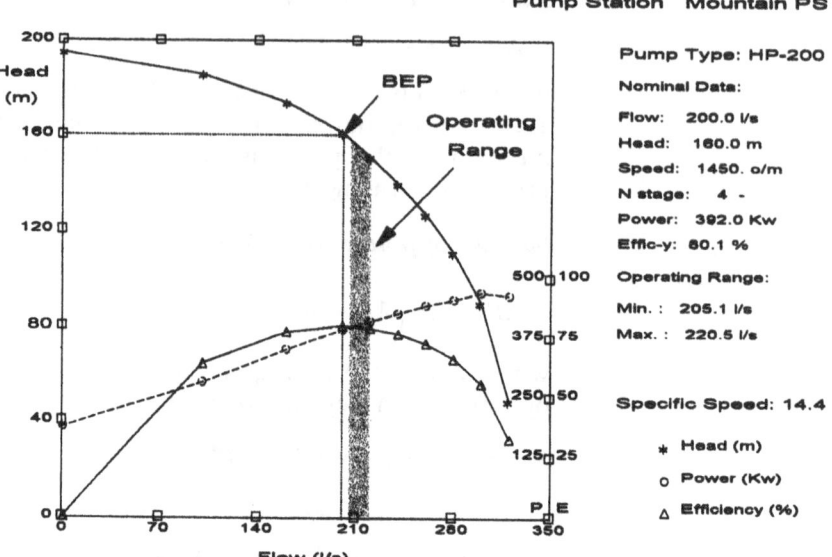

Figure 7.8 Mountain PST, pump No. 1.

Figure 7.9 Mountain PST, Unit I, theoretical and real efficiency.

maximum efficiency of all three pumps in the Old PST is low, some 15% below the expected, theoretical values, while the average efficiency is still lower. The case for replacement of these pumps appears to be very strong.

The corresponding results for the last important station – Main Booster PST – are shown in Fig. 7.13. They are somewhat

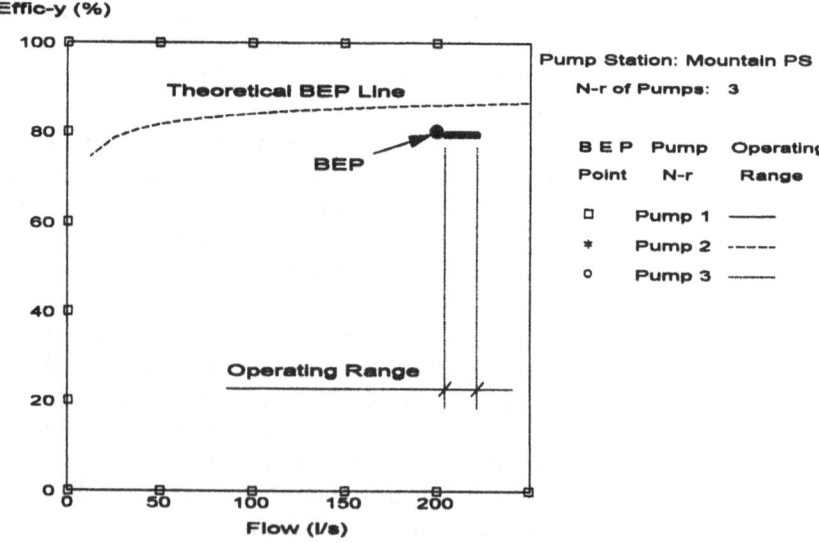

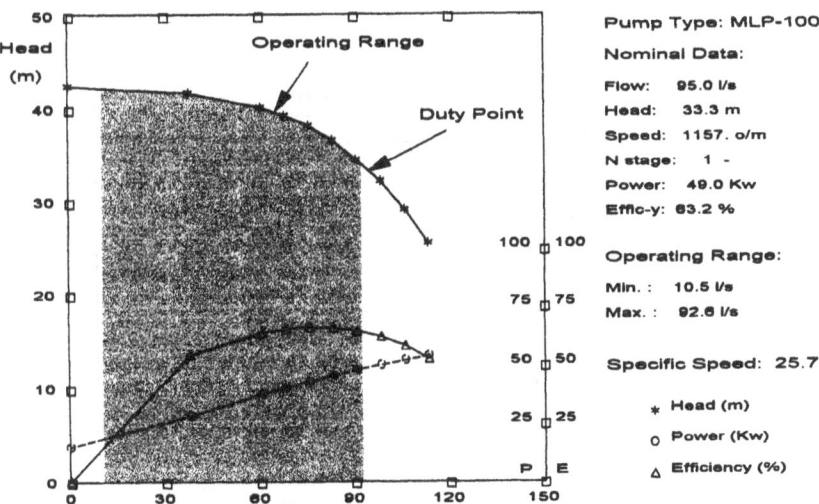

Figure 7.10 Old PST, small pumps.

better than those for Old PST, but worse than for Mountain PST – which is presumably newer.

However, the final conclusions cannot be made without quantifying the possible losses and gains; to do that, one must take the tariff into account.

Electricity tariffs vary in complexity throughout the world, tending to reflect a variety of political, economic and practical

Figure 7.11 Old PST, large pump.

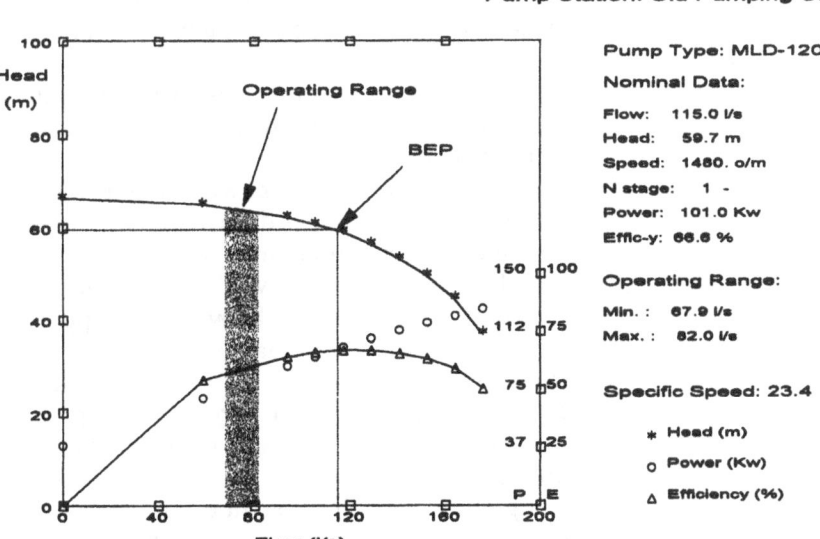

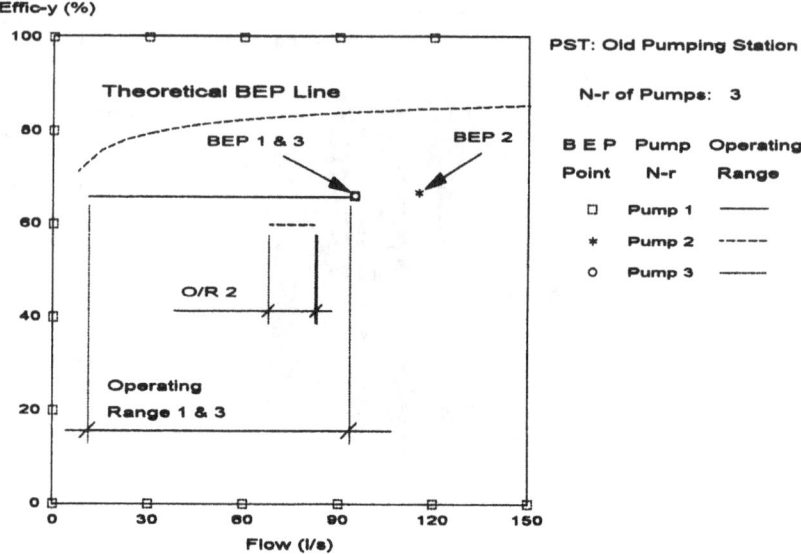

Figure 7.12 Old PST, theoretical and real efficiency.

factors of the particular country or region. One factor that must be common to all power producers is the high level of capital investment required for power plants, whatever the source of energy, be it oil, coal, hydroelectric power, etc. This being the case, all of them should be interested in spreading the load on their systems as evenly as possible throughout the day, to minimise peak demands, which have to be met by creating adequate generating capacity. Tariff structures with any degree of sophistication will normally, therefore, offer incentives to customers to reduce peak time use and increase off-peak use of energy.

Our example is based on tariff structures in the United Kingdom, but similar exercises could be carried out for other tariff structures.

For this example, the power supplier's tariffs are essentially simple: charges are based on both total energy consumed and maximum power used in a given period of time (usually a month). The rates vary with the season, day of the week and time of day. The contract between the supplier and the customer includes forecasts of maximum power to be used by the customer, with penalties for exceeding the contract, etc. (just as a bank may charge for exceeding an agreed overdraft limit!).

Theoretically, the user can compute the bill in advance – but nobody does, to the best of the authors' knowledge. It is too complicated and the user does not have all the relevant data (energy used, maximum power) at hand, even for a single site. If

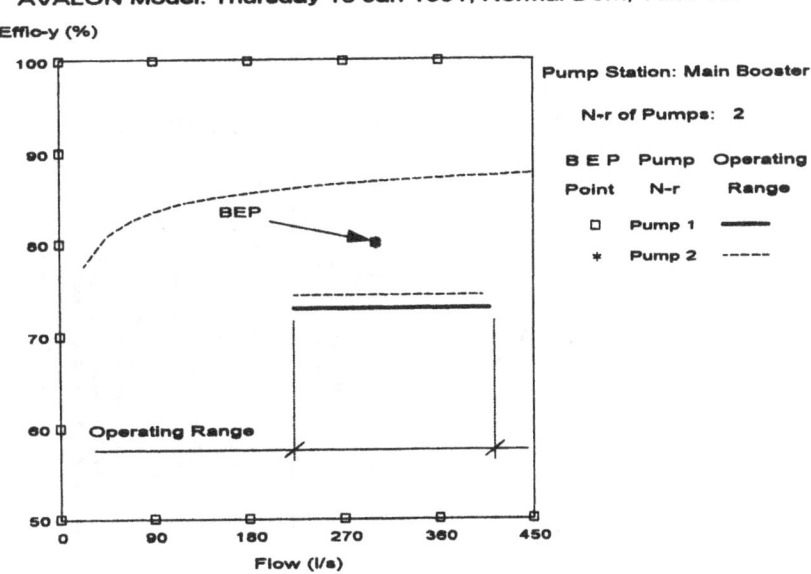

Figure 7.13 Main Booster PST, theoretical and real efficiency.

there are several sites (and usually there are), the task is really hopeless, so all that can be done is to wait patiently for the bill. The user then puts the new data into tables, computes the specific use of energy per cubic metre of water delivered to customers and the specific prices of water (p m^{-3}) and of energy (p kWh^{-1}), makes graphs showing variations of either, etc. – but cannot easily tell when and why the cost was incurred, much less how to reduce it.

The costs could be displayed in graphical form, as has been done in Figs 7.14 and 7.15 for the first pump in Mountain PST and Old Pumping Station respectively. The graphs contain the following information:

- Power (kW) used by the pump
- Daily rate – price of energy (p kWh^{-1})
- Costs (GBP h^{-1}) – every bar represents the cost of energy used in that hour

Maximum demand charges are not shown here because they are related to the site – not to the individual units – and embrace a full month, not a single day.

The prospects for savings are meagre: the pump did operate continuously through the period of 'night'. (Curiously, the public in general believe that the duration of the average night is equal to that of the average day – but for electric companies 'night' is just 7 hours and 'day' is 17 hours, all the year round!)

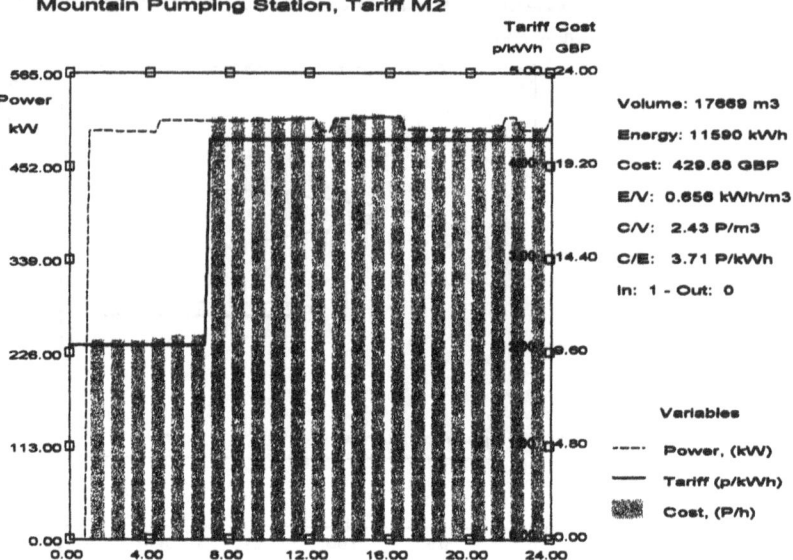

Figure 7.14 Mountain PST, Unit I, energy costs.

Figure 7.15 Old PST, Unit I, energy costs.

Note that the tariff in question does not include special high rates in the early evening hours.

However, there are other tariffs called 'Seasonal-Time-Of-Day' tariffs (STOD) that offer some benefits (e.g. no charge on maximum power used), but charge heavily for energy used during peak hours (usually between 16:00 and 19:00, in winter

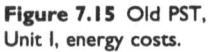

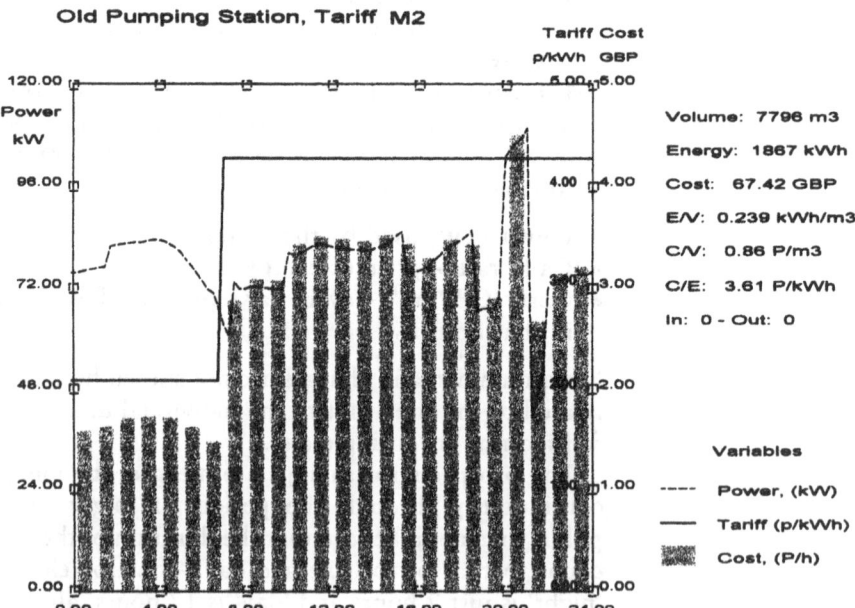

months). Then the rate is 6–7 times higher than average, to induce users to switch off their pumps (if they can afford to).

In some cases the pump must operate at that time (for instance, to keep the pressure high enough or to prevent emptying of a small reservoir), but in many cases the user can alter the control policy and stop pumps during this critical period. The potential savings could be estimated as follows.

Savings in costs ΔC_3 (GBP)

$$\Delta C_3 = (r_{PR} - r_{ave}) \sum_{\text{peak hours}} P\Delta t \tag{7.6}$$

where r_{PR} (p kWh^{-1}) is peak hours rate, r_{ave} (p kWh^{-1}) is the average rate, P (kW) is the power used by the pump and t (h) is time; note that summation is done only for the period of peak rate (from 16:00 to 19:00 in this case), naturally.

The quick appraisal   The analysis shows where money is being wasted, by examining the smallest item in the system – a single pump. Ideally, the user should analyse the behaviour of every pump and take appropriate action (i.e. change control policy or replace pumps). As there are many pumps in the system, this may seem to be 'looking for the needle in a haystack'. A small effort in organising the results will make the job easier. The potential savings are first computed for each pump and then summed for pumping stations. The information needed is:

- rated values;
- operational data – time, volume discharged, energy used, total cost, maximum power;
- possible savings on all three accounts – optimum efficiency (compared to the achievable maximum), average efficiency (compared to BEP) and average rate (compared to peak rate).

Figures 7.16 and 7.17 show results for a pump in Mountain PST and Old PST. A summary of results is shown in Fig. 7.18. Possible daily savings are of the order of 62 GBP, 6.1% of costs. The figure is quite reasonable. Note that the costs for maximum power used are not included. Another point is that the savings might not be achievable – three-quarters of savings are due to relatively low BEP. Note also that, although pumps in Old Pumping Station operate in an unfavourable range, the possible savings are almost insignificant in the total. The advice is clearly to replace older pumps with newer units, which will have a higher efficiency.

Another option is to change the tariff. Instead of this rather flat tariff, which includes charges for maximum power, try one

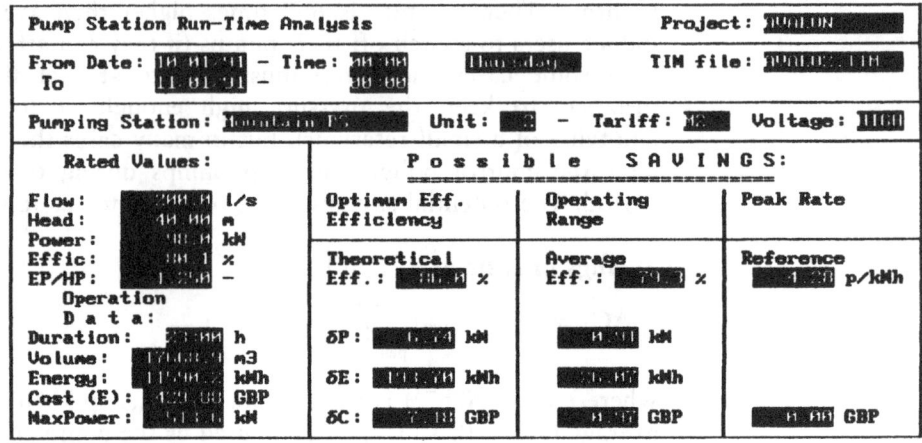

Figure 7.16 Possible savings, Unit II, Mountain PST.

that takes into account the time of the day during winter (so-called STOD tariff) and charges nothing for maximum power used.

The simulation run was repeated with the alternative tariff; hydraulics remained unchanged, of course, but the costs were dramatically changed – see the summary in Fig. 7.19. Note the sharp increase in total costs (compare with Fig. 7.18); the main difference is in the third column of possible savings. Obviously the pumps did operate during peak-rate hours – and the penalty is extremely high. But even if this item is deleted, the total daily costs are still higher than in the previous case (about 1276 GBP

Figure 7.17 Possible savings, Unit I, Old PST.

```
 Pump Station Run-Time Analysis                    Project: ▓▓▓▓▓▓
 From Date: ▓▓ ▓▓ ▓▓ – Time: ▓▓ ▓▓     ▓▓▓▓▓▓▓▓    TIM file: ▓▓▓▓▓▓ ▓▓▓
 To         ▓▓ ▓▓ ▓▓ –        ▓▓ ▓▓
 Pumping Station: ▓▓▓ ▓▓▓▓▓▓▓ ▓▓   Unit: ▓   – Tariff: ▓▓▓   Voltage: ▓▓▓▓

   Rated Values:            P o s s i b l e    S A V I N G S :
                            ==================================================
 Flow:     ▓▓▓ ▓ l/s    Optimum Eff.        Operating          Peak Rate
 Head:     ▓▓▓ ▓▓▓ m    Efficiency          Range
 Power:    ▓▓▓ ▓ kW
 Effic:    ▓▓ ▓ x       Theoretical         Average            Reference
 EP/HP:    ▓ ▓▓▓ –      Eff.: ▓▓▓▓▓ x       Eff.: ▓▓▓▓ x       ▓ ▓▓▓ p/kWh
   Operation
   D a t a :
 Duration: ▓▓ ▓▓ h      δP: ▓▓ ▓▓ kW        ▓ ▓▓ kW
 Volume:   ▓▓▓▓ m3
 Energy:   ▓▓▓ ▓▓ kWh   δE: ▓▓▓ ▓▓ kWh      ▓▓▓ ▓▓ kWh
 Cost (E): ▓▓ ▓▓ GBP
 MaxPower: ▓▓▓ ▓ kW     δC: ▓▓▓ ▓▓ GBP      ▓ ▓▓ GBP            ▓ ▓▓▓ GBP

 Next Pump (y/n)?
```

```
┌─────────────────────────────────────────────────────────────────────────────┐
│ System Run-Time Analysis (All PST)                    Project:  AVALON        │
├─────────────────────────────────────────────────────────────────────────────┤
│ From Date: 10/01/91 - Time: 00:00    Thursday         TIM file: AVALO2.TIM    │
│ To         11/01/91 -        00:00                                             │
├──────────────────┬───────┬────────┬────────┬────────┬═══Possible Savings═══   │
│ Pumping Station  │ Time  │ Volume │ Energy │ Cost   │MaxETA┬OpRange┬PeakRate   │
│                  │ hh:mm │  m3    │  kWh   │ GBP    │ GBP  │ GBP   │ GBP       │
│ Main Booster PS  │ 13:15 │19720.7 │ 4182.6 │ 179.0  │14.65 │ 14.19 │ 0.00      │
│ Old Pumping St   │ 24:00 │ 7999.8 │ 1940.9 │  70.6  │20.20 │  1.17 │ 0.00      │
│ Mountain PS      │ 24:00 │26466.7 │17440.7 │ 628.2  │10.54 │  1.21 │ 0.00      │
│ High Zone Boost  │ 10:45 │ 1163.7 │  182.3 │   7.6  │ 0.00 │  0.15 │ 0.00      │
│ Borehole No.1    │ 24:00 │ 3690.0 │ 1014.7 │  39.0  │ 0.00 │  0.00 │ 0.00      │
│ Deep Well Pump   │ 00:45 │ 1260.0 │  686.7 │  30.6  │ 0.00 │  0.00 │ 0.00      │
│ Tower Booster    │ 18:45 │ 2734.4 │  410.2 │  17.1  │ 0.00 │  0.00 │ 0.00      │
│ Borehole No.2    │ 24:00 │ 4320.0 │ 1188.0 │  43.6  │ 0.00 │  0.00 │ 0.00      │
│                                                                               │
│ T o t a l:          67355.4  27046.1  1015.74   45.39   16.72   0.00          │
└─────────────────────────────────────────────────────────────────────────────┘
```

Figure 7.18 Tariff M2, total possible savings.

against 1016 GBP). Does this mean that the proposal should be rejected? No, at least not before a thorough analysis of causes and effects.

The point is that the old control policy was developed over years to suit tariff M2, leaving little room for improvement – just over 6%. For a different tariff, a *new* control policy should be developed, taking into account its particular features. When an optimum is found – a local one – the two policies can be compared and a decision taken.

Figure 7.19 Tariff M3, total possible savings.

In this case the effort should be directed to preventing the pumps operating during peak-rate hours, if this requirement is

```
┌─────────────────────────────────────────────────────────────────────────────┐
│ System Run-Time Analysis (All PST)                    Project:  AVALON        │
├─────────────────────────────────────────────────────────────────────────────┤
│ From Date: 10/01/91 - Time: 00:00    Thursday         TIM file: AVALO3.TIM    │
│ To         11/01/91 -        00:00                                             │
├──────────────────┬───────┬────────┬────────┬────────┬═══Possible Savings═══   │
│ Pumping Station  │ Time  │ Volume │ Energy │ Cost   │MaxETA┬OpRange┬PeakRate   │
│                  │ hh:mm │  m3    │  kWh   │ GBP    │ GBP  │ GBP   │ GBP       │
│ Mountain PS      │ 24:00 │26466.7 │17440.7 │1275.3  │21.41 │  2.40 │ 581.77    │
│ Main Booster PS  │ 13:15 │19720.7 │ 4182.6 │ 370.4  │30.14 │ 27.94 │ 158.35    │
│ Old Pumping St   │ 24:00 │ 7999.8 │ 1940.9 │ 131.0  │37.58 │  1.84 │  52.63    │
│ High Zone Boost  │ 10:45 │ 1163.7 │  182.3 │  18.4  │ 0.00 │  0.35 │   9.39    │
│ Borehole No.1    │ 24:00 │ 3690.0 │ 1014.7 │  77.1  │ 0.00 │  0.00 │   0.00    │
│ Deep Well Pump   │ 00:45 │ 1260.0 │  686.7 │  89.7  │ 0.00 │  0.00 │   0.00    │
│ Tower Booster    │ 18:45 │ 2734.4 │  410.2 │  34.8  │ 0.00 │  0.00 │   0.00    │
│ Borehole No.2    │ 24:00 │ 4320.0 │ 1188.0 │  82.1  │ 0.00 │  0.00 │   0.00    │
│                                                                               │
│ T o t a l:          67355.4  27046.1  2078.81   89.13   32.54  802.13         │
└─────────────────────────────────────────────────────────────────────────────┘
```

compatible with the safety of supply. The flows provided by the three main pumping stations can be seen in Fig. 7.20. Note that flows are high during the critical period, 16:00 to 19:00, when the highest costs were incurred. This might be expected. Peak demand for power, certainly by domestic customers, will almost always coincide with peak demand for water.

Figure 7.21 shows water levels in High Level Reservoir and the discharge of Mountain Pumping Station feeding this reservoir. The profile reaches its maximum at 08:00, falls to a minimum around 16:00 and then starts its rise towards the midnight target level. Note that only two of the three installed pumps were used, to keep the maximum power charge at a lower level.

Obviously a different control policy is needed, which will stop the pumps during peak-rate hours without jeopardising supply. After a few simulation runs, this was achieved. Figure 7.22 shows flows through the pumping stations, operating under the new policy. All pumping has stopped during the period of the peak charge. Flow through the Main Booster PST is gravity flow through the station bypass pipe (Fig. 7.23).

Has the new policy ensured that supply is maintained? Figure 7.24 shows operation of Mountain PST and changes in High Level Reservoir under the new policy – note the difference (compared to Fig. 7.21): the water level is kept high until 16:00, then falls rapidly because all the pumps have been stopped. However, under the new policy the reservoir is full at 16:00

Figure 7.20 Old policy – Mountain, Main Booster and Old PSTs.

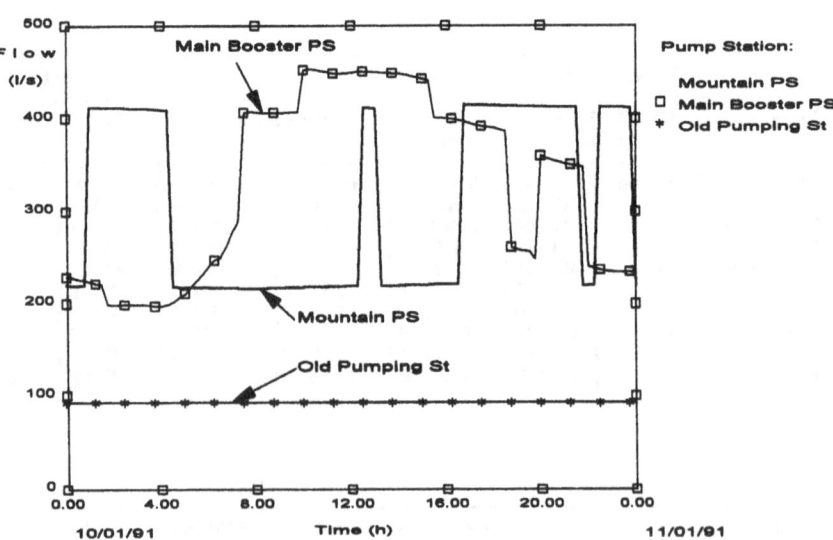

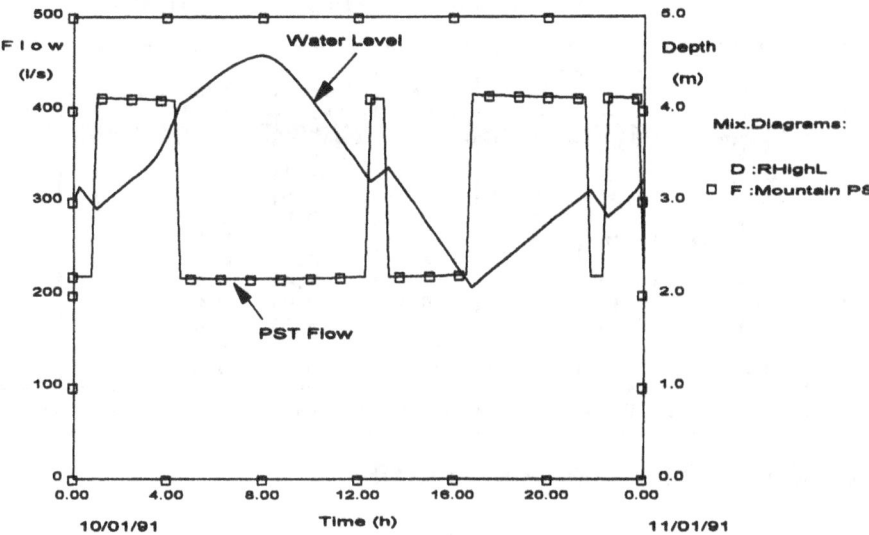

AVALON Model: Thursday 10 Jan 1991, Normal Dem, Tariff M2

Figure 7.21 Profile control – Mountain PST and High Level Reservoir.

Figure 7.22 New policy – Mountain, Main Booster and Old PSTs.

hours and has sufficient water to maintain supply during the critical period. Although the level in the reservoir falls to a slightly lower level than under the previous regime, there are still adequate emergency reserves.

Figures 7.25 to 7.27 show costs for one pump from each of the three pumping stations. The results are summarised in Fig. 7.28. The possible savings amount to 72.46 GBP, or some 6.3%

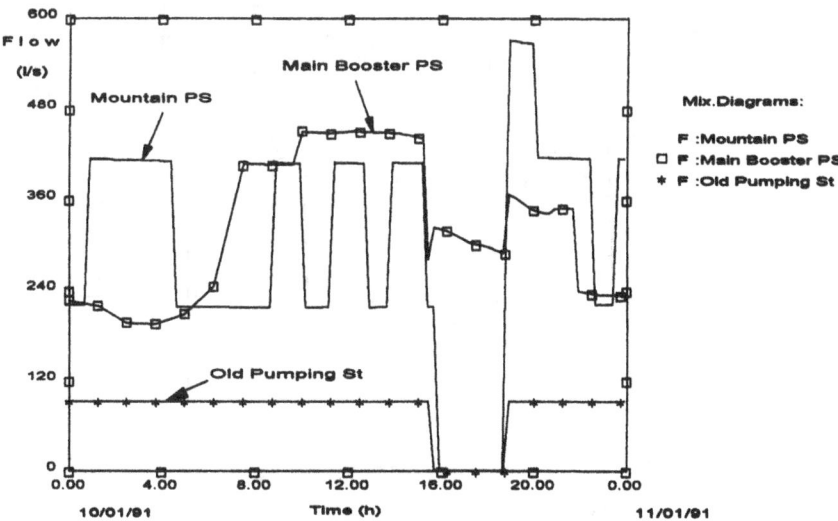

AVALON Model: Thursday 10 Jan 1991, Tariff M3, New CONTROL

```
W E S N E T :AVALON Model: Thursday 10 Jan 1991, Tariff M3, New CONTROL
            Time Interval :10/01/91 00:00 - 11/01/91 00:00

    PUMP STATION:Main Booster PS    No.of Units: 2   Max.Power: 700.0 kVA
```

Date	Time	Flow (l/s)	Head (m)	Power (kW)	En.Loss (kWh)	VarSpeed (rpm)	Status on/off
10/01/91	15.00	443.05	54.56	394.4	0.0	0.	11
10/01/91	15.15	441.45	54.60	393.7	0.0	0.	11
10/01/91	15.30	281.70	-0.09	0.0	0.0	0.	00
10/01/91	15.45	322.92	-0.12	0.0	0.0	0.	00
10/01/91	16.00	320.69	-0.12	0.0	0.0	0.	00
10/01/91	16.15	319.10	-0.11	0.0	0.0	0.	00
10/01/91	16.30	316.05	-0.11	0.0	0.0	0.	00
10/01/91	16.45	311.83	-0.11	0.0	0.0	0.	00
10/01/91	17.00	307.94	-0.11	0.0	0.0	0.	00
10/01/91	17.15	304.86	-0.10	0.0	0.0	0.	00
10/01/91	17.30	301.67	-0.10	0.0	0.0	0.	00

```
   Total- Flow: 27.03 Ml     Energy: 3502. kWh  Cost: 100.0 GBP Loss:   0. GBP
```

Line: 60

Figure 7.23 Operation of Main Booster Station – new policy.

Figure 7.24 New profile – Mountain PST and High Level Reservoir.

of daily costs, but the more important point is how these results fare in comparison with the earlier ones (Fig. 7.18).

The total daily costs are now 1156.30 GBP, which is above the 1015.74 GBP from the first run. This is not the end, because the new tariff does not incur additional costs, while the old tariff also charges for the maximum power used. The three pumping stations use about 1545 kW (Mountain PS 1000 kW, Main Booster 400 kW and Old PST 145 kW) at 7.32 GBP kW^{-1} –

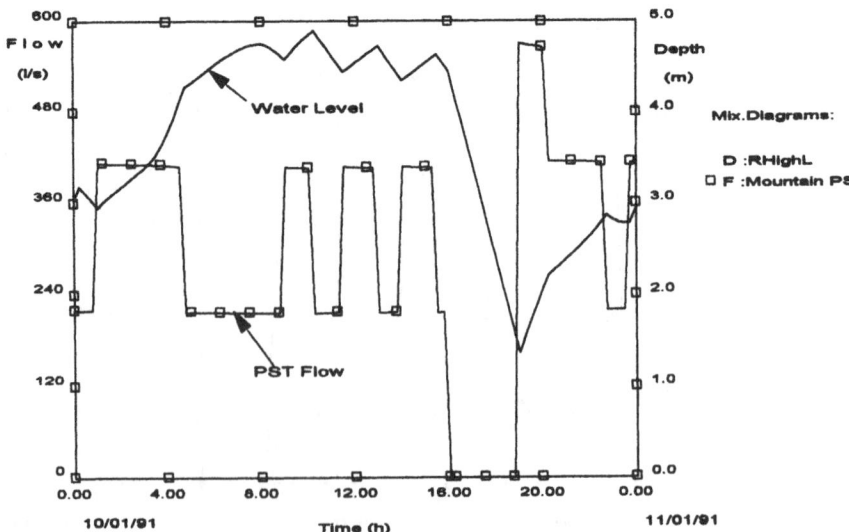

AVALON Model: Thursday 10 Jan 1991, Tariff M3, New CONTROL

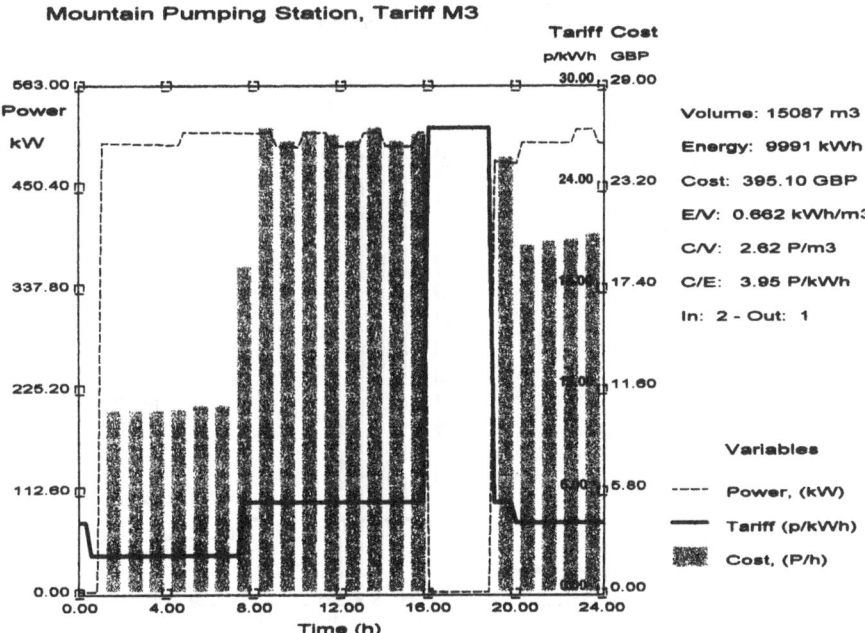

Figure 7.25 New control policy, Unit I, Mountain PST.

monthly. This sum should be reduced to one day; the additional cost is then

$$\Delta cost = 7.32 \times 1545/31 = 364.82 \, GBP$$

The total daily costs for the old tariff are therefore actually 1380.56 GBP (= 1015.74 + 364.82), which is 16.2% higher than the corresponding costs for the new tariff (1156.30 GBP). Note, however, that these additional costs apply only during the four winter months, and not in the remaining eight months. On the other hand, the high rate applies only for working days (Monday–Friday). More investigation is needed before the final decision is made, but it seems that the new tariff might be the better option.

This simple case illustrates the complexity of such studies even in relatively simple cases. It cannot be done without a well calibrated model. The results are valid only for a short time, before something is changed: demands, network, tariffs – anything. It is clear that such analysis is best made by staff, rather than by hired specialists; the advantage of the first group is that they know the system and have the responsibility for it.

7.5 Modernisation of control systems

Once the 'go ahead' for modernisation is given, the water company has to solve numerous problems, the most obvious being as follows.

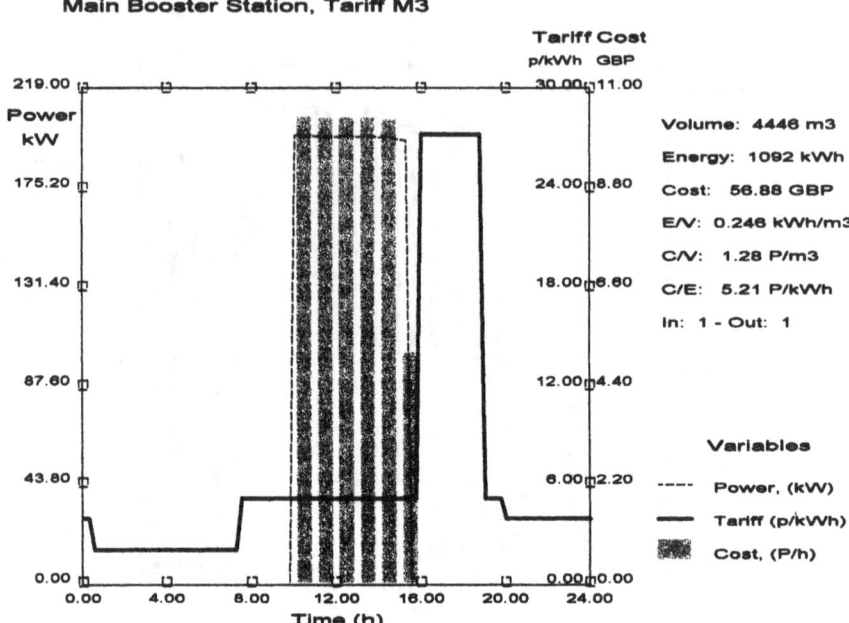

Main Booster Station, Tariff M3

Figure 7.26 New
control policy, Unit I,
Main Booster PST.

- Financing the project
- Obtaining import licences, if required
- Reconciling different technical standards and regulations
- Staff training
- Redeployment or recruitment of hardware and software maintenance staff
- Supply of spare parts
- Upgrading the technology and knowledge

Fundamental questions to be answered are the following.

- Will the control system use a telemetry system or data-loggers or both?
- How can the project be phased in? How much of the new technology can staff absorb whilst maintaining service within acceptable limits?
- What will happen if the new technology fails (such radical innovations often have teething problems)? Can staff run two systems in parallel?
- Can the staff take over from the contractors completely – and when?
- Is the new technology adequate for dealing with all emergencies in this system?
- How can human problems within the company be overcome?

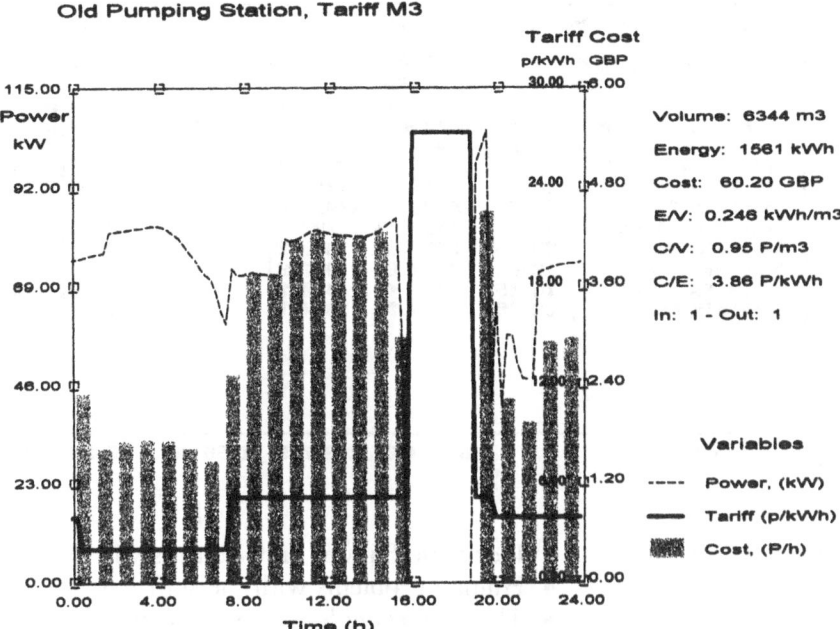

Figure 7.27 New control policy, Unit I, Old PST.

- Can the company hire new specialists and how can they merge them with the old hands?

Modernisation brings forward human problems, which are far from minor inconveniences, involving three groups of people, i.e. the staff, the manager(s) and the contractors.

1 The staff will quickly perceive the following:
   - Old values, skills, local knowledge and expertise are in danger of being lost or diminished, and the status of many individuals reduced (this may be an unjustified fear, but no less debilitating for that). It is important that water supply skills and knowledge are merged with the new technology, not overwhelmed by it.
   - New skills are needed, many of them quite unfamiliar – such as familiarity with computers.
   - New people will come in and acquire an important role within the company.
   - Re-learning and changing well established habits is a painful process – the more so for older hands.
2 The manager(s) in charge of the project is (are) beset with many problems:
   - The transition period is expensive and complicated – especially if two systems have to be run in parallel for a while.

System Run-Time Analysis (All PST)				Project: AVALON		
From Date: 10/01/91 – Time: 00:00 Thursday				TIM file: AVALO4.TIM		
To 11/01/91 – 00:00						
Pumping Station	Time hh:mm	Volume m3	Energy kWh	Cost GBP	Possible Savings		
					MaxETA GBP	OpRange GBP	PeakRate GBP
Main Booster PS	11:00	16347.6	3582.4	100.0	15.23	11.73	0.00
Old Pumping St	20:45	6916.5	1705.1	69.5	20.43	3.02	0.00
Mountain PS	21:00	25546.2	17018.0	667.9	11.32	0.99	0.00
High Zone Boost	10:45	1163.5	182.2	10.4	0.00	0.35	9.39
Borehole No.1	20:30	3357.0	923.2	40.7	0.00	0.00	0.00
Deep Well Pump	12:00	1728.0	941.8	90.9	0.00	0.00	0.00
Tower Booster	10:45	2734.4	410.2	35.0	0.00	0.00	0.00
Borehole No.2	20:30	3690.0	1014.7	45.1	0.00	0.00	0.00
T o t a l:		61483.2	25857.5	1156.30	46.98	16.09	9.39

Figure 7.28 New control policy – the summary.

- Inevitable failures and setbacks will happen.
- Human problems with the staff have to be dealt with.
- The new technology and procedures have to be adopted as they come.

3 The contractors should not forget the following:
- Cooperation with local staff is essential.
- It can be very rewarding to collect and evaluate local experience (see the simple case below) – and build safeguards into the new system.
- Local staff should not be pushed too hard and over their limits.

These difficulties are quite common (see Chase and Jones, 1993; Ormsbee and Lansey, 1994).

Once started, modernisation is irreversible; it develops its own momentum and is difficult, if not impossible, to reverse. The only questions are how long it will take and how much it will cost.

Now, a tip: Phase the project carefully and succeed at each step before moving to the next. It is always good to start with loggers, train local staff and mobilise a hard core of dedicated specialists before starting a full telemetry system – which is far more expensive and complicated. Introducing GIS also needs careful consideration (see Scherer and Phebey, 1995). Further advice and guidelines may be found elsewhere (Brammer and Schulte, 1993; Chase and Ormsbee, 1993; Gotoh et al., 1993; Halpern and Pascal, 1987; Huntington, 1984, 1993; Nguyen, 1994; Rance et al., 1993a,b; Robertson, 1993; Schulte and Malm, 1993; Snoxell, 1991, 1993a,b).

A simple case is included here to show how modernisation is not exactly a one-way transfer of knowledge and expertise. The case is based on the operation of a small system that was successfully optimised (see Jowitt et al., 1988), but could easily have involved clashes between old and new.

Figure 7.29a shows variations of demand during a typical day. Variations of water level in the main reservoir are shown in Fig. 7.29b. Both the observed water levels and results of the optimisation procedure are shown. Note that optimisation will provide higher levels in the morning hours and lower in the evening – around 8 p.m.

These data permit computation of reserve storage of water, expressed as a reserve time T_R, the period for which the stored water can feed demand in an emergency when the inflow is completely stopped. The result is shown in Fig. 7.29c. Note that the usual practice is to keep T_R almost constant over the day, between 10 and 12 hours of average demand. Optimisation will generally provide even longer T_R – except in the afternoon and evening where it drops below 9 hours.

The difference appears to be small to the analyst, but it might be important to the operators responsible for maintaining a reliable supply, reducing the safety margin at a particularly sensitive time of day or in a vulnerable system. It might be regarded as an unacceptable risk. If the operators are unable to express their fears openly, quantify the adverse effects of the proposed policy or even discuss the change with the interloping 'policy whizz-kid', they can still offer passive resistance and refuse cooperation, much to the surprise of the computer scientist. With a two-way communication, such problems could be easily overcome, to the benefit of all.

Finally a few tips for those planning the modernisation:

- The level of local technology and expertise cannot be changed too much in one step.
- Even the most modern systems need good maintenance (more than old ones) and this implies considerable costs and effort.
- The concept of 'war games' for senior executives might be very useful and illuminating.
- Over-riding concern should be *safety of supply* – not the technological breakthrough.
- Do not forget the ever-present need for reappraisal and improvements, given the speed of changes in the modern world.
- Modernisation will be found to be irreversible – once started, it cannot be abandoned.
- Black-box solutions are unacceptable. The staff will reject them or, even worse, might follow them blindly until the first

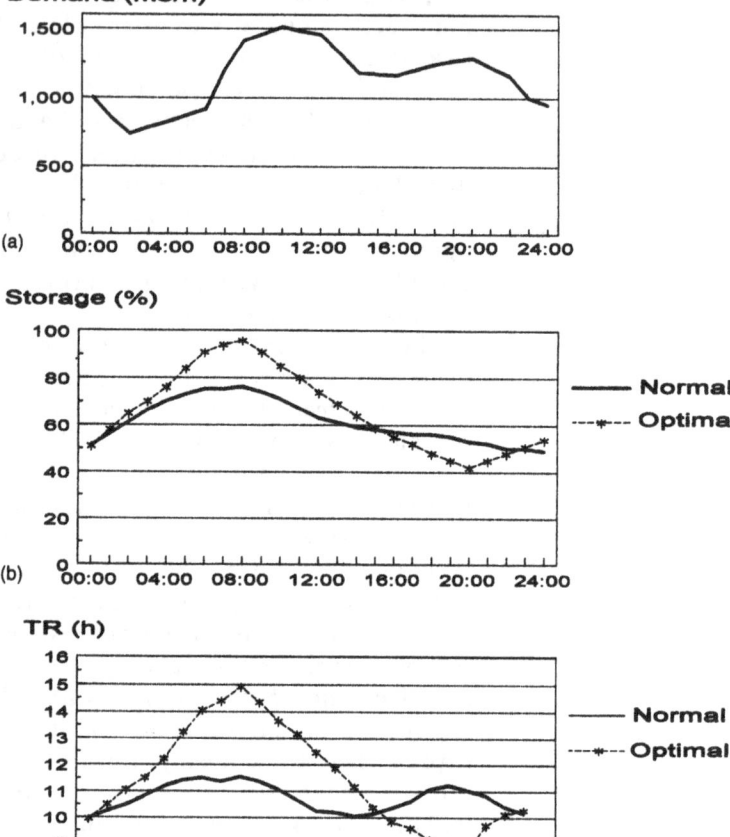

Figure 7.29 Water demand: (a) the demand; (b) two control policies; (c) reserve time.

major disaster. Staff need to understand the principles of what is going on around them, if not all the technical detail.

- Small alterations and improvements are the sure way forward. A big leap could (and probably would) end in a disaster.
- Progress is not possible without the active participation of the water supply system staff. They know the system better than anyone else. Simulation exercises provide excellent 'middle ground' where the staff can cooperate with computer scientists on an equal footing, to produce better software for full automatic control of the water supply system or its parts.
- How can one optimise a system without knowing it well? Give the credit to all experts, but trust only your own staff – after all *they* will have to run the system.

- Water supply systems are *open* systems. All possible inputs cannot be monitored and controlled, and a measure of 'educated guessing' will always be needed – hence the value of experience.

7.6 Models and operational management – cases from real life

Example 1

Ormsbee and Lansey (1994) have summarised experiences of optimal control of water supply systems, using the latest available data. Their conclusions were that a typical control system consists of: (a) hydraulic network model, (b) demand forecast model, and (c) optimal control model, all linked to the real system through SCADA (supervisory control and data acquisition).

1 *Hydraulic network models* vary from a simple mass balance, through regression models, simplified hydraulic model (linearised equations) to the full hydraulic model with all main facilities and trunk mains.
2 *Demand forecast models* have attracted less attention and consequently are less developed, so additional work is needed in that area. At present, forecasts are made using historical data – spatial and temporal.
3 *Optimal control models* are based on mathematical methods such as linear programming (LP), dynamic programming (DP), integer programming (IP), mixed-integer programming (MIP), linear-quadratic programming (LQP), non-linear programming (NLP), etc. The time horizon is usually limited to a day or two, and weekly and seasonal factors have to be included through constraints. The aim is to:
   - decrease water volume pumped,
   - decrease total system head (and pressure-related system losses),
   - increase pump efficiency by better selection of units,
   - use storage to keep the operating range close to BEP,
   - maximise off-peak (cheaper rate) pumping,
   - keep the number of starts and stops within reasonable limits.

Decision variables are either *direct*, e.g. pump operating time, or *indirect*, e.g. reservoir levels or pump discharge. It seems that there is no general solution to this problem; each case is unique, requiring specialist knowledge and custom-made software (nothing 'off the shelf'). Cooperation of staff is essential, but must be won – it is not granted. More research work is still required, also practical experience.

Several cases of implementation of these ideas in real life have been reported in the last decade, and some of the best-known examples are briefly described below.

Example 2

Morita and Arakawa (1989) describe a case study. It is a gravity-fed area supplied from a single reservoir; the network has four PRVs and 18 control points where pressure and demand are monitored; there is no pumping, no sources and no transfers.

The intention is to keep pressures close to the lower limit of the prescribed range, almost constant, avoiding over- and under-pressures and surges. Demand is predicted for 24 hours ahead, using operational data. A knowledge base is established, with a model capable of reacting in an intelligent way, taking into account previous experience and demand forecasts. The need for constant surveillance by people is thus removed.

Example 3

Alla (1985) states that a mathematical model of the western suburb of Paris, with a population of one million, has been developed and put to operational use. The production capacity of this system is 400 Ml d^{-1}, total length of water mains around 2000 km, and there are 40 reservoirs with a total storage of 80 Ml. The system is covered by telemetry, so the control centre has access to some 50 main facilities. The aim is to achieve optimal control over this large and complex system. The procedure consists of the following activities:

- daily prediction of consumption in the next 24 hours, using historical data and weather forecasts;
- hydraulic modelling of several sub-networks;
- discrete dynamic programming of the whole system, taking sub-networks as macro nodes – the program takes into account peculiarities of electricity tariffs, marginal costs of treatment of water and individual pump characteristics;
- examination of the solution and repeating the previous two steps if it is not feasible in one sub-network.

The end-result is an operational plan for the next day. It is implemented and carefully watched by comparing the data coming from main facilities to the control centre. If there is a significant disagreement, then the optimisation procedure is reactivated. The benefits for the water company are as follows.

- Reduced costs of electrical energy, with a return period of some 5–6 years approximately.
- Better understanding of how the system really operates. Some weak points were identified and fixed and general knowledge has increased.
- Development of a demand prediction program is an asset in its own right.

Example 4

A similar approach was adopted by Thames Water Utilities for control of London's water supply system (see Burnell et al., 1993; Rance et al., 1993a; Davies, 1993). A simplified network model for the whole system has been developed. It gives an overall picture by linking detailed models of several sub-networks. The point is that, although hydraulic relationships are non-linear, the regime implications are linear – hence linear programming could be applied for optimisation. Therefore, the control model has a simple balance for water treatment plants, main trunks and the London Water Ring Main. Three program modules – Demand Prediction, Trunk Scheduling and Network Modelling – share a common database. Special care was given to the development of a user-friendly GUI (graphical user interface) for operators. Individual demand areas are modelled using local knowledge.

The work starts after midnight, with computation of the demand in the next few days, and more closely for the next 24 hours. An iterative procedure is then used to determine the optimum solution that is feasible, trying at the same time to reduce pump switches wherever possible. The solution is then passed to local operational centres for further analysis. There the proposed plan is tested on detailed mathematical models. One of these models (the South-West Area Model) has 797 nodes, 1088 links, 15 demand zones, 15 local pumping stations, 21 trunk pumps, 67 control valves and 39 transfers, which is a rather large model by today's standards. Any comment or request is referred back to the control centre for review and reappraisal. It might lead to repetition of the whole procedure if a serious discrepancy has been discovered.

This system is applied for day-to-day operation of London's network. It is expected that it will cut down operational costs to a certain measure and also increase the security of supply.

Example 5

Chase and Jones (1993) have described a tailor-made computer-based control system developed for the city of Plano, Texas, USA. The population there is close to 100 000; the system consists of two pressure zones, four pumping stations and nine reservoirs, but no water treatment plants – it relies on imported water.

The system is based on SCADA covering the main facilities. The control model was developed using hydraulic relationships, tested and implemented. The operators communicate with model and data through a Windows-based interface – a user-friendly GUI. SCADA provides operational data converted into ASCII data files and transferred to the model for various simulation and optimisation runs. The costs are computed assuming that the tariff system is simple. The operator can change loading and boundary conditions. The decision is put into effect through SCADA again – in the reverse direction – to control pumps.

The system could also be used for off-line training of operators. Despite its sophistication – or perhaps because of it – there were difficulties with local staff. The authors deserve credit for their frankness in revealing these difficulties, such as: '... the system is operated by individuals who, quite frankly, can be somewhat intimidated and suspicious of computers ...'. Everyone in this field has experienced similar problems, which are bound to appear as long as *persons outside the water company have to develop and implement operational models*. However, it is clear that valuable experience and knowledge about real systems has been captured and put to use.

Example 6

Another case is reported by Schulte and Malm (1993). This is a new system designed and built for DuPage Water Commission of Elmhurst, Chicago, Illinois, USA. Its capacity is 185 million US gallons per day (approx. 700 Ml d^{-1}), total length of water mains is 145 miles (233 km), with pipe diameter ranging from 12 to 72 inches (300 to 1800 mm). Storage capacity is 92.5 million US gallons (347 Ml), approximately 12 hours of reserve. The system has 64 delivery points in an area of 650 km^2, through 19 remotely controlled valves. The ratio of peak to average demand can go up to 1.7, which means that the reserve of storage may fall to just 4 hours during peak demand days. All facilities are covered by telemetry.

To control such a huge and important system, SCADA was integrated with the model ('Hydraulic Simulator') for:

- training, before commissioning,
- supervisory control, after commissioning.

The system monitors water balance, on-line, and updates automatically forecasted demands. Water quality modelling (chlorine concentration) will be implemented in the future.

Example 7

Cubillo and Mercier (1993) report a joint research programme based on Paris, France, and Madrid, Spain. The aim is to control water quality in their distribution systems, using mathematical modelling and direct monitoring.

Madrid's network supplies a population of 4.6 million, and has 6100 km of pipes. The development of the model will take about 1.5 years and may have upwards of 100 000 nodes. As for Paris, the model will cover only part of the Greater Paris Area with population of 4 million and 8400 km of pipes. The pilot area selected for the study has a demand of $865 \text{ m}^3 \text{ d}^{-1}$ and population of 6120.

The benefits of this project will be a better understanding of water quality changes in the network, as well as predicting consequences of different treatment processes and other related processes. This should be achieved through modelling of conservative (inert) and non-conservative (active) substances. The entire piping system should be included and dynamic calculations used to describe these complicated phenomena. The system is GIS-based. Experimental work and calibration were in progress at the time of writing.

Example 8

Further progress in Paris is reported by Nguyen (1994). He describes the new control centre for the City of Paris, started in 1987. The centre provides:

- real-time data processing, and
- off-line hydraulic modelling,

for 11 pumping stations and 85 remotely controlled valves. Information about the systems comes as 600 digital signals (pressure, flow, WQ) and 5400 binary signals from the network. The time step can be 2.5, 5, 10, 20 or 60 min, but usually is

10 min. Operational data are transferred via the Ethernet network and stored in an Oracle database.

The model covers all major facilities and pipes above 300 mm size. Therefore, it is a so-called skeleton model. It has:

- 646 sections (pipes),
- 484 nodes,
- 18 pumps,
- 82 valves,
- 18 water tanks,
- 18 districts,
- 186 'laws' – presumably daily diagrams – adjusted every 10 min.

Simulation is automatically performed every 10 min, off-line, and gives predicted values of flows, water level in reservoirs, etc. Ten minutes later the predicted values are compared to the real data; if there are significant differences, an alarm is raised.

Example 9

Lansey and Awumah (1994) report another case study for the city of Austin, Texas, USA. A system for on-line optimisation of pump operations has been developed and implemented. The hydraulic model is based on regression equations, which in turn are obtained through several simulations on the mathematical model of the water supply system. The optimal solution is found by using dynamic programming (DP) with adequate constraints. The numbers of switches between 'on' and 'off' could be controlled. At present, this method can be applied on systems with only one or two reservoirs. The cost of energy is minimised, but other costs – such as maximum power charge – have to be accounted for through constraints, indirectly.

It was estimated that the savings could amount to a maximum 9%; this was achieved in days when the reservoir has been kept full all the time for reasons known to the operator (probably a 'play safe' policy).

7.7 Leads for beginners

At the end of this chapter, a few leads for beginners:

- Models should be made by local staff. Who knows the system better?
- Modern tools – such as GUI – remove earlier difficulties and fears because they are increasingly user-friendly, and few people should be intimidated by computers.

- Simulation is better for real-life use than optimisation based on numerical techniques unknown to the staff.
- 'All-mains' models are coming, and they will be based on GIS data – but not before a whole new software interface layer has been developed to reduce the size of the model.
- Operational data are now being acquired in large quantities. They are very reliable and should be used confidently throughout the water company.
- There may be no immediate, obvious benefit, dramatic cuts in operational costs, etc., from the introduction of modelling. Nevertheless, modelling is here to stay. It will improve overall knowledge about the system and expertise of the staff. How can one quantify *that*?

Appendix A
Standard interface formats[*]

Rajko Cavor

A.1 Introduction

Modern programs for network modelling have been designed to accept input from many different types of system (e.g. telemetry, GIS), and from any manufacturer of each different type. They achieve this 'open systems' approach by the use of standard interface formats. These are formats specified for inputting logger data (SLI), telemetry data (STI) and geographical data (SGI). In principle, it is much simpler for systems manufacturers to produce output in this standard format than for the general-purpose program to read dozens of different formats.

In this appendix, some details of each of the three standard interfaces are given. For further information, and advice on how to produce the formats from your systems, please contact your distributor. The three standard interfaces are:

- STI – Standard Telemetry Interface
- SLI – Standard Logger Interface
- SGI – Standard Geographical Interface

A.2 Standard Telemetry Interface (STI)

To establish the exchange and use of telemetry data outside the real-time telemetry computer system, there are two groups of problems to be solved by the telemetry manufacturers and the end-users.

Manufacturers

The problem is to provide controlled, formatted output of telemetry data at regular time intervals, to a named file, and a file transfer mechanism to a PC environment. Bearing in mind that all telemetry systems have extensive archive facilities, this requirement is basically archiving telemetry data in a different file format.

The user must specify – when purchasing – an independent, configurable and readable source of data. This will certainly require changes in the real-time software, rather than post-

---

[*] We are grateful to Rajko Cavor for writing this appendix, and to Aquaware Systems Ltd for their kind permission to reproduce the SGI specification.

installation surgery on the telemetry database, which could endanger vital telemetry functions, and even sometimes make extraction of data to the outside world impossible.

The user should adopt the view that data collected through the telemetry system are one of the company's most valuable assets in managing and operating the water supply system. Consequently, data must be made accessible, and useful for day-to-day analysis, and not locked up in a specialised telemetry system.

Users

The problem is to provide a PC link (PC file-server), connected to the telemetry computer via a direct communications link, or through a network workstation. This link will handle the regular transfer from telemetry, and the copying of files to the file-server directory structure.

Once data are transferred to a PC, they can be manipulated and used by many different software packages, or within WESNET's integrated management environment, for both modelling and calibration as well as for various analytical tasks.

STI Format

The Standard Telemetry Interface hardware and software organisation and the directory structure are shown in Fig. A.1. The format consists of a telemetry root directory, and 31 rolling archive directories, each containing a number of daily telemetry data files written to a PC at regular, prescribed time intervals. The actual data files are named after the date, month, hour and minute (DDMMhhmm.dat), and written to the appropriate day-in-the-month directory, which enables a simple sorting and archiving mechanism.

The format of the data files is a simple ASCII, comma-separated format, with *date and time* written in the first record, followed by a stream of records, each containing the *unique item ID* (from the telemetry system) and the *reading*. The *reading* for each item can be averaged over the transfer time interval, or can be the last reading from the outstation (in most cases it is the only one within the interval anyway). For reservoirs, flows and pressures, a 15 minute time interval has usually proven satisfactory because of the low speed of change and inertia of a water supply system.

Telemetry configurator

The most important link in the whole structure is the telemetry configurator, extracted from the telemetry database with all relevant data for analysis of the telemetry data by external users. This configurator can be used by the telemetry system as a look-up table when extracting and sending out data, as well as being used on the PC side for data recognition, filtering and sorting.

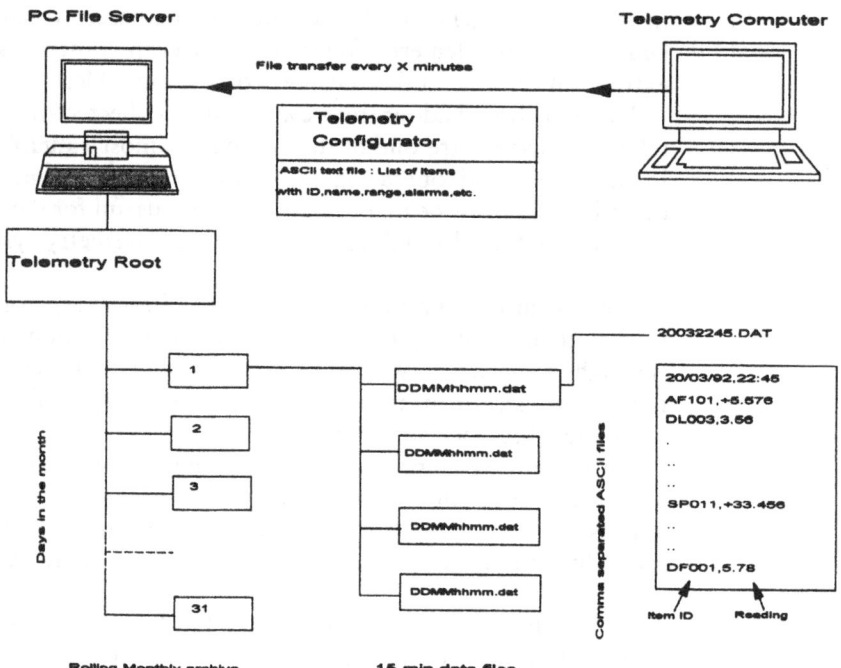

The procedure for creating and updating this configuration is of major importance for the use and validation of telemetry data. It must be worked out as part of the regular updating procedure of the telemetry database (e.g. after a specified number of changes in the telemetry database, the configurator should be written out and transferred automatically to the PC link).

The configurator is also of crucial importance for building a user-friendly system for external users, who are not usually familiar with internal telemetry structures, cumbersome IDs and the number of parameters (minimum, maximum, units, etc.) attached to each telemetry item. The configurator can also provide a strong cross-checking link between telemetry instrumentation, local measuring equipment and local knowledge about the water supply system.

A.3 Standard Logger Interface (SLI)

File details

The Standard Logger Interface file format is basically an ASCII, CSV (comma-separated value) format (extension: CSV, PRN). Key words in the file are used to denote the record type as follows:

- '*text*' – comment line (no limit to the number of these records);

- '*ch*' – logger channel, which should be followed by the channel number, the type of measurement (key words 'flow' and 'pressure') and the units; channel number also locates the reading within the data record ('ch', 2, means that the readings will be second in the data record);
- '*time*' – followed by start date (dd/mm/yy), start time (hh:mm), logging intervals, units (hour, min, sec) and the number of readings following;
- *data record* – no key words, comma-separated readings, no restrictions on range, or number of decimal places.

The suggested file extension for these files is *.PRN, which will make files directly readable for spreadsheet packages.

Standard Logger Interface – an example (SLI.PRN)

'text', 'Title: Standard Logger Interface – CSV file format'
'text', 'Site: Parkstone Road, Poole, Dorset'
'text', 'Note: two channel, flow and pressure measurement'
'ch', 1, 'flow', 'l/s'
'ch', 2 'pressure', 'm'
'time', 17/03/93, 11:45, 15.0, min, 31
46.17,16.51
43.59,16.70
42.91,16.90
42.97,16.83
43.40,16.93
44.39,16.90
43.22,17.10

A.4 Standard Geographical Interface (SGI)

Preliminary remarks

The interface between GIS and general-purpose software, including programs for network modelling, is at two levels:

- downloading graphics only, via the CGM format;
- The SGI – an ASCII format download, containing critical elements of network geometry (e.g. pipes, nodes).

When using the SGI, it is worth noting that some editing is usually necessary on the GIS platform, since most GIS databases contain *all* geographic details of the network. Much of this data will be irrelevant when building a hydraulic model of the system. One solution is the creation of a 'hydraulic layer' on the GIS. This layer would contain data that affects the hydraulic performance of the network, and would be downloaded to the general-purpose software via the SGI.

The full specification of the SGI developed by Aquaware Systems Ltd is reproduced here with their kind permission. (Please note that Aquaware Systems own the copyright to the following material relating to their Standard Geographical Interface software. Any enquiries for permission to reproduce this appendix in whole or in part should be addressed to: Aquaware Systems Ltd, 73 Glenferness Avenue, Bournemouth BH3 7ES, UK.) The specification has been slightly modified so that some nomenclature conforms with the main text of this book.

Standard Geographical Interface (SGI) is an object-oriented, expandable data structure that uses ASCII CSV (comma-separated value) format and assumes node/link representation of the physical network. SGI acts as a system-independent data transfer format providing a link between GIS/mapping systems and network modelling packages. Physical network connectivity and attributes are defined via nodes and links lists (nodes list must be declared first in full). The format of node and link elements/objects and all fields within the object record section are described below.

Version 3 of SGI introduces streamlining of version 2.1.1, additions to model geometry definition and, as a major change, a set of new extensions to link the network model to the control and operations data and the concept of the 'Ready-to-Run' WESNET32 models.

Network model features

1 *Nodes*

- *Ordinary nodes*
  Information defining an ordinary node
  [Node]   Node Element
  node = Label, X, Y, Z, Area Code, Text, Colour, Alignment, Text Size

Label	Unique ID up to eight characters
X	X coordinate of the node in metres
Y	Y coordinate of the node in metres
Z	Node elevation (ground level) in metres AOD (Above Ordnance Datum)
Area Code	Area ID up to eight characters
Text	Node long name or comment – 16 characters
Colour	Node colour index 1 to 255

Alignment   Node centred quadrant for related text
             display
Text Size   Node text size in relative units 1 to 99

comment = Any text
gis_key = GIS/Database key

Records comment and gis_key are optional.

- *Reservoirs*
Information defining a reservoir
[Reservoir] Node Element
Label, X, Y, Z, Area Code, Text, Colour, Align-
  ment, Text Size

All these items are similar to those defining an ordinary
node (see above).

data = RBL, TWL, Max. Control Depth

RBL   Reservoir bottom level (metres
                       AOD)
TWL   Top water level (metres AOD)
Max. Control   Maximum depth used by con-
Depth   trol devices (FLV, etc.)

Volume = Di, Vi, Di+1, Vi+1, ...,

Di, Vi   Depth (m), volume (m^3) co-
                       ordinates

At least one such pair should be specified (maximum
depth, maximum volume), but a reservoir with a complex
shape may require several pairs of coordinates to define
fully the relationship between depth and volume.

- *Transfer nodes*
Information defining a transfer node
[Transfer] Node Element
Label, X, Y, Z, Area Code, Text, Colour, Align-
  ment, Text Size

Similar to ordinary nodes (see above).

- *Fixed heads*
Information defining a fixed head node
[Fixed Head] Node Element
Label, X, Y, Z, Area Code , Text, Colour, Align-
  ment, Text Size

Similar to ordinary nodes (see above).

2 *Links*

- *Pipes*
  Information defining a pipe
  [Pipe] Link Element
  pipe = Node_L1, Node_L2, Lp, Dp, Fp, FF, LL,
         Age, Material index, Text, Colour

Node_L1	Unique node label of the first node
Node_L2	Unique node label of the second node
Lp	Model pipe length (metres)
Dp	Pipe diameter (millimetres)
Fp	Friction factor (–)
FF	Friction formula identifier, two characters, one of the following three options:

          CW   Colebrook–White
          HW   Hazen–Williams
          DW   Darcy–Weissbach

LL	Coefficient of local loss (–)
Age	Year of construction (eg. 1996)
Material index	Code selected from Table A.1
Text	Pipe long name/comment, up to 16 characters
Colour	Pipe colour index 1 to 255

  route = Xpi, Ypi, Xpi+1, Ypi+1, ...

Xpi, Ypi	Pair of coordinates (metres) describing changes in pipe direction

  comment = Any text
  gis_key = GIS/Database key

  The route record can be created with any number of coordinate pairs attached to a pipe element. If route records are not present, end node coordinates will be used *but* end node coordinates should not be entered with the pipe data as they will be identified from the node list.

  Records comment and gis_key are optional.

- *Pumping stations*
  Information defining a pumping station
  [PST] Link Element
  Node_L1, Node_L2, Lp, Dp, Fp, FF, LL, Age, Material index, Text, Colour

  route = Xpi, Ypi, Xpi+1, Ypi+1, ...

  All this information is similar to that required by a pipe (see above).

Table A.I Pipe material table

Material index	Description	
1	AK	Alkathene
2	AC	Asbestos cement
3	BR	Brick
4	CI	Cast iron
5	SI	Spun (grey) iron
6	CO	Concrete
7	CSB	Concrete segments bolted
8	CSU	Concrete segments unbolted
9	CC	Concrete box culvert
10	DI	Ductile iron
11	GRC	Glass reinforced concrete
12	GRP	Glass reinforced plastic
13	PSC	Plastic/steel composite
14	GRP	(Unplasticised) polyvinyl chloride
15	PE	Polyethylene
16	RPM	Reinforced plastic matrix
17	ST	Steel
18	VC	Vitrified clay

```
data = Duty Head, No. of Pumps, Bypass Flag
       (1/0)

power = Tariff index, Voltage index, Power
        Consum., E/H ratio, Power Coeff.
        (CosFi)

unit = UnitID, Duty Flow, Max. Speed, Min.
       Speed, Local Loss, No. of Stages,
       Suct. Level
```

- *Valves*
  Information defining a valve
  ```
  [Valve] Link Element
  pipe = Node_L1, Node_L2, Lp, Dp, Fp, FF, LL,
         Age, Material index, Text, Colour,
         XV, YV
  ```

All this information is similar to that required for a pipe (see above). The entry of the coordinates of the actual location of the valve on the pipe, XV and YV, is optional.

```
route = Xpi, Ypi, Xpi+1, Ypi+1, ...
```

Information similar to that required for pipe route (see above).

```
data = Valve Type index, Diameter, Loss When
       Fully Open, Control Type index
```

Valve Type index	Index of valve types (see Table A.2)
Diameter	Internal diameter in milli- metres
Local Loss	Coefficient when fully open (–)
Control Type index	Index of valve control types (see Table A.3)

The function of 'status flag' in previous versions is now assumed by Control Type index. Direction of flow is always assumed to be from node 1 to node 2.

- *Meter*
  Information defining a meter
  [Meter] Link Element
  pipe = Node_L1, Node_L2, Lp, Dp, Fp, FF, LL,
          Age, Material index, Text, Colour,
          XV, YV

All this information is similar to that required for a pipe (see above). The entry of the coordinates of the actual location of the valve on the pipe, XV and YV, is optional.

route = Xpi, Ypi, Xpi+1, Ypi+1, ...

Table A.2 Valve construction types

Valve Type index	Description
1	Plug valve (default)
2	Ball valve
3	Butterfly valve
4	Gate valve
5	Needle valve
6	Other (spare)

Table A.3 Valve control types

Control Type index	Description	
0	MOV	Manually operated valve
1	FLV	Float valve (on res. inlet)
2	NRV	Non-return valve
3	THV	Throttled valve
4	PRV	Pressure-reducing valve
5	PSV	Pressure-sustaining valve
6	PCV	Pressure-controlled valve
7	TCV	Time-controlled valve
8	FCV	Flow-controlled valve

Information similar to that required for pipe route (see above).

```
data = Type index, Diameter, Loss When Open
```

3 Pumps (individual unit characteristics)
```
[Pump]
data = UnitID, Nominal Flow, Rated Speed, Suc-
       tion Diameter, Pressure Diameter
curve = q1, h1, p1, q2, h2, p2, ...
```

4 Demands
```
[Demand]
node = NodeID, Leakage(l/s), Nominal Pres-
       sure(m)
user = NodeID, Category ID, No. Properties,
       Spec. Consum, Direct Demand(l/s)
```

5 Background maps
Creates list of maps (in CGM format) that cover network model geographical area.

```
[Maps]  CGM maps
map = File path
```

Control and operations features

The following records, `control`, `link`, `wq`, `timetable`, are applicable to *all* network elements (node, pipe, reservoir, PST, transfer node, fixed head node, valve etc.) to describe object functional features and linkage to other systems (loggers, telemetry).

1 Control
```
control = Report Flag, Control Node/Pipe ID,
          Set Value 1, Set Value 2, ..., Set
          Value n
```

Report Flag	0 = omit, 1 = include element in Save/Report List
Control Node/ Pipe ID	Node label (max. 8 chars) or two nodes on pipe labels (max. 16 chars given as Node1 Node2 string)
Set Value 1, ..., Set Value n :	Control set values (e.g. pressure, depth, flow,...)

Tables A.4–A.6 show the control element (node or pipe) and set values for all valve and pumping station modes of operation.

2 *Link to external data sources*

link = External Link ID, Data Source index,
          Manufacturer index, Source File

External Link ID	Unique ID (max. 8 chars) of logger or telemetry device
Data Source index	1 = logger, 2 = telemetry
Manufacturer index	Index of manufacturers (see Table A.7)
Source File	Full path (include filename) to data file

3 *Water quality (nodes only)*

Describes initial and additional concentrations of simulated substance at any node and for any time-related decay rate.

wq = Init.Concentration, Add.Concentration,
          Decay Rate

Init.Concentration	At node (mg l^{-1})
Add. Concentration	Injected at node (mg l^{-1})
Decay Rate	Substance- or source dependant decay rate coefficient (1 day^{-1})

Table A.4   Valve control parameters

Valve control type	Control node/ pipe	Set value 1	Set value 2	Set value 3	Set value 4
MOV	–	opening (%)	–	–	–
FLV	reservoir	inlet level (m)	control depth range (m)	throttling type 0/1[a]	backfeed pipe 0/1[b]
NRV	–	–	–	–	–
THV	–	opening (%)	–	–	–
PRV	node	set pressure (m)	–	–	–
PSV	node	set pressure (m)	–	–	–
PCV	node	init. opening (%)	pressure 1	opening 1 ↑	opening 1 ↓
TCV	–	–	–	–	–
FCV	pipe	init. opening (%)	flow 1 ($l\ s^{-1}$)	opening 1	–

[a] FLV throttling type: 0 = continuous within depth range, 1 = on/off action.
[b] FLV backfeed pipe: 0 = none, 1 = yes.

Table A.5 Pumping stations control parameters

	FLOW	Plan	AUTO	PROF[a]
Control node/pipe	node	node	node	node
Set value 1	1	2	3	4
Set value 2	task ID[b]	task ID	task ID	task ID
Set value 3	max. power	max. power	max. power	max. power
Set value 4	low press. on	low press. on	pump 1 P on	pump 1 status[a]
Set value 5	high press. off	high press. off	pump 1 P off	pump 1 D on
Set value 6			pump 2 P on	pump 1 D off
Set value 7			pump 2 P off	pump 2 D on
Set value 8				pump 2 D off
⋮				
Set value n				

[a] PST in Profiler mode uses deviations from target level/pressure to switch pumps on and off.
[b] Task ID: 0 = raw water PST, 1 = source PST, 2 = booster PST.
[c] Pump status (initial): 0 = not available, 1 = stand-by, 2 = running.

Table A.6 Other nodes and pipes control parameters

Network element	Control node/pipe	Set value 1	Set value 2
Reservoir	–	initial depth (m)	–
Fixed head	–	total head	–
Transfer	–	flow sign −1/+1	–
Pipe	–	open/close 0/1	–

Table A.7 Manufacturers' index

Index	Loggers	Telemetry
1	Spectrascan	STI
2	Radcom / SLI	
3	Technolog / SLI	
4	SLI logger	

4 Real-time control profiles – timetable record

`Timetable` record provides for direct import of real-time control profiles for:

- pumping stations (FLOW, PLAN, PROF)
- control valves (PRV/PSV, TCV)
- fixed head with variable profile
- transfer node with variable profile

Typically this record will be repeated within one element section (e.g. PST) to cover all changes of control profile during the simulation period. If profile is constant value, then only one (initial date/time) record will be necessary. The control profiles are shown in Tables A.8–A.10.

```
timetable = Date, Time, Control_1, Control_2,
            ..., Control_n
```

Date	dd/mm/yy
Time	hh:mm
Control_1...n	Array of control values (flow, pressure, depth, speed, etc.)

Table A.8 Pumping stations control profiles

PST control type	Control I	Control 2
FLOW	target flow (l s^{-1})	–
PLAN	pump statuses[a]	Var. speed (rpm)[b]
PROF	target pressure(m)	–

[a] All units statuses as string 001100.
[b] Var. speed only for PSTs with active var. speed unit.

Table A.9 Valve control profiles

Valve control type	Control I
PRV	set pressure (m)
PSV	set pressure (m)
TCV	set opening (%)

Table A.10 Other control profiles

Network element	Control I
Fixed head	total head (m AOD)
Transfer	+/− transfer flow (l s^{-1})

Appendix B
Friction loss formulae

The most popular of these is the Darcy–Weissbach formula. Other widely used equations are the Hazen–Williams and Manning formulae.

B.1 Darcy–Weissbach formula

The Darcy–Weissbach formula states that

$$\Delta H_{\text{fric}} = \lambda \frac{Lv^2}{2gd} \tag{B.1}$$

where λ (–) is the friction coefficient, L (m) and d (m) are the pipe's length and internal diameter, respectively, v (m s^{-1}) is the velocity of the water and g ($= 9.806\ 65$ m s^{-2}) is the acceleration due to gravity. Unfortunately, λ is not a constant but an implicit function of several factors, notably:

- the Reynolds number, $Re = vd/\nu$, where ν is the kinematic viscosity of the fluid (m^2 s^{-1});
- the internal roughness of the pipe's wall, k.

The variation of λ in relation to Re and k is explained in many textbooks, see Streeter and Wylie (1979) for instance.

There are five distinctly different flow regimes, as follows.

1 *Laminar flow,* where

$$\lambda = 64/Re \tag{B.2}$$

2 *Transient zone between laminar and turbulent flow* where λ is not well defined; several exponential formulae such as the one below were proposed:

$$\lambda = A + \frac{B}{Re^\kappa} \tag{B.3}$$

where A, B and κ are empirical constants.

3 *Turbulent flow where the laminar sublayer is thicker than the average height of internal roughness k* and the resistance is caused by frictional forces only:

$$\frac{1}{\sqrt{\lambda}} = -2\ \log_{10}\left(\frac{2.51}{Re\sqrt{\lambda}}\right) \tag{B.4}$$

4 *Turbulent flow where the sublayer cannot cover all internal peaks* so the losses are partly caused by friction and partly by

vortices formed behind the peaks; this is the well known Colebrook formula:

$$\frac{1}{\sqrt{\lambda}} = -2 \ \log_{10}\left(\frac{2.51}{Re\sqrt{\lambda}} + \frac{k}{3.71d}\right) \tag{B.5}$$

5 *Fully developed turbulent flow* where the laminar sublayer is very thin, so the losses are mainly caused by vortices:

$$\frac{1}{\sqrt{\lambda}} = -2 \ \log_{10}\left(\frac{k}{3.71d}\right) \tag{B.6}$$

The values of internal roughness k, for different pipe materials (cast iron, ductile iron, steel, asbestos cement, polyethylene or PVC) are given in pipe manufacturers' literature (see, for instance, Miller, 1978) – for brand new pipes only.

These equations for λ are very accurate, being the result of many very precise experiments in several laboratories. The problem is that λ is a function of Reynolds number Re, i.e. fluid velocity, which is not known at the beginning of the computation. Further unknowns are the actual roughness of the pipe, k, and the internal diameter of the pipe, d, after several years of service when the net flow area could be significantly reduced due to deposits and encrustation.

Equivalent surface roughness data for different pipe materials are given in Table B.1 (from Bhave, 1991). These values will increase with age.

B.2 Hazen–Williams formula

The Hazen–Williams formula exists in many different forms, one of which is:

Table B.1 Equivalent surface roughness

Pipe material	Roughness k (mm)
Riveted steel	0.90–9.00
Concrete	0.30–3.00
Wooden stave	0.20–0.90
Cast iron	0.26
Galvanised iron	0.15
Asphalted cast iron	0.13
Commercial steel, wrought iron	0.05
Uncoated asbestos cement, PVC pipes with waviness, prestressed concrete	0.04
PVC	0.0021
Drawn tubing of aluminium, brass, copper, lead, glass and plastic	0.0015
Coated asbestos cement, spun cement-lined, spun bitumen-lined	0.0015

Table B.2 Hazen–Williams coefficient C for new pipes

	C value for various pipe diameters				
Pipe material	75 mm (3 inch)	150 mm (6 inch)	300 mm (12 inch)	600 mm (24 inch)	1200 mm (48 inch)
Uncoated cast iron	121	125	130	132	134
Coated cast iron	129	133	138	140	141
Uncoated steel	142	145	147	150	150
Coated steel	137	142	145	148	148
Wrought iron	137	142	–	–	–
Galvanised iron	129	133	–	–	–
Coated spun iron	137	142	145	148	148
Uncoated asbestos cement	142	145	147	150	–
Coated asbestos cement	147	149	150	152	–
PVC pipes with waviness	142	145	147	150	150
Concrete	69–129	79–133	84–138	90–140	95–141
Prestressed concrete	–	–	147	150	150
Spun cement-lined; spun bitumen-lined, PVC, brass, lead, copper	147	149	150	152	153
Newly scraped mains	109	116	121	125	127
Newly brushed mains	97	104	108	112	115

Table B.3 Effect of age on coated cast-iron pipes

		C value for various pipe diameters				
No.	Degree of corrosive attack	75 mm (3 inch)	150 mm (6 inch)	300 mm (12 inch)	600 mm (24 inch)	1200 mm (48 inch)
	Pipes 30 years old					
1	slight	100	106	112	117	120
2	moderate	83	90	97	102	107
3	appreciable	59	70	78	83	89
4	severe	41	50	58	66	73
	Pipes 60 years old					
1	slight	90	97	102	107	112
2	moderate	69	79	85	92	96
3	appreciable	49	58	66	72	78
4	severe	30	39	48	56	62
	Pipes 100 years old					
1	slight	81	89	95	100	104
2	moderate	61	70	78	83	89
3	appreciable	40	49	57	64	71
4	severe	21	30	39	48	54

$$\Delta H = \frac{10.675}{d^m} L \left(\frac{Q}{C}\right)^\kappa \tag{B.7}$$

where Q (m^3 s^{-1}) is the flow, and the two empirical constants are $\kappa = 1.852$ and $m = 4.8704$.

The coefficient C is, in a way, the pipe capacity. Engineers in the USA and the UK are familiar with this formula and can estimate the value of C quite accurately both for old and for new pipes. More details may be found in Streeter and Wylie (1979) or Bhave (1991). The C values for clean new pipes of different materials are given in Table B.2 (after Bhave, 1991). These values will decrease with the pipe's age, as indicated in Table B.3.

The relationship between λ and C is:

$$\lambda = \frac{1014.2}{C^{1.852} d^{0.0184} Re^{0.148}} \tag{B.8}$$

B.3 Manning formula

The Manning formula is usually expressed as:

$$\Delta H = \frac{n^2 v^2 L}{R_h^{4/3}} \tag{B.9}$$

where v (m s^{-1}) is the velocity, L (m) is the pipe length, R_h (m) is the hydraulic radius, equal to $d/4$ for round-section pipes, and n is Manning's roughness coefficient (Table B.4).

By introducing all constants equation (B.9) becomes:

$$\Delta H = \frac{n^2 L}{(d/4)^{4/3}} \left(\frac{Q}{\pi d^2/4}\right)^2 = 10.293 \frac{n^2 Q^2 L}{d^{16/3}} \tag{B.10}$$

which might be more practicable.

The relationship between λ and Manning's coefficient n is

$$\lambda = 124.6 \frac{n^2}{d^{1/3}} \tag{B.11}$$

For instance, if $n = 0.012$ and $d = 300$ mm $= 0.300$ m, the value of λ will be 0.0268.

Table B.4 Manning's coefficient for different pipe materials

Pipe material	Manning's coefficient n
PVC	0.008–0.011
Brass, copper, glass, lead, prestressed concrete	0.009–0.012
Concrete	0.010–0.017
Wooden stave	0.011–0.013
Welded steel	0.012–0.013
Coated cast iron	0.012–0.014
Uncoated cast iron	0.013–0.015
Galvanised iron, riveted steel	0.015–0.017

Appendix C
Local (minor) loss coefficients

Local (minor) losses are caused by valves, bends, reductions, sleeves and other fittings. Usually, they are not as important as friction losses. The expression for such a loss is:

$$\Delta H = \zeta \frac{v^2}{2g} \tag{C.1}$$

Table C.1 Minor losses

Type of fitting	ζ value	Type of fitting	ζ value
ENTRY LOSSES		*Angle branches*	
Sharp-edged entrance	0.50	Flow in line	0.35
Re-entrant entrance	0.80	Line-to-branch/Branch-to-line:	
Slightly rounded entrance	0.25	30.0°	0.40
Bell-mouthed entrance	0.05	45.0°	0.60
Foot valve and strainer	2.50	90.0°	0.80
INTERMEDIATE LOSSES		*Sudden enlargements*	
Elbows		4:5 inlet:outlet diameter	0.15
R/D = 0.5, 22.5°	0.20	3:4	0.20
R/D = 0.5, 45.0°	0.40	2:3	0.35
R/D = 0.5, 90.0°	1.00	1:2	0.60
		1:3	0.80
Close radius bends		1:5 and over	1.00
R/D = 1 approx., 22.5°	0.15		
R/D = 1 approx., 45.0°	0.30	*Sudden contractions*	
R/D = 1 approx., 90.0°	0.75	5:4 inlet:outlet diameter	0.15
		4:3	0.20
Long radius bends		3:2	0.30
R/D = 2 to 7, 22.5°	0.10	2:1	0.35
R/D = 2 to 7, 45.0°	0.20	3:1	0.45
R/D = 2 to 7, 90.0°	0.40	5:1 and over	0.50
Sweeps		*Tapers*	
R/D = 8 to 50, 22.5°	0.05	Flow to small end	negligible
R/D = 8 to 50, 45.0°	0.10	Flow to large end:	
R/D = 8 to 50, 90.0°	0.20	4:5 inlet:outlet diameter	0.03
		3:4	0.04
Mitre elbows		1:2	0.12
2 piece, 22.5°	0.15		
2 piece, 30.0°	0.20	*Valves*	
2 or 3 piece, 45.0°	0.30	Gate valve, fully open	0.12
2 piece, 60.0°	0.65	Gate valve, ¼ closed	1.00
3 piece, 60.0°	0.25	Gate valve, ½ closed	6.00
2 piece, 90.0°	1.25	Gate valve, ¾ closed	24.0
3 piece, 90.0°	0.50	Globe valve	10.0
4 piece, 90.0°	0.30	Right angle valve	5.00
		Reflux valve	1.00
Tees		Butterfly valve	0.30
Flow in line	0.35		
Line-to-branch/Branch-to-line:		**EXIT LOSSES**	
sharp-edged	1.20	Sudden enlargement	1.00
radiused	0.80	Bell-mouthed outlet	0.20

where ζ (−) is the loss coefficient. Its value can be found in manuals − see Idel'chik (1979) or Miller (1978) for instance. A few values are given in Table C1.

Appendix D
Pumps

D.1 Specific consumption of energy

The use of energy per volume of water delivered is equal to

$$e = \frac{E}{V} = \frac{\rho g H_p}{3600\,000\eta}$$

in kWh m^{-3}. For example, a pump providing the head of 60 m, with 73.5% efficiency, will use

$$e = (1000 \times 9.81 \times 60)/(3600\,000 \times 0.735) = 0.222 \text{ kWh m}^{-3}$$

Note that flow, Q, is not directly included in this simple equation. Table D.1 might help for quick computations. The values in Table D.1 should be increased by 20–30% to take into account wire-to-water efficiency (additional mechanical and electrical losses in the coupling, bearings and electric motor).

Table D.1 Specific consumption of energy

Pump flow (l s^{-1})	Efficiency (%)	Pump head (m)					
		30	50	70	100	125	150
1	34	0.242	0.400	0.961	0.800	1.001	1.202
2	44	0.186	0.310	0.433	0.619	0.774	0.929
3	50	0.164	0.272	0.381	0.544	0.681	0.817
5	58	0.141	0.235	0.329	0.469	0.587	0.705
10	66	0.124	0.206	0.289	0.414	0.516	0.619
20	72	0.114	0.189	0.265	0.379	0.473	0.568
30	75	0.109	0.181	0.255	0.363	0.454	0.545
50	78	0.105	0.175	0.244	0.350	0.436	0.524
100	81	0.101	0.168	0.236	0.336	0.420	0.504
200	83	0.099	0.164	0.230	0.328	0.410	0.492
Larger units	85	0.096	0.160	0.224	0.320	0.401	0.481

D.2 Characteristics

The data for five typical pumps are given in Tables D.2–D.7. The notation is:

- relative flow $v = Q/Q_{BEP}$
- relative head $h = H_p/H_{p,\,BEP}$
- relative power $\beta = P/P_{BEP}$
- relative efficiency $\eta' = \eta/\eta_{BEP}$

BEP is best efficiency point. The data are taken from the references.

Appendix D

Table D.2 Centrifugal pump $n_q = 21$

v	h	β	η'
0.00	1.110	0.343	0.000
0.20	1.108	0.462	0.480
0.40	1.100	0.596	0.738
0.60	1.080	0.735	0.882
0.80	1.050	0.872	0.963
0.90	1.027	0.935	0.988
1.00	1.000	1.000	1.000
1.10	0.958	1.058	0.996
1.20	0.904	1.113	0.975
1.30	0.843	1.166	0.940
1.40	0.762	1.213	0.879
1.50	0.670	1.260	0.798
1.60	0.530	1.300	0.652
1.70	0.330	1.335	0.420
1.80	0.000	1.360	0.000

This information could be used to reconstruct the data for a centrifugal pump if the following data are known for the BEP (nominal) operating point:

- flow $Q_0 = 28\ \mathrm{l\ s^{-1}}$
- head $H_p = 63$ m
- power $P = 22.8$ kW
- efficiency $\eta = 76\%$
- rotational speed $n = 2930$ rpm

The specific speed is equal to

Table D.3 Operational characteristics – an estimate

Flow Q ($\mathrm{l\ s^{-1}}$)	Head H_p (m)	Power P_h (kW)	Efficiency (%)
0.0	69.9	7.8	0.0
5.6	69.8	10.5	36.5
11.2	69.3	13.6	56.1
16.8	68.0	16.8	67.0
22.4	66.2	19.9	73.2
25.2	64.7	21.3	75.1
28.0	63.0	22.8	76.0
30.8	60.4	24.1	75.7
33.6	57.0	25.4	74.1
36.4	53.1	26.6	71.4
39.2	48.0	27.7	66.8
42.0	42.2	28.7	60.6
44.8	33.4	29.6	49.6
47.6	20.8	30.4	31.9
50.4	0.0	31.0	0.0

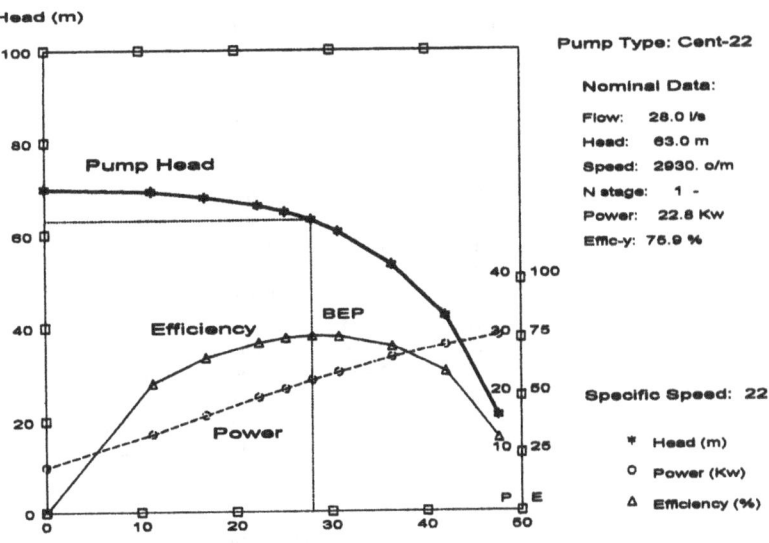

Figure D.1 Estimated
characteristics.

Table D.4 Centrifugal pump n_q = 42

v	h	β	η'
0.0	1.220	0.486	0.000
0.20	1.208	0.548	0.441
0.40	1.180	0.662	0.713
0.50	1.160	0.723	0.802
0.60	1.138	0.782	0.873
0.80	1.082	0.896	0.966
0.90	1.044	0.950	0.989
1.00	1.000	1.000	1.000
1.10	0.937	1.043	0.988
1.20	0.866	1.086	0.957
1.30	0.785	1.128	0.905
1.40	0.684	1.162	0.824
1.50	0.550	1.194	0.691
1.60	0.310	1.223	0.406
1.68	0.000	1.237	0.000

$$n_q = \frac{n\sqrt{Q}}{H_p^{0.75}} = \frac{2930\sqrt{0.028}}{63^{0.75}} = 21.9 \cong 22$$

which means that this is a centrifugal pump, so the data from
Table D.2 could be used. The result is given in Table D.3. This
characteristic is shown in Fig. D.1 below. The result is very
convincing, so the data could be used until better information is
obtained by direct measurements on the unit in question. Data
for other pumps with different specific speeds are given in Tables
D.4–D.7

Table D.5   Centrifugal pump $n_q = 64$

v	h	β	η'
0.00	1.353	0.650	0.000
0.20	1.308	0.665	0.393
0.40	1.256	0.720	0.698
0.50	1.225	0.760	0.806
0.60	1.192	0.805	0.888
0.80	1.108	0.906	0.978
0.90	1.060	0.955	0.999
1.00	1.000	1.000	1.000
1.10	0.903	1.035	0.960
1.20	0.786	1.060	0.890
1.30	0.645	1.086	0.772
1.40	0.445	1.107	0.563
1.50	0.200	1.123	0.267
1.56	0.000	1.127	0.000

Table D.6   Centrifugal Pump $n_q = 106$

v	h	β	η'
0.00	1.717	1.186	0.000
0.20	1.582	1.100	0.288
0.40	1.450	1.050	0.552
0.50	1.384	1.031	0.671
0.60	1.314	1.013	0.778
0.80	1.176	1.003	0.938
0.90	1.097	1.001	0.986
1.00	1.000	1.000	1.000
1.10	0.860	0.999	0.947
1.20	0.673	0.990	0.816
1.30	0.445	0.954	0.606
1.40	0.160	0.876	0.256
1.45	0.000	0.808	0.000

Table D.7   Mixed-flow pump $n_q = 169$

v	h	β	η'
0.00	1.866	0.995	0.000
0.20	1.738	0.963	0.361
0.40	1.575	0.993	0.634
0.50	1.487	1.003	0.741
0.60	1.396	1.010	0.829
0.80	1.212	1.013	0.957
0.90	1.110	1.011	0.988
1.00	1.000	1.000	1.000
1.10	0.855	0.970	0.970
1.20	0.660	0.880	0.900
1.30	0.406	0.695	0.759
1.40	0.080	0.400	0.280
1.42	0.000	0.300	0.000

D.3 Electric motor speed

The data in Table D.8 correspond to squirrel cage motors.

Table D.8 Electric motor speed

Pole pairs	Synchronous speed (rpm)	Power (kW)			
		0.7 to 2.2	3.0 to 7.5	11 to 22	30 to 75
Europe: frequency f = 50 Hz					
1	3000	2820	2855	2910	2935
2	1500	1405	1420	1445	1475
3	1000	930	945	960	975
4	700	–	–	720	735
USA: frequency f = 60 Hz					
1	3600	3380	3425	3490	3520
2	1800	1685	1700	1730	1770
3	1200	1115	1130	1145	1170
4	900	–	–	860	880

D.4 Operational difficulties

The pumps sometimes operate less than perfectly. The signs of trouble are:

- The pump cannot provide the same flow as before.
- Consumption of energy rises.
- Noise and vibrations are stronger.
- Start-ups fail very often.
- Frequent trip-offs happen without apparent reason.
- Bearings are too hot.

A total breakdown is now a real possibility and something should be done before it is too late. A few tips are the following.

- Collect evidence and talk with the local staff.
- Make a list of possible malfunctions.
- Check the obvious – personally.
- Have a good look around. In many cases the cause of trouble is not the pump itself but something else.
- Do not make conclusions prematurely.

The hydraulic causes for pump malfunctions are:

- suction pipe is not air-tight;
- losses in the suction system are now greater due to something, like clogging of the inlet pipe;
- the pump is forced to operate in the low-flow range, which might be unstable;

- the impeller is partly blocked (clogging?) or damaged (cavitation?);
- the check valve is damaged or broken completely;
- the throttling valve is partially closed (somebody's mistake?);
- the hydraulic thrust is not properly balanced;
- the pressure on the suction side is too low, due to some changes in the system;
- the pressure on the delivery side is too high/too low, again due to changes outside the pumping station;
- another pumping station interferes with this one, etc.

The mechanical causes might be:

- bearings are of poor quality or worn out;
- the rotating parts (impeller, shaft, clutch) are not well balanced;
- the seal rings are worn out;
- the loading of flanges is excessive, etc.

Most of these problems are caused by poor installation and/or maintenance.

Appendix E
Valve characteristics

Manufacturers will provide data for their products, including the values of loss coefficient ζ or its equivalent, like the flow coefficient k_v, which is

Table E.1   Fully open valve coefficients

D_n		Butterfly valves			Ball valves		
mm	inch	k_v	ζ	k_D	k_v	ζ	k_D
50	2	110	0.828	0.740	480	0.044	0.979
65	2.5	210	0.649	0.779	750	0.043	0.980
80	3	330	0.603	0.790	1300	0.030	0.985
100	4	610	0.431	0.836	2300	0.030	0.985
125	5	1000	0.392	0.848			
150	6	1500	0.361	0.857	5400	0.028	0.986
200	8	2700	0.352	0.860	10000	0.026	0.987
250	10	4300	0.339	0.864	16000	0.024	0.988
300	12	6600	0.298	0.878	24000	0.023	0.989
350	14	8900	0.304	0.876	31400	0.024	0.988
400	16	11500	0.310	0.874	43000	0.022	0.989
450	18	15000	0.292	0.880	57000	0.020	0.990
500	20	18800	0.284	0.883	73000	0.019	0.991
600	24	27600	0.273	0.886			
700	28	38600	0.258	0.891			
800	32	51500	0.248	0.895			

Table E.2   Partially open valves (I)

Angle (deg)	Ball valve		Butterfly valve		
	ζ	k_D	k_v	ζ	k_D
90 open	0.05	0.976	6600	0.298	0.878
85	0.13	0.941			
80	0.26	0.891	5985	0.363	0.856
75	0.55	0.803			
70	0.95	0.716	3486	1.069	0.695
65	1.50	0.632			
60	2.35	0.546	1841	3.833	0.455
55	3.40	0.477			
50	5.2	0.402	1183	9.283	0.312
45	7.6	0.341			
40	12	0.277	661	29.73	0.180
35	18	0.229			
30	28	0.186	394	83.68	0.109
25	47	0.144			
20	83	0.109	197	334.7	0.055
15	185	0.073			
10	670	0.039			
0	∞	0	0	∞	0

Table E.3 Partially closed valves (II)

Opening (%)	Gate valve ζ	Gate valve k_D	Ball valve ζ	Ball valve k_D	Butterfly valve ζ	Butterfly valve k_D	Needle valve ζ	Needle valve k_D
100	0	1.000	0.20	0.912	0.40	0.845	2.35	0.546
90	0.06	0.971	0.41	0.842	0.50	0.816	2.60	0.527
80	0.17	0.925	1.01	0.705	0.82	0.741	2.90	0.506
70	0.44	0.833	2.90	0.506	1.70	0.609	3.45	0.474
60	0.98	0.711	8.0	0.333	4.65	0.421	4.35	0.432
50	2.06	0.572	22.5	0.206	12.5	0.272	5.70	0.386
40	4.60	0.423	66	0.122	34	0.169	8.40	0.326
30	10.0	0.302	235	0.065	102	0.099	13.0	0.267
20	35.0	0.167	2250	0.021	400	0.050	25	0.196
10	120	0.091	10000	0.010	5000	0.014	97	0.101
0	∞	0	∞	0	∞	0	∞	0

$$k_v = \frac{Q^*}{31.6}\sqrt{\frac{\rho}{\Delta p^*}} \tag{E.1}$$

where Q^* is expressed in $m^3\ h^{-1}$ and Δp^* in bar. By introducing SI units, $m^3\ s^{-1}$ and Pa, we get

$$\frac{\Delta p}{\rho} = 10^5\left(\frac{3600\ Q}{31.6\ k_v}\right)^2 = \zeta\frac{v^2}{2} = \frac{\zeta}{2}\left(\frac{Q}{A_n}\right)^2 \tag{E.2}$$

or

$$\zeta = 26\left(\frac{10^4 A_n}{k_v}\right)^2 \tag{E.3}$$

We shall now consider an example: the valve $D_n = 600\ \text{mm}$ has $k_v = 27\ 600$; therefore $A_n = 0.600^2\pi/4 = 0.283\ m^3$ and $\zeta = 26\ (10000 \times 0.283/27\ 600)^2 = 0.273$.

Another widely used coefficient is

$$k_D = \frac{1}{\sqrt{1+\zeta}} \tag{E.4}$$

which is a unitless flow coefficient.

Data for fully open butterfly and ball valves are given in Table E.1. Values for partially closed valves of different types are given in Tables E.2 and E.3. These data could be used as default values, but of course no effort should be spared to get real data from the manufacturer.

Appendix F
Water consumption patterns

Real data can be obtained only by direct measurements; attempts to provide 'typical patterns' are not likely to succeed due to interference of many factors, largely unknown at present. Even when data are available, the patterns are difficult to discern. In many cases, the data are irregular, like in the case shown in Fig. F.1. Note that both the base demand and the sudden jumps are unrelated to the day in the week, or to the time of day. Obviously the process is automated and the level of demand is dictated by its requirements, unknown outside the dairy. The useful information from such data is only the range of flows: the maximum demand and the minimum night flow.

In other cases the patterns are recognisable, as the remaining examples (Figs F.2–F.81) will illustrate, taken in the UK recently. The data are included here for orientation only, and should not be used uncritically in other countries and different climatic conditions. The samples cover just a few days, usually a week, but a picture could be formed nevertheless.

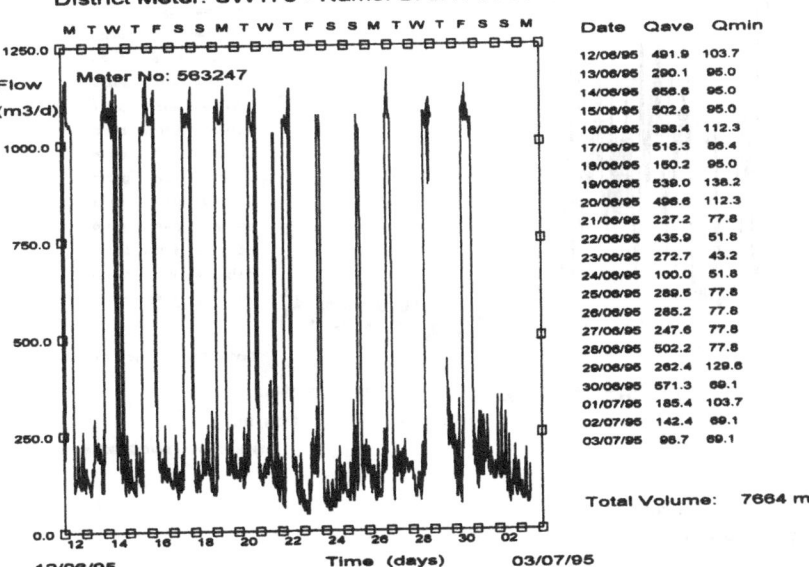

Figure F.1 Irregular pattern – a dairy.

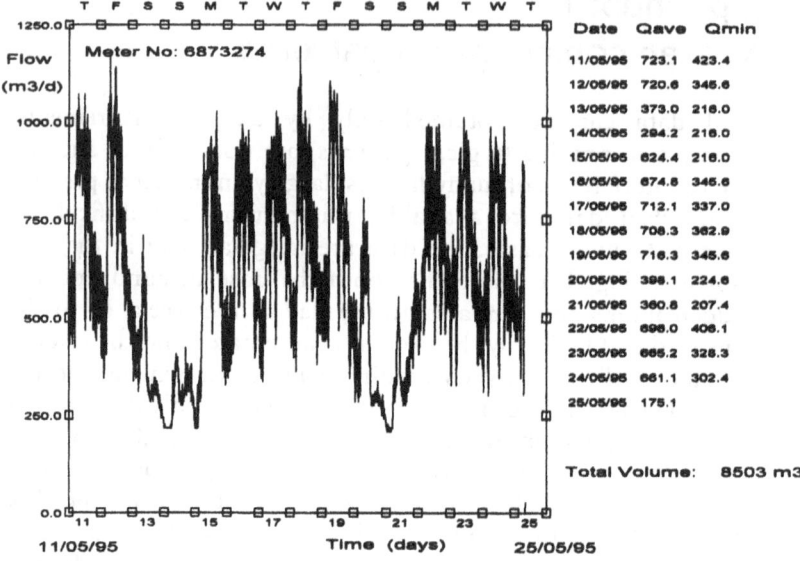

Figure F.2 Chocolate
factory – normal
operation.

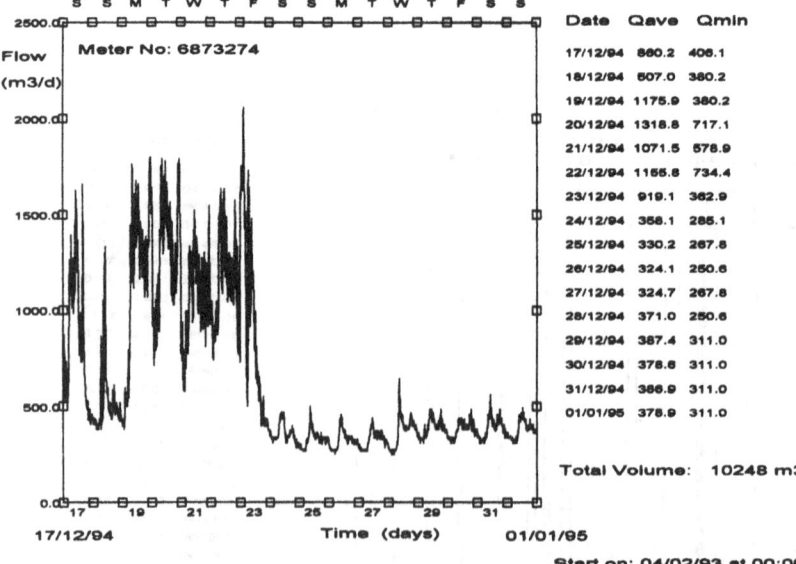

Figure F.3 Chocolate
factory – before
Christmas.

Data for Customer Meter: AW358 - Name: CHOCOLATE FACTORY

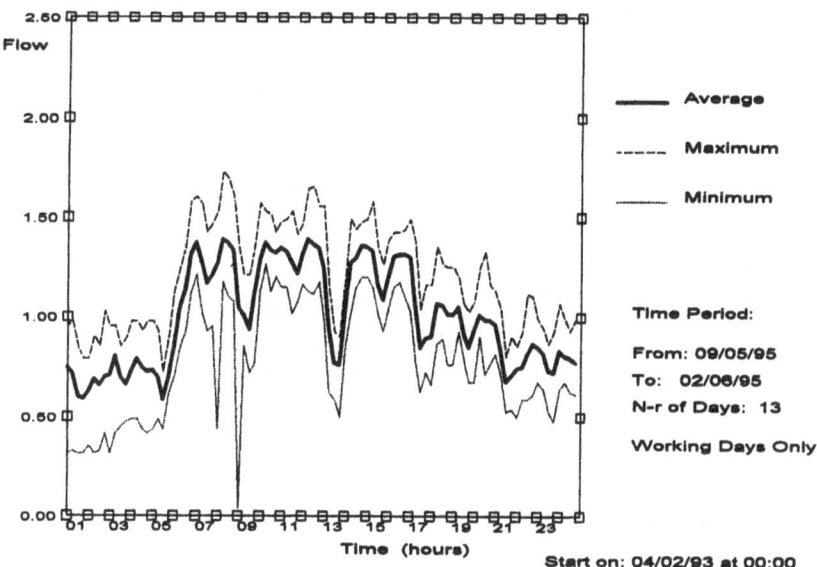

Figure F.4 Chocolate
factory – workdays
pattern.

Data for Customer Meter: AW358 - Name: CHOCOLATE FACTORY

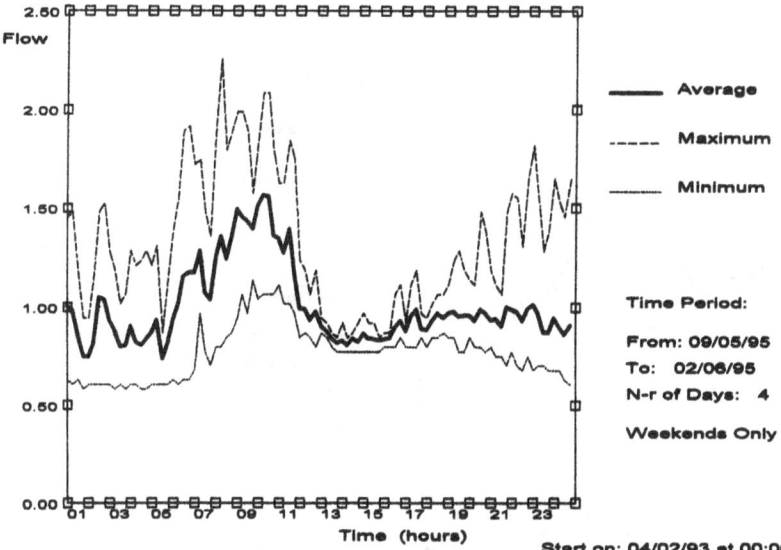

Figure F.5 Chocolate
factory – weekend days.

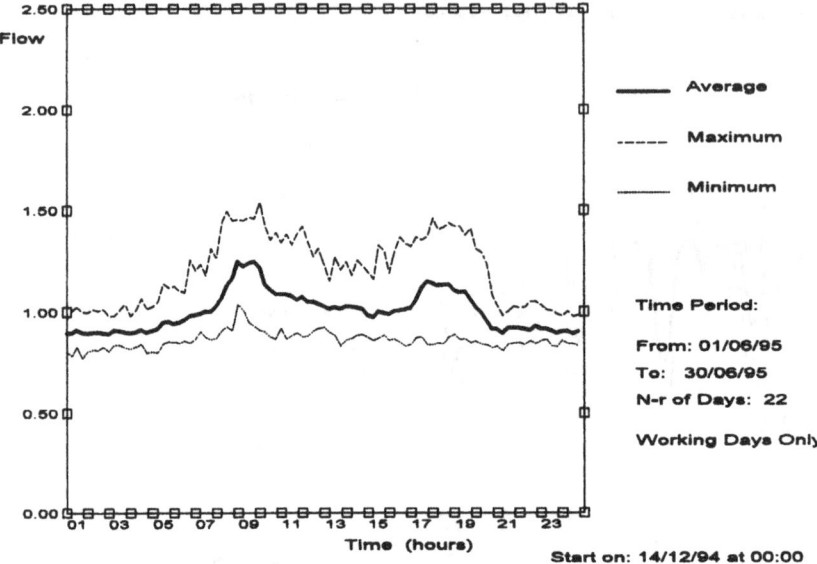

Figure F.6 Chemical
factory – workdays.

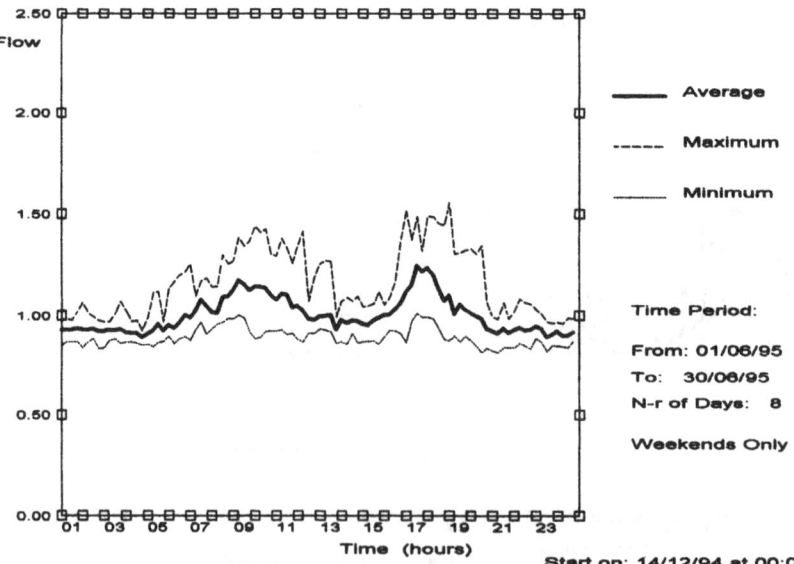

Figure F.7 Chemical
factory – weekends.

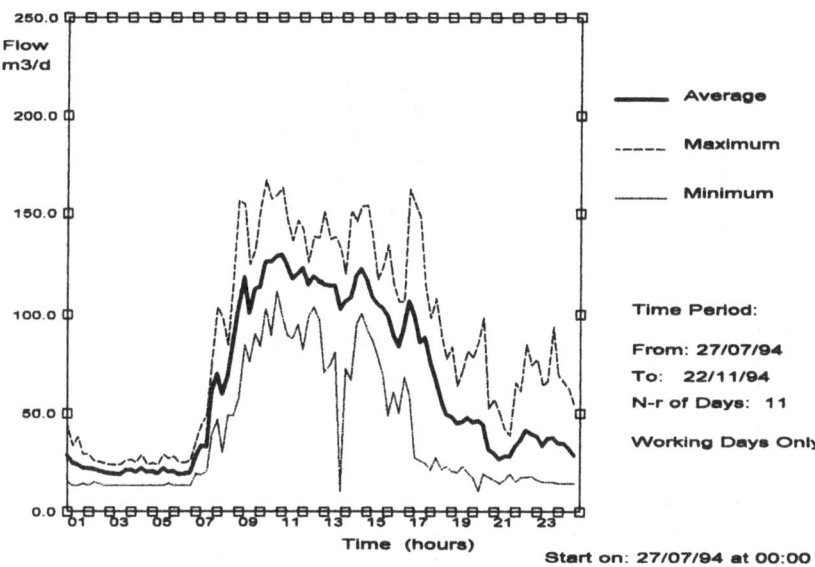

Figure F.8 Electronics
industry – workdays.

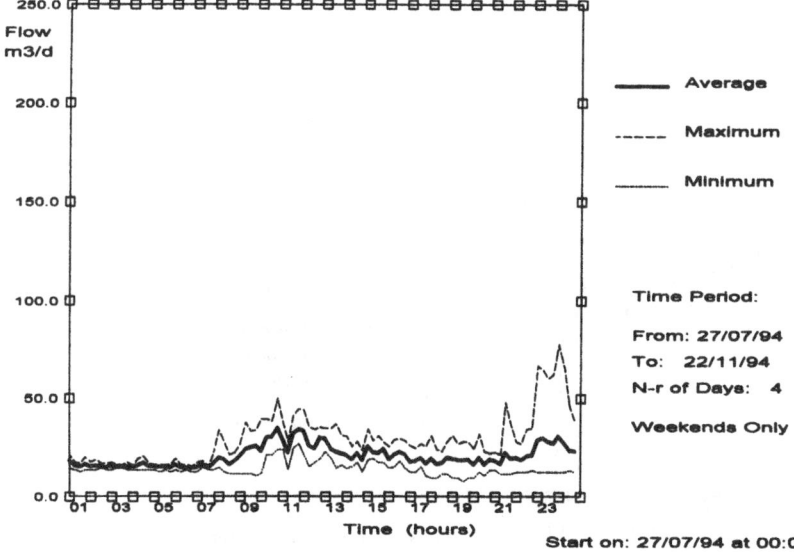

Figure F.9 Electronics
industry – weekends.

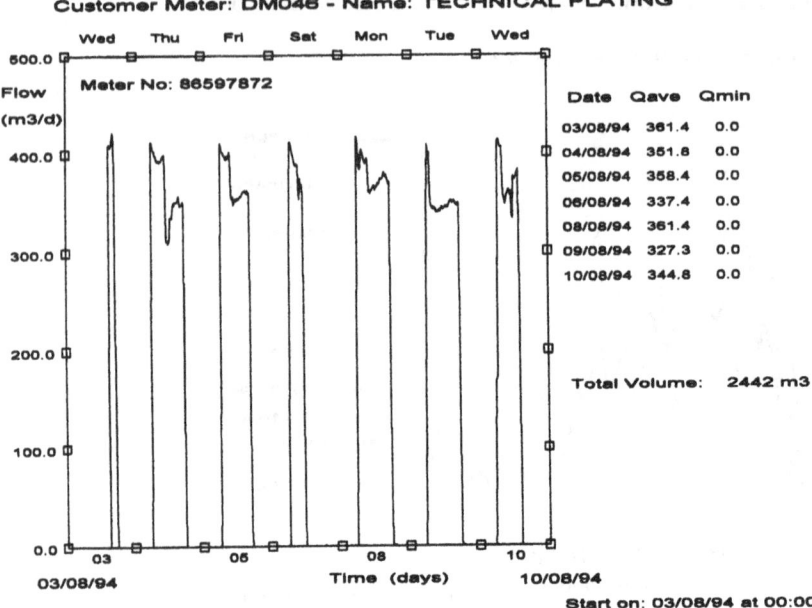

Figure F.10 Industrial
plant – the data.

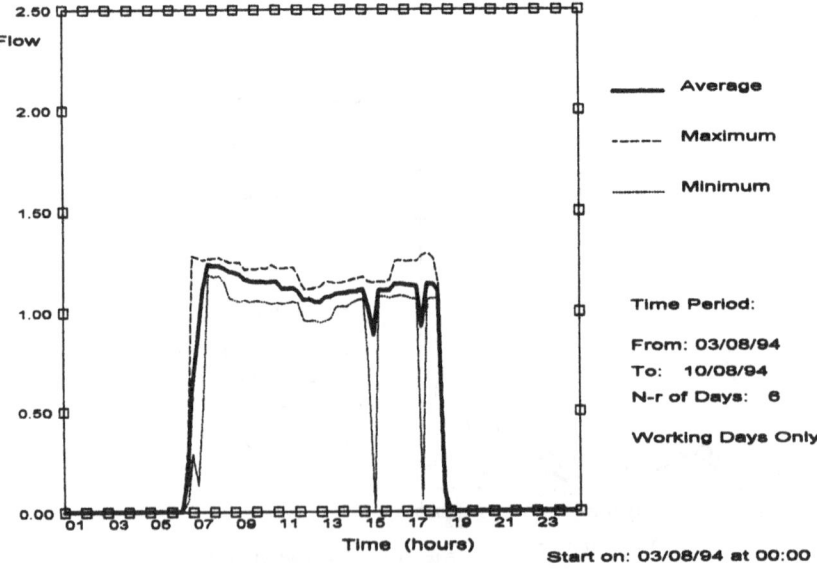

Figure F.11 Industrial
plant – workdays.

Customer Meter: DM013 - Name: THE BREWERY

Date	Qave	Qmin
28/10/94	436.6	103.7
29/10/94	229.6	109.4
30/10/94	175.2	103.7
31/10/94	367.9	123.8
01/11/94	385.5	115.2
02/11/94	410.5	103.7
03/11/94	435.8	149.8
04/11/94	358.3	106.6
05/11/94	235.4	100.8
06/11/94	168.3	118.1
07/11/94	401.7	97.9
08/11/94	391.7	0.0

Total Volume: 3997 m3

Start on: 28/10/94 at 00:00

Figure F.12 A brewery
– the data.

Data for Customer Meter: DM013 - Name: THE BREWERY

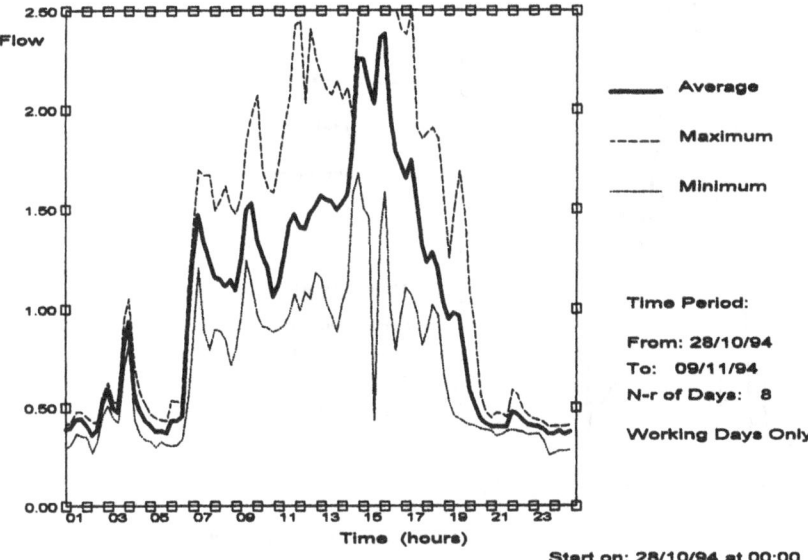

——— Average

- - - - Maximum

——— Minimum

Time Period:

From: 28/10/94
To: 09/11/94
N-r of Days: 8

Working Days Only

Start on: 28/10/94 at 00:00

Figure F.13 A brewery
– the workdays pattern.

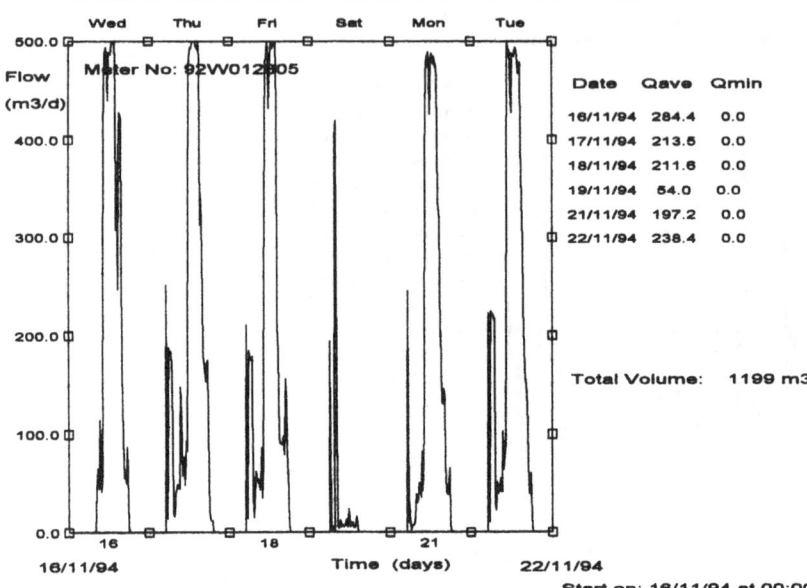

Figure F.14 A bacon
factory – the data.

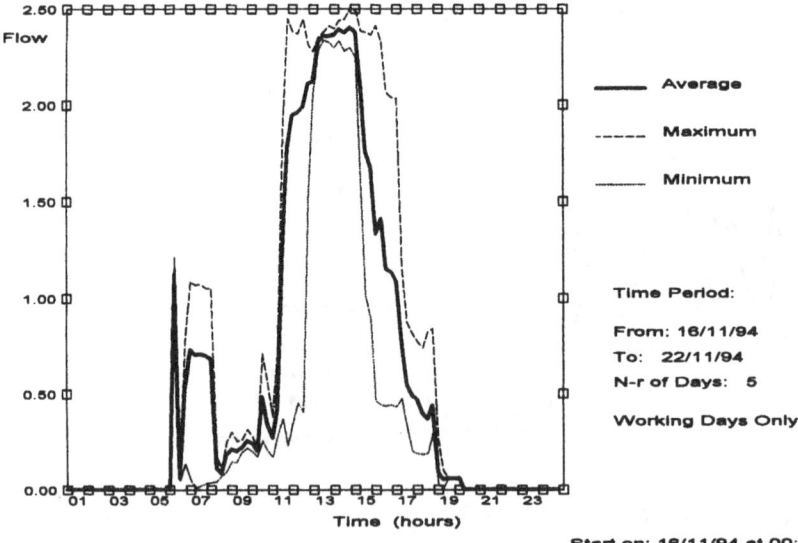

Figure F.15 A bacon
factory – the workdays
pattern.

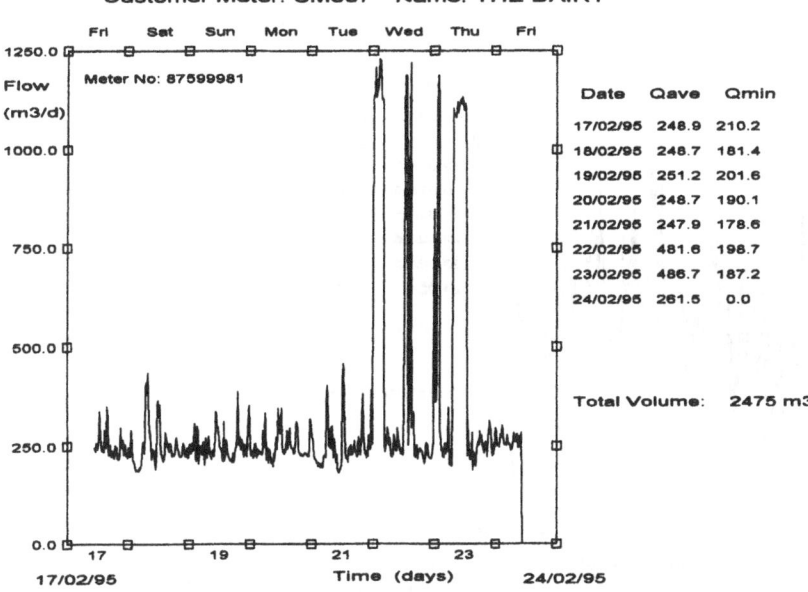

Figure F.16 A dairy –
the data.

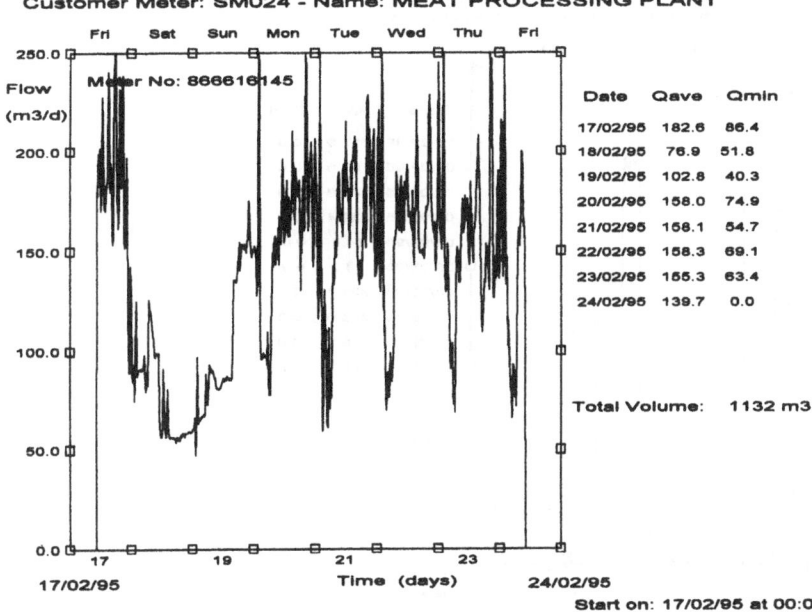

Figure F.17 A meat
processing plant – the
data.

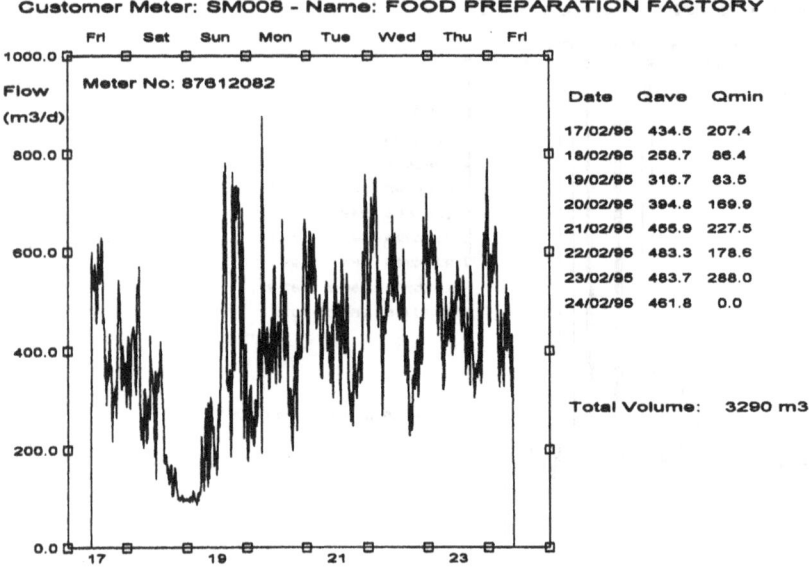

Figure F.18 A food preparation factory – the data.

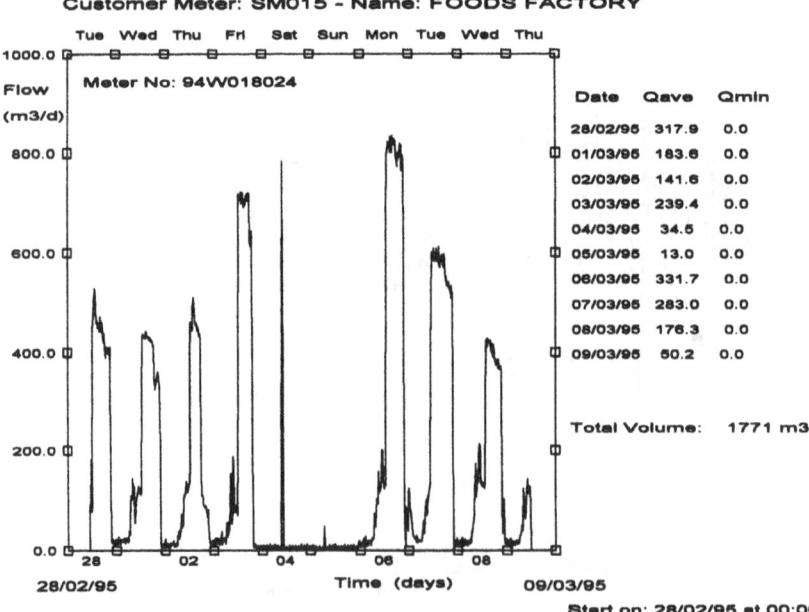

Figure F.19 A food producing factory.

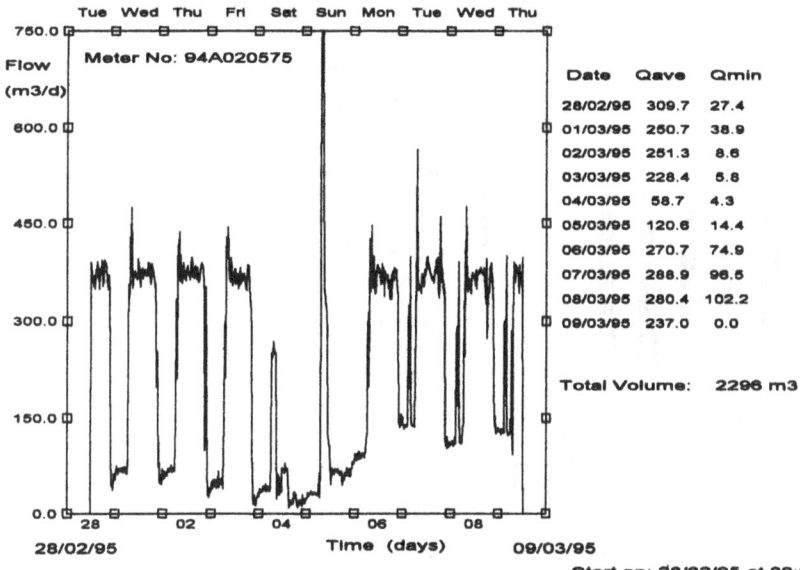

Figure F.20 Food
production – the data.

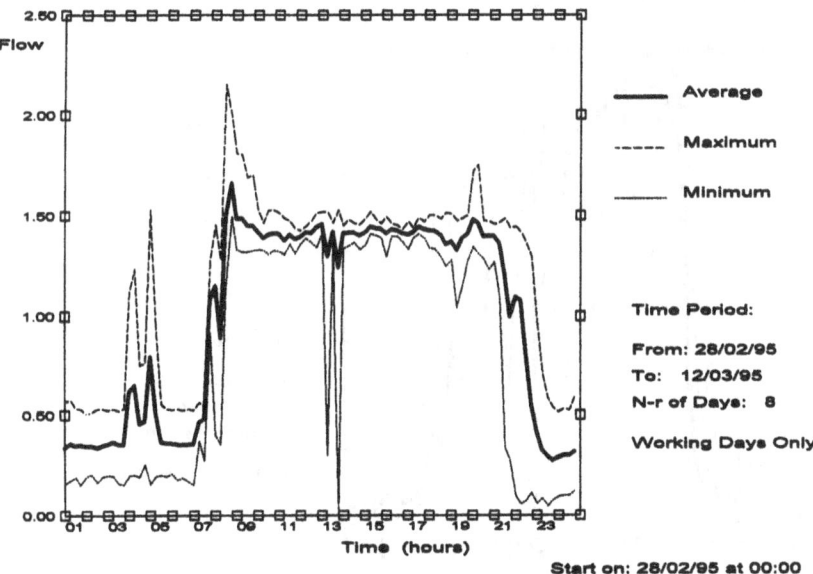

Figure F.21 Food
production – the work-
days pattern.

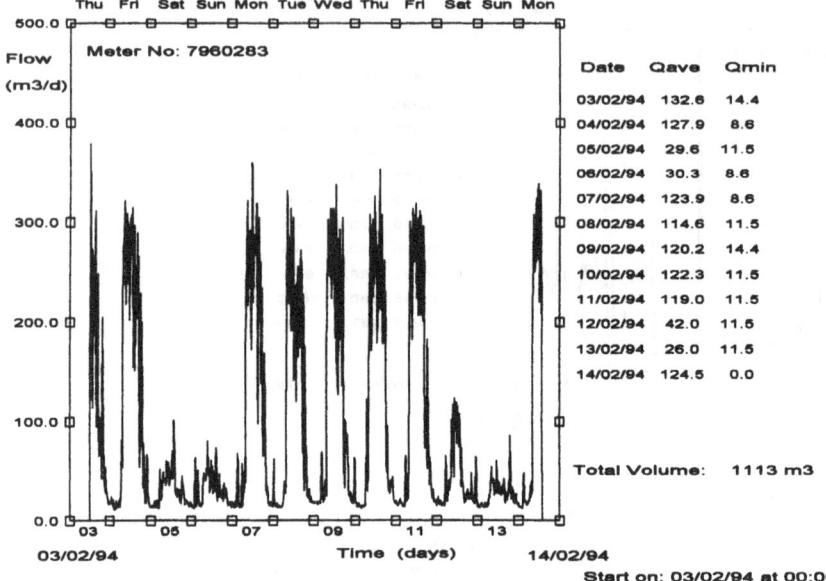

Figure F.22 A hospital –
the data.

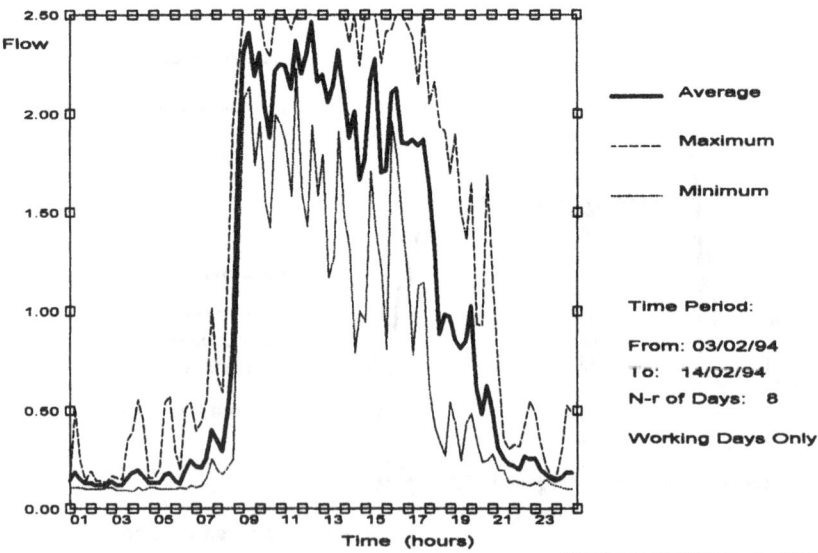

Figure F.23 A hospital –
the workdays pattern.

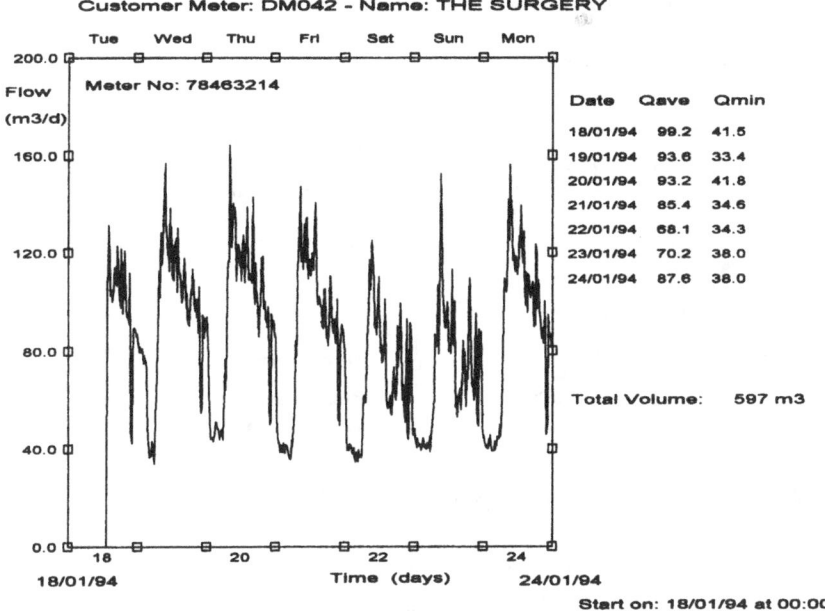

Figure F.24 A surgery –
the data.

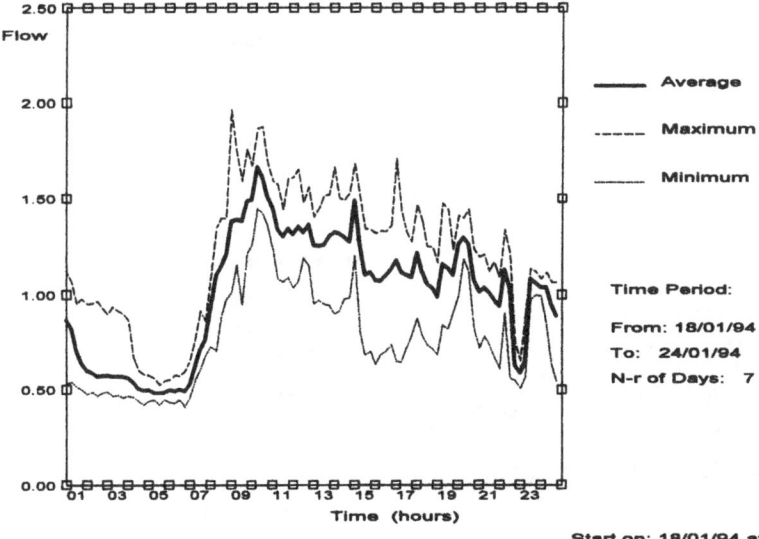

Figure F.25 A surgery –
the all-days pattern.

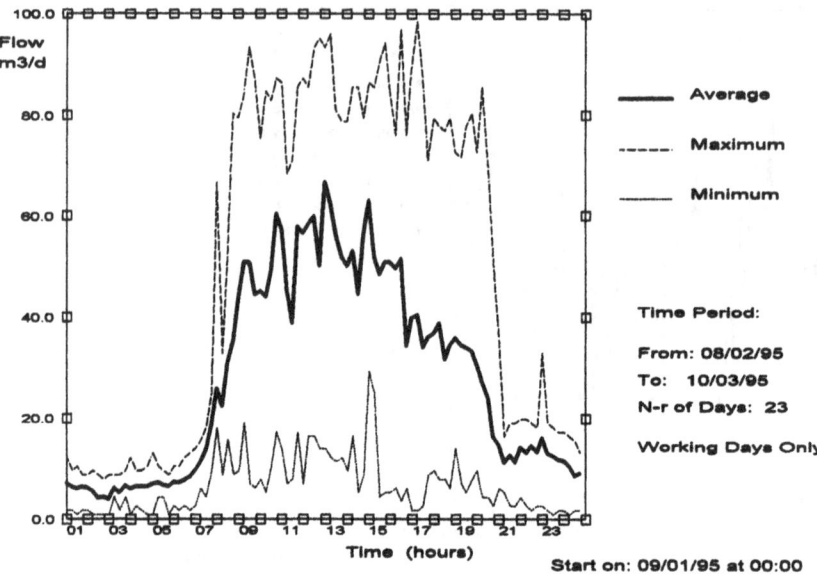

Figure F.26 Hospital
pattern – workdays.

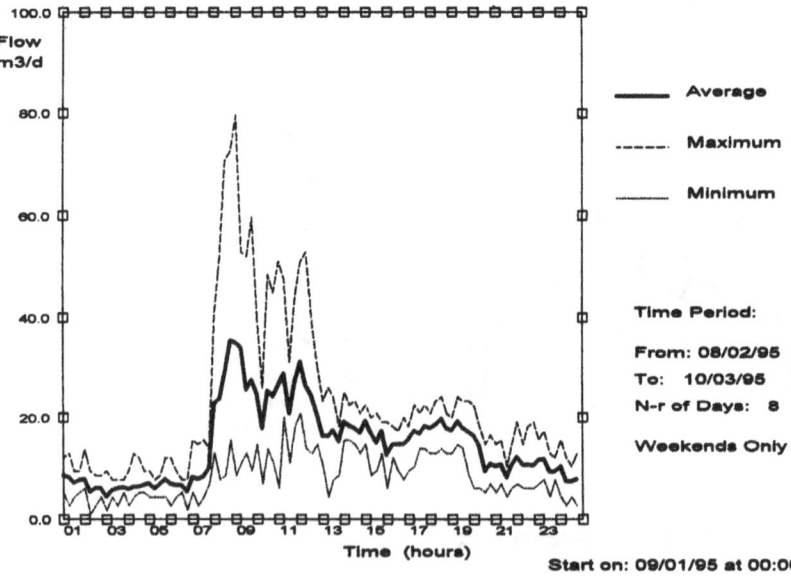

Figure F.27 Hospital
consumption on week-
ends.

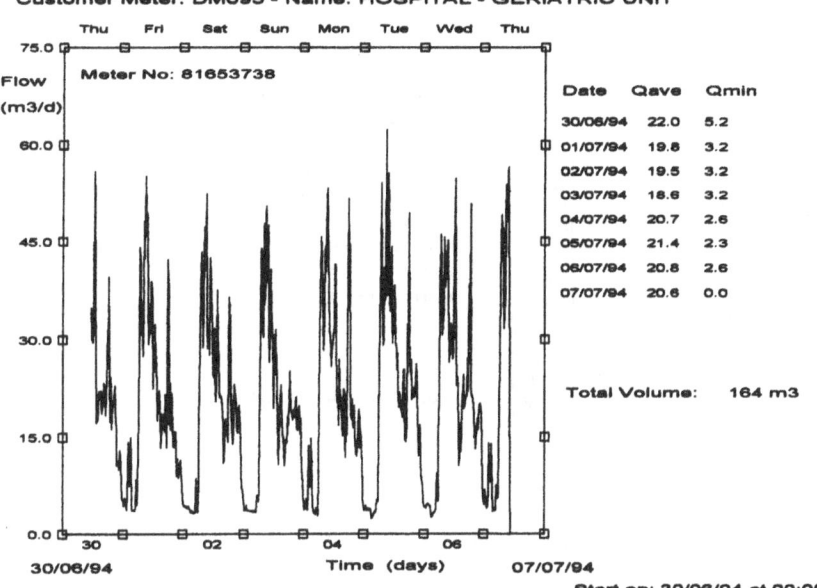

Figure F.28 A unit in a hospital – the data.

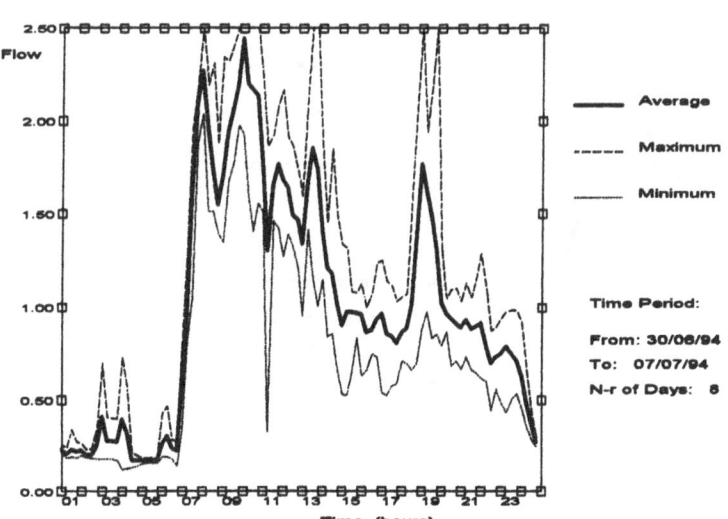

Figure F.29 A unit in a hospital – the pattern.

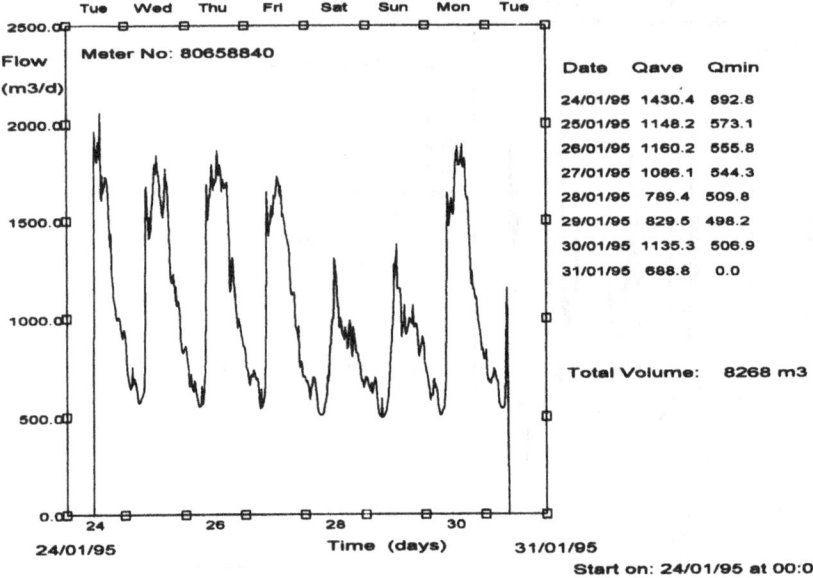

Figure F.30 A university
– the data.

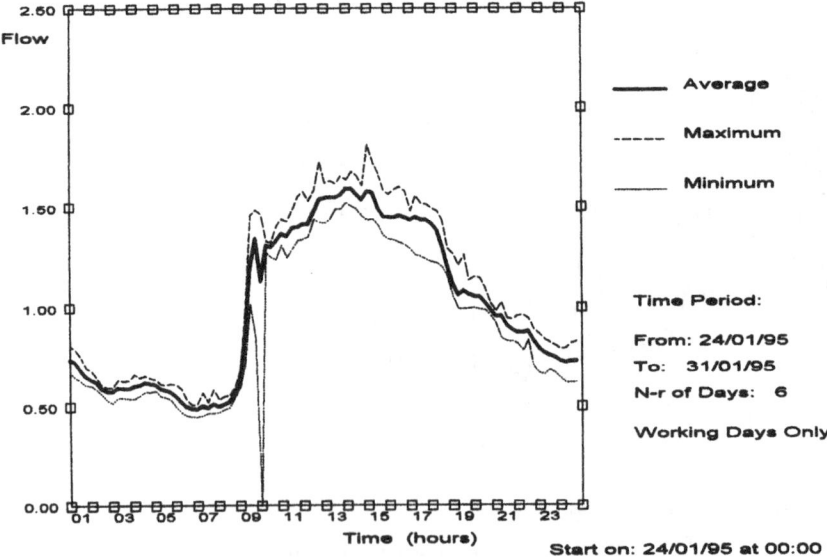

Figure F.31 A university
– the workdays pattern.

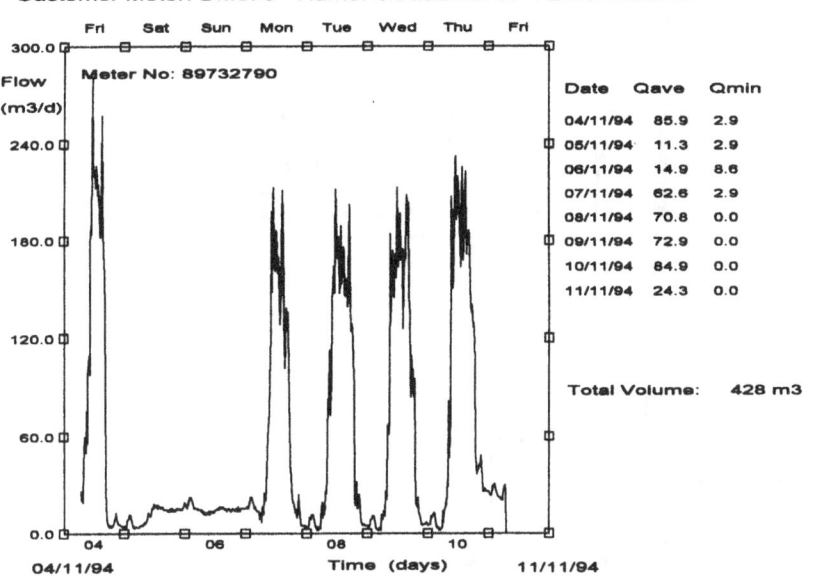

Figure F.32 A college –
the data.

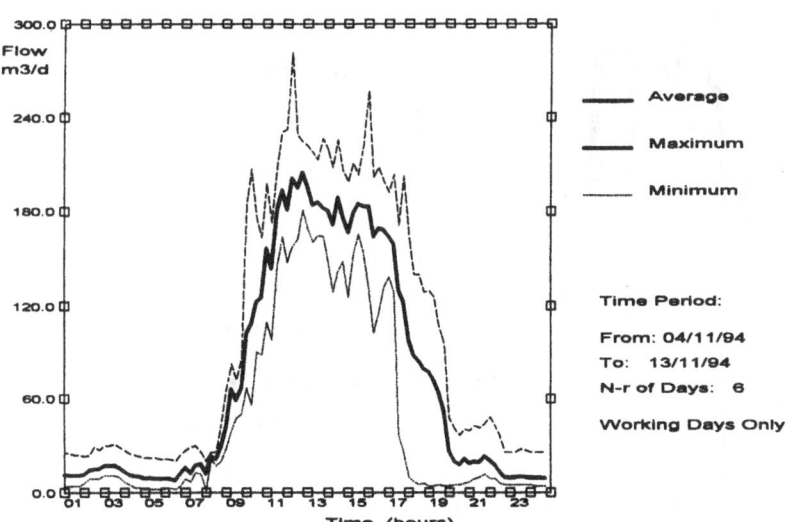

Figure F.33 A college –
workdays pattern.

Customer Meter: SM031 - Name: TECHNICAL COLLEGE

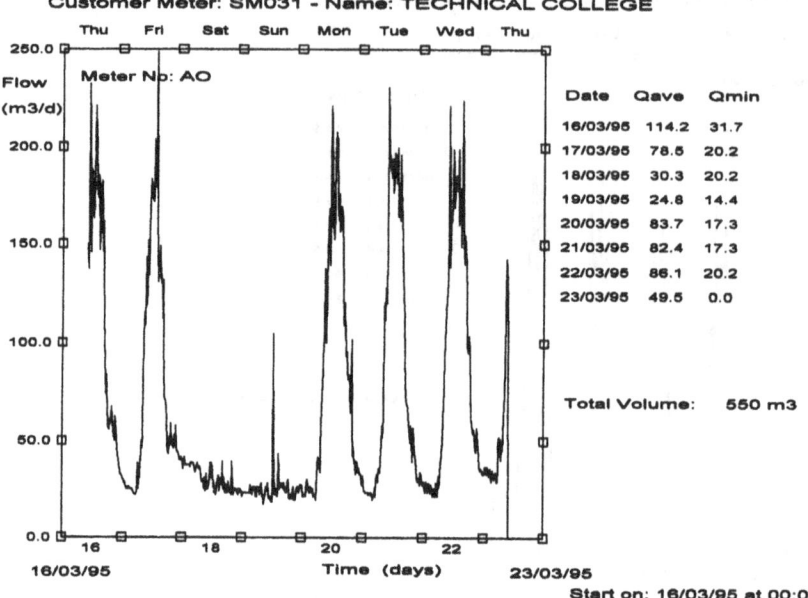

Date	Qave	Qmin
16/03/95	114.2	31.7
17/03/95	78.5	20.2
18/03/95	30.3	20.2
19/03/95	24.8	14.4
20/03/95	83.7	17.3
21/03/95	82.4	17.3
22/03/95	86.1	20.2
23/03/95	49.5	0.0

Total Volume: 550 m3

Start on: 16/03/95 at 00:00

Figure F.34 Another college – the data (II).

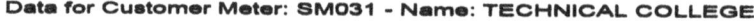

Data for Customer Meter: SM031 - Name: TECHNICAL COLLEGE

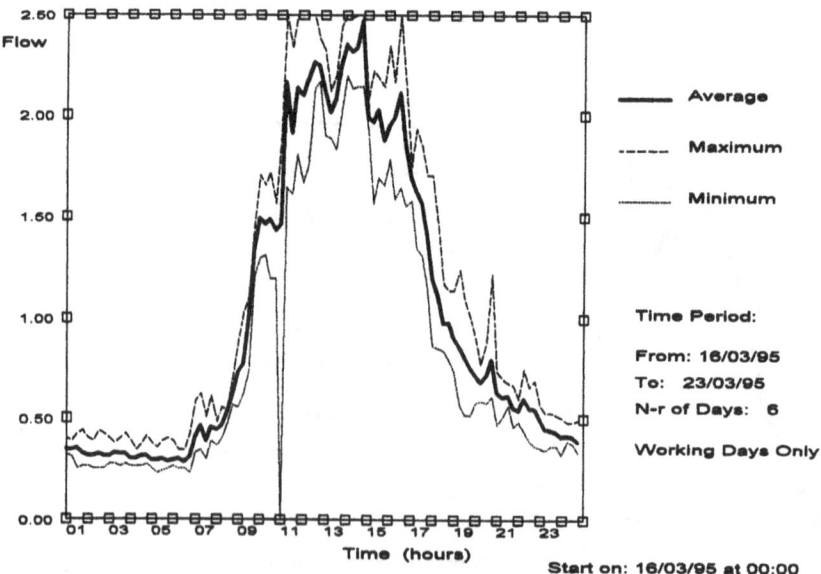

——— Average

----- Maximum

——— Minimum

Time Period:

From: 16/03/95
To: 23/03/95
N-r of Days: 6

Working Days Only

Start on: 16/03/95 at 00:00

Figure F.35 The college – workdays pattern (II).

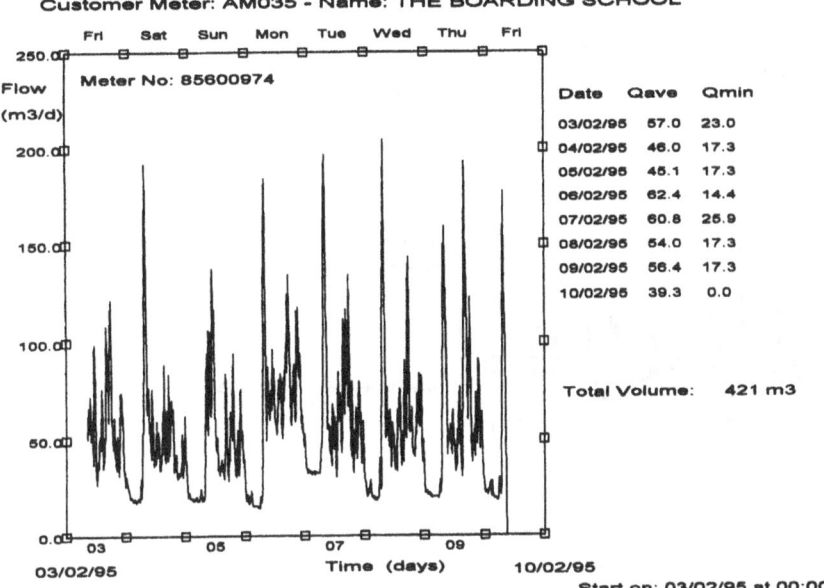

Figure F.36 A boarding school – the data.

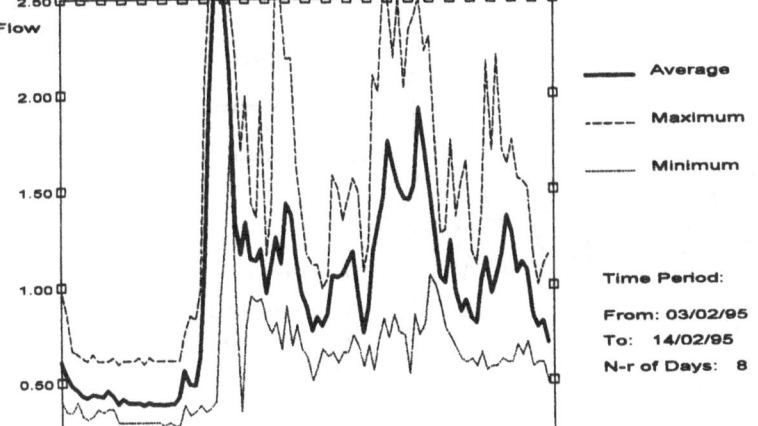

Figure F.37 The boarding school pattern.

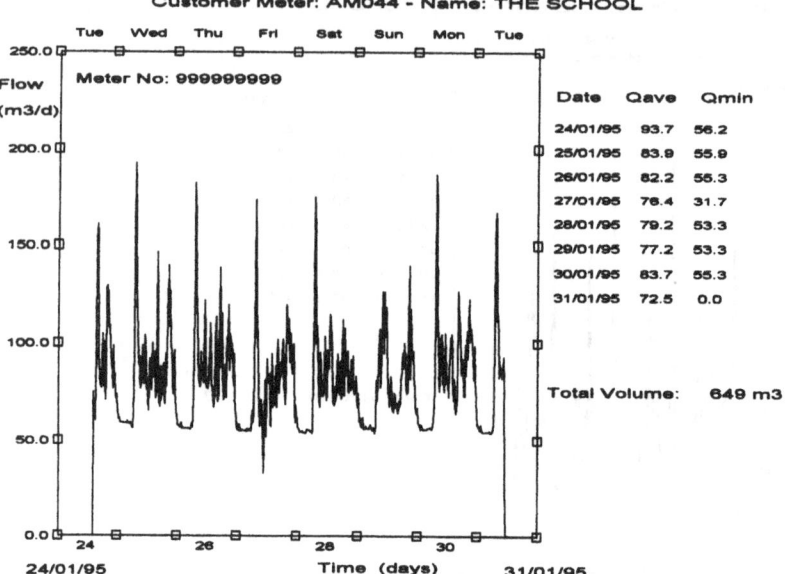

Figure F.38 A school –
the data.

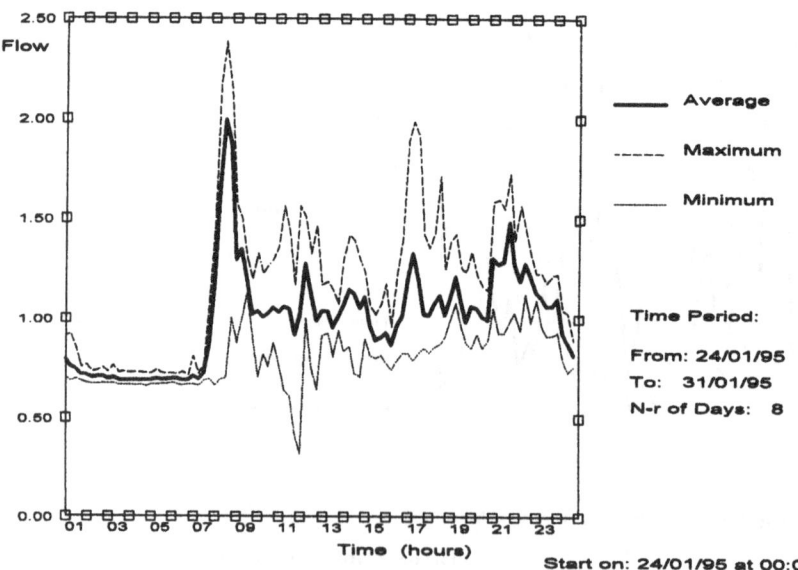

Figure F.39 A school –
the all-days pattern.

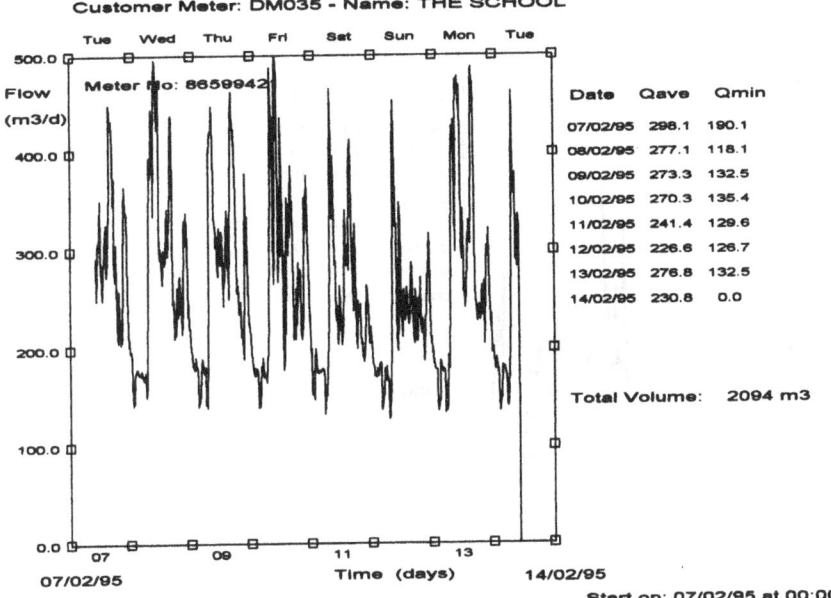

Figure F.40 Another school – the data (II).

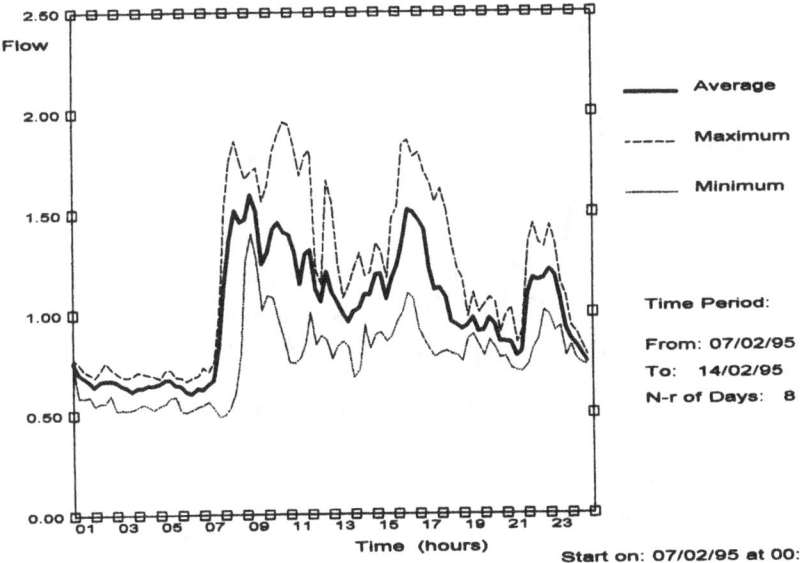

Figure F.41 The school pattern (II).

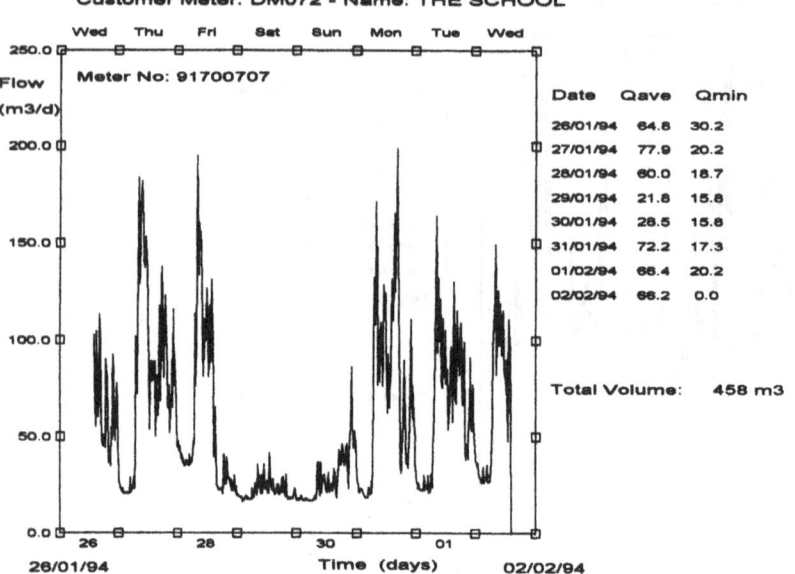

Figure F.42 A primary
school – the data.

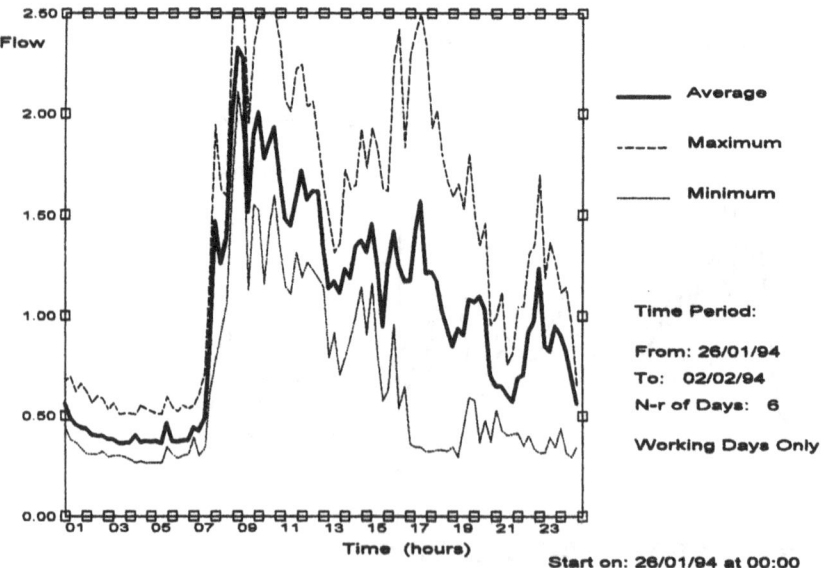

Figure F.43 The
primary school – the
workdays pattern.

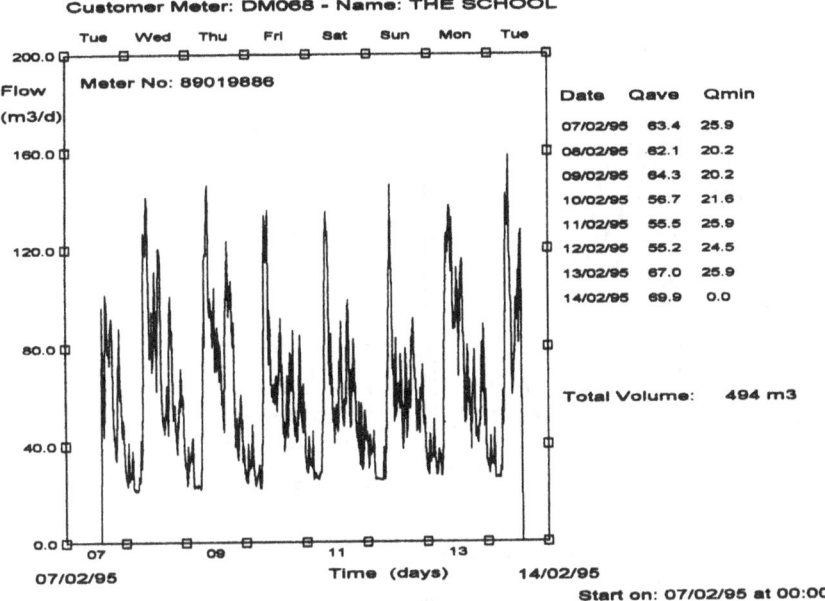

Figure F.44 A boarding
school – the data.

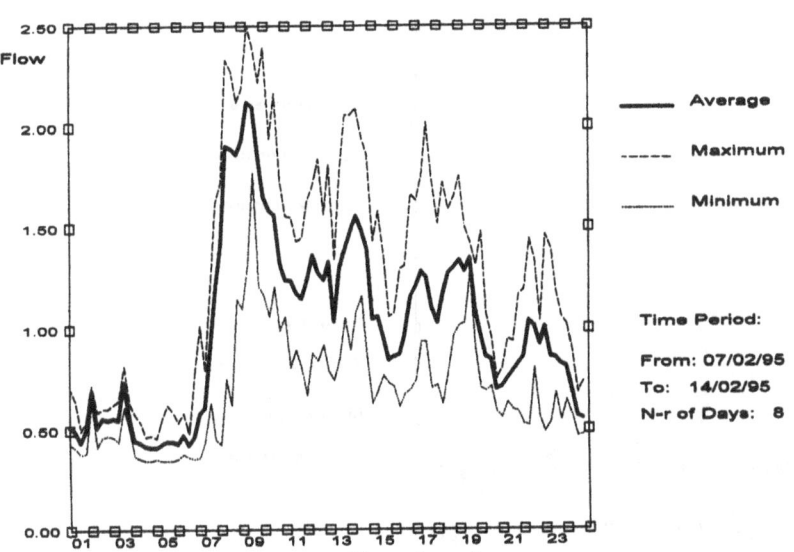

Figure F.45 The boarding school pattern.

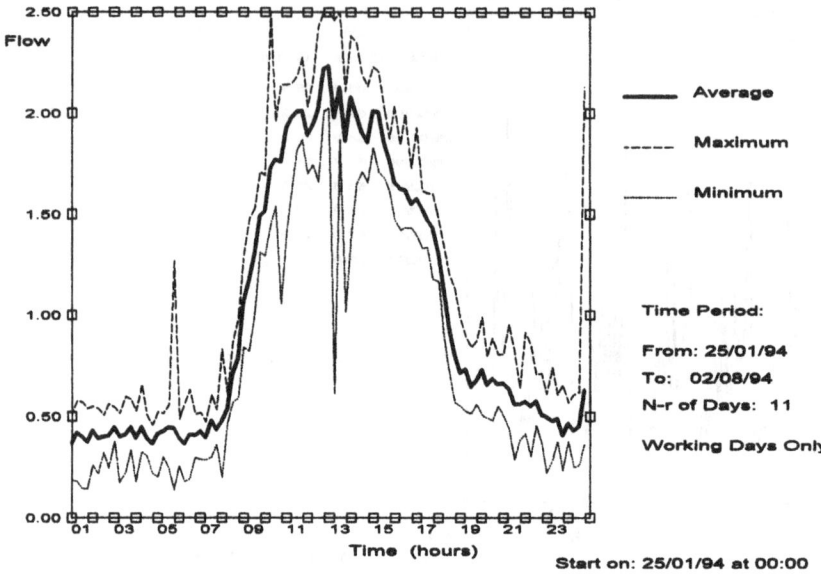

Figure F.46 A bank –
workdays.

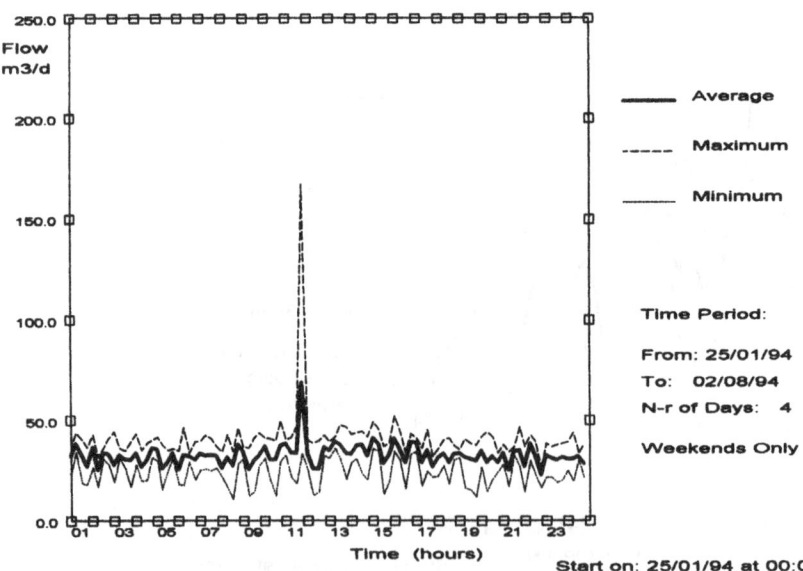

Figure F.47 A bank –
weekend data.

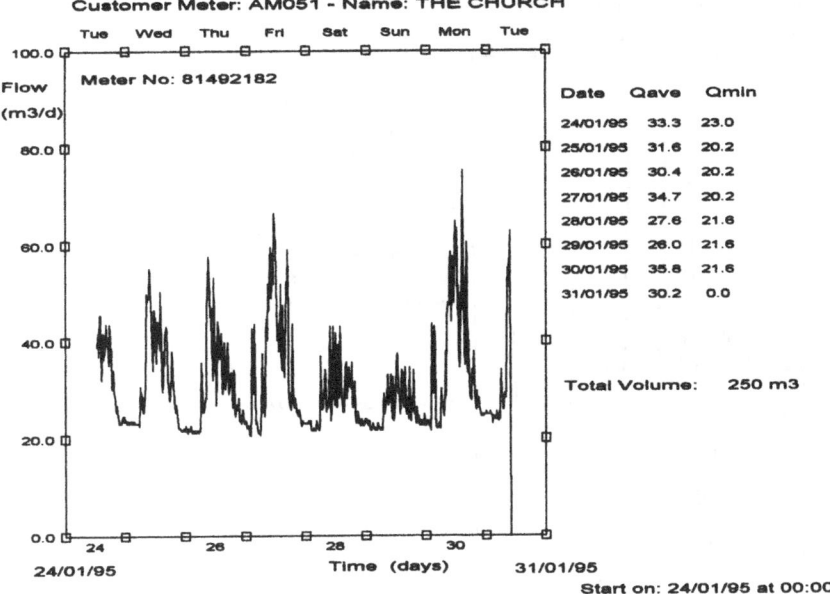

Figure F.48 A church —
the data.

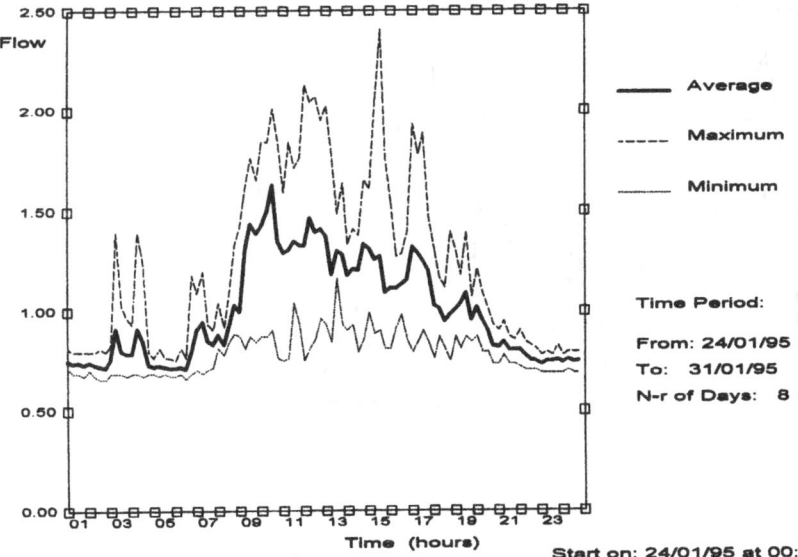

Figure F.49 The church
pattern.

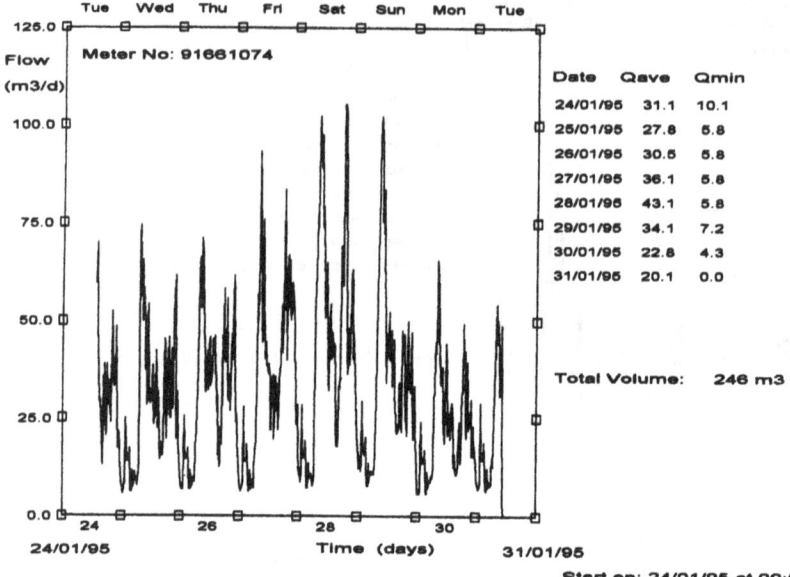

Figure F.50 A hotel –
the data.

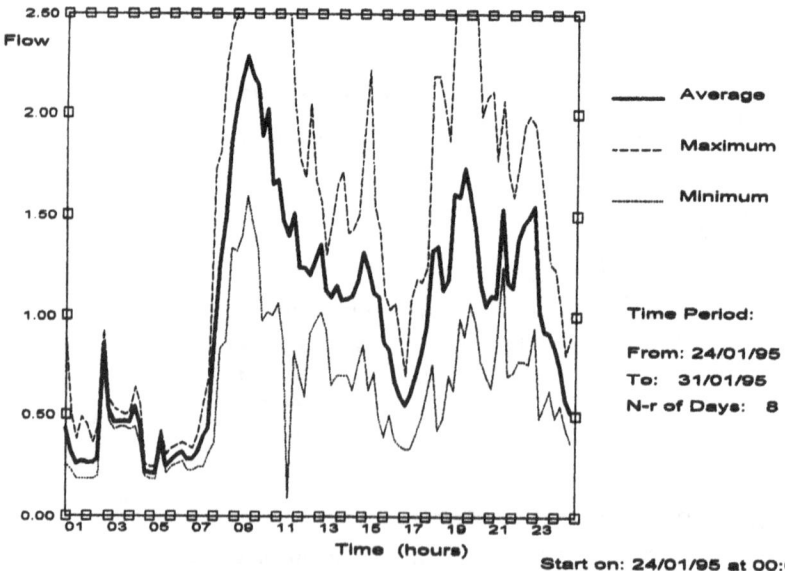

Figure F.51 The hotel
all-days pattern.

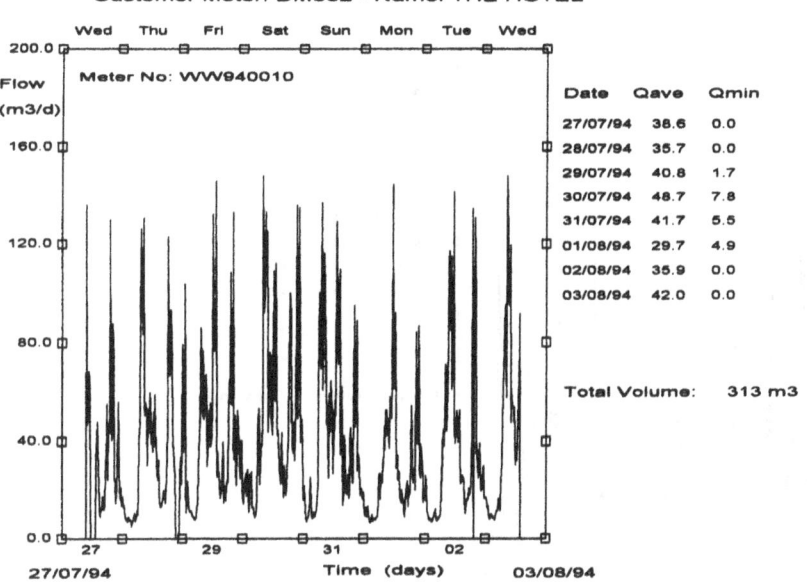

Figure F.52 Another hotel – the data (II).

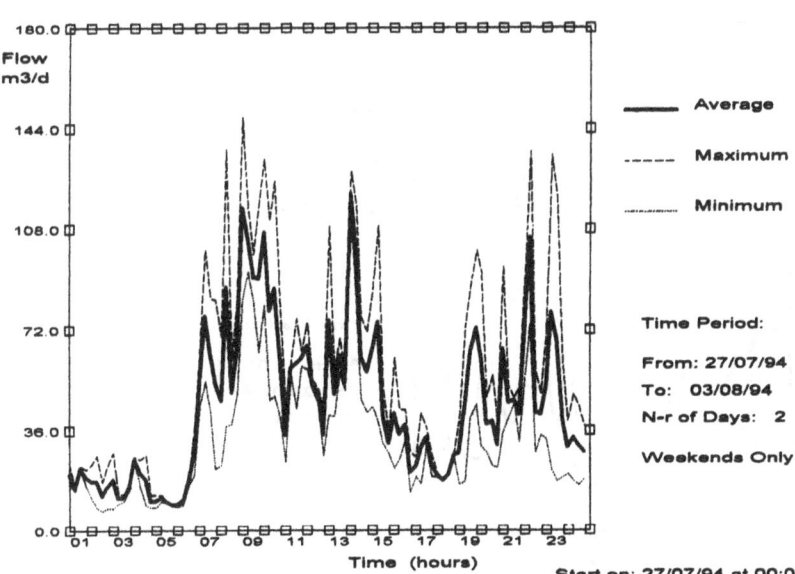

Figure F.53 The hotel weekends pattern (II).

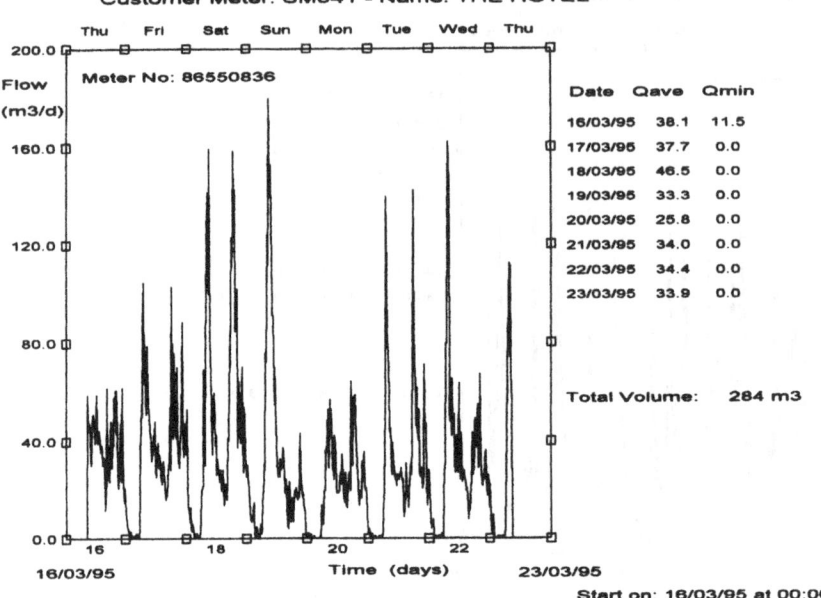

Figure F.54 Another
hotel – the data (III).

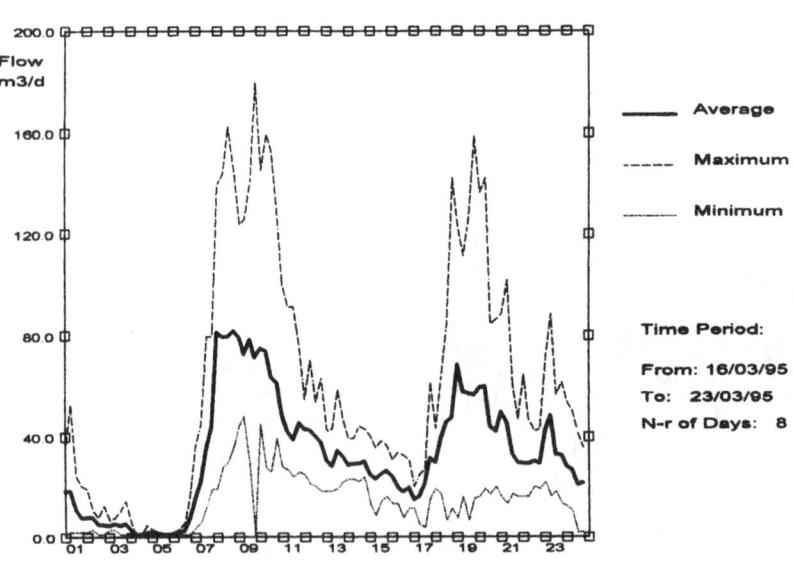

Figure F.55 The hotel
all-days pattern (III).

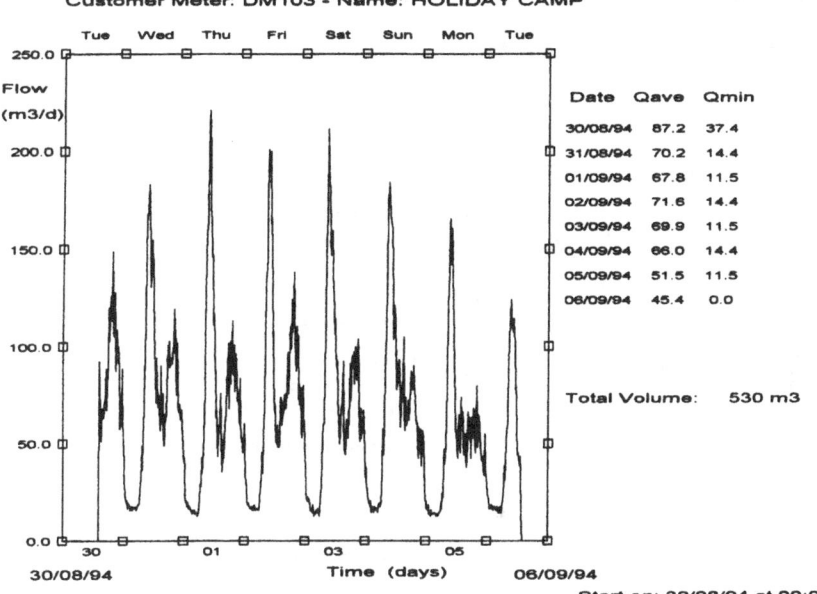

Figure F.56 A holiday
camp – the data.

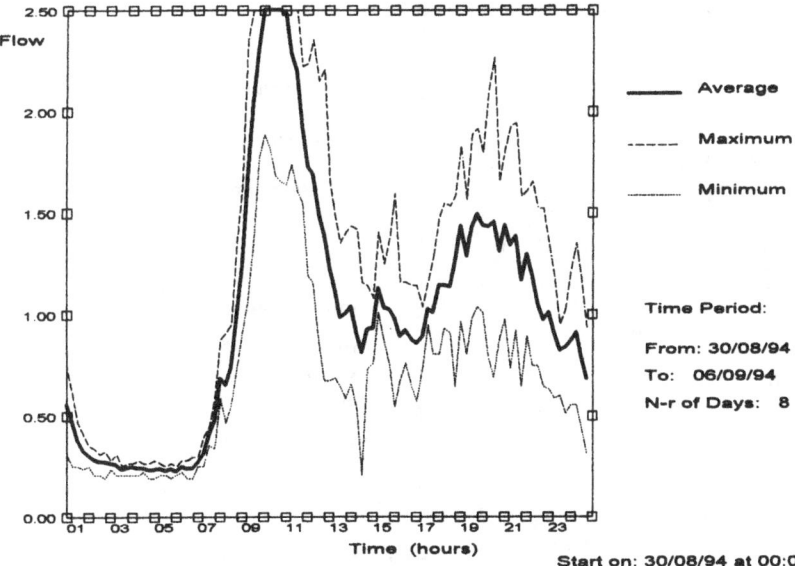

Figure F.57 The holiday
camp pattern.

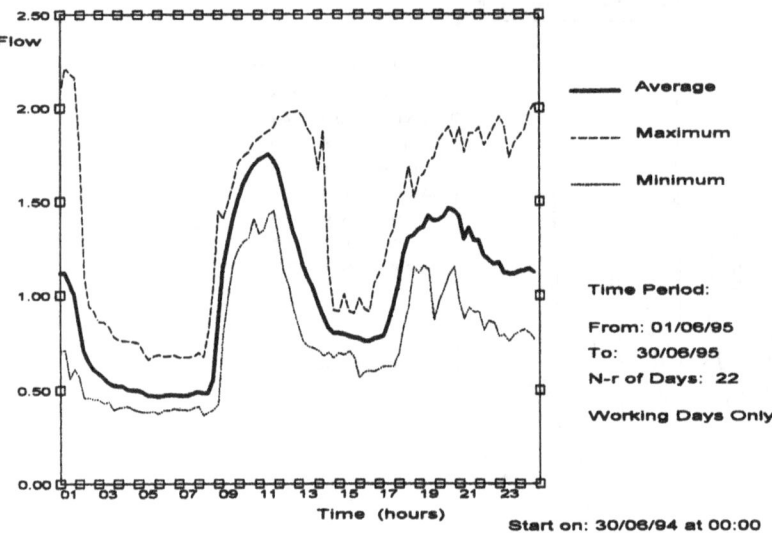

Figure F.58 Holiday
park workdays pattern.

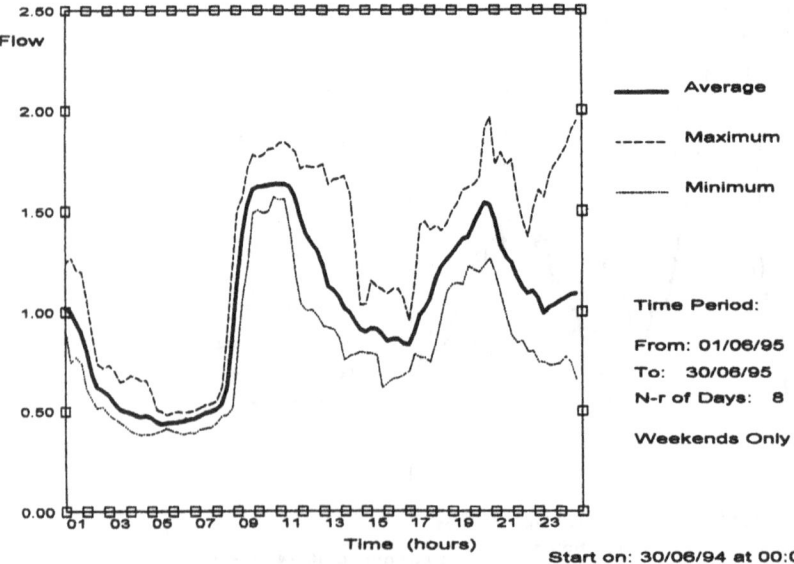

Figure F.59 Holiday
park weekend pattern.

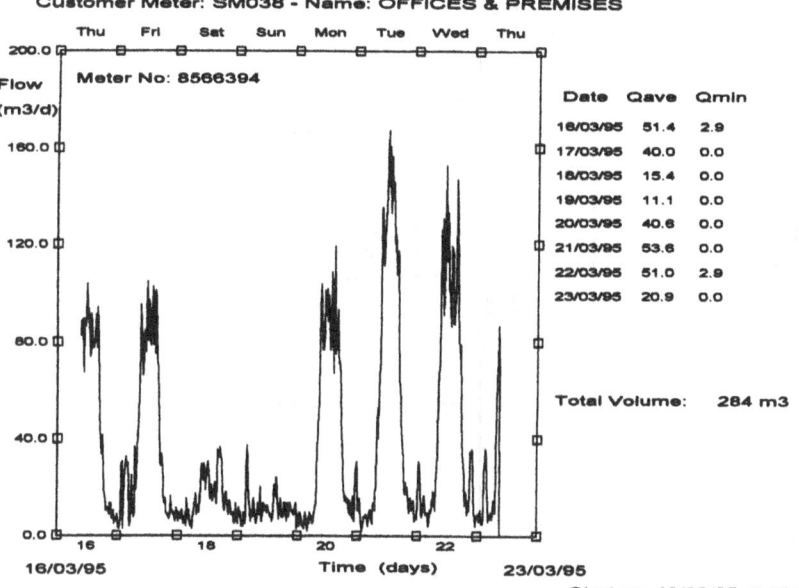

Figure F.60 Offices and
premises – the data.

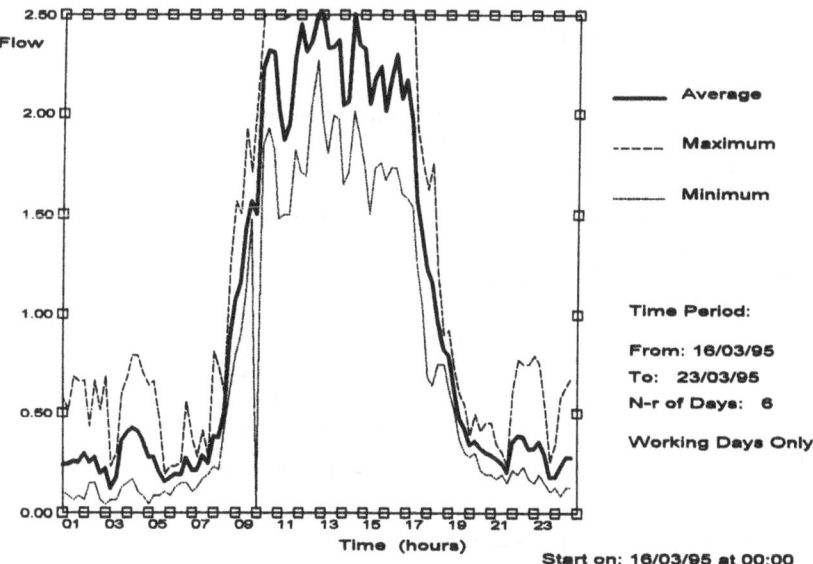

Figure F.61 Offices and
premises – the workdays
pattern.

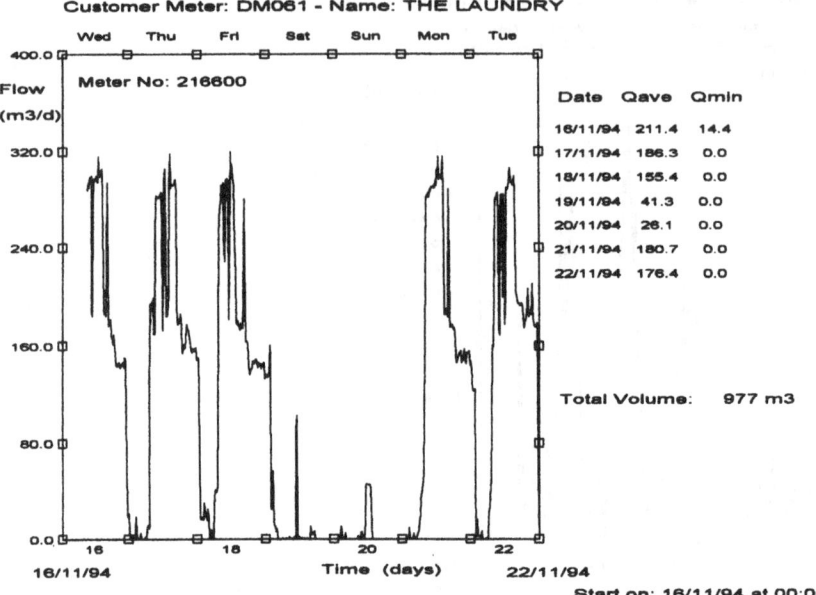

Figure F.62 A laundry –
the data.

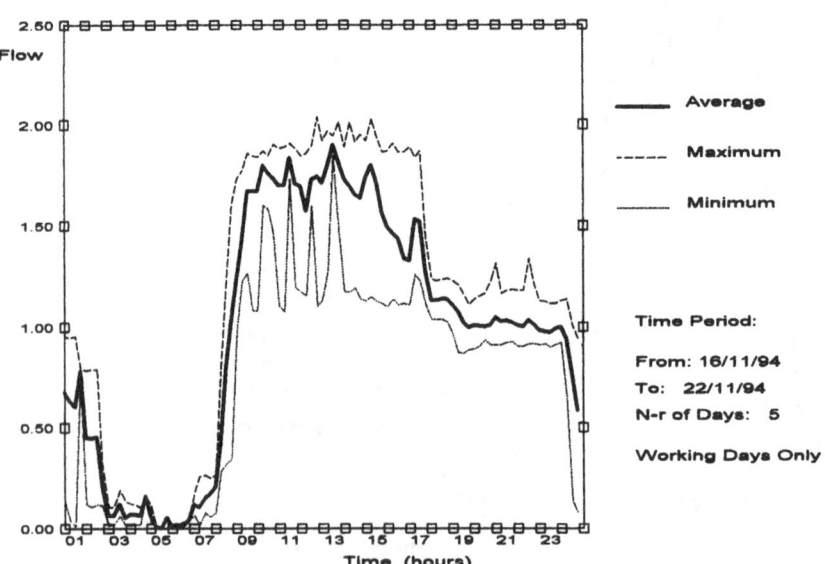

Figure F.63 The laundry
pattern.

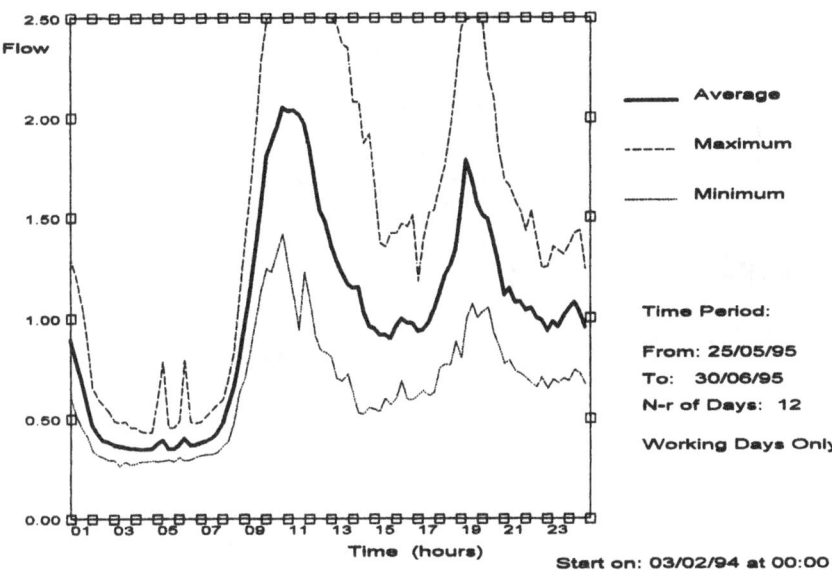

Figure F.64 A caravan
park workdays pattern.

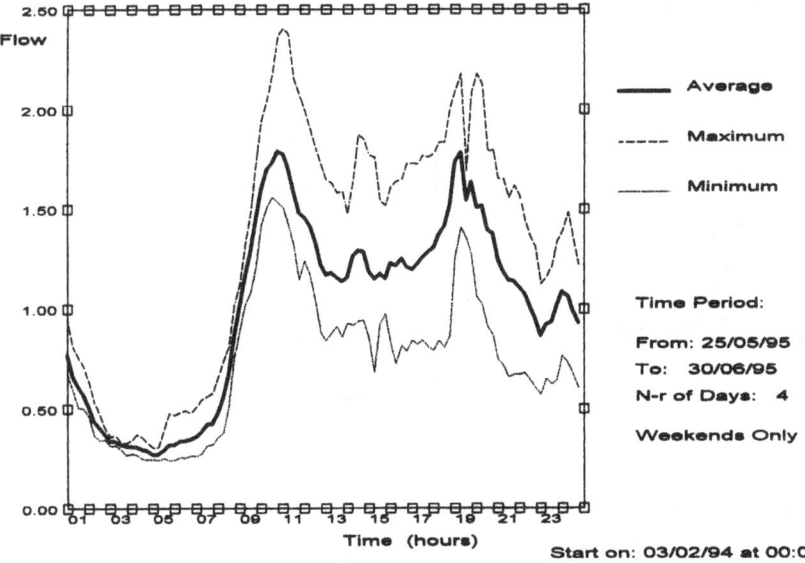

Figure F.65 A caravan
park weekend pattern.

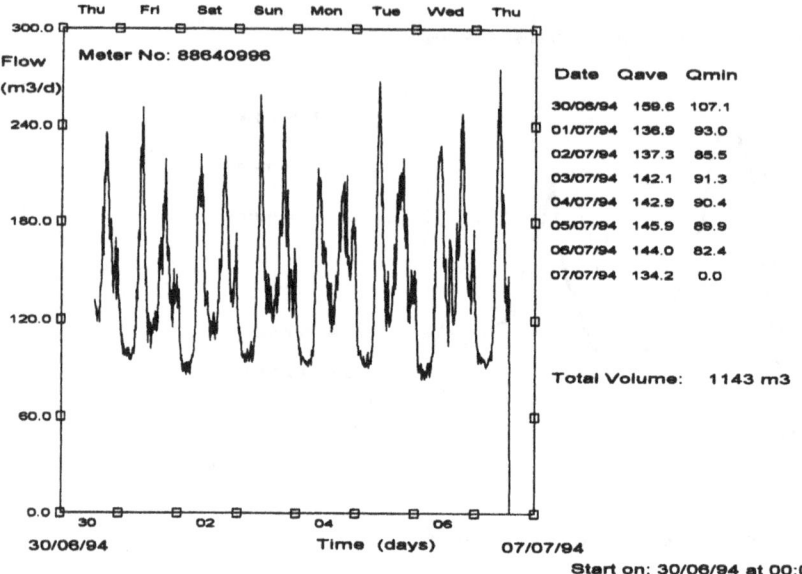

Figure F.66 A holiday
park – the data.

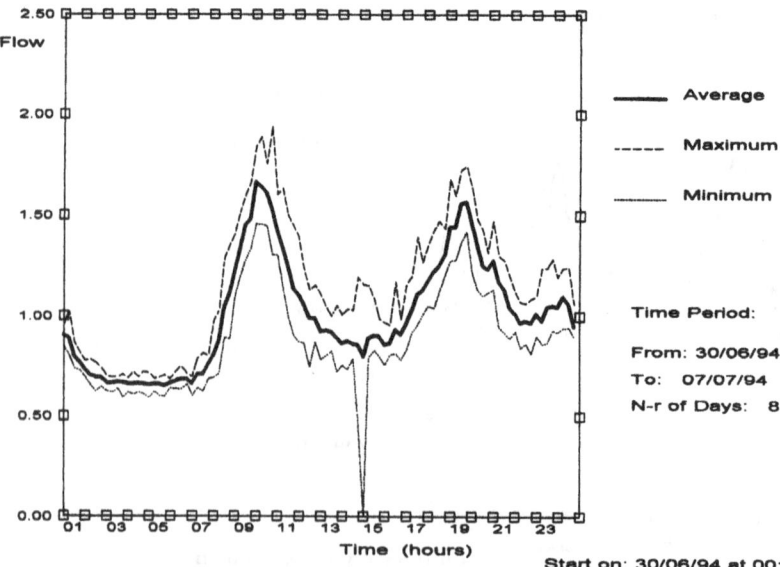

Figure F.67 The holiday
park pattern.

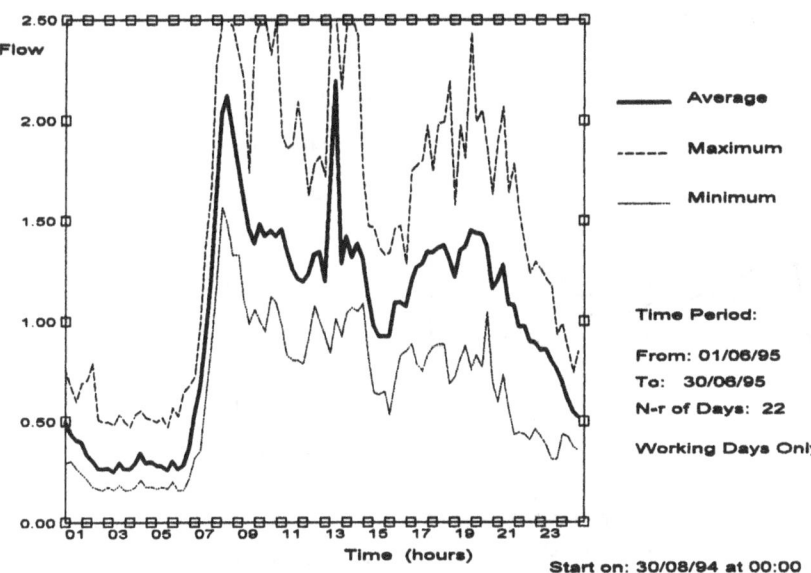

Figure F.68 A military
camp – the workdays
pattern.

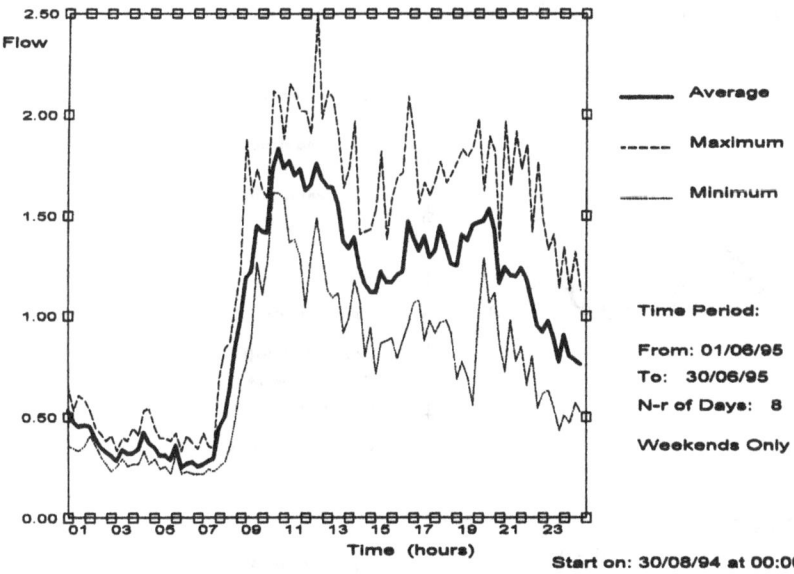

Figure F.69 The
military camp weekend
pattern.

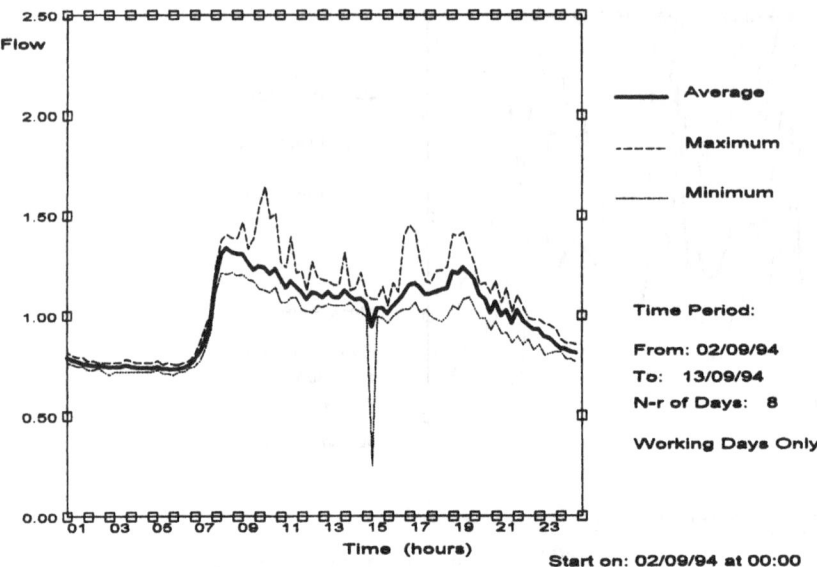

Figure F.70 The
military barracks –
the workdays pattern.

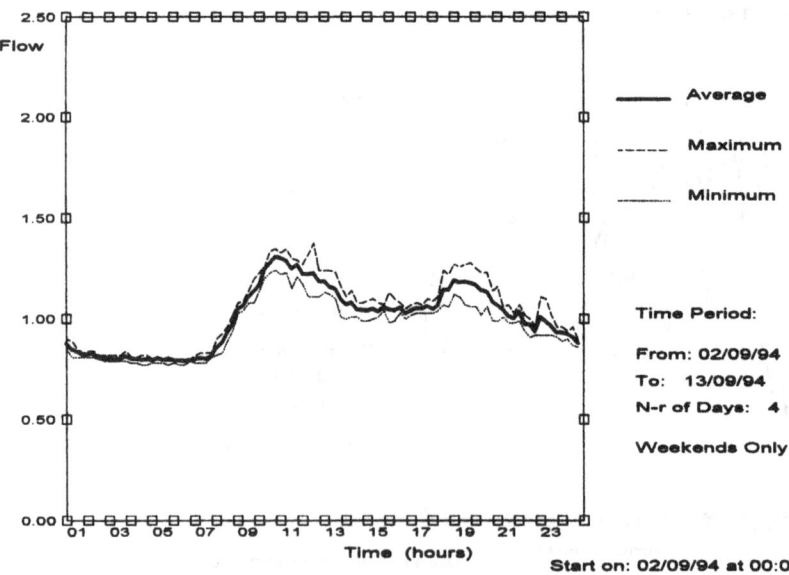

Figure F.71 The mili-
tary barracks – the
weekend pattern.

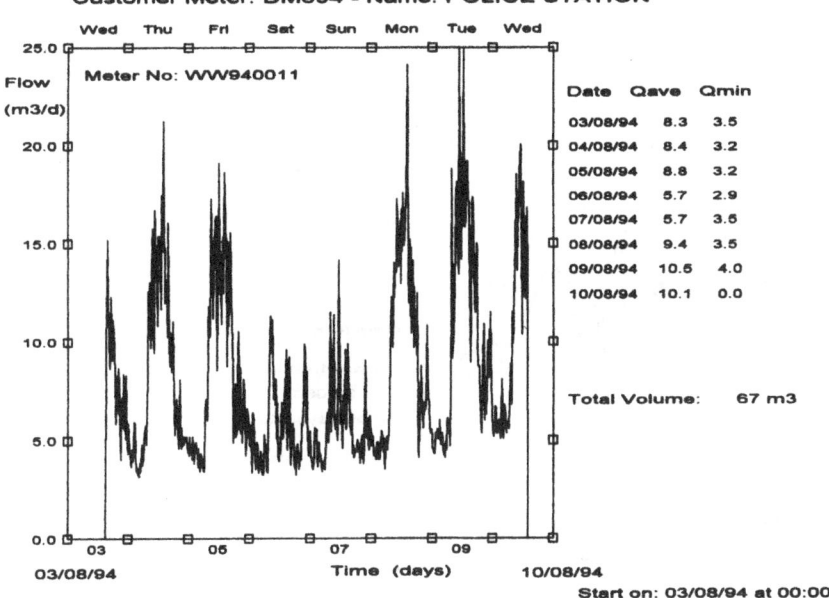

Figure F.72 A police
station – the data.

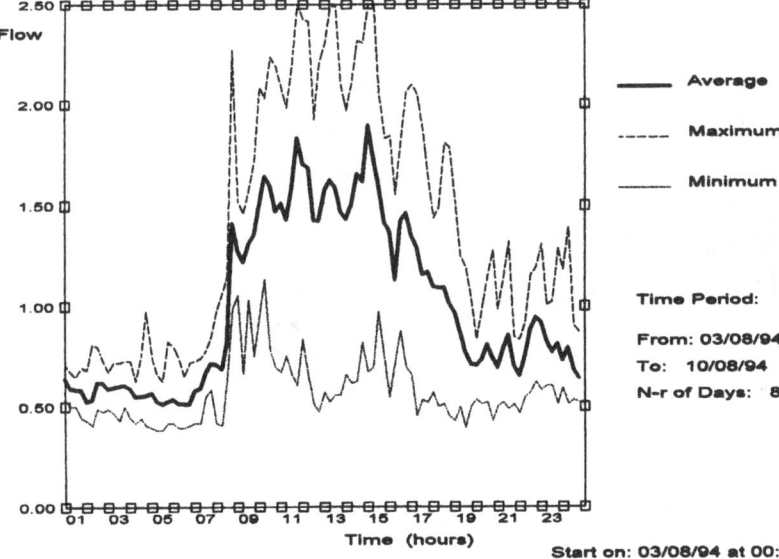

Figure F.73 The police
station pattern.

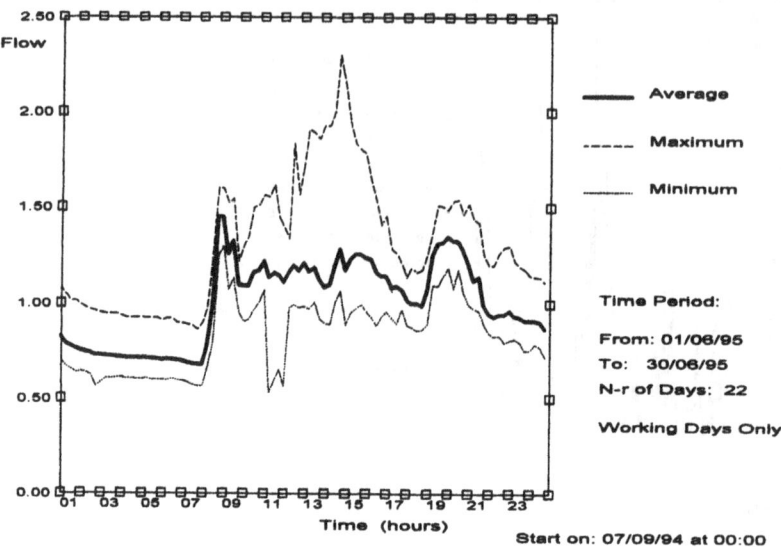

Data for Customer Meter: DM014 - Name: YOUTH CUSTODY CENTR

Figure F.74 A youth custody centre – workdays pattern.

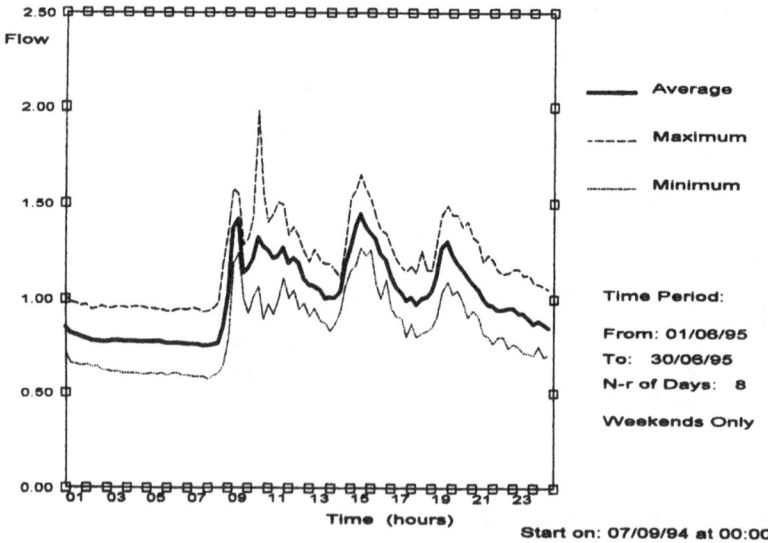

Data for Customer Meter: DM014 - Name: YOUTH CUSTODY CENTRE

Figure F.75 The youth custody centre weekend pattern.

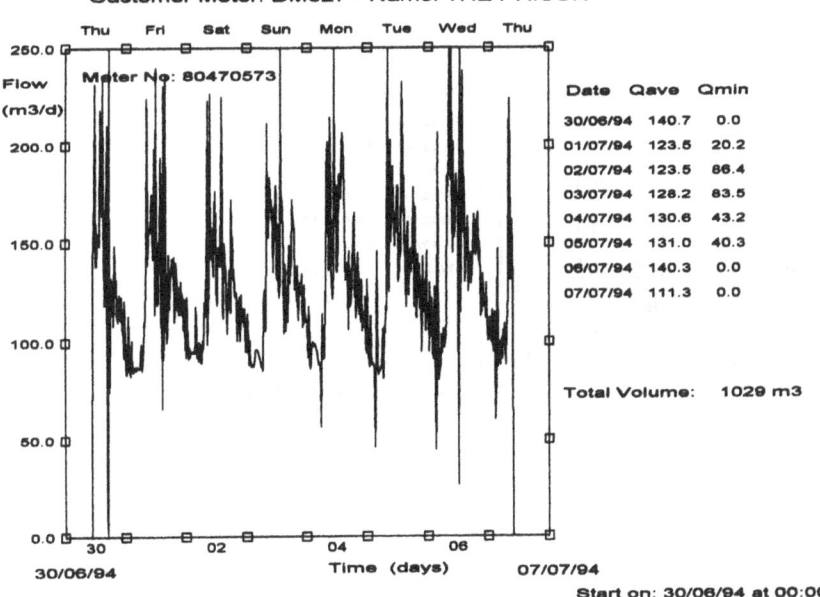

Figure F.76 A prison –
the data.

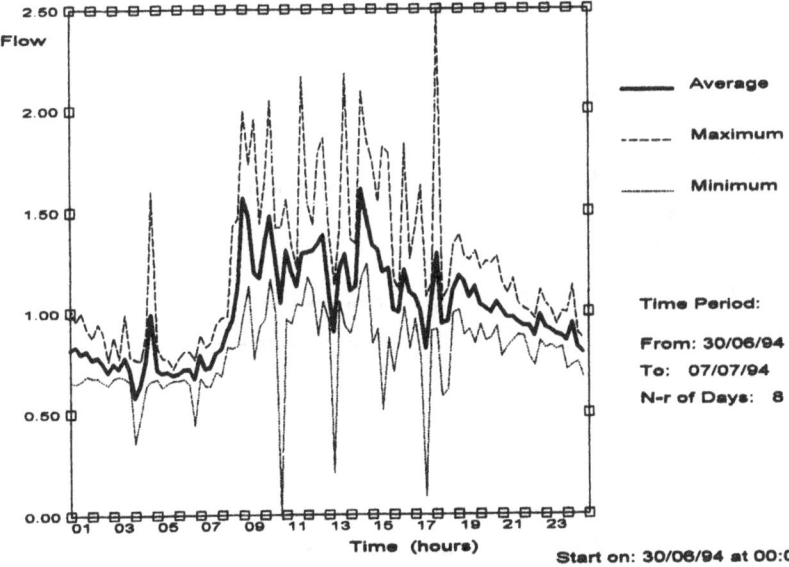

Figure F.77 The prison
pattern.

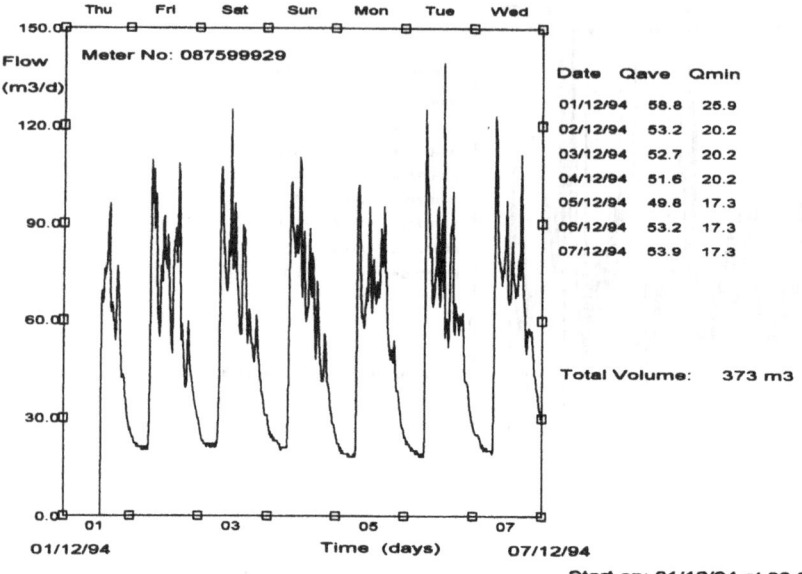

Figure F.78 Another
prison – the data (II).

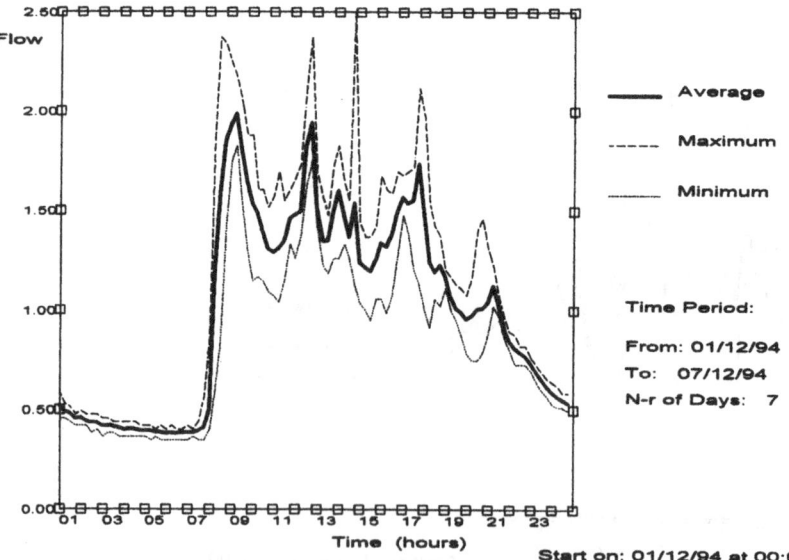

Figure F.79 The prison
pattern (II).

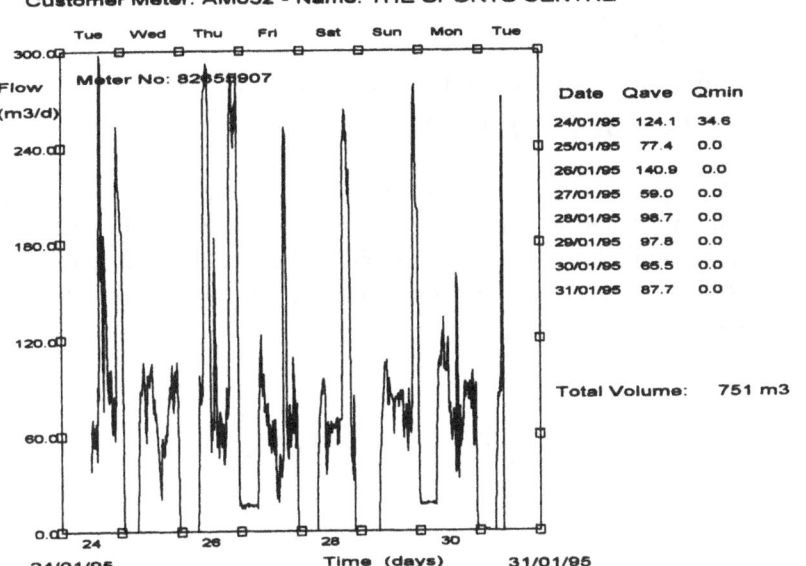

Figure F.80 A sports centre – the data.

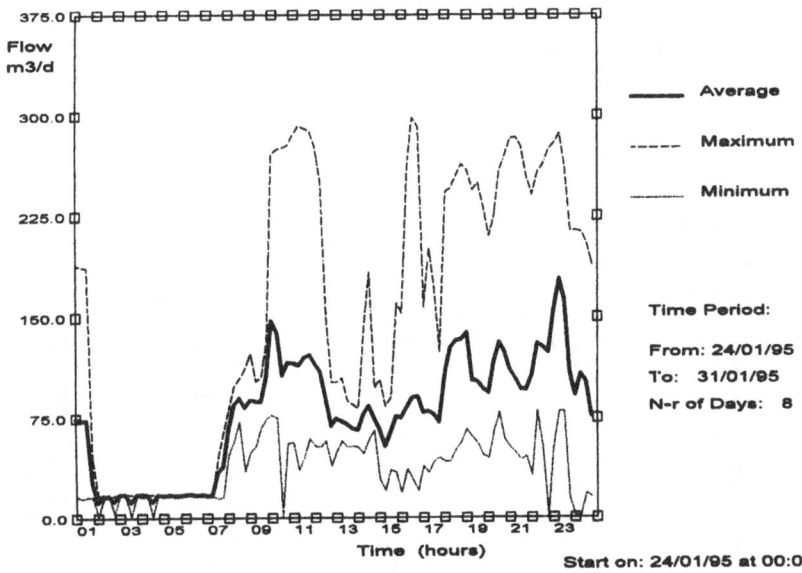

Figure F.81 Sports centre pattern.

Appendix G
Manual computations – a test case: three-loop system

Figure G.1 A system with three loops.

The layout of the three-loop system is shown in Fig. G.1, and data are listed in Table G.1. The simplified network ready for the calculations is shown in Fig. G.2. Table G.2 gives the computation sheet. The remaining stages are shown in Figs G.3–G.8.

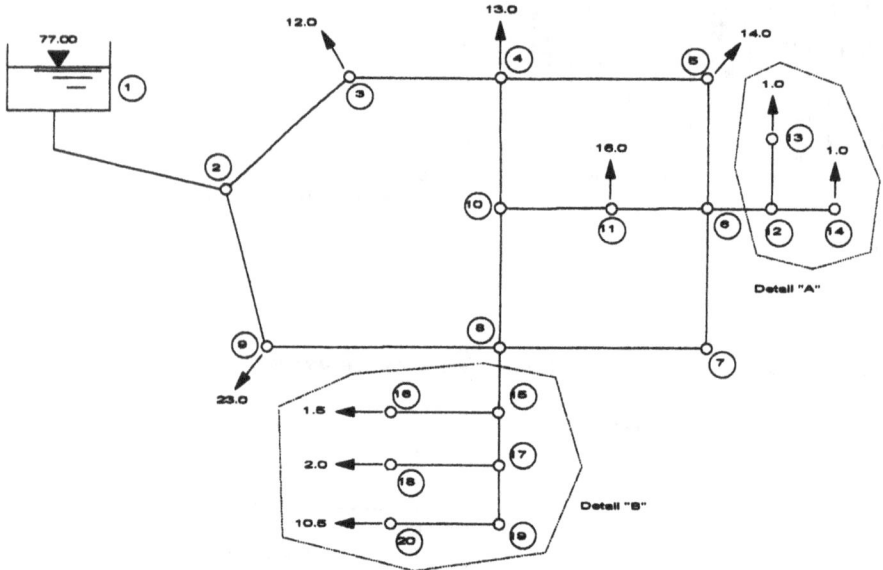

Table G.1 The data

Line (from–to)	Length (m)	Diameter (mm)	λ	K
1–2	6195	400	0.020	1000
2–3	2205	300	0.020	1500
3–4	1182	250	0.020	2000
4–5	768	250	0.020	1300
5–6	910	200	0.020	4700
6–7	639	200	0.020	3300
7–8	1063	250	0.020	1800
8–9	1470	300	0.020	1000
9–2	1617	300	0.020	1100
4–10	1182	250	0.020	2000
8–10	882	300	0.020	600
10–11	532	250	0.020	900
11–6	310	200	0.020	1600

Table G.2 The computation sheet

Loop	Line	K	Q^0 (l s⁻¹)	Δh^0 (m)	$2K|Q^0|$	Q^1 (l s⁻¹)	Δh^1 (m)	$2K|Q^1|$	Q^2 (l s⁻¹)	Δh^2 (m)
I	2–3	1500	42.0	2.646	126.0	40.49	2.459	121.5	40.57	2.469
	3–4	2000	30.0	1.800	120.0	28.49	1.623	114.0	28.57	1.632
	4–10	2000	0	0	0	2.44	0.012	9.76	3.50	0.025
	10–8	600	0	0	0	−8.14	−0.040	9.77	−9.79	−0.058
	8–9	1000	−29.0	−0.841	58.0	−30.51	−0.931	61.02	−30.43	−0.926
	9–2	1100	−52.0	−2.974	114.4	−53.51	−3.150	117.7	−53.43	−3.140
Total for loop I				0.631	418.4		−0.027	433.7		0.002
II	4–5	1300	17.0	0.376	44.2	13.05	0.221	33.93	12.07	0.189
	5–6	4700	3.0	0.042	28.2	−0.95	−0.004	8.93	−1.93	−0.018
	6–11	1600	16.0	0.410	51.2	5.42	0.047	17.34	2.71	0.012
	11–10	900	0	0	0	−10.58	−0.101	19.04	−13.29	−0.159
	10–4	2000	0	0	0	−2.44	−0.012	9.76	−3.50	−0.025
Total for loop II				0.828	123.6		0.151	89.0		−0.001
III	10–11	900	0	0	0	10.58	0.101	19.04	13.29	0.159
	11–6	1600	−16.0	−0.410	51.2	−5.42	−0.047	17.34	−2.71	−0.012
	6–7	3300	−15.0	−0.742	99.0	−8.37	−0.231	55.24	−6.64	−0.145
	7–8	1800	−15.0	−0.405	54.0	−8.37	−0.126	30.13	−6.64	−0.079
	8–10	600	0	0	0	8.14	0.040	9.77	9.79	0.058
Total for loop III				−1.557	204.2		−0.263	131.5		−0.019

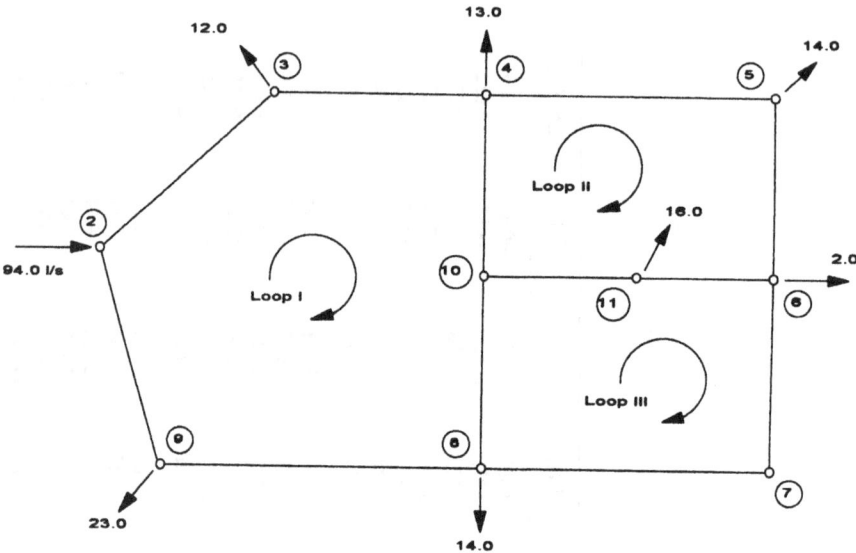

Figure G.2 Network
prepared for calculation.

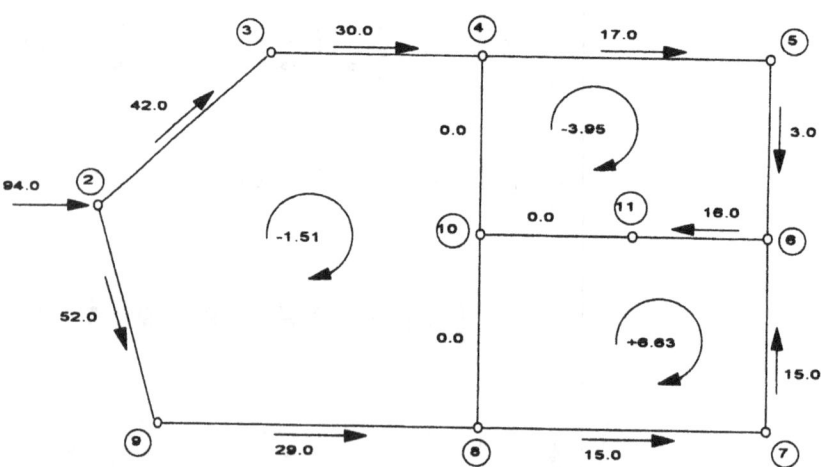

Figure G.3 Initial approximation and the first
correction.

$$
\begin{array}{ccc} q_1 & q_2 & q_3 \end{array}
$$

$$
\begin{bmatrix} 418.40 & 0.00 & 0.00 \\ 0.00 & 123.60 & -51.20 \\ 0.00 & -51.20 & 204.20 \end{bmatrix} = \begin{bmatrix} -0.631 \\ -0.828 \\ +1.557 \end{bmatrix}
$$

$$
q_1 = -0.00151
$$

$$
q_2 = -0.00395
$$

$$
q_3 = +0.00663
$$

Figure G.4 The system of equations (I).

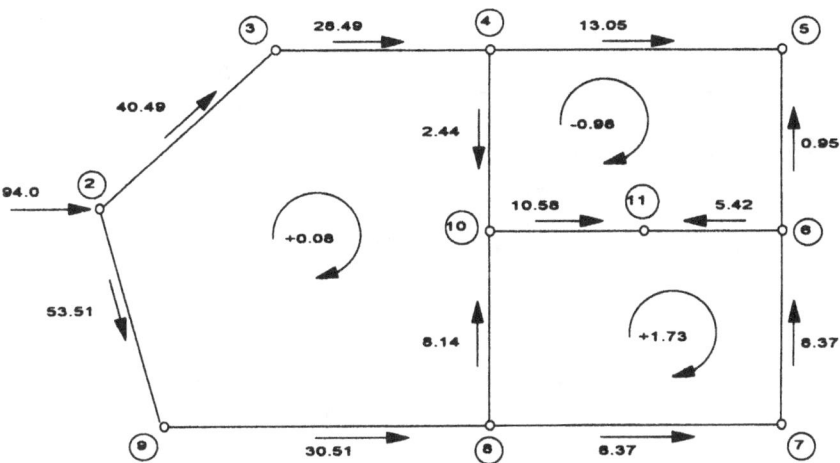

Figure G.5 The second approximation and corrections.

$$
\begin{array}{ccc}
q_1 & q_2 & q_3
\end{array}
$$

$$
\begin{bmatrix}
433.70 & -9.76 & -9.77 \\
-9.76 & 89.00 & -36.38 \\
-9.77 & -36.38 & 131.52
\end{bmatrix}
=
\begin{bmatrix}
+0.027 \\
-0.151 \\
+0.263
\end{bmatrix}
$$

$$q_1 = +0.00008$$

$$q_2 = -0.00098$$

Figure G.6 The system
of equations (II).

$$q_3 = +0.00173$$

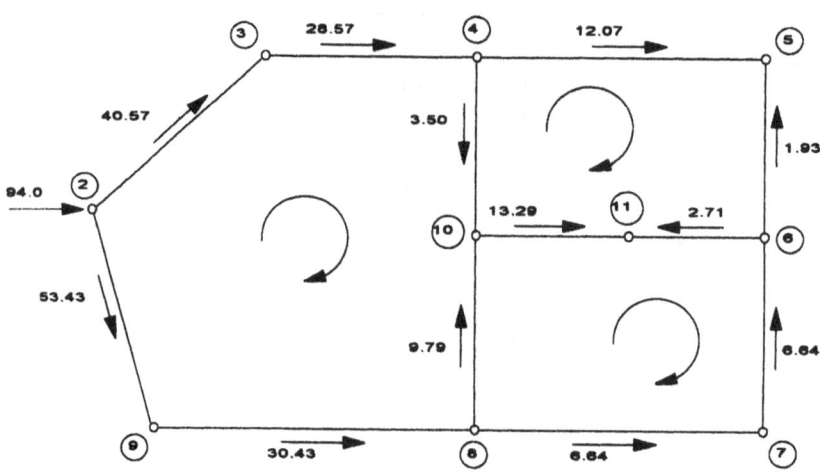

Figure G.7 The third
approximation.

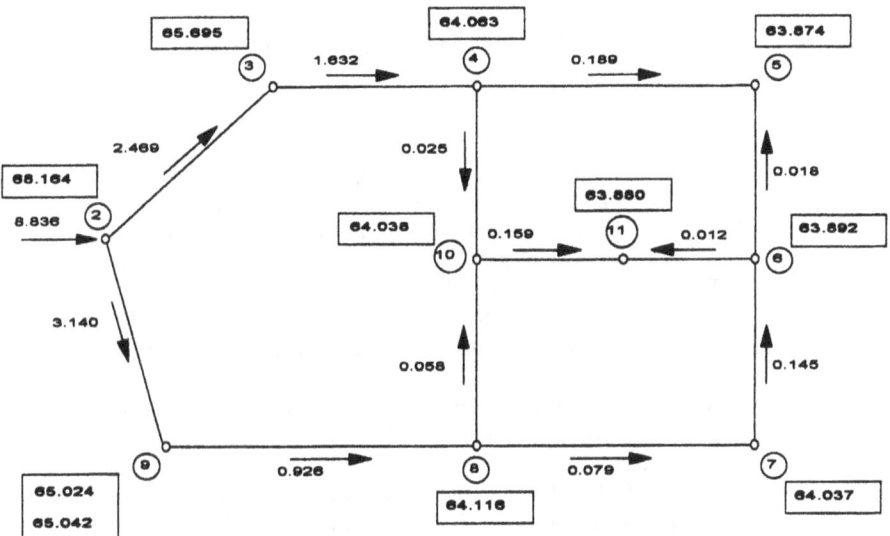

Figure G.8 Heads and head losses in the network.

References and further reading

ABB (1989): 'Speed Control of Pumps with SAMI Frequency Converters', ABB Drives, Asea Brown Boveri, Helsinki, Finland.

Ainsworth, K. and S. Ross (1994): 'Dealing with Customers' Leakage', in 'Managing Leakage', eds S.G. Bessey et al., UK Water Industry, Water Research Centre, Marlow, UK.

Al Hames, H. (1987): 'Safe Drinking Water Supply by Automation - Example of the City of Hamburg', Water 2000 Symposium, Nice, France, pp. 9-3-1 to 9-3-8.

Alla, P. (1985): 'Optimal Operations of Large Water Supply Networks', AQUA, Journal of Water Supply Research and Technology, UK, No. 6, pp. 320–324.

Alla, P. (1988): 'Information and Expert Systems in Water Supply', International Report 7, Proceedings of XVII Congress of the International Water Supply Association, Brazil.

Archibald, G. (1994): 'Estimating Unmeasured Water Delivered', in 'Managing Leakage', eds S.G. Bessey et al., UK Water Industry, Water Research Centre, Marlow, UK.

Bailey, R.J., P.K. Jolly and R.F. Lacey (1986): 'Domestic Water Use Patterns', WRC Report TR 225, Water Research Centre, Marlow, UK.

Barker, R.A. (1993): 'Network Analysis at Ground Level', in 'Integrated Applications in Water Supply', ed. B. Coulbeck, Vol. 2, John Wiley & Sons, pp. 3–15.

Barrufet, A. (1985): 'Study on Peak and Average Demand Coefficients Which are Interrelated' (French), AQUA, Journal of Water Supply Research and Technology, UK, No. 6, pp. 316–319.

Beal, D. (1987): 'The Implementation of Information Technology to the Year 2000 - A Realistic Approach', Water 2000 Symposium, Nice, France, pp. 311–317.

Beal, D. (1988): 'Integrated Distribution Management Systems', AQUA, Journal of Water Supply Research and Technology, UK, No. 5, pp. 267–276.

Bessey, S.G. (1985): 'Progress in Pressure Control', AQUA, Journal of Water Supply Research and Technology, UK, No. 6, pp. 325–330.

Bessey, S.G. (1990): 'Water Savings from Flow Modulated Pressure Control', Best Practice Programme, A/23, Department of Energy, UK.

Bessey, S.G. and R. Garrett (1994): 'Summary Report', in 'Managing Leakage', eds S.G. Bessey et al., UK Water Industry, Water Research Centre, Marlow, UK.

Bessey, S.G. and A. Lambert (1994): 'Reporting Comparative Leakage Performance', in 'Managing Leakage', eds S.G. Bessey et al., UK Water Industry, Water Research Centre, Marlow, UK.

Bessey, S.G., et al. (1994): 'Managing Leakage', UK Water Industry, Water Research Centre, Marlow, UK.

Bhave, P.R. (1991): 'Analysis of Flow in Water Distribution Networks', Technomic, Lancaster, USA.

Bland, P. and J.C. Townend (1987): 'An Integrated Approach to Telemetry - North West Water Authority - A Regional Telemetry Scheme', Water 2000 Symposium, Nice, France, pp. 9-2-1 to 9-2-4.

Bojarski, J. and M. Walden (1996): 'PLC vs. DCS - Finding the Best Value in Process Control', in 'Productivity Through Innovation', 1996 Computer Conference Proceedings, April 21-24, 1996, Chicago, American Water Works Association, Denver, USA, pp. 41–49.

Bos, M. and P.A. Jarrige (1989): 'Mathematical Modelling of Water Distribution Networks Under Steady-State Conditions - Recent Developments and Future Projects' (French), AQUA, Journal of Water Supply Research and Technology, UK, Vol. 38, pp. 352–357.

Bost, J.-F. (1986): 'State of the Art of Microtechnology in Water Services', General Report 2, Proceedings of XVI Congress of the International Water Supply Association, Special Subject No. 4, Rome, Italy.

Boulos, P.F., T. Altman, R.W. Bowcock, A.K. Dhingra and F. Collevati (1994): 'An Explicit Algorithm for Modelling Distribution System Water Quality with Applications', Proceedings of 2nd International Conference on Water Pipeline Systems, ed. D.S. Miller, BHR Group Conference Series, Publication No. 10, London, UK, pp. 405–423.

Boulos, P.F., J.E. Heath, D.H. Freiberg, J.J. Ro and M. Meyer (1996): 'Water Distribution Network Management: A Fully Integrated Approach', in 'Productivity Through Innovation', 1996 Computer Conference Proceedings, April 21–24, 1996, Chicago, American Water Works Association, Denver, USA, pp. 17–27.

Brammer, L.F. and A.M. Schulte (1993): 'A Joint US and UK Approach to Water Supply and Distribution Modelling', Journal of the Institution of Water and Environmental Management, UK, Vol. 7, No. 5, pp. 471–480.

Buchweitz, G. (1988): 'Control of Waterworks: State of the Art Today and Future Developments', AQUA, Journal of Water Supply Research and Technology, UK, No. 5, pp. 254–260.

Burnell, D., J. Race and P. Evans (1993): 'The Trunk Scheduling System for the London Water Ring Main', in 'Integrated Applications in Water Supply', ed. B. Coulbeck, Vol. 2, John Wiley & Sons, pp. 203–217.

Capener, A. and B. Ratcliffe (1994): 'Managing Water Pressure', in 'Managing Leakage', eds S.G. Bessey et al., UK Water Industry, Water Research Centre, Marlow, UK.

Carr, R.J. and A.B. Doherty (1993): 'A Cost-Benefit Analysis of Including All Distribution Mains in Operational Models', in 'Integrated Applications in Water Supply', ed. B. Coulbeck, Vol. 2, John Wiley & Sons, pp. 33–46.

Castensson, R.G. (1989): 'Urban Water Resources Supply Conflicts', Proceedings of the NATO Urban Water Resources Advanced Research Workshop, Isle of Man, UK, pp. 18–27.

Caves, J.L. and T.C. Earl (1979): 'Computer Applications - A Tool for Water Distribution Engineers', Journal of the American Water Works Association, USA, pp. 230–235.

Cesario, L. (1995): 'Modeling, Analysis, and Design of Water Distribution Systems', American Water Works Association, Denver, USA.

Chase, D.V. and G.L. Jones (1993): 'Lessons Learned from Application of Optimal Control Technology', in 'Integrated Applications in Water Supply', ed. B. Coulbeck, Vol. 2, John Wiley & Sons, pp. 377–394.

Chase, D.V. and L.E. Ormsbee (1993): 'Computer-Generated Pumping Schedules for Satisfying Operational Objectives', Journal of American Water Works Association, Vol. 85, No. 7, pp. 54–61.

CHW (1986): 'The Process Control Handbook 1986–87', CHW Roles and Associates Ltd., Cambridge, UK

Clark, R.M. (1994): 'Modelling Water Quality Changes and Contaminant Propagation in Drinking Water Distribution Systems: A US Perspective', AQUA, Journal of Water Supply Research and Technology, UK, Vol. 43, No. 3, pp 133–143.

Cleverly, G.J. and D.J. Algeo (1988): 'The Use of Network Analysis Techniques in the Development and Optimisation of a Large UK Urban Water Distribution System', AQUA, Journal of Water Supply Research and Technology, UK, Vol. 37, No. 2, pp. 67–72.

Cohen, J. (1990): 'The Development of a Dynamic Calculation Model for Drinking-Water Networks', AQUA, Journal of Water Supply Research and Technology, UK, Vol. 39, No. 3, pp. 172–187.

Cohen, J. and P.M. de Visser (1992): 'The Development and Application of Monitoring Systems for the Distribution of Drinking Water', AQUA, Journal of Water Supply Research and Technology, UK, Vol. 41, No. 6, pp. 352–359.

Cornish, R.J. (1940): 'The Analysis of Flow in Networks of Pipes', Paper No. 5219, Journal of the Institution of Civil Engineers, UK.

Cosgriff, G.O., et al. (1985): 'Interactive Computer Modelling: Monitoring and Control of Melbourne's Water Supply System', Water Resources Research, Vol. 21, No. 2, pp. 123–129.

Coulbeck, B. and C.H. Orr (1988): 'Computer Applications in Water Supply', Vol. 2: 'Systems Optimisation and Control', Research Studies Press, Herts, UK.

Coulbeck, B. and C.H. Orr (1989): 'Computer Programs for Real-Time Optimised Control of Water Distribution Systems', HYDROCOMP '89: Computational Modelling and Experimental Methods in Hydraulics, Elsevier Applied Science, London, 1989, pp. 289–300.

Cross, H. (1936): 'Analysis of Flow in Networks of Conduits or Connectors', University of Illinois, Bulletin No. 286, USA.

Croucher, P. (1989): 'Communications and Networks, A Handbook for the First-time Users', Sigma Press, Wilmslow, UK.

Cubillo, F. and M. Mercier (1993): 'An Approach to Water Quality Modelling in the Paris and Madrid Distribution Networks', Water Supply, Vol. 11, Nos. 3/4, pp. 377–385.

Cullen, N. (1987): 'Computing - Not Yet Half-Way There', Water 2000 Symposium, Nice, France, pp. 3-3-1 to 3-3-12.

Davies, M. (1993): 'Area Control Modelling', in 'Integrated Applications in Water Supply', ed. B. Coulbeck, Vol. 2, John Wiley & Sons, pp. 189–202.

Davis, A.L. and R.W. Jeppson (1979): 'Developing a Computer Program for Distribution Systems Analysis', Journal of the American Water Works Association, Vol. 71, No. 5.

Davoust, J.L. (1986): 'Application of Data Processing to Remote Control of Potable Waterworks and Dependent Networks', Colloque Eau et Informatique, Paris, France, pp. 385–393.

De Moyer, R., H.D. Gilman and M.Y. Goodman (1973): 'Dynamic Computer Simulation and Control Methods for Water Distribution Systems', General Electric Co., USA.

Dickenson, C. (1988): 'Pumping Manual' (8th edn), The Trade and Technical Press, Morden, Surrey, UK.

Donachie, R.P. (1974): 'Digital Program for Water Network Analysis', Transactions of the ASCE, Journal of the Hydraulics Division, Vol. 100, HY-3, pp. 393–403.

Dzadey, N. and G. Price (1996): 'Development of a Strategic Operational Hydraulic Model for Integration with Melbourne Water Management System', in 'Productivity Through Innovation', 1996 Computer Conference Proceedings, April 21-24, 1996, Chicago, American Water Works Association, Denver, USA, pp. 245-249.

Elton, A., L.F. Brammer and N.S. Tansley (1994): 'Recent Advances in Modelling Water Quality in Distribution Network', Special Subject 7: 'Managing Water Distribution Systems', Water Supply, IWSA, Vol. 12, Nos. 1/2, pp. SS 7-5 to 7-9.

Epp, R. and A.G. Fowler (1970): 'Efficient Code for Steady-State Flows in Networks', Transactions of the ASCE, Journal of the Hydraulics Division, Vol. 96, HY-1, pp. 43–56.

Fallside, F. (1977): 'On-line Control of a Water Supply System', International Workshop on Instrumentation and Control for Water and Wastewater Treatment and Transport Systems, Paper No. 40, London-Stockholm, pp. 6-40-1/10.

Fallside, F. (1981): 'Systems Engineering and Microelectronics in Water Supply', IWES - Microelectronics in Water Industry, UK, pp. 17–41.

Fontaine, J. (1983): 'Towards Integral Automation in Water Distribution' (French), AQUA, Journal of Water Supply Research and Technology, UK, No. 6, pp. 302–312.

Fowles, G. (1993): 'Flow, Level and Pressure Measurement in the Water Industry', Butterworth-Heineman, Oxford, UK.

Frischherz, H. and M. Sacher (1984): 'Function and Choice of Pumping Machinery', Proceedings of the XV Congress of International Water Supply Association, Special Subject 14, Monastir, Tunisia.

Gagnon, J.-L. and P.T. Bowen (1996): 'Supply Safety and Quality of Distributed Water', in 'Productivity Through Innovation', 1996 Computer Conference Proceedings, April 21–24, 1996, Chicago, American Water Works Association, Denver, USA, pp. 287–294.

Garrett, P. (1994): 'This Year's Model for Sherwood', Water Bulletin, Water Services Association, UK, No. 589, 21 January 1994, pp. 8.

Germanopoulos, G. and P.W. Jowitt (1989): 'Leakage Reduction by Excess Pressure Minimisation in a Water Supply Network', Proceedings of Institution of Civil Engineers, UK, Part 2, Vol. 87, pp. 195–214.

Gessler, J. (1981): 'Analysis of Pipe Networks', Chapter 4 in 'Closed-Conduit Flow', eds M.H. Chaudhry and V. Yevjevic, Water Resources Publications, Littleton, USA, pp. 61–99.

Gilman, H.D., M.Y. Goodman and R.V. Metkowski (1971): 'Mathematical Modelling for Water Distribution Systems', Office of the Water Resources Research, US Department of the Interior, USA.

Gilman, H.D., M.Y. Goodman and R. De Moyer Jr (1972): 'Replication Modelling for Water Distribution Control', Proceedings of the 29th Annual Conference of AWWA, Chicago, USA.

Glasbrook, D. (1993): 'The Practical Application of GIS, and its Use for AMP2', IWO Conference 1993, Cardiff, UK.

Gotoh, K., J.K. Jacobs, S. Hosoda and R.L. Gerstberger (1993): 'Instrumentation and Computer Integration of Water Utility Operations', American Water Works Association Research Foundation and Japan Water Works Association, Denver, USA.

Grombach, P. (1986): 'Centralised Monitoring and Control of Water Supply System of the City of Zurich: An Example of the Systematic Optimisation of Operation by Computer', Colloque Eau et Informatique, Paris, France, pp. 216–225.

Gwynne, M., M. Hilton and G. Parkes (1996): 'Realising the Benefits of Telemetry Data', Proceedings of the Conference 'Knowledge, Information and Data 1996', BHR Group, Cranfield, UK, pp. 105–111.

Haddon, M. (1994): 'Ringing the Changes for Meter Reading', Water Bulletin, Water Services Association, UK, No. 605, pp. 9–11.

Haddon, M. (1995): 'Dial M for Meter', Water Bulletin, Water Services Association, UK, No. 650, 21 April 1995, pp. 13.

Halpern, O. and O. Pascal (1987): 'Water Distribution System for the 3rd Millennium' (French), Water 2000 Symposium, Nice, France, pp. 1-1-1/14.

Hamam, Y.M. and A. Brameller (1971): 'Hybrid Method for the Solution of Piping Networks', Proceedings of IEEE, Vol. 118, No. 11, pp. 1607–1612.

Hoogsten, K.J. and J.T. van der Zwan (1985): 'Distribution System Modelling', AQUA, Journal of Water Supply Research and Technology, UK, No. 5, pp. 247–251.

Hosho, T. and N. Fukui (1983): 'Application of Speed Control for Water Supply System Pumps', Hitachi Review, Japan, Vol. 32, No. 1, pp. 29–32.

Howard, K., B. Rosindale and R. Thomas (1993): 'Implementation of a SCADA and a Decision Support System to Provide Close Pressure Control', in 'Integrated Applications in Water Supply', ed. B. Coulbeck, Vol. 2, John Wiley & Sons, pp. 235–248.

Huntington, R. (1979): 'Resurrection of Water Supply Distribution - New Life for a Faithful Servant', 84th Summer General Meeting and Conference of Water Supply Engineers, Bournemouth, UK.

Huntington, R. (1984): 'Improvement of Cost and Performance of Distribution Systems', Proceedings of XV Congress of International Water Supply Association, Special Subject No. 9, Monastir, Tunisia.

Huntington, R. (1986): 'Effects of ICA on Mechanical and Electrical Maintenance', Conference of Water Supply Engineers, UK.

Huntington, R. (1990): 'Wessex Water's Integrated Water Distribution Management System', Proceedings of the 5th IAWPRC Workshop, Japan.

Huntington, R. (1993): 'Updating and Development of a Regional Telemetry Scheme', 6th IAWQ Workshop on Instrumentation, Control and Automation of Water and Wastewater Treatment and Transport System, Canada.

Idel'chik, I.E. (1979): 'Memento des Pertes de Charge' (traduit du russe), Eyrolles, Paris, France.

Isaacs, L. and K.G. Miles (1980): 'Linear Theory Methods for Network Analysis', Transactions of the ASCE, Journal of the Hydraulics Division, Vol. 106, HY-7, pp. 1191–1201.

Jeffery, J. and C.R. Taylor (1993): 'Development in Leakage Control at North Surrey Water Ltd - A Case Study', AQUA, Journal of Water Supply Research and Technology, UK, Vol. 42, No. 4, pp. 223–232.

Jeppson, R.W. and A.L. Davis (1976): 'Pressure Reducing Valves in Pipe Network Analysis', Transactions of the ASCE, Journal of the Hydraulics Division, Vol. 102, HY-7, pp. 987–1001.

Johnson, I., R. Casey and R. Turner (1993): 'The Operational Management System for London Water', in 'Integrated Applications in Water Supply', ed. B. Coulbeck, Vol. 2, John Wiley & Sons, pp. 281–301.

Jones, R. (1984): 'Utilisation of Mathematical Models for Simulation of Water Distribution Networks Operation' (French), SOGREAH, Grenoble, France.

Jowitt, P.W., R.T. Garrett, S.C. Cook and G. Germanopoulos (1988): 'Real-Time Forecasting and Control for Water Distribution', in 'Computer Applications in Water Supply', Vol. 2: 'Systems Optimisation and Control', eds B. Coulbeck, and C.H. Orr, Research Studies Press, Herts, UK.

Joyner, M. and G. Wright (1996): 'Distribution Network Management - Norwich and Beyond', in 'Productivity Through Innovation', 1996 Computer Conference Proceedings, April 21–24, 1996, Chicago, American Water Works Association, Denver, USA, pp. 239–243.

Kado, M. and H. Itoh (1987): 'A Satisfactory Water Service Expert', Water 2000 Symposium, Nice, France, pp. 2-1-1/20.

Kanbayashi, T., et al. (1977): 'Computer Control and Operation Information System for Large Water Supply Systems', International Workshop on Instrumentation and Control for Water and Wastewater Treatment and Transport Systems, Paper No. 41, UK, London-Stockholm, pp. 6-41-1/7.

Karney, B.W. and D. McInnis (1992): 'Efficient Calculations of Transient Flow in Simple Pipe Networks', Transactions of the ASCE, Journal of the Hydraulics Division, Vol. 118, No.7, pp. 1014–1030.

Kashiwagi, M. and K. Matsumoto (1980): 'Full Automation for Managing Modern Water Supply Systems', Hitachi Review, Japan, Vol. 29, No. 5, pp. 229–232.

Kootattep, S. and H. Aya (1985): 'Appropriate Method of Distribution Networks Analysis for Developing Countries', AQUA, Journal of Water Supply Research and Technology, UK, No. 6, pp. 311–315.

Kordic, M., et al. (1976): 'Mathematical Modelling of Regional Water Supply Systems in Yugoslavia', Journal of the American Water Works Association, Vol. 69, No. 4, pp. 203–205.

Kroon, J.R. and M.W. Spiteri (1993): 'Automated Network Model Data Reduction and Submodel Management', in 'Integrated Applications in Water Supply', ed. B. Coulbeck, Vol. 2, John Wiley & Sons, pp. 173–185.

KSB (1975): 'KSB Centrifugal Pumps Lexicon', Klein, Schanzlin & Becker Co., Frankenthal, Germany.

Lambert, A. (1994): 'Interpreting Measured Night Flows', in 'Managing Leakage', eds S.G. Bessey et al., UK Water Industry, Water Research Centre, Marlow, UK.

Lansey, K.E. and K. Awumah (1994): 'Optimal Pump Operations Considering Pump Switches', Journal of Water Resources Planning and Management, ASCE, Vol. 120, No. 1, pp. 17–35.

Lawson, S. (1996): 'Pulling Down the Cost of Telemetry', Proceedings of the Conference 'Knowledge, Information and Data 1996', BHR Group, Cranfield, UK, pp. 99–104.

Lonsdale, P.B. (1984): 'Automatic Control of Pressure in Distribution Systems', Wessex Water plc, Bristol, UK.

Maidment, D.R. and S.P. Miaou (1985): 'Transfer Function Models of Daily Urban Water Use', Water Resources Research, Vol. 21, No. 4, pp. 425–432.

Martin, L. and M. Farley (1994): 'Leakage Management Techniques, Technology and Training', in 'Managing Leakage', eds S.G. Bessey et al., UK Water Industry, Water Research Centre, Marlow, UK.

Matsumoto, K., et al. (1977): 'Control Models for the Computer Operation of Large Water Supply Systems', International Workshop on Instrumentation and Control for Water and Wastewater Treatment and Transport Systems, Paper No. 78, London-Stockholm, pp. 10-78-1/5.

McCann, B. (1989): 'Network Analysis - the Core of Distribution Management', Water Bulletin, No. 357, Water Services Association, UK.

McNaught, C. (1993): 'Water Down Energy Costs', Water and Waste Treatment, Vol. 37, No. 7, pp. 38 and 44.

McPherson, M.B. (1975): 'System Operation, Control and Monitoring; Part I: Concept of Integrated Automatic Operational Control', ASCE Urban Water Resources Research Programme, Technical Memorandum No. 15.

Miller, D.S. (1978): 'Internal Flow Systems', BHRA Fluid Engineering Series, UK.

Mitchell, V. (1993): 'London's Telemetry System', Water and Waste Treatment, Vol. 37, No. 7, pp. 26.

Monin, L.A. (1984): 'Function and Choice of Pumping Machinery', Proceedings of XV Congress of International Water Supply Association, Special Subject No. 14, Monastir, Tunisia.

Monro, M. and A. Martin (1992): 'Water Distribution Network Analysis', Pipes and Pipelines International, March/April, pp. 37–39.

Morita, S. and T. Arakawa (1989): 'Water Distribution Control System', Water Nagoya '89, ASPAC IWSA (1S-22-B22), pp. 626–635.

Muguruma, K. (1995): 'Control and Management of Water Distribution Systems in Tokyo', Water Supply, IWSA, Vol. 13, Nos. 3/4, pp. 19–24.

Nandy, B. and L. Owen (1993): 'Knowledge Modelling, Operational Simulation of Water Supplies Through Interactive Visualisation', in 'Integrated Applications in Water Supply', ed. B. Coulbeck, Vol. 2, John Wiley & Sons, pp. 261–278.

Nerenberg, R. and R. Butterworth (1996): 'Using Chicago's Distribution/Tunnel Model for Practical Regional Decisions', in 'Productivity Through Innovation', 1996 Computer Conference Proceedings, April 21–24, 1996, Chicago, American Water Works Association, Denver, USA, pp. 233–237.

Nguyen, B. (1994): 'Automatic Operation of the Water Distribution of the City of Paris', IWSA Regional Conference, Zurich, Switzerland, pp. 311–317.

Nguyen, B. and A. Montiel (1994): 'On-line Quality Control in Distribution Networks', IWSA Regional Conference, Zurich, Switzerland, pp. 299–309.

Obradovic, D. (1973): 'Analysis of Water Distribution Systems', Proceedings of the 8th Yugoslav International Symposium on Information Processing, Informatica 1973, Paper D-18, Bled, Yugoslavia.

Obradovic, D. (1989): 'Modernisation of Urban Water Supply Systems', Proceedings of the NATO Urban Water Resources Advanced Research Workshop, Isle of Man, UK, pp. 150–165.

Obradovic, D. (1993): 'Mathematical Modelling of Water Supply Systems', IHE - International Institute for Infrastructural, Hydraulic and Environmental Engineering, Delft, The Netherlands.

Obradovic, D. and A. Filip (1986): 'The Value of Data Banks: The Case of Belgrade Water Supply System', Colloque Eau et Informatique, Paris, France, pp. 111–126.

Obradovic, D. and M. Kordic (1984): 'Mathematical Modelling of Water Supply Systems in Yugoslavia', Proceedings of XV Congress of the International Water Supply Association, Special Subject No. 9, Monastir, Tunisia.

Obradovic, D. and M. Kordic (1986): 'Studying a Disastrous Situation Before it Actually Happens', Proceedings of XVI Congress of the International Water Supply Association, Special Subject No. 4, Rome, Italy.

Obradovic, D., M. Kordic and D. Jakic (1984): 'Mathematical Modelling of an Existing Regional Water Conveyance System',

International Conference Hydrosoft '84, Portoroz, Yugoslavia, pp. 5-43/55.

Olner, P. (1985): 'Telemetry Control Systems and Integrated Data Systems', in 'Instrumentation and Control of Water and Waste Water Treatment and Transport Systems', ed. R.A.R. Drake, Pergamon Press, Oxford and New York.

Onoda, H. (1995): 'Satisfying Control and Management for Service Systems by an Automatic Water Meter Reading Method', Water Supply, IWSA, Vol. 13, Nos. 3/4, pp. 149–154.

Ormsbee, L.E. and K.E. Lansey (1994): 'Optimal Control of Water Supply Pumping Systems', Journal of Water Resources Planning and Management, ASCE, Vol. 120, No. 2, pp. 237–252.

Ormsbee, L.E. and D.J. Wood (1986): 'Explicit Pipe Network Calibration', Journal of Water Resources Planning and Management, ASCE, Vol. 112, No. 2, pp. 166–182.

Pocock, A. (1992): 'Metering - Today's Technology', Water Bulletin, UK, No. 521, 14 August 1992, pp. 19.

Poveda, M.G. (1988): 'Automation and Control: General Concepts and Their Application in Water Supply and Waste Disposal Systems', AQUA, Journal of Water Supply Research and Technology, UK, No. 5, pp. 249–253.

Prasifka, D.W. (1988): 'Water-Supply Planning, Issues, Concepts and Risks', Van Nostrand Reinhold, New York, USA.

Quevedo, J. and G. Cembrano (1986): 'Water Demand Forecasting Through Time-Series Analysis', Colloque Eau et Informatique, Paris, France, pp. 308–317.

Rance, J.P., P.L.M. Bounds, S.T. Tennant and B. Ulanicki (1993a): 'Integration of a Network Simulator into On-line Control System for London's Water Networks', in 'Integrated Computer Applications in Water Supply', ed. B. Coulbeck, John Wiley & Sons, New York, USA.

Rance, J.P., N. Britton, B. Ulanicki, J. Grummant and S. Chen (1993b): 'Computer Aided Network Planning for London's Water System', in 'Integrated Applications in Water Supply', ed. B. Coulbeck, Vol. 2, John Wiley & Sons, pp. 47–59.

Rao, H.S. and W.B. Don Jr (1977): 'Extended-Period Simulation of Water Systems', Transactions of the ASCE, Journal of the Hydraulics Division, part A, Vol. 103, HY-2, pp. 97–108, part B, Vol. 103, HY-3, pp. 281–294.

Ray, C. (1996): 'Maximising Control Over Your Distribution System by Optimising Network Management', Conference on 'Cost-Effective Management of Water Pipelines and Networks', WSA, London, UK.

Rees, D.H. (1996): 'Data Rich But Information Poor', Knowledge, Information and Data Conference 1996, SFK Technology and East Midlands Branch of CIWEM, Buxton, Derbyshire, UK.

Ridley, W.F. (1980): (Chairman of the Technical Working Group on Waste of Water) 'Leakage Control Policy and Practice', National Water Council, London, UK.

Robertson, J. (1993): 'Optimum Network Management Through Integration of Operations Processes', in 'Integrated Applications in Water Supply', ed. B. Coulbeck, Vol. 2, John Wiley & Sons, pp. 317–329.

Rossman, L.A. (1993): 'EPANET Users Manual', Drinking Water Division, Risk Reduction Engineering Laboratory, Environmental Protection Agency, Cincinnati, Ohio, USA.

Santoni, A., et al. (1987): 'EXPERT Systems - New Tools for Water Treatment', Water 2000 Symposium, Nice, France, pp. 6-1-1/8.

Scherer, P. and T. Phebey (1995): 'Geographical Information Systems', AQUA, Journal of Water Supply Research and Technology, UK, Vol. 44, No. 3, pp. 118–124.

Schulte, A.M. and A.P. Malm (1993): 'Integrating Hydraulic Modelling and SCADA Systems for System Planning and Control', Journal of American Water Works Association, Vol. 85, No. 7, pp. 62–66.

Shamir, U. and C.D. Howard (1968): 'Water Distribution Systems Analysis', Transactions of the ASCE, Journal of the Hydraulics Division, Vol. 94, HY-1, pp. 219–234.

Shore, D.G. (1986): 'A Method for Using Computerized Customer Information in Water Distribution Network Models', Journal of Institution of Water Engineers and Scientists, UK, Vol. 40, Issue 2, pp. 131–138.

Skrentner, R.G. (1996): 'Keeping Up With Technology: Planned Obsolescence', in 'Productivity Through Innovation', 1996 Computer Conference Proceedings, April 21–24, 1996, Chicago, American Water Works Association, Denver, USA, pp. 309–313.

Slipper, M.J. (1985): 'The Application of ICA to Water Supply and Distribution Management', AQUA, Journal of Water Supply Research and Technology, UK, No. 5, pp. 279–284.

Snoxell, J.D. (1991): 'New Technology in the Water Industry - Making It Stick in the Real World', Proceedings of the Pipeline Industry Guild, UK.

Snoxell, J.D. (1993a): 'Network Management in Water Supply', International Conference 'Computer Applications in Water Supply and Distribution', De Montfort University, Leicester, UK, pp. 331–345.

Snoxell, J.D. (1993b): 'The Use and Value of GIS in the UK Water Industry; Experience in Wessex Water Services Ltd', IWSA Specialised Conference on GIS, Lyons, 1992.

Snoxell, J.D. (1994): 'Management of a Widespread Multi-Plant Water Supply System', IWSA Regional Conference, Zurich, 1994.

Snoxell, J.D., D. Obradovic and R. Cavor (1989): 'The Use of Real-Time Telemetry and Modelling in Conjunction for the Operational Management of Water Supply Systems', Computational Modelling and Experimental Methods in Hydraulics - Hydrocomp '89, Dubrovnik, Elsevier Applied Science, London, 1989, pp. 301–310.

Sorensen, C. (1996): 'Applying Networking Technology to Water Utility SCADA Systems', in 'Productivity Through Innovation', 1996 Computer Conference Proceedings, April 21–24, 1996, Chicago, American Water Works Association, Denver, USA, pp. 7–15.

Spectrascan (1992): 'Integrated Water Management System, Operating Manual', Biwater Operations Ltd, Spectrascan Products Division, Waterlooville, UK.

Stephenson, D. (1985): 'Continuous Simulation of Flow in Pipe Networks', AQUA, Journal of Water Supply Research and Technology, UK, No. 5, pp. 258–262.

Streeter, V.L. and E.B. Wylie (1979): 'Fluid Mechanics' (7th edn), McGraw-Hill, USA.

Swains, P.W. (1996): 'Bigger and Better: The Case for Large, Fully Calibrated, All-Mains Dynamic Network Models', in 'Productivity Through Innovation', 1996 Computer Conference Proceedings, April 21–24, 1996, Chicago, American Water Works Association, Denver, USA, pp. 151–164.

Swamy, S.A. (1986): 'Unaccounted-for Water and Leak Detection', Proceedings of XVI Congress of the International Water Supply Association, Rome, Italy.

Takagi, T., et al. (1983): 'Distribution Network Control for Water Supply System', Hitachi Review, Japan, Vol. 32, No. 5, pp. 259–264.

Tarquin, A.J. and J. Dowdy (1989): 'Optimal Pump Operation in Water Distribution', Transactions of the ASCE, Journal of the Hydraulics Division, Vol. 115, No. 2, pp. 158–168.

Technolog (1990): 'GPS - General Purpose Software for NEWLOG Data Loggers, User Guide to Operation', Technolog Ltd, Cromford, UK.

Trifunovic, N. (1995): 'Water Transport and Distribution', Lecture Notes, IHE - International Institute for Infrastructural, Hydraulic and Environmental Engineering, Delft, The Netherlands.

Trow, S. and M. Hall (1994): 'Setting Economic Leakage Targets', in 'Managing Leakage', eds S.G. Bessey et al., UK Water Industry, Water Research Centre, Marlow, UK.

Tustin, J.R. (1994): 'Economics of Leakage in a Water Supply System', presented to Royal Society of Civil Engineers, February 1994, London, UK.

Tustin, J.R. (1996): 'Development of an Information Technology System to Provide Efficient Management of Leakage from a Water Supply Company Network', Conference on 'Cost-Effective Management of Water Pipelines and Networks', WSA, London, UK.

Twort, A.C., F.M. Law, F.W. Crowley and D.D. Ratnayama (1994): 'Water Supply' (4th edn), Edward Arnold, Hodder Headline Group, London, UK.

van der Zwan, J.T. (1988): 'Computer Utilization in Distribution', AQUA, Journal of Water Supply Research and Technology, UK, No. 1, pp. 9–13.

van der Zwan, J.T. (1990): 'Water Transport and Distribution', International Institute for Hydraulic and Environmental Engineering, Delft, The Netherlands.

Walker, R.S. (1988): 'Water Industry Expert Systems Club' (Final Report), Water Research Centre, Swindon, UK.

Walski, T.M. (1983): 'Technique for Calibrating Network Models', Journal for Water Resources Planning and Management, ASCE, USA, Vol. 109, Issue 4, pp. 360–372.

Walski, T.M. (1984): 'Analysis of Water Distribution Systems', Van Nostrand Reinhold, New York, USA.

Walski, T.M. (1986): 'Case Study: Pipe Network Model Calibration Issues', Journal of Water Resources Planning and Management, ASCE, USA, Vol. 112, No. 2, pp. 238–249.

Walski, T.M. (1993): 'Tips for Saving Energy in Pumping Operations', Journal of American Water Works Association, Vol. 85, No. 7, pp. 49–53.

Warren, R. and M. Langley (1996): 'Practical Experience in the Integration of GIS and Network Modelling Systems', Proceedings of the Conference 'Knowledge, Information and Data 1996', BHR Group, Cranfield, UK.

Wessex Water (1993a): 'WESNET Technical Reference Guide', Aquaware Systems Ltd, 73 Gelenfurness Avenue, Bournemouth, BH3 7ES, UK.

Wessex Water (1993b): 'WESNET User's Guide', Aquaware Systems Ltd, Bournemouth, UK.

Whyte, M.W. (1989): 'A New Approach in Establishing Water Industry Standards of Service', Proceedings of the NATO Urban Water Resources Advanced Research Workshop, Isle of Man, UK, pp. 100–116.

Wilde, A. (1996): 'Real-time Modelling - Just a Pipedream?' Proceedings of the Conference 'Knowledge, Information and Data 1996', BHR Group, Cranfield, UK.

Williams, M.F. (1996): 'Accurately Analysing Your Network by Advanced Network Modelling Techniques', Conference on 'Cost-Effective Management of Water Pipelines and Networks', WSA, London, UK.

Wood, D.J. and C.O. Charles (1972): 'Hydraulic Network Analysis Using Linear Theory', Transactions of the ASCE, Journal of the Hydraulics Division, Vol. 98, HY-7, pp. 1157–1170.

WRC Reports: 'Pump Scheduling in Water Supply', Water Research Centre, Technical Reports-Engineering, TR 232, Swindon, UK.

Yates, M. (1989): 'A Meter for Pump Efficiency Measurement', World Pumps, British Pump Manufacturers Association, UK.

Young, J. and M. Valmores (1996): 'Distribution System Modeling for Development of Water Quality Based Operating Procedures', in 'Productivity Through Innovation', 1996 Computer Conference Proceedings, April 21–24, 1996, Chicago, American Water Works Association, Denver, USA, pp. 275–280.

Zarghamee, M.S. (1971): 'Mathematical Modelling for Water Distribution Systems', Transactions of the ASCE, Journal of the Hydraulics Division, Vol. 97, HY-1, pp. 1–14.

Glossary

Abstraction Licence	Permission given to a UK water company by the national Environment Agency to abstract from a natural source of water to a certain level, daily and/or annually
affinity laws	The relationships between rotational speed of pumps and discharge, head and power respectively (also called similarity laws)
all-mains model	Model embracing all pipes and facilities of the real system (*opp.* skeleton model)
AMR (automatic meter reading)	The electronic reading of customers' meters from a remote central station
AOD	Above Ordnance Datum
area	A part or a subdivision of the territory covered by a water supply system; could be further subdivided into zones
authorised water use	Unmetered but legitimately drawn water for fire fighting, street washing, mains flushing, sewer cleaning, fountains, parks, etc.
AUTO control	Control policy for a pumping station: its pumps operate according to variations of water level (level-controlled pumps) in the control reservoir or of pressure (pressure-controlled pumps) in the control node
Availability Charge	Charge levied on a water company for the permanent provision of power up to the agreed value (given in kVA)
BEP	Best efficiency point (of a pump)
billing system	The system that takes data from customers' meters to produce bills and track payments
booster PST	A pumping station in a water distribution system that is used to increase ('boost') pressure in the mains
calibration (of model)	Adjustment of various coefficients in the model (e.g. friction loss coefficient) in order to get as good an agreement between the model and the real system as possible
check valve	see NRV (non-return valve)
CIS (customer information system)	The system that manages, processes and controls information on a utility's customers; it supports and controls interaction with the customers, most notably meter reading and billing

communication pipe	Pipe from a distribution main to the stopcock
consumer	A person or a company supplied with water by the water utility (*syn.* customer)
consumption	A volume of water taken from the water supply system into a consumer's installation
contact tank	A reservoir designed to allow a period of time for free chlorine to be in contact with water, prior to its delivery into the public water supply system
contamination	Any introduction into water of micro-organisms, chemicals, wastes or waste water that may make water unfit for use
control node	A particular point within the water supply system where pressure is measured and transmitted to a pumping station or control valve as the control signal
control pipe	A particular pipe within the water supply system where flow is measured and transmitted to a main valve as the control signal
control policy	A set of rules determining how a pumping station or a control valve should be run under changing conditions
control signal	Current value of pressure or water level (at a control node) or flow (in a control pipe) which causes changes in the corresponding pumping station or control valve according to the pre-set rules
control valve	A valve within the distribution network that operates according to a given set of rules (*see* FCV, FLV, LCV, NRV, PCV, PRV, TCV)
cos ψ	Ratio between active and total electric power (usually close to 0.90); also called power factor
customer service	System that develops and generates a response to a customer's problems or enquiries
database	An assembly of mutually related graphical and/or non-graphical data sets stored and manipulated by a specific program
data-logger	A device that captures, converts and stores data such as flows and pressures
DBMS (database management system)	A software package providing capability for storage and retrieval of information
DCS (distributed control system)	A control system in which several digital processors (usually microcomputers) are directly connected to the controlled devices

DDC (direct digital control)   A control system in which a digital controller (usually a microcomputer) is directly connected to the controlled devices

demand   The volume of water that has to be put into a supply system to satisfy the requirements of customers (*see* consumption) plus leakage and other waste

demand diagrams   Diagrams showing variations of water demand during a given period of time (a day, a year) for a particular class of users (domestic, hotels, industry, etc.)

demand district   An area of the distribution system that is (or can be) isolated from the rest of the system by closing a small (usually) number of valves ('district valves')

distribution system   The pipes and fittings in a water supply system that are used to deliver water locally to the end-users

district meter   A device for measuring the quantity of water passing into or leaving a district

district metering   Monitoring of water flows into and out of part of the distribution system to determine this system integrity and the presence of leaks

efficiency   Ratio of useful output power to total input power

EMR (electronic meter reading)   The use of hand-held data entry terminals or similar devices to increase the productivity of meter readers

Ethernet   A local area network on which computers transmit and receive data over the same common medium

excessive demand   Quantity of water used due to unexpected circumstances (like pipe bursts, fighting major fires, etc.)

FCV (flow-controlled valve)   A valve whose opening changes according to variations of flow in its control pipe

file-server   A device that provides common information and database access to users on the computer network

fittings   Includes pipes (other than mains), taps, cocks, valves, ferrules, water meters, cisterns, water closets and other similar apparatus used in connection with the supply and use of water

fixed-head node   A point within the water supply system where water level/pressure is time-invariant (like storage basins, lakes, rivers, large pools and reservoirs)

fixed-speed pump   Any pump driven by a constant-speed motor (e.g. asynchronous electric motor)

FLOW control   Control policy for a pumping station: its discharge will vary in time according to a given rule

flowmeter   An instrument that measures flow through a pipe (*see* water meter)

FLV (float valve)   A device that prevents overspilling from a reservoir or a water tower

GIS (geographical information system)   A computer system designed for relating any physical feature, description, event, or environment to a geographical location through the integrated use of computer graphics and database systems

GUI (graphic user interface)   Software and hardware designed to simplify the user interface with a computer system

HDET (hand-held data entry terminal)   A portable device used to record meter readings in the field, which are then transferred electronically to the billing system

HGL (hydraulic grade line)   Heads in the system

incrustation   The forming of dense solids as a crust on the inside surface of a pipe

LAN (local area network)   An interconnected group of computers in a relatively small geographic area

LCP (level-controlled pumps)   Pumps that are switched 'on' or 'off' according to variations of water level in their control reservoir/water tower

LCV (level-controlled valve)   Its opening changes according to variations of water level in a given reservoir or water tower

load allocation   The process of assigning water consumption to nodes in a network model

low suction pressure protection   A device that automatically switches off a pump (or pumps) whenever suction pressure drops below a set level (to prevent cavitation and entrainment of air)

mathematical model   Description of a real system (a water supply system) by mathematical means in order to get an entity that behaves more or less like the original

maximum power charge   Charge for the maximum power used for longer than 15 min during the current month

metered consumption   Quantity of water, delivered to the consumer, is registered on a water meter installed in the service connection

metering system	The equipment and process to determine the customer's consumption of water
MIU (meter interface unit)	An electronic device responsible for communication between the utility office and the meter; it reads the meter when given a signal to do so and transmits information to the utility
MNF (minimum night flow)	Minimum water flow during the 'night': usually over a period of two hours any time between midnight and 7:00 a.m.
model calibration	*see* calibration (of model)
monthly billed tariffs	*see* tariffs
network analysis	Mathematical modelling of a system (a water supply system)
node	A point in the water supply system where either pressure (e.g. a reservoir) or inflow or outflow (e.g. user connections) is known; *see* fixed-head node, ordinary node, transfer node, variable-head node)
NRV (non-return valve)	A device that prevents backflow; also known as a check valve
OMR (off-site meter reading)	A system in which a meter reader, either walking or driving, obtains a reading from the vicinity of the meter without directly contacting it
on–off valve	A valve used for either fully open or shut-off service
ordinary node	Any node that is not a fixed-head, variable-head or transfer node (e.g. pipe junction, service connections)
over-pressure protection	A device that stops all pumps in a pumping station whenever pressure in the network exceeds a given limit
PCP (pressure-controlled pumps)	*see* AUTO control
PCV (pressure-controlled valve)	The opening of the valve changes according to variations of pressure in a given point (control node)
peak demand	The maximum load placed upon the distribution system during a day (lasting at least 15 min) in a given zone or total water supply system
pipe, communication	That part of a service pipe that is vested in the supplier (i.e. the water company's responsibility)
pipe, service	The pipe connecting the customer's premises to the distribution system; comprises communication pipe and supply pipe

pipe, supply	That part of a service pipe that is vested in the customer (i.e. the customer's responsibility)
PLAN control	Control policy for a PST: its pumps operate according to the schedule (*syn.* time-controlled pumps)
PLC (programmable logic controller)	A device that includes PID control and many functions of remote terminal units
power factor	*see* cos ψ
pressure control	Reduction of hydraulic pressure at service connections to the acceptable minimum (say, 15 m) in all operating regimes
pressure-related demand	Relationship (empirical) between consumption of water and service pressure in the same node
pressure-related losses	Relationship (empirical) between local losses of water (leakage mostly) and service pressure in the same node
PROF control	Control policy for PST: its pumps operate in such a way that water level in its control reservoir follows a prescribed time-variable pattern, i.e. profile
PRV (pressure-reducing valve)	Device that keeps local pressure in a given point (*syn.* control node) as close as possible to the set value, which can be a constant or vary in time
PST	Pumping station
pump, fixed-speed	*see* fixed-speed pump
pump, rated value of	Discharge, head and power at the pump's best efficiency regime, its rotational speed being equal to the nominal value
pump scheduling	Planning the operation of available pumps in order to satisfy both the demand and the economic criteria
pump status	A pump may be either in operation or idle or out of order
pump, throttled	The valve on the pressure side of this pump is partially closed to decrease its discharge
pump, variable-speed	*see* variable-speed pump
Q-H-P curve	Pump characteristics: flow Q, head H, power P
regulating range of a FLV	Range of water level in a reservoir/water tower where the influence of a float valve (FLV) is felt (0.5 m or so)

reverse logic (in AUTO control)	An unusual control policy for pumps in AUTO control mode: units are switched 'on' when water level in the control reservoir exceeds a given mark, and 'off' when water level drops sufficiently low (note that this is a common control policy for sewage pumping stations)
RTU (remote terminal unit)	A key element in a supervisory control and data acquisition system that provides the interface between the master station and field devices such as pumps, valves and instruments
SCADA (supervisory control and data acquisition)	Remote control of devices such as pumps and valves, and remote acquisition of data such as flows, levels and pressures
section valve	A valve installed in a pipeline to shut it off for the purpose of inspection or repair
service pipe	Pipe connecting the customer to the distribution system, with or without a water meter
service pressure	Pressure measured at consumer's connection
similarity laws	Alternative name for affinity laws
simulation	Imitation of a real event or sequence of events on an adequate mathematical model
skeleton model	A model containing a subset of all pipes in the system (*opp.* all-mains model)
skeletonization	The practice of deleting certain minor pipes and facilities from the system model in order to create a more manageable model that is still accurate enough
snapshot	A steady state of the water supply system (as if it were 'frozen')
specific consumption	Specific consumption of energy is usually given in kilowatt-hours per cubic metre ($kWh\ m^{-3}$); specific consumption of water is usually given in litres per capita per day or litres per property per day
supply site charge	A lump sum to be paid for each facility that uses energy from the public electricity system
target flow	Discharge to be delivered by a particular pumping station as planned (constant or time-varying)
tariffs	Charges levelled on a water company for energy and maximum power used in an agreed time period (usually a month)

telemetry data	Data gathered by a telemetry system (remote sensing, transfer and processing) about flows, water levels and pressures within the system
TCP (time-controlled pumps)	*see* PLAN control
TCV (time-controlled valve)	The opening of such a valve follows a time-varying plan prepared in advance
transfer node	A node where a known quantity of water is either delivered to the water supply system or taken out of it
transmission mains	The pipes in a water supply system that are used to transport water from sources to areas of demand
trunk main	A larger pipe with few branches used mainly for transport and less for distribution of water
TWL	Top water level of a reservoir/water tower
two-speed valve	A level-controlled valve with two sets of rules: one for the case when water level rises and a different one when water level falls
UFW (unac-counted-for water)	Water taken from the system, but which cannot be attributed to any known user; i.e. difference between a water utility's production and its sales to consumers; it can be 'authorised' or 'unauthorised'
unauthorised water use	Losses of water due to leakage, overflowing reservoirs, illegal tapping, theft, inaccurate meters, incorrect meter reading, etc.
unmetered consumption	Volume of water taken by consumers other than through metered connections
user category	The users of water are classified into several categories with similar patterns of water use, i.e. domestic, hotels, hospitals, etc.
valve	A device to control water flow by throttling, and/or to stop the flow altogether; *see* FCV, FLV, LCV, NRV, PCV, PRV, TCV
variable-head node	A reservoir
variable-speed pump	A pump whose speed can be regulated within a certain range (usually between 70 and 100% of the nominal speed)
volume–elevation curve	Description of the shape of a water tower by giving several points: depth in metres (from the bottom) and volume in cubic metres

water consumption	Quantity of water taken by consumers from the distribution system
water demand	Quantity of water needed to cover both the real consumption and losses of water within a zone in a given period of time
water losses	Water lost from the system due to leakage, seepage, spilling and other causes, before reaching the consumers
water main	Pipe owned and used by the water company to make a supply of water available to customers
water meter	A device installed in a consumer's connection to measure the quantity of water delivered to that customer (*see* flowmeter)
WL	Water level (in a reservoir/water tower)
Zone	A part of an area; could be further subdivided into demand districts

Index

Page numbers in **bold** indicate main entries